Solid Mechanics and Its Applications

Volume 218

Aims and Scope of the Series

The fundamental questions arising in mechanics are: *Why?*, *How?*, and *How much?* The aim of this series is to provide lucid accounts written by authoritative researchers giving vision and insight in answering these questions on the subject of mechanics as it relates to solids.

The scope of the series covers the entire spectrum of solid mechanics. Thus it includes the foundation of mechanics; variational formulations; computational mechanics; statics, kinematics and dynamics of rigid and elastic bodies: vibrations of solids and structures; dynamical systems and chaos; the theories of elasticity, plasticity and viscoelasticity; composite materials; rods, beams, shells and membranes; structural control and stability; soils, rocks and geomechanics; fracture; tribology; experimental mechanics; biomechanics and machine design.

The median level of presentation is to the first year graduate student. Some texts are monographs defining the current state of the field; others are accessible to final year undergraduates; but essentially the emphasis is on readability and clarity.

More information about this series at http://www.springer.com/series/6557

David Wagg · Simon Neild

Nonlinear Vibration with Control

For Flexible and Adaptive Structures

Second Edition

Springer

David Wagg
Department of Mechanical Engineering
University of Sheffield
Sheffield
UK

Simon Neild
Department of Mechanical Engineering
University of Bristol
Bristol
UK

ISSN 0925-0042
ISBN 978-3-319-10643-4
DOI 10.1007/978-3-319-10644-1
ISSN 2214-7764 (electronic)
ISBN 978-3-319-10644-1 (eBook)

Library of Congress Control Number: 2014949364

Springer Cham Heidelberg New York Dordrecht London

Printed on acid-free paper

Springer is part of Springer Science+Business Media (www.springer.com)

Preface

Identifying, modelling and controlling nonlinear vibrations is becoming increasingly important in a range of engineering applications. This is particularly true in the design of flexible structures such as aircraft, satellites, bridges, sports stadia and other tall/slender structures. There are also applications in the areas of robotics, mechatronics, micro-electro-mechanical systems (MEMS) and non-destructive testing (NDT) and related disciplines such as structural health monitoring (SHM).

In the majority of cases, the trend is towards lighter structures, increased flexibility and other higher levels of performance requirements. It is increasingly common for structures to have integrated actuator and sensor networks to carry out tasks such as limiting unwanted vibrations, detecting damage and in some cases changing the shape of the structure. These types of structures have become known as *smart structures* (sometimes called adaptive or intelligent structures). They are often made of new composite materials and their ability to perform multiple tasks means that these types of smart structures are *multifunctional*.

Nonlinear behaviour in structural dynamics arises naturally from a range of common material and geometric non-linearities. By their nature, these structures are typically made up of highly flexible continuous elements such as beams, cables and plates. They are also required to operate in a dynamic environment and, as a result, understanding the vibration behaviour of the structures is critically important.

The focus of this book is first to give a comprehensive treatment of nonlinear multi-modal structural vibration problems, and then to show how (a limited set of) control techniques can be applied to such systems. The emphasis is on continuous structural elements with relatively simple geometry, which enables a range of analytical and approximate techniques to be presented, without the need for extensive numerical simulation. It should be emphasized that there is no attempt to provide a comprehensive treatment of nonlinear control techniques in this book. Instead, a limited set of control approaches which apply to problems of *vibration control* are presented.

The aim was to make the book accessible to the reader with some background knowledge in linear vibration. The book falls into two main parts. The first five chapters have been developed from lecture notes taught at masters level, and example problems are included at the ends of Chaps. 2–4. The second half of the book, Chaps. 5–8, has more of a research emphasis, with case studies and research examples shown where appropriate.

Chapters 1–3 contain introductory material on nonlinear vibration phenomena and control methods for nonlinear vibration. Chapter 4 introduces the approximate techniques such as harmonic balance, and perturbation methods which can be used for analysis of nonlinear vibration problems. The topic of modal analysis for nonlinear structures is discussed in detail in Chap. 5. In particular, normal form analysis is used to model multi-modal vibration response for nonlinear structures.

Then each of Chaps. 6–8 is dedicated to a particular type of structural element. Chapter 6 is focused on beams, Chap. 7 on cables and Chap. 8 on plates and shells. In these chapters, a selection of nonlinear vibration case studies is presented. Discussions of control methods are also included where appropriate.

Since the first edition of this book was published, there has been continued interest in modelling and controlling nonlinear vibrations. The main change we have introduced is that the normal form methods described in Chaps. 4 and 5 are now based on the second order form of the governing equations. This is a more natural approach for structural dynamics problems and has other advantages that are explained in the relevant sections. In addition to this, many more minor additions have been made to update the text.

This book has only been possible with the generous help and support of many colleagues and collaborators. In particular, we would like to acknowledge the work of Andres Arrieta Diaz, Andrea Cammarano, Alicia Gonzalez-Buelga, Tom Hill, Irina Lasar, Xuanang Liu, Julian Londono, Nihal Malik, Claire Massow, Jack Potter, Alex Shaw and Zhengfan Xin who carried out some of the original work, which is presented in this book. For informed discussion on the scope of the book and feedback on the draft manuscript, we would like to thank Nick Alexander, Alex Carrella, Mike Davies, David Ewins, Peter Gawthrop, Peter Green, Dan Inman, Irfan Khan, Bernd Krauskopf, Steve Shaw, Lawrie Virgin, Paul Weaver and Keith Worden. We would also like to thank Series Editor, Graham Gladwell, for his detailed technical comments on the draft manuscript. In addition, we are very grateful to Paul Neild, who meticulously proofread the manuscript of the first edition and to Tom Hill, who kindly produced Figs. 5.6–5.12 and 7.10–7.13 for the second edition. Finally, we would like to thank Tom Spicer from Springer for his help and support.

Contents

Chapter 1
Introduction to Nonlinear Vibration and Control

Abstract The performance requirements of flexible structures are continually increasing. Often structures are required to have integrated control and sensor systems to carry out tasks such as limiting unwanted vibrations, detecting damage and in some cases changing the shape of the structure. These types of structures have become known as *smart structures* (sometimes called adaptive or intelligent structures). The ability to perform multiple tasks means that the smart structure is *multifunctional*. By their nature, these structures are typically highly flexible and are required to operate in a dynamic environment. As a result, the vibration behaviour of the structure is of critical importance. Not only is vibration important, it is often nonlinear, due to a range of effects which naturally arise in flexible structural dynamics. Applying control to the structure to limit unwanted vibration and to effect any shape changes also requires detailed knowledge of the vibration characteristics. This chapter introduces the basic ideas of nonlinear vibration and control, which will be used in later chapters to underpin the analysis of more complex structural elements.

1.1 Vibration of Flexible Structures

Vibrations occur in a wide range of structural and mechanical systems when the system is shaken (or suddenly disturbed) by an external force. Vibration typically takes the form of a series of cyclic movements, or oscillations, in the structure. The time-scale of the cyclic motion is important because it relates to the *frequency* of the oscillation. Cyclic oscillations in the form of approximate sine waves are one of the most common types of vibration response. In this case, the frequency of the oscillation, f, can be related to the time period of the oscillation, T (seconds), such that $f = 1/T$ (cycles/second), as shown schematically in Fig. 1.1a.

For vibrations to occur, the structure needs to have a restoring force, which returns the structure towards its resting position when disturbed by an external force. This property makes the structure behave as a *flexible* body which flexes (or stretches) when pushed or pulled, as opposed to a rigid body which is assumed to be inflexible. Flexibility is common to the majority of structural elements such as rods, beams, plates, shells, membranes, cables, shafts, etc. which are used throughout structural, mechanical and aerospace engineering.

D. Wagg and S. Neild, *Nonlinear Vibration with Control*,
Solid Mechanics and Its Applications 218, DOI 10.1007/978-3-319-10644-1_1

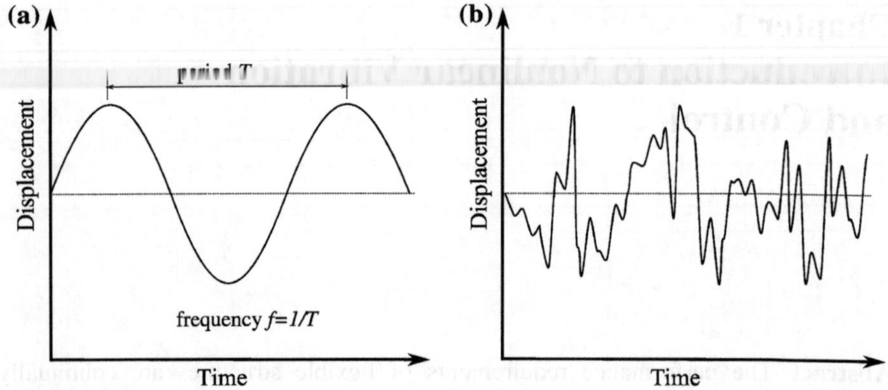

Fig. 1.1 Basic nonlinear vibration definitions: **a** period and frequency for a harmonic response, **b** a non-periodic vibration response

Flexible physical structures naturally possess two other properties which are important for vibration. First, they have a mass distribution which depends on their shape and material make-up. When the structure vibrates, the movement of mass in the structure leads to inertia forces being generated. Secondly, the material typically has some material damping (or energy dissipation mechanism), which has the effect of reducing the magnitude of oscillation. The exact nature of the physical damping mechanism is usually much more difficult to model than either the inertia or restoring forces. However, for most structural and mechanical systems of interest, the physical damping will be relatively small.

So, the forces involved in a flexible body vibration problem are (i) the inertia forces, $\mathbf{F}_I$, which are related to the mass distribution in the structure, (ii) restoring forces, $\mathbf{F}_R$, related to the material properties and shape of the structure, sometimes referred to as *stiffness* (or spring) forces, (iii) damping or dissipation forces, $\mathbf{F}_D$, and (iv) excitation or external input forces, $\mathbf{F}_E$. When modelling vibrations, it is conventional to write the governing equations of motion in a form where the external excitation force is on the right-hand side and the physical system forces are on the left-hand side, such that

$$\mathbf{F}_I + \mathbf{F}_D + \mathbf{F}_R = \mathbf{F}_E. \tag{1.1}$$

Each of the force terms in Eq. (1.1) can be considered to be a vector which corresponds to the forces at a series of discrete points on the structure. The displacements (relative to the initial resting position) at the N discrete points are represented by the vector $\mathbf{x} = [x_1, x_2, \ldots, x_N]^T$, where N is the number of *degrees-of-freedom* in the structural model. Then the velocities are $\dot{x}_i = \mathrm{d}x_i/\mathrm{d}t$ for each degree of freedom $i = 1, 2, 3, \ldots N$ and the accelerations $\ddot{x}_i = \mathrm{d}^2x_i/\mathrm{d}t^2$ for $i = 1, 2, 3, \ldots N$, where t is time. In terms of vectors, velocity is $\dot{\mathbf{x}} = \mathrm{d}\mathbf{x}/\mathrm{d}t$ and acceleration $\ddot{\mathbf{x}} = \mathrm{d}^2\mathbf{x}/\mathrm{d}t^2$.

For linear vibration problems, the inertia forces are represented by $\mathbf{F}_I = \boldsymbol{M}\ddot{\mathbf{x}}$ where $\boldsymbol{M}$ is an $N \times N$ mass matrix, the dissipation forces are given by $\mathbf{F}_D = \boldsymbol{C}\dot{\mathbf{x}}$

where C is the $N \times N$ damping matrix and the restoring forces are $\mathbf{F}_R = K\mathbf{x}$, where K is the $N \times N$ stiffness matrix. Substituting these relationships into Eq. (1.1) gives

$$M\ddot{\mathbf{x}} + C\dot{\mathbf{x}} + K\mathbf{x} = \mathbf{F}_E. \tag{1.2}$$

Equation (1.2) is the fundamental governing equation for linear vibration of structural and mechanical systems, where $\mathbf{F}_E$ is the dynamic forcing vector.

There is a range of external forces that can be applied, the most common of which for vibration problems is a continuous single-frequency sine wave force called *harmonic* forcing. When this type of continuous, harmonic forcing is applied, the initial response motion is *transient*, and after some length of time the response is said to be at *steady-state*. The length of time after which transient motion becomes steady-state depends on the problem being considered. With harmonic forcing, and if the restoring force is linear-elastic, as in Eq. (1.2), the steady-state vibrations will be harmonic with the same frequency as the forcing.

Nonlinear vibration problems are simply those which cannot be modelled by Eq. (1.2). For example, it is possible that the restoring force is not linear such that $\mathbf{F}_R \neq K\mathbf{x}$. When this (or other nonlinear effects are present), a range of *non*-harmonic nonlinear vibration phenomena can occur when the system is harmonically forced. Often the response is *periodic*[1] and so for the single-frequency harmonic input, the system response contains multiple components each with a different frequency. However even relatively small nonlinear effects (often called weak nonlinearity) can lead to very complex nonlinear dynamics, as shown schematically in Fig. 1.1b.

1.2 Causes of Nonlinear Vibration

There are many physical phenomena which lead to nonlinear vibration problems. Some of the most common causes of nonlinearity are discussed below.

1.2.1 Material Properties

The constitutive relationships for any material, expressed as stress-strain or force-displacement relationships, are typically nonlinear. In Fig. 1.2a a typical stress-strain relationship for an axially-loaded rod made of a non-ferrous alloy, such as aluminium, is shown. A linear relationship exists between stress and strain, $E = \sigma/\varepsilon$ up to an *elastic limit* (or limit of proportionality), where E is the Young's modulus of the material. This is the linear elastic region of material behaviour where stress, σ, can be assumed to be linearly proportional to strain, ε, or force proportional to displacement—a relationship that is often referred to as Hooke's law. A similar elastic

[1] Periodic is more general than harmonic, it has a repeating pattern but this repeating pattern is not necessarily limited to a single-frequency sine wave.

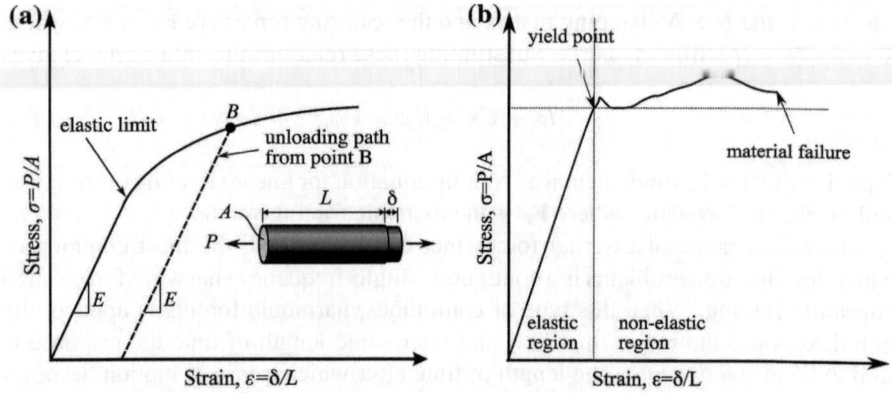

Fig. 1.2 Examples of material nonlinearity, showing **a** typical stress-strain relationship for an axially-loaded rod made of a non-ferrous alloy such as aluminium, **b** typical stress-strain for low carbon steel materials

region exists for the behaviour of steel shown in Fig. 1.2b. In the elastic region the material can be loaded and unloaded repeatedly along the same stress-strain (force-displacement) line, or loading path.

Beyond the elastic limit, metals typically yield and then exhibit a region of non-elastic behaviour. The first behaviour after yielding is typically *plastic* material behaviour. In this regime, permanent displacements can occur and unloading paths are not necessarily the same as the loading paths. In some cases, unloading in the non-elastic region leads to a switch back into linear behaviour, but with some residual strain. This is shown in Fig. 1.2a, where the material is loaded from zero up to point B and then unloaded. This type of behaviour is known as *material hysteresis*. Some materials, such as some types of rubberized bands, have no discernible elastic region, and exhibit hysteresis when loaded and unloaded, as shown schematically in Fig. 1.3a. The loop formed by the loading and unloading curves is often referred to as the hysteresis loop.

When repeatedly loaded and unloaded, in a cyclic manner, some structural materials can exhibit a progressive reduction in the maximum point on the loading curve, such that for each cycle, the hysteresis loop appears to rotate slightly, as shown schematically in Fig. 1.3b. This type of behaviour can occur during the progressive failure of a material which is undergoing cycling loading, and is sometimes known as *cyclic softening*. In some materials the opposite effect occurs, which is a form of *cyclic hardening*. In this type of behaviour the material is said to have a form of *memory*, meaning that the behaviour at any time, t, depends on the history of the loading the material has experienced.

A class of materials that exhibit a different type of memory is *shape memory alloys (SMAs)*. With these materials, temperature can be used to switch between the austenitic state and the martensite state of the material, and each state has a different shape associated with it. As a result, the SMA appears to remember and be able to switch between the shapes in the two different states. This type of shape-change

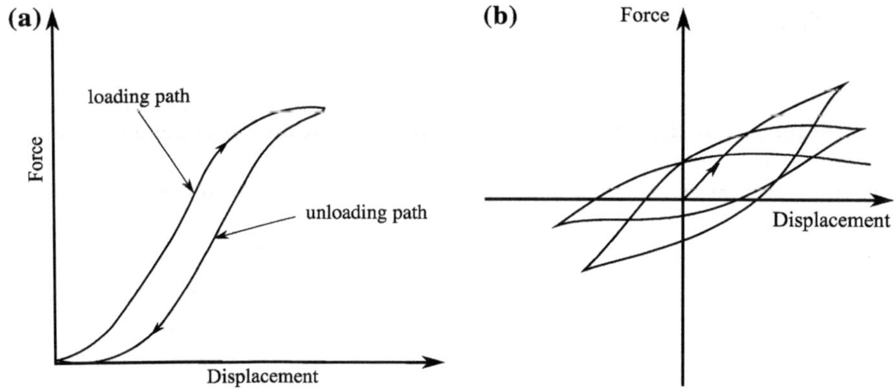

Fig. 1.3 Typical force-displacement behaviour: **a** for a visco-elastic material such as a rubberized (bungee cord or similar) band, **b** a schematic representation of cyclic softening behaviour

behaviour can be used either for direct shape-change applications, or to create an SMA actuator.

It is important to realize that the material behaviour shown in Fig. 1.3 is actually quite dependent on the rate at which the loading is applied. This *rate dependence* of material behaviour can be very significant for nonlinear vibration problems. The simplest rate-dependent behaviour is where the force is directly proportional to the velocity, for example $force = constant \times \dot{x}$. This type of relationship is often used to approximate the resistance of a fluid-filled damper element in linear vibration theory, where it is often referred to as *viscous* damping. This was the relationship used in Eq. (1.2), where the dissipative forces were assumed to be a linear coefficient matrix multiplied by the system velocities $\mathbf{F}_D = C\dot{\mathbf{x}}$. Some viscous behaviour is more accurately modelled using a nonlinear relationship. For example, some acoustic damping applications have a viscous damping which is closer to a power law function of the form $force = constant \times (\dot{x})^n$, where typically values are $2 \leq n \leq 3$.

A material which has a stress–strain relationship that can be modelled by a combination of viscous and elastic behaviour is said to be visco-elastic. This type of material typically has a hysteretic behaviour such as that shown in Fig. 1.3. An important observation in respect of these materials is that the energy dissipated during a loading and unloading cycle is proportional to the area of the hysteresis loop.

Other considerations include whether the material behaves in an isotropic and/or homogeneous manner. However, making such assumptions is often invalid—with composite materials, for example. In addition, the presence of residual stresses, damage, wear or other environmental factors can have a significant effect on material behaviour.

1.2.2 Geometric Nonlinearity

Geometric nonlinearity occurs when the nonlinear effect is due entirely to the geometry of the problem under consideration. A simple example is that of a pendulum, which has an approximately linear behaviour for very small angles of displacement, but an inherently nonlinear behaviour when the angular displacements become larger. This effect occurs because the pendulum restoring force contains a $\sin\theta$ term. Figure 1.4a shows a pendulum bob with mass m, pivoted at point O and constrained to rotate in a circular plane by a rigid, light rod of length l (which is assumed to be massless). If the pendulum is disturbed from the resting (downwards) position, gravity provides a restoring force of $mg\sin\theta$ tangential to the circular arc. Assuming no damping in the pivot and applying Newton's second law to the pendulum bob along the tangent to the circular arc of motion, we find that $ml\ddot{\theta} = -mg\sin\theta$ (noting that the length along the circular arc is $s = l\theta$, so that the acceleration is $\ddot{s} = l\ddot{\theta}$). Dividing by ml gives the equation of motion for the undamped, unforced pendulum as

$$\ddot{\theta} + \frac{g}{l}\sin\theta = 0.$$

It is interesting to note that because both the inertia force and the restoring force depend on mass, m has been scaled out of the equation of motion.[2] It is usual to set $\frac{g}{l} = \omega^2$ so that the frequency, ω, is a function of the pendulum length l (assuming gravity g remains constant). Approximating $\sin\theta \approx \theta - \frac{\theta^3}{3!} + \frac{\theta^5}{5!} + \cdots$, we find the crudest (or lowest order) approximation is that $\sin\theta \approx \theta$. This approximation is valid for *small* pendulum oscillations, but as the oscillations become larger the nonlinear terms become significant. This type of approximation is a form of *linearisation*, and is used extensively throughout the study of nonlinear vibration. In this simple example it is possible for us to compute values of θ to decide exactly what "small" means—see for example Jordan and Smith (1999). In more complex systems, the exact limits of the smallness approximation are often not so clear, and this should be kept in mind when assessing the validity of expanded nonlinear approximations.

Large deflections of beams and columns are other common sources of geometric nonlinearity in structural dynamics. When a beam or column is acted upon by a load, it will deflect into a deformed shape with a certain curvature. The basic modelling assumption for beam bending is that the bending moment in the beam is proportional to the curvature. To obtain a linear equation of motion for beam vibration, it is usual to assume that the deflections are small, such that nonlinear terms arising from curvature can be neglected—this is discussed in detail in Sect. 6.2—but it means in effect that the beam is very close to being straight. Any deviations from this will lead to geometric nonlinearities and errors in the modelling process, if these nonlinearities are not accounted for.

[2] Strictly speaking, this is only true in a vacuum. In low density fluids the behaviour is very close, but in liquids the effects of mass become significant, Neill et al. (2007).

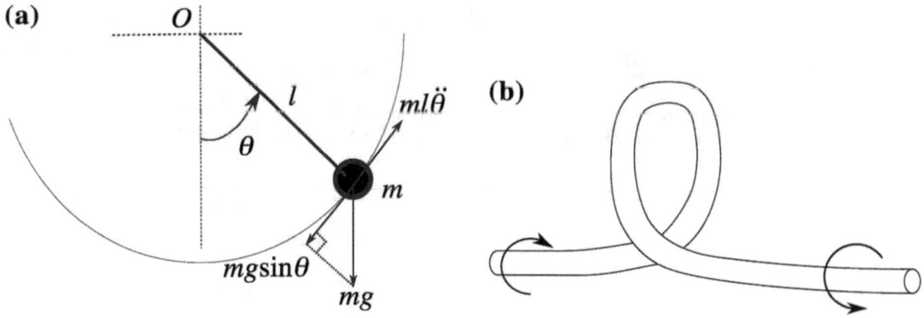

Fig. 1.4 Geometric nonlinear behaviour: **a** a simple pendulum and **b** localization in a twisted rod

At very high levels of deformation, elastic structural elements can develop *localized* effects. For example, when a rod is twisted repeatedly, a form of localized buckling can occur in which the rod forms a loop at some point along its length, as shown in Fig. 1.4b. These types of effects have been observed in physical engineering applications such as drillstrings (used in the oil industry), pipelines and cables, textile yarns and, more recently, supercoiling and packing of DNA molecules and other biological applications. Further details are given in Thompson and Champneys (1996) and van der Heijden (2008).

1.2.3 External Forces and Constraints

Nonlinearity can be caused by external forces acting on a linear system. An important example is the interaction of elastic bodies with aerodynamic forces, known as *aeroelasticity*, that forms an important branch of structural vibration, especially for aerospace structures and slender civil engineering structures subject to wind loading. A simple case of so-called steady aeroelasticity, applied to a single-degree-of-freedom aerofoil, is shown in Fig. 1.5a. In this example, as the air flows past the aerofoil with velocity U, the aerofoil pitches upwards with positive angle of attack, θ. Here $M(U, \theta)$ is the total aerodynamic moment which is a function of both U and θ. The rotational spring, k_θ, provides a restoring moment which is trying to return θ to the $(\theta = 0)$ resting position. Using the small angle approximation, we can assume that the equation of motion for the aerofoil is approximately linear, such that $\ddot{\theta} + k_\theta \theta = M(U, \theta)$. However, the total aerodynamic moment, M is a function of both the flow velocity, U, and the induced angle of attack, θ. Wind tunnel experiments on sections similar to that in Fig. 1.5a show that, as M increases, there comes a point where it overcomes the restoring moment, and the system becomes *statically unstable* leading to large and damaging oscillations. This point of static instability is usually referred to as *divergence*, and has a dynamic counterpart called *flutter*, which is extremely dangerous for aeroelastic structures. In fact, for the majority of

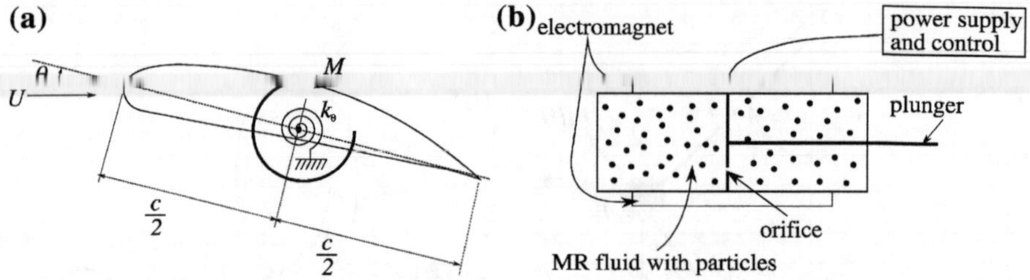

Fig. 1.5 Nonlinear behaviour from external forces: **a** a simple aerofoil and **b** a magnetorheological damper system

aeroelastic structures, $\boldsymbol{M}$ is a nonlinear function of angle of attack, θ, and flutter is a classic case of a static equilibrium position becoming unstable when the system is subjected to certain dynamic excitations. This loss of stability occurs via a Hopf bifurcation—discussed in Chap. 2.

Magnetic forces acting on otherwise linear elastic structural elements can also be the cause of nonlinear behaviour: for example, magnetic forces acting on a linear elastic cantilever beam, as discussed in Chap. 6. Externally applied magnetic forces have come into widespread use in the form of *magneto-rheological dampers*, or MR dampers. These damping systems use a special fluid which contains metallic particles. When a magnetic field is applied to the fluid, the rheological properties change significantly. A cross-section through a damper containing MR fluid is shown schematically in Fig. 1.5b. In this example, the magnetic field is generated using an electromagnet which is controlled using an input voltage from a control system. These types of dampers are used extensively in automotive applications for semi-active suspension systems. They are also being used increasingly for seismic engineering applications, to damp out unwanted vibrations in bridges and other structures.

The effect of the magnetic force is to make the MR fluid behave like a viscoelastic material with parameters which vary as the magnetic flux varies. A related phenomenon occurs in electrorheological fluid, which can alter its viscosity when an electric current is passed through it.

For systems that are manufactured at the micro scale, such as micro-electro-mechanical systems (MEMS), or even smaller at the nano scale, inter-molecular forces such as the van der Waals force can become significant external nonlinear force terms.

Constraints occur when structural elements are not allowed to move freely in all directions. An example of a constrained system is stretching in beams, rods and cables. This occurs when the structural element is clamped at both ends. For example, when a beam is constrained like this, any load which causes a deflection will stretch the beam, inducing an axial tension force. This occurs even when the deflection is small such that it can be assumed that no geometric nonlinearity is present. This example is discussed further in Chap. 6.

1.2.4 Freeplay, Backlash, Impact and Friction

Nonlinear behaviour can be caused by the interaction of structural elements. For example, it is often impossible to manufacture things without some degree of tolerance, particularly when two or more components interact. This leads to a degree of "looseness" or freeplay between components. The example shown in Fig. 1.6a shows a pair of rotating spur gears. For effective meshing, the gears cannot be mounted too closely, so a small amount of freeplay, or backlash, exists. When rotating at operating speed, the gear teeth can lose contact and then impact against each other again. It is also possible for the gear teeth to oscillate in a repeated series of impacts. This can sound like *rattle* in gearboxes, especially when idling. Spur gears are often attached to flexible shafts, and although the gears can usually be modelled as rigid bodies, the shaft vibration can require multi-modal modelling techniques.

Other examples can occur, often because of manufacturing tolerances: for example an aileron that can move a small amount when set at zero flap angle, shown in Fig. 1.6b. This means that, with a certain aerodynamic forcing, it is possible for the aileron to vibrate to and fro in the freeplay zone, often with an accompanying "buzzing" sound. In the freeplay zone, the movement in the aileron cannot be controlled. Freeplay is sometimes also referred to as a deadzone in other similar applications.

When vibration occurs where motion-limiting constraints are present, repeated impacts can occur. This type of behaviour is sometimes called *vibro-impact* motion. The classic example is shown in Fig. 1.7a, where a vertically clamped cantilever beam, which is sinusoidally forced, has a motion-limiting constraint (the impact stop) near the tip, see Moon and Shaw (1983). As the beam vibrates, it has an impact each time the beam hits the constraint. Although the beam has linear dynamics, the nonlinear effect of the impact makes the problem nonlinear. A common way to model this type of impact system is to assume that the velocity of the beam at impact is reversed and reduced by a coefficient of restitution, r, such that $\dot{u}(b,t)_{after} = -r\dot{u}(b,t)_{before}$, where $u(b,t)$ is the transverse displacement of the beam at the point of impact. To make this type of model work, it is usually assumed that only the first vibration mode of the beam is significant and that the impact is effectively instantaneous.

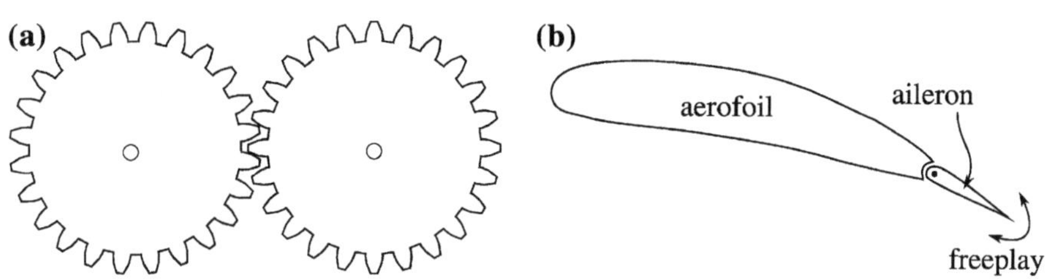

Fig. 1.6 Nonlinear behaviour from freeplay and backlash: **a** rotating spur gears and **b** an aerofoil with freeplay in the aileron

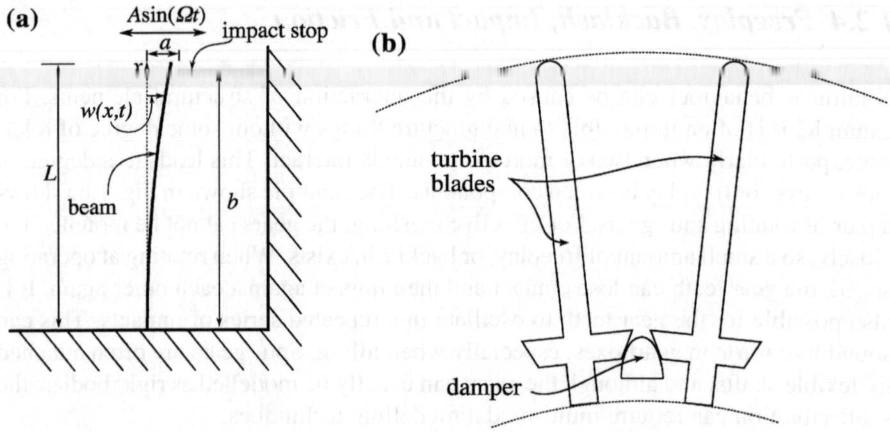

Fig. 1.7 Nonlinear behaviour from impact and friction: **a** a sinusoidally forced cantilever beam subjected to a motion-limiting constraint (the impact stop) near the tip, **b** a friction damper used to reduced vibration in turbine blades

When structural elements are already in contact, friction occurs along the contact surface. There are two behaviour states: either *stick* when there is no relative movement, or *slip* when the elements slide. Repeated stick and slip cycles can induce vibration in flexible structural elements. This is what happens in bowed musical instruments as the bow is drawn across the string. In Fig. 1.7b a nonlinear vibration example is shown with a friction damper used to reduce vibration in turbine blades. This type of damper is often parabolic in shape and uses dry friction in between the turbine blades to limit their vibration while the turbine is rotating at high speed.

Friction effects are very significant in machines. Bearings for rotating shafts usually consist of rolling elements, which have different characteristics from sliding friction. The inclusion of lubricants makes modelling more complex, and study is usually carried out at microscopic scale using tribology.

Freeplay, backlash, impact and friction nonlinearities are often modelled using non-smooth assumptions. Non-smooth assumptions allow relatively simple models to be used for phenomena which are otherwise complex, such as vibration and impact combined. For example, the assumption that an impact is represented by a sudden jump in velocity via the coefficient of restitution rule allows vibro-impact problems to be considerably simplified. In fact, this is only realistic if the time of impact is very small compared to the time-scales being used for the dynamic modelling. Looking at the impact in a higher resolution reveals that the process is strongly nonlinear, but not non-smooth. In general, this applies to all the underlying physical processes in the (macro-scale) mechanical domain, and this should be kept in mind when using non-smooth assumptions. In electrical systems, quantum mechanics or MEMS, non-smooth behaviour may be closer to the underlying physics. Detailed accounts of non-smooth modelling techniques are given by Brogliato (1999) and di Bernardo et al. (2007).

1.2.5 Control and Delay

Generally, control forces can be added to Eq. (1.1) so that

$$\mathbf{F}_I + \mathbf{F}_D + \mathbf{F}_R = \mathbf{F}_E + \mathbf{F}_C,$$

where $\mathbf{F}_C$ is a vector of control forces (note that the sign of the vector $\mathbf{F}_C$ is taken as positive here, but feedback control inputs normally need to be negative to make the system stable). Just like the nonlinear external forces discussed in Sect. 1.2.3, the control forces, $\mathbf{F}_C$, are functions of the system states, $\mathbf{x}$, $\dot{\mathbf{x}}$, so that $\mathbf{F}_C = \mathbf{F}_C(\mathbf{x}, \dot{\mathbf{x}})$. In many cases $\mathbf{F}_C$ is a linear function of $\mathbf{x}$ and $\dot{\mathbf{x}}$, and as a result can be thought of as analogous to additional linear stiffness and damping terms. However, some control methods are inherently nonlinear. For example, adaptive control uses displacement and velocity signals from the system to produce a control signal. However, during this process the signals are multiplied together, which leads to quadratic nonlinear terms in $\mathbf{F}_C$. So even when applied to linear systems, this type of controller can lead to nonlinear dynamic behaviour.

Hybrid controllers use discrete *switchings* between different regimes of operation. The effect of the switches can lead to non-smooth dynamic behaviour, even if all different regimes of operation are linear. Examples include sky-hook control, which will be discussed in Chap. 6.

Introducing delays into dynamical systems leads to nonlinear effects. Delays can be fixed, variable, parameter-dependent or state-dependent. Delays can occur in machine cutting systems where the object being cut is rotating, such as on a lathe. In these systems, the new surface depends on the cut from the previous rotation. So if the delay from the previous rotation is τ, then the system is dependent on the state now, $x(t)$, and the state at time τ in the past, $x(t-\tau)$. Including $x(t-\tau)$ terms in the governing differential equation leads to *delay differential equations* or DDEs.

Some laser systems which receive optical feedback are naturally modelled as systems of DDEs (Krauskopf 2005). Having actuators in structural systems can also introduce delay effects. The actuator takes a certain time to respond to an input signal, which can be modelled as a fixed or parameter-dependent delay.

Delays in vibration systems can have the effect of acting like *negative* damping. Positive damping takes energy out of the system whereas negative damping effectively adds energy. If the negative damping effects outweigh the positive damping in the system, the system can become unstable, which is normally very undesirable. For this reason, delay effects in vibrating systems should be treated with care.

1.3 Mathematical Models for Vibration

Having models of vibration which accurately capture physical behaviour enables engineers to predict future behaviour of the system being considered. It also allows control mechanisms to be designed, for example, when the levels of vibration need to be actively reduced. It is important to remember that the mathematical models are only an approximation to reality. Verifying mathematical models against real physical behaviour, from either laboratory-based experiments or in situ observations, is an essential part of the design process.

1.3.1 Linear Vibration Modelled Using Sine Waves

For linear systems, exciting the system with a sine wave results in the system responding with sine waves. Considering the linear equation of motion, Eq. (1.2), when there is only one degree of freedom, gives

$$m\ddot{x} + c\dot{x} + kx = F_e, \tag{1.3}$$

where m, c and k are the scalar parameters of the system, and F_e is the forcing. This is the ordinary differential equation governing a single-degree-of-freedom lin-

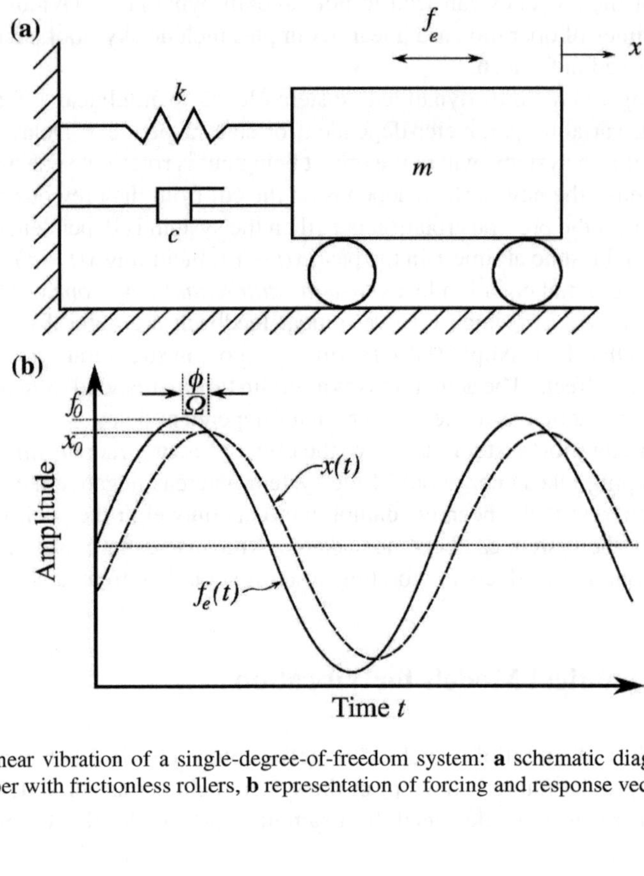

Fig. 1.8 Linear vibration of a single-degree-of-freedom system: **a** schematic diagram of mass-spring-damper with frictionless rollers, **b** representation of forcing and response vectors

ear oscillator, with viscous damping and forcing, which is shown schematically in Fig. 1.8a.

In general, ordinary differential equations of this type have two parts to the solution. The solution to the homogeneous equation (i.e. Eq. (1.3) with $F_e = 0$) corresponds to the transient behaviour of the oscillator, which depends on the initial displacement $x(t = 0)$ and velocity $\dot{x}(t = 0)$ (sometimes called the complementary function). The second part of the solution is the particular solution for the forcing function F_e being applied to the system, and this solution corresponds to the steady-state (i.e. long-term) behaviour of the system. In fact the transient solution dies out quickly, so the particular solution is usually the one of primary interest.

If the forcing is sinusoidal, then the response is also sinusoidal. However the sinusoidal response will typically not be the same as the excitation. It would be expected that both the *amplitude*[3] of the response sine wave and the *phase* (i.e. time difference) will be different from the input signal. So we would expect that if the forcing is $F_e = F_0 \sin(\Omega t)$, the displacement response would be, $x = X_r \sin(\Omega t - \phi)$, where F_0 is the amplitude of the forcing input, X_r is the amplitude of the displacement response, Ω is the forcing frequency and ϕ is the phase shift between the two sine waves, as shown in Fig. 1.8b. Note that Ω is in the form of a circular (or angular) frequency, $\Omega = 2\pi f_f$, where f_f is the forcing frequency in cycles per second, and $f_f = 1/T_f$, where T_f is the period of the forcing in seconds.

An alternative way of representing the sine waves is to use complex functions. The forcing sine wave can be written as

$$F_e = \frac{F_0}{2\mathrm{i}}(\mathrm{e}^{\mathrm{i}\Omega t} - \mathrm{e}^{-\mathrm{i}\Omega t}), \tag{1.4}$$

where F_0 is real. The response sine wave needs to include a phase delay, ϕ, but X_r must be real. So the response sine wave is written as

$$x = \frac{X}{2\mathrm{i}}\mathrm{e}^{\mathrm{i}\Omega t} - \frac{\overline{X}}{2\mathrm{i}}\mathrm{e}^{-\mathrm{i}\Omega t}, \tag{1.5}$$

where X is a complex constant and $\overline{X}$ is the complex conjugate of X. The use of complex function X allows both amplitude and phase information to be included in the response sine wave. The amplitude function, X_r, is the modulus of X and the phase, ϕ, is the argument of X.

A further advantage of using the complex formulation is that when Eqs. (1.4) and (1.5) are substituted into Eq. (1.3), it is more straightforward to compare coefficients of the time-dependent terms (i.e. the exponential terms) than if the phase had been included explicitly in Eq. (1.5). Carrying out the substitution of Eqs. (1.4) and (1.5) into Eq. (1.3) and comparing coefficients of the $\mathrm{e}^{\mathrm{i}\Omega t}$ and $\mathrm{e}^{-\mathrm{i}\Omega t}$ terms, we find the relationships

---

[3] Amplitude is sometimes used to denote only the magnitude of displacements. Throughout this text it will be used to imply a *magnitude* of the quantity under discussion, be it force, velocity, acceleration or displacement.

$$(k - m\Omega^2 + \mathrm{i}c\Omega)X = F_0, \quad (1.6)$$

$$(k - m\Omega^2 - \mathrm{i}c\Omega)\overline{X} = F_0 \quad (1.7)$$

which can also be written in the alternative formulation of

$$\left(1 - \left(\frac{\Omega}{\omega_n}\right)^2 + i2\zeta\frac{\Omega}{\omega_n}\right)kX = F_0, \quad (1.8)$$

$$\left(1 - \left(\frac{\Omega}{\omega_n}\right)^2 - i2\zeta\frac{\Omega}{\omega_n}\right)k\overline{X} = F_0, \quad (1.9)$$

where $\omega_n = \sqrt{k/m}$ is the (undamped) *natural frequency* and $\zeta = c/2m\omega_n$ is the damping ratio.

To find the amplitude of response, X_r, and the phase lag, ϕ, Eqs. (1.8) and (1.9) (or Eqs. (1.6) and (1.7)), are separated into real and imaginary parts. This is done by first writing

$$\frac{kX}{F_0} = \frac{1}{\left(1 - \left(\frac{\Omega}{\omega_n}\right)^2 + i2\zeta\frac{\Omega}{\omega_n}\right)}, \quad (1.10)$$

$$\frac{k\overline{X}}{F_0} = \frac{1}{\left(1 - \left(\frac{\Omega}{\omega_n}\right)^2 - i2\zeta\frac{\Omega}{\omega_n}\right)}. \quad (1.11)$$

Multiplying the numerator and denominator of the right-hand sides of Eqs. (1.10) and (1.11) by the complex conjugate of the denominators, gives

$$\frac{kX}{F_0} = \frac{\left(1 - \left(\frac{\Omega}{\omega_n}\right)^2\right) - i2\zeta\frac{\Omega}{\omega_n}}{\left(1 - \left(\frac{\Omega}{\omega_n}\right)^2\right)^2 + 4\left(\zeta\frac{\Omega}{\omega_n}\right)^2},$$

$$\frac{k\overline{X}}{F_0} = \frac{\left(1 - \left(\frac{\Omega}{\omega_n}\right)^2\right) + i2\zeta\frac{\Omega}{\omega_n}}{\left(1 - \left(\frac{\Omega}{\omega_n}\right)^2\right)^2 + 4\left(\zeta\frac{\Omega}{\omega_n}\right)^2},$$

from which we can show that for both X and $\overline{X}$ the magnitude (or amplitude) of the complex vectors is

$$X_r = |X| = \left(\frac{F_0}{k}\right)\frac{1}{\sqrt{\left(1 - \left(\frac{\Omega}{\omega_n}\right)^2\right)^2 + 4\zeta^2\left(\frac{\Omega}{\omega_n}\right)^2}}. \quad (1.12)$$

As there are two complex vectors, there are two distinct values for the phase delay, namely

$$\arg(X) = \arctan\left(\frac{-2\zeta\frac{\Omega}{\omega_n}}{\left(1-\left(\frac{\Omega}{\omega_n}\right)^2\right)}\right) = -\phi, \tag{1.13}$$

$$\arg(\overline{X}) = \arctan\left(\frac{2\zeta\frac{\Omega}{\omega_n}}{\left(1-\left(\frac{\Omega}{\omega_n}\right)^2\right)}\right) = \phi. \tag{1.14}$$

The phase delay, ϕ, is taken as a solution of arctan, which goes smoothly from $\phi = 0$ at $\Omega = 0$ to $\phi = \pi$ as Ω becomes much larger than ω_n. These functions define the response of the linear system as

$$\begin{aligned} x =& -\mathrm{i}\frac{X_r}{2}\mathrm{e}^{-\mathrm{i}\phi}\mathrm{e}^{\mathrm{i}\Omega t} + \mathrm{i}\frac{X_r}{2}\mathrm{e}^{\mathrm{i}\phi}\mathrm{e}^{-\mathrm{i}\Omega t} \\ =& -\mathrm{i}\frac{X_r}{2}\mathrm{e}^{\mathrm{i}(\Omega t-\phi)} + \mathrm{i}\frac{X_r}{2}\mathrm{e}^{-\mathrm{i}(\Omega t-\phi)} \\ =& \frac{X_r}{2\mathrm{i}}(\mathrm{e}^{\mathrm{i}(\Omega t-\phi)} - \mathrm{e}^{-\mathrm{i}(\Omega t-\phi)}) \\ =& X_r\sin(\Omega t-\phi), \end{aligned} \tag{1.15}$$

which corresponds to two counter-rotating complex vectors in the complex plane.

These vectors always have equal and opposite imaginary parts, with the result that the sum of the two vectors is a sine wave in the real plane, as shown schematically in Fig. 1.9.

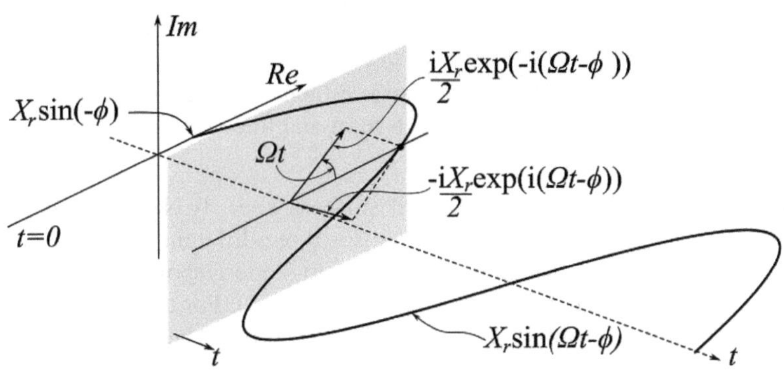

Fig. 1.9 Linear vibration response of a single-degree-of-freedom system showing how the two counter-rotating complex response vectors form a sine wave

In many texts on linear vibration, this type of analysis is performed by assuming $x = X\mathrm{e}^{\mathrm{i}\Omega t}$, where X is a complex constant and where, for a sine wave forcing, the imaginary parts of the complex forcing and response functions are taken, and for cosine forcing the real parts are used, which in this case gives more directly the result that

$$\frac{X}{F_0} = \frac{1/k}{\left(1-\left(\frac{\Omega}{\omega_n}\right)^2 + i2\zeta\frac{\Omega}{\omega_n}\right)} = H(\mathrm{i}\Omega),$$

where $H(\mathrm{i}\Omega)$ is defined as the (receptance) complex frequency response function (FRF) for the system. However, for nonlinear vibration, it is essential to use both the $\mathrm{e}^{\mathrm{i}\Omega t}$ and $\mathrm{e}^{-\mathrm{i}\Omega t}$ terms, otherwise important *cross-coupling* terms will not appear in the analysis. This will be discussed further in Sect. 1.3.2.

From Eq. (1.12), two special cases can be noted. First when $\Omega = 0$, the response is $X_r = \frac{F_0}{k}$ which corresponds to the static force displacement relationship $x_s = \frac{F_0}{k}$, when there is no dynamic excitation. The second special case occurs when $\Omega = \omega_n\sqrt{1-2\zeta^2}$, which corresponds to the maximum value of the displacement response for the linear system, given by Eq. (1.12). The shape of the response is in the form of a *resonance* peak, with resonance occurring exactly at $\Omega = \omega_n\sqrt{1-2\zeta^2}$. The concept of resonance is of major importance for the study of both linear and nonlinear vibration, especially when damping in the system is small. The reason can be seen from Eq. (1.12), where at resonance

$$x_{0,max} = \left(\frac{F_0}{k}\right)\frac{1}{\sqrt{4\zeta^2\sqrt{1-\zeta^2}}}, \quad \text{for } \zeta < \frac{1}{\sqrt{2}}$$

so that the smaller the damping ratio, ζ, the larger the maximum response, $x_{0,max}$ (see Example 1.1). Lightly damped resonances usually lead to larger than desired displacements in most structural systems. The difficulties of controlling this type of behaviour will be outlined in Sect. 1.4 and forms the core challenge of *active vibration control*. An example of linear resonance is considered next.

Example 1.1 Linear resonance

Problem Compute the amplitude and phase resonance curves for a linear oscillator with mass $m = 1$ kg and stiffness $k = 16$ N/m and three different damping values $c = 0.5, 1.0, 4.0$ N s/m.

Solution When the mass $m = 1$ kg and stiffness $k = 16$ N/m, the natural frequency is $\omega_n = \sqrt{16/1} = 4$ rads/s. Then for three different values of damping, $c = 0.5, 1.0, 4.0$ N s/m, the ratio of dynamic to static response amplitude X_r/x_s and phase ϕ can be plotted from Eqs. (1.12) and (1.13). For the three c values the corresponding damping ratios are $\zeta = 0.0625, 0.125, 0.5$. The results are plotted in Fig. 1.10. The ratio X_r/x_s is often referred to as the *dynamic amplification factor*. Note that as Ω increases the phase lag, ϕ, gets larger, and, when resonance occurs (at frequency $\Omega = \omega_n\sqrt{1-2\zeta^2}$) the phase lag is exactly $\phi = \pi/2$. ■

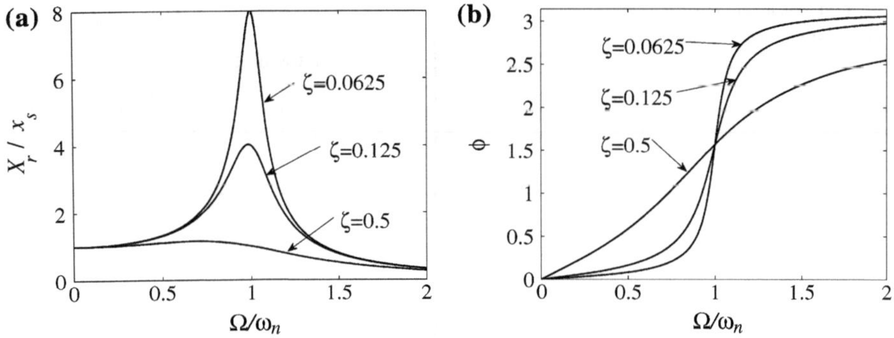

Fig. 1.10 Linear vibration response of a single-degree-of-freedom system showing amplitude and phase functions

Finally, in this section we note that the original equation of motion, Eq. (1.3) can be written as

$$\ddot{x} + 2\zeta\omega_n\dot{x} + \omega_n^2 x = \frac{F_e}{m}, \tag{1.16}$$

by using the definitions of natural frequency, $\omega_n = \sqrt{k/m}$, and damping ratio, $\zeta = c/2m\omega_n$. This is an important form for two reasons. First, the parameters of the system ω_n and ζ can very often be estimated from experiments without knowing m, c, or k. Second, linear systems with more than one degree of freedom can be transformed into a system of equations with an identical form to Eq. (1.16)—discussed further in Sect. 1.3.3.

1.3.2 Nonlinear Vibration Modelled Using Sine Waves

Now consider the nonlinear equation of motion

$$\ddot{x} + 2\zeta\omega_n\dot{x} + \omega_n^2 x + \alpha x^3 = F\sin(\Omega t), \tag{1.17}$$

where $F = F_0/m$. This equation includes a cubic displacement term, and is usually referred to as the Duffing oscillator. In structural engineering the Duffing oscillator is used to model nonlinear force displacement behaviour such as *hardening* and *softening*. An example of typical force displacement behaviour is shown in Fig. 1.11. The physical mechanisms which cause this type of behaviour will be discussed in more detail in Chap. 6.

The forcing is harmonic and can be represented as $F\sin(\Omega t) = \frac{F}{2\mathrm{i}}(\mathrm{e}^{i\Omega t} - \mathrm{e}^{-i\Omega t})$, where F is real. Assuming the response is also harmonic, and using the exponential form found in the linear example Eq. (1.15), we find

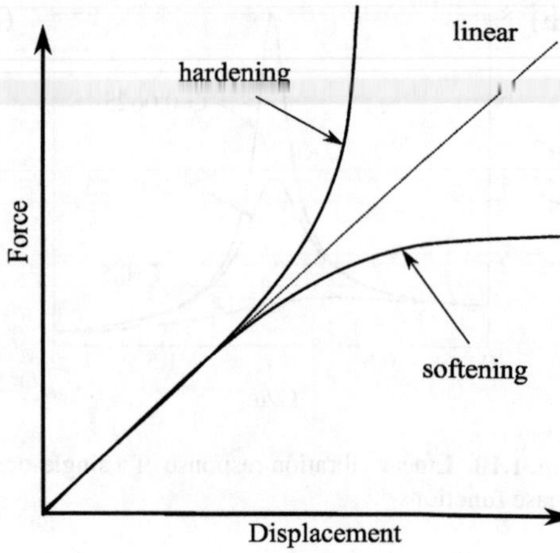

Fig. 1.11 Nonlinear force-displacement behaviour

$$x = \frac{X_r}{2\mathrm{i}}(\mathrm{e}^{\mathrm{i}(\Omega t-\phi)} - \mathrm{e}^{-\mathrm{i}(\Omega t-\phi)}), \tag{1.18}$$

where X_r and ϕ are the amplitude and phase (for the linear case, when $\alpha = 0$, these values are given by Eqs. (1.12) and (1.13) respectively).

The main difference from the previous linear example is the presence of the cubic term. To investigate what happens to the assumed solution, Eq. (1.18), when substituted into the governing equation, (1.17), consider

$$\begin{aligned} x^3 &= \left(\frac{X_r}{2\mathrm{i}}\right)^3 (\mathrm{e}^{\mathrm{i}(\Omega t-\phi)} - \mathrm{e}^{-\mathrm{i}(\Omega t-\phi)})^3, \\ &= \frac{X_r^3}{8\mathrm{i}^3}(\mathrm{e}^{\mathrm{i}3(\Omega t-\phi)} - 3\mathrm{e}^{\mathrm{i}2(\Omega t-\phi)}\mathrm{e}^{-\mathrm{i}(\Omega t-\phi)} + 3\mathrm{e}^{\mathrm{i}(\Omega t-\phi)}\mathrm{e}^{-\mathrm{i}2(\Omega t-\phi)} - \mathrm{e}^{-\mathrm{i}3(\Omega t-\phi)}), \\ &= \frac{X_r^3}{8\mathrm{i}^3}(\mathrm{e}^{\mathrm{i}3(\Omega t-\phi)} - 3\mathrm{e}^{\mathrm{i}(\Omega t-\phi)} + 3\mathrm{e}^{-\mathrm{i}(\Omega t-\phi)} - \mathrm{e}^{-\mathrm{i}3(\Omega t-\phi)}), \\ &= \frac{X_r^3}{4}(3\sin(\Omega t-\phi) - \sin(3(\Omega t-\phi))), \end{aligned} \tag{1.19}$$

from which it can be seen immediately that the cubic term has a vibration response at both the forcing frequency, Ω, and three times the forcing frequency, 3Ω. This is contrary to the original assumption that the response, x, is a single-frequency sine wave with frequency Ω.[4]

This illustrates that, for a nonlinear system with a single-frequency harmonic forcing, a response can be expected at frequencies other than just the input frequency. In this example, the effect of the cubic nonlinearity is to *generate* an additional harmonic response at three times the forcing frequency. The harmonic generation occurs

[4] Note that this result can be found more directly using a sine wave substitution. However, the exponential form is useful for normal form analysis in Chaps. 4 and 5.

because of the additional cross-coupling terms which appear in the cubic expansion. If the method were repeated with an assumed solution including both $\sin(\Omega t)$ and $\sin(3\Omega t)$, the cross-coupling terms would generate harmonics at $5, 7$ and 9Ω. In general, the use of the term harmonic in this context means an additional frequency response related to the lowest (or fundamental) forcing frequency, which is Ω in this case.

Now consider simplifying the problem by setting the forcing and damping to zero, $F = \zeta = 0$, which results in a governing equation of the form

$$\ddot{x} + \omega_n^2 x + \alpha x^3 = 0. \tag{1.20}$$

As the system is undamped, any non-zero initial conditions, $x(0)$ and $\dot{x}(0)$, will result in periodic motion. Now assume a solution $x = X_r \sin(\omega_r t)$, where ω_r is the frequency of the response motion. Note that as there is no forcing sine wave to compare with, the phase term in the response will be assumed (for the moment) to be zero.[5] Substituting this assumed solution into Eq. (1.20), using the result of Eq. (1.19) we find

$$-\omega_r^2 X_r \sin(\omega_r t) + \omega_n^2 X_r \sin(\omega_r t) + \frac{X_r^3}{4}(3\sin(\omega_r t) - \sin(3\omega_r t)) = 0.$$

Gathering the coefficients of the harmonic terms, $\sin(\omega_r t)$ and $\sin(3\omega_r t)$, gives

$$\left[(\omega_n^2 - \omega_r^2)X_r + \frac{3\alpha X_r^3}{4}\right]\sin(\omega_r t) - \frac{\alpha X_r^3}{4}\sin(3\omega_r t) = 0. \tag{1.21}$$

To satisfy this expression, the coefficients of the harmonic terms must equal zero, which can only be true for the $\sin(3\omega_r t)$ if either α or X_r is zero. To get round this, assume (at least for now) that the $\sin(3\omega_r t)$ term can be ignored, in which case Eq. (1.21) becomes

$$[(\omega_n^2 - \omega_r^2)X_r + \frac{3\alpha X_r^3}{4}]\sin(\omega_r t) \approx 0,$$

leading to

$$(\omega_n^2 - \omega_r^2)X_r + \frac{3\alpha X_r^3}{4} \approx 0,$$

from which an approximate value for the response frequency can be obtained, which is

$$\omega_r^2 \approx \omega_n^2 + \frac{3\alpha X_r^2}{4} \rightsquigarrow \omega_r \approx \omega_n\sqrt{1 + \frac{3\alpha X_r^2}{4\omega_n^2}}. \tag{1.22}$$

[5] In fact, if $\dot{x}(0) \neq 0$ this assumption becomes invalid.

Equation (1.22) implies that the frequency of the response, ω_r, is a function of the amplitude of response,[6] X_r. This *amplitude dependence* is another key difference between nonlinear vibration problems and linear ones (see Example 1.2). This analysis, although simplified, is a basic form of the *harmonic balance* method. One can think of having to balance all the individual harmonic terms, after substituting the assumed trial solution in the equation of motion. A more detailed discussion of harmonic balance will be given in Chap. 4, Sect. 4.2. Now, a Duffing oscillator example is considered.

Example 1.2 Duffing oscillator

Problem Compute time-series responses for a Duffing oscillator given by Eq. (1.17) with the following parameters $\omega_n = 4\,\text{rad/s}$, $\zeta = 0.0625$ and $F = 16\,\text{N/m}$. Consider two cases; (a) when $\alpha = 1.0\,\text{N/kg m}^3$ $\Omega = 5\,\text{rads/s}$ and (b) $\alpha = 16\,\text{N/kg m}^3$, $\Omega = 1.75\,\text{rads/s}$. What type of steady-state behaviour is observed?

Solution The time-series responses are computed using numerical integration which is described in more detail in Example 2.1, in Chap. 2. Figure 1.12 shows the computed results for the two cases. In Fig. 1.12a, the system is being forced close to linear resonance and α is small compared to ω_n^2. The resulting displacement response, x, is close to being harmonic (at least to the naked eye). In Fig. 1.12b, the forcing frequency is at a third of linear resonance, and the nonlinear parameter, α is the same magnitude as ω_n^2. In this case the response is clearly non-harmonic, and contains at least two frequency components. ■

1.3.3 Multiple Degrees-of-Freedom

In the preceding two subsections, systems with only a single degree of freedom were considered. Most flexible structures exhibit multiple degrees-of-freedom. One of the main challenges of vibration modelling is to capture this type of behaviour accurately. For control, damping out unwanted vibrations in a system where there are multiple possible modes of behaviour is also a major challenge. Here, an introduction to the effect of multiple degrees-of-freedom is briefly given, starting with a linear multi-degree-of-freedom system.

Recall Eq. (1.2),

$$M\ddot{\mathbf{x}} + C\dot{\mathbf{x}} + K\mathbf{x} = \mathbf{F}_E, \tag{1.23}$$

which represents the governing equation for linear vibration of structural and mechanical systems, where $\mathbf{F}_E$ is the dynamic forcing vector. The matrices M, C and K are not diagonal in general, so the equations are coupled. However, if a transformation

[6] In many texts on nonlinear vibration, α is assumed to be small, in which case the square root term can be approximated using $(1+a)^{1/2} \approx (1+a/2)$ to give $\omega_r \approx \omega_n[1 + \frac{3\alpha X_r^2}{8\omega_n^2}]$.

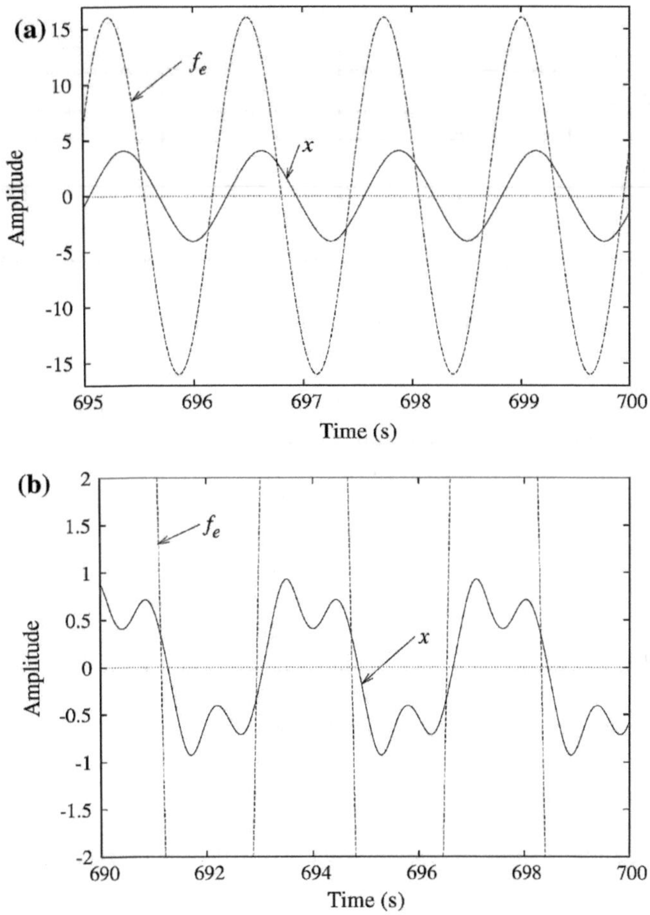

Fig. 1.12 Nonlinear vibration response of a single-degree-of-freedom Duffing oscillator with $\omega_n = 4\,\text{rads/s}$, $F_0 = 16\,\text{N}$, $\zeta = 0.0625$ and **a** $\alpha = 1.0\,\text{N/kg}\,\text{m}^3$, $\Omega = 5\,\text{rads/s}$, **b** $\alpha = 16$, $\Omega = 1.75\,\text{rads/s}$ (Note at this scale the forcing sine wave is only partially shown.)

could be found which replaced M, C and K with equivalent diagonal matrices, the problem could be simplified.

To do this, the vector of displacements, $\mathbf{x}$, is transformed by making the substitution, $\mathbf{x} = \Psi\mathbf{q}$ to give

$$M\Psi\ddot{\mathbf{q}} + C\Psi\dot{\mathbf{q}} + K\Psi\mathbf{q} = \mathbf{F}_E.$$

Pre-multiplying by Ψ^T results in

$$\Psi^T M\Psi\ddot{\mathbf{q}} + \Psi^T C\Psi\dot{\mathbf{q}} + \Psi^T K\Psi\mathbf{q} = \Psi^T\mathbf{F}_E.$$

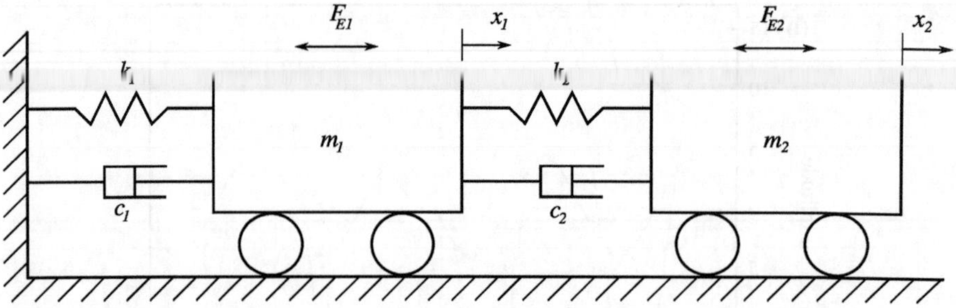

Fig. 1.13 A two-degree-of-freedom system with linear springs and viscous dampers

What is required is that Ψ be chosen so that $\Psi^T M\Psi$, $\Psi^T C\Psi$ and $\Psi^T K\Psi$ are simultaneously diagonalised. This can be achieved if M, C and K satisfy certain conditions (discussed in more detail in Sect. 5.2).

Setting $\mathbf{F}_E = 0$ and assuming a sinusoidal solution for each x_i, we can find the system eigenvalues and eigenvectors (an example is shown below, and a more in depth discussion of the eigenvalue problem is given in Chap. 5). The N eigenvalues relate to the N natural frequencies of the system and the *modal matrix* Ψ is formed such that the columns contain the eigenvectors. As eigenvectors are non-unique, Ψ is normalized such that $\Psi^T M\Psi = I$, where I is the identity matrix. Then $\Psi^T K\Psi = [\omega_{ni}^2]$, becomes a diagonal matrix containing the squared natural frequencies.

To diagonalise the C matrix, as well as the M and K matrices, a further assumption is needed. A sufficient (but not necessary) condition is that the damping matrix, C, is linearly proportional to the mass and stiffness matrices defined as $C = \alpha M + \beta K$, where α and β are scalar constants. This type of *proportional* damping is often known as *Rayleigh* damping after (Rayleigh 1894a, b) who was first to use this method.[7] With proportional damping $\Psi^T C\Psi = [2\zeta_i\omega_{ni}]$, which is a diagonal matrix with ζ_i as the *modal* damping coefficient. This results in N equations of the form

$$\ddot{q}_i + 2\zeta_i\omega_{ni}\dot{q}_i + \omega_{ni}^2 q_i = \psi_i^T \boldsymbol{F}_E, \tag{1.24}$$

where ψ_i^T is the ith row of Ψ^T. Equation (1.24) has exactly the same form as the single-degree-of-freedom system, Eq. (1.16), and can be solved in the same way. In fact, this type of linear system has the special property that each of the equations can be solved separately and the results added together (or superimposed) to get the total response. This *principle of superposition* is a key part of the analysis of linear vibrating systems.

Example 1.3 A two-degree-of-freedom linear system

Problem Find the undamped natural frequencies and mode-shapes for a two-degree-of-freedom oscillator with lumped masses $m_1 = m_2 = 1$ kg coupled by springs with

[7] In fact this can be extended to include additional terms formed from combinations of M and K, known as extended Rayleigh damping. See Clough and Penzien (1993) for a detailed discussion.

stiffness $k_1 = k_2 = 16\,\text{N/m}$ and dampers $c_1 = c_2 = 0.5\,\text{N s/m}$, as shown schematically in Fig. 1.13. Then compute the steady-state amplitude and phase response for $0 \leq \Omega \leq 10$ for the case when the forcing functions $F_{E1} = 16\sin(\Omega t)\text{N}$ and $F_{E2} = 0$ are applied.

Solution For an oscillator of the type shown schematically in Fig. 1.13, the mass and stiffness matrices are given by

$$M = \begin{bmatrix} 1 & 0 \\ 0 & 1 \end{bmatrix}, \quad K = \begin{bmatrix} 32 & -16 \\ -16 & 16 \end{bmatrix}.$$

For small damping, the natural frequencies and mode-shapes can be found by setting $C = \mathbf{F}_E = 0$ and substituting the trial solution $\mathbf{x} = \mathbf{x}_0 \sin(\omega t)$ into Eq. (1.23), where $\mathbf{x}$ and $\mathbf{x}_0$ are both 2×1 vectors. This leads to an eigenvalue problem of the form $(-\omega^2 M + K)\mathbf{x}_0 = 0$, which for the nontrivial solution (i.e. the case when $\mathbf{x}_0 \neq 0$) requires that $|-\omega^2 M + K| = 0$. This gives

$$\begin{vmatrix} 32 - \omega^2 & -16 \\ -16 & 16 - \omega^2 \end{vmatrix} = 0. \tag{1.25}$$

Solving Eq. (1.25) gives natural frequencies $\omega = 2.46\,\text{rad/s} = \omega_{n1}$ and $\omega = 6.45\,\text{rad/s} = \omega_{n2}$. Substituting each of these values in turn gives values for the two eigenvectors $\mathbf{x}_{01} = [-0.526, \ -0.851]^T$ and $\mathbf{x}_{02} = [-0.851, 0.526]^T$, which correspond to the vibration *mode-shapes* for the system. In this example, $\mathbf{x}_{01}$ corresponds to a vibration mode where the masses are in phase with each other, and $\mathbf{x}_{02}$ corresponds to a vibration mode where the masses are out of phase with each other.

The resulting behaviour (with damping included) is shown in Fig. 1.14 when the forcing functions $F_{E1} = 16\sin(\Omega t)\text{N}$ and $F_{E2} = 0$ are applied. The two masses have clear resonances when the forcing frequency is at the two natural frequency values such that $\Omega = 2.46\,\text{rad/s}$ and $\Omega = 6.45\,\text{rad/s}$. At $\Omega = 2.46\,\text{rad/s}$, the masses are oscillating in phase with an amplitude ratio given by $\mathbf{x}_{01}$ and at $\Omega = 6.45\,\text{rad/s}$ the masses are almost π out of phase, with a ratio given by $\mathbf{x}_{02}$ (note that in the figure the maximum positive amplitudes per period are shown, so the sign difference in $\mathbf{x}_{02}$ is not evident). Just after $\Omega = 6.45\,\text{rad/s}$ there is a 2π jump down in the phase of mass two as it lags more than 2π behind the forcing frequency. ■

Superposition does not generally hold for nonlinear systems, and so this powerful technique cannot be extended from linear directly to nonlinear. As discussed in Sect. 1.3.2, even for a single-degree-of-freedom nonlinear system, many different and complex responses are possible. This makes generalizations very difficult and the approach to nonlinear vibrations tends to require a problem-specific approach.

However, there are many aspects of linear vibration which can be used in the study of nonlinear vibrations. This is most obvious when using linear approximations to the nonlinear system, which is a core technique, usually carried out in a small (local) region of the system parameters. Having linearized the system in a region of interest, we can perform an eigen analysis to give information on the system dynamics.

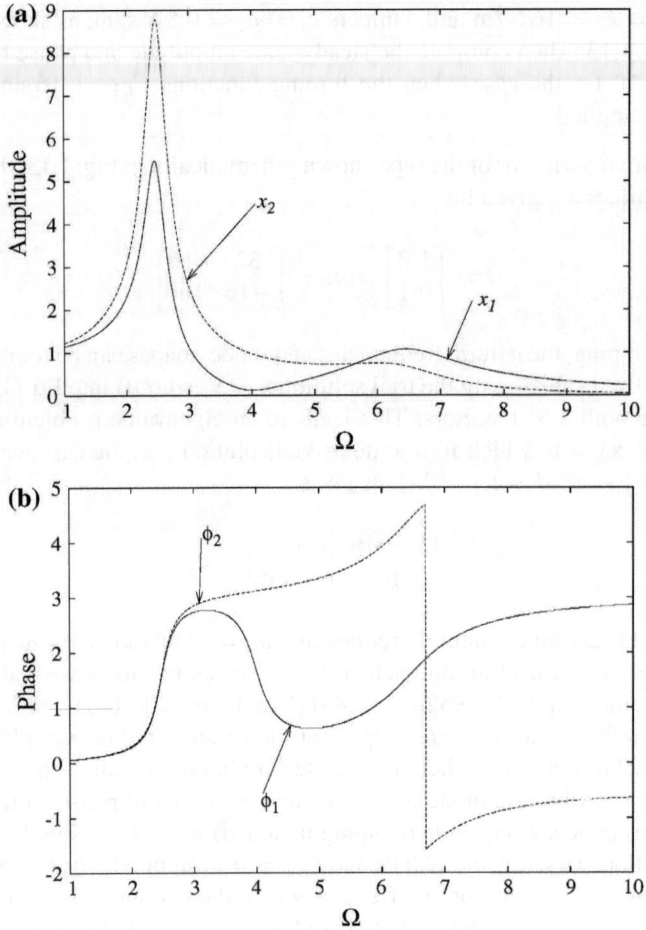

Fig. 1.14 Linear vibration response of a two-degree-of-freedom example: **a** shows the *maximum* amplitude of response per forcing period for steady-state vibration, for each mass, and **b** shows the phase difference between the forcing function F_{E1} and each of the masses

For nonlinear mechanical and structural systems, the concepts of resonance and physical modes of vibration are as important as they are in linear problems. In many cases the nonlinear effects can be modelled as a small *perturbation* of an underlying linear system. This will be discussed in detail in Chaps. 2–4.

1.4 Control of Nonlinear Vibrations

In Sect. 1.2.5 it was noted that control forces can be added to a vibrating system in order to control the behaviour of the system. One of the primary reasons to add control is to reduce unwanted vibrations. Of course, vibrations can be reduced using dampers and/or other damping treatments. In this case damping systems without any active control forces are called *passive* vibration-reduction devices. A classical tuned mass damper (see for example Den Hartog (1934)) is one kind of passive vibration-reduction device. Using a damper which has a passive capability in conjunction with active control forces is called *semi-active* control. An alternative type of semi-active control is a passive device which has parameters that can be altered to improve performance during operation. The final type of vibration reduction is to use a system which has only active control forces, with no passive capability. This is called simply *active control*. Active control is able to generate motion rather than just resist it (and so can put energy into the system), and as a result can also potentially render the system unstable.

The most useful form of control for low frequency vibration applications[8] is *feedback control*. Feedback control uses information from the system response in order to change the control input, $\mathbf{F}_C$, to achieve an improved behaviour. To monitor the response of the system, measurements are taken using sensors to record particular system variables such as displacements, accelerations or forces. These measurements are then *fed back* and used to update the control forces via a *control algorithm*. The control algorithm is designed to give an optimized system performance based on the requirements of the designer.

A block diagram of a typical feedback control system is shown in Fig. 1.15. This type of block diagram shows the flow (or logic) of the feedback system, from an input *demand*,[9] signal to a system response *output*. The key point to note is that part of the system response is measured and used by the control algorithm inside the controller block. In the example shown, the output from the system is subtracted from the demand signal[10] to form an error, $e(t) = r(t) - x(t)$, which is used by the controller in order to give an improved response.

The system to be controlled is usually referred to as the *plant* in control terminology. If there is just one input demand and one response output from the plant, the control system is said to be a *single-input, single-output* (SISO) system. If there is more than one input or output, the control system is said to be *multiple-input, multiple-output* (MIMO).

[8] For higher frequencies, feedforward control often becomes more appropriate for linear systems, see for example Fuller et al. (1996). Many control approaches use a combination of feedback and feedforward control. A discussion of this for nonlinear systems is given by Slotine and Li (1991).

[9] Sometimes known as *setpoint* or *reference* this is the desired system output.

[10] Note the negative feedback. As a general rule, positive feedback will cause instability.

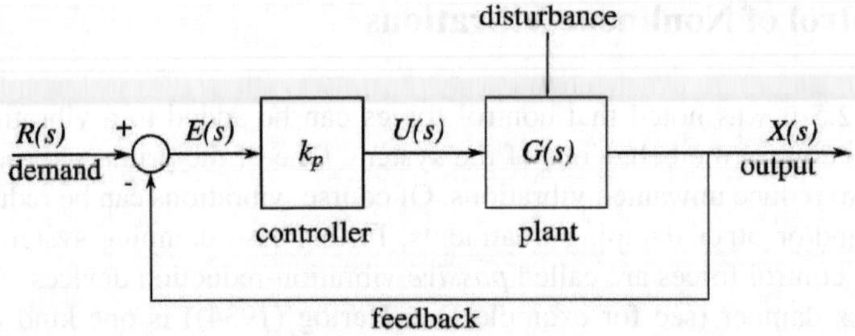

Fig. 1.15 Feedback control block diagram

A control system with a feedback path is often referred to as a *closed-loop* system. Conversely a system without a feedback path can be referred to as an *open-loop* system, either with or without feedforward control.[11]

Adding feedback introduces issues regarding the stability of the combined controller and system behaviour. Feedback controllers can destabilize a system in certain situations, and the design and application of these types of systems is a large subject in its own right. The problem is usually split into two parts. First, design the system to be stable in a perfect (no noise) environment. This is the stability problem. Second, design the stable system to be as robust as possible to noise and other disturbances (shown in Fig. 1.15 acting on the plant) and to uncertainty which may occur under operating conditions. This is the robustness problem.

1.4.1 Feedback Control of Linear Systems

To control the behaviour of a linear system, control forces can often be added to Eq. (1.1) so that

$$M\ddot{\mathbf{x}} + C\dot{\mathbf{x}} + K\mathbf{x} = \mathbf{F}_E + \mathbf{F}_C,$$

where $\mathbf{x}$ represents the displacements and $\mathbf{F}_C$ is the vector of control forces. This type of second-order differential equation can either be analysed as it is or put into a first-order form, which in some cases can make the analysis easier. To develop a first-order form[12] let $\boldsymbol{x} = [\mathbf{x}^T, \dot{\mathbf{x}}^T]^T$, $F_E = 0$ and $\mathbf{F}_C = P\mathbf{u}$ such that

$$\dot{\boldsymbol{x}} = A\boldsymbol{x} + B\mathbf{u}, \tag{1.26}$$

[11] Feedforward control is a type of controller that does not use feedback from the system being controlled.

[12] Notice that there is a subtle difference between $\boldsymbol{x}$, which is the $2N \times 1$ state vector, and $\mathbf{x}$, which is the $N \times 1$ displacement vector. This is used to maintain (as far as possible) notation conventions from control engineering, nonlinear dynamics and structural vibration.

where

$$A = \begin{bmatrix} 0 & I \\ M^{-1}C & M^{-1}K \end{bmatrix}, \quad B = \begin{bmatrix} 0 \\ M^{-1}P \end{bmatrix},$$

and where $\mathbf{u} = \{u_1, u_2, \ldots, u_N\}^T$ is the vector of control signals u_i, I is the identity matrix and P is a constant matrix representing the control mechanism/hardware. Equation (1.26) represents the first-order or *state space form* of the controlled system. It is normally combined with an output equation $\mathbf{y} = \bar{C}\boldsymbol{x}$, which represents the case when the state vector, $\boldsymbol{x}$, cannot be directly observed, and instead $\mathbf{y}$ are the observations from the sensors.[13]

As an example, consider the single-degree-of-freedom (SDOF) equation of motion, Eq. (1.3), when there is a single control input, u, gives

$$m\ddot{x} + c\dot{x} + kx = f_c = pu, \tag{1.27}$$

where x is the displacement of the mass m and p is a scalar constant which can be thought of as a gain. Taking the Laplace transform[14] of Eq. (1.27), and assuming zero initial conditions we find

$$X(s) = \frac{p}{(ms^2 + cs + k)} U(s) = G(s)U(s),$$

where s is the Laplace variable, and $G(s)$ is the *transfer function* for the single-degree-of-freedom system given in Eq. (1.27). Assuming that the control task is to make the displacement $x(t)$ follow a predetermined reference signal $r(t)$ such that $x(t) \rightarrow r(t)$ in the steady state, we define the error as $e(t) = r(t) - x(t)$. Now the error can be used as feedback, so that when $e \neq 0$, some control effort is applied to the system as shown in Fig. 1.15.

Using the Laplace transform variables shown in Fig. 1.15, $U(s) = k_p E(s) = k_p(R(s) - X(s))$, and by using the logic of the block diagram in Fig. 1.15 (assuming no disturbance), we can show that the steady-state relationship for the closed-loop system is

$$\frac{X(s)}{R(s)} = \frac{G(s)k_p}{1 + k_p G(s)}.$$

For this system the only control parameter to be selected is k_p. The process of choosing the best k_p is often referred to as *control design*. For the design process, the Laplace parameter s is related to a generalized frequency parameter, ω, by the relationship $s = \mathrm{i}\omega$. Then the relationship between the input function $R(\mathrm{i}\omega)$ and the output $X(\mathrm{i}\omega)$ is governed by $G(\mathrm{i}\omega)k_p/(1 + k_p G(\mathrm{i}\omega))$, which is referred to as the *closed-loop transfer function*. Here it is denoted as $L(\mathrm{i}\omega)$.

---

13 To avoid confusion with the damping matrix, $\bar{C}$ is used as the control output matrix.

14 Note that the convention of writing the Laplace transform of a variable as a capital letter is used here.

It should be noted from Fig. 1.15 that the system has negative feedback. However, for sinusoidal signals, should the output signal become phase shifted far enough, the effect will be the same as positive feedback and the system will become unstable. For this to happen, the amplitude of the closed-loop transfer function must be $|L(i\omega)| \geq 1$ and the phase $\arg(L(i\omega)) \geq -\pi$. The reason that only half a wavelength is required is that the negative feedback inverts the signal to look like the input signal after only π phase lag. These conditions can be derived mathematically and are known as the *Nyquist stability criterion*. The use of this criterion is a fundamental design technique for linear control systems—see Goodwin et al. (2000) for details. For a particular k_p value, the frequency can be varied from $0 \leq \omega \leq \omega_{max}$ to see if $L(i\omega)$ remains stable.

The next example uses these ideas for a single-input, single-output system.

Example 1.4 Linear control

Problem For the controlled oscillator given by Eq. (1.27), find the stability of the closed-loop transfer function when mass $m = 1$ kg, stiffness $k = 16$ N/m, damping is $c = 0.5$ and $p = 1$.

Solution With the mass and stiffness values, the natural frequency $\omega_n = \sqrt{16/1} = 4$ rads/s. Then $G(s) = 1/(s^2 + 0.5s + 16)$, which has *poles*[15] of $-0.25 \pm i3.99$. The poles are in the left-hand side of the complex plane, which means that the uncontrolled (or open-loop) system is stable.

For this example, the poles for the controller system (the closed-loop poles) are found from the poles of

$$L(s) = \frac{k_p}{s^2 + 0.5s + 16 + k_p}, \tag{1.28}$$

which gives poles of $s_{1,2} = -0.25 \pm 0.5\sqrt{4(16 + k_p) - 0.25}$, which for positive k_p are always complex and in the left-hand plane, and therefore stable. Note also that the effect of k_p is analogous to adding stiffness to an oscillator. Substituting $s = i\omega$ gives

$$L(i\omega) = \frac{k_p}{16 + k_p - \omega^2 + i0.5\omega}. \tag{1.29}$$

Multiplying top and bottom of Eq. (1.29) by $(16 + k_p - \omega^2 - i0.5\omega)$ allows the real and imaginary parts to be found, from which

$$|L(i\omega)| = \frac{k_p}{\sqrt{(16 + k_p - \omega^2)^2 + (0.5\omega)^2}}, \tag{1.30}$$

and

[15] Poles are the complex roots of the denominator of $G(s)$.

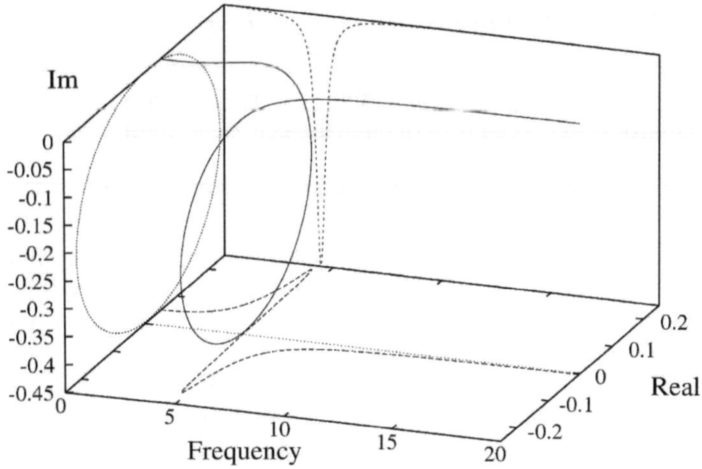

Fig. 1.16 Plot of the complex frequency response function of Eq. (1.28) with $k_p = 4$ for the frequency range $0 \leq \omega \leq 20$ with projections onto the complex, real-frequency and imaginary-frequency planes, following Ewins (2000)

$$\arg(L(\mathrm{i}\omega)) = \arctan\left(\frac{0.5\omega}{(16 + k_p - \omega^2)}\right). \tag{1.31}$$

Note the strong similarities with the analysis which led to Eqs. (1.12) and (1.13). In fact, except for the presence of the control parameter, k_p, the two sets of expressions are analogous. The Nyquist stability criterion says that the ω values at which instability occurs are when $|L(\mathrm{i}\omega)| = 1$, so computing Eq. (1.30) when $k_p = 4$ gives two points at which $|L(\mathrm{i}\omega)| = 1$. Of these, the value closest to the instability point is $\omega \approx 4.82\mathrm{rad/s}$, from which $\arg(L(\mathrm{i}4.82)) \approx 2.5$ (using Eq. (1.31) plus π). So in this example the system is stable by a margin of $\pi - 2.5 \approx 0.641$ radians or 37°. This stability margin is called the *phase margin*. ■

The plot of the complex frequency response function of Eq. (1.28) is shown in Fig. 1.16, with three projections onto other planes. Note that the projection onto the complex plane is a circle,[16] while the projections onto the frequency planes show peaks and inverted peaks.

Frequency-based analysis is a powerful tool for the examination and design of linear vibrations and linear control. However, nonlinearity complicates the frequency response to such an extent that nonlinear control design requires a different approach. This is introduced next.

---

[16] In fact, for viscous damping, this is only approximately circular. See Ewins (2000) for a more detailed discussion of the properties of these functions.

1.4.2 Feedback Control of Nonlinear Systems

A direct analogy between linear and nonlinear control systems can be made in that nonlinear control systems can be written in the general form of

$$\dot{\boldsymbol{x}} = \mathbf{f}(\boldsymbol{x}) + \mathbf{g}(\boldsymbol{x})u, \tag{1.32}$$

where $\mathbf{f}$ is the nonlinear system function, $\mathbf{g}$ is the nonlinear controller function, $\boldsymbol{x}$ is the state vector $\boldsymbol{x} = [\mathbf{x}^T, \dot{\mathbf{x}}^T]^T$ and u is the control signal. This is a nonlinear version of Eq. (1.26). When system states cannot be observed directly, $\mathbf{y} = \mathbf{h}(\boldsymbol{x})$ is used, where $\mathbf{h}$ is the nonlinear output function.

In some situations, the nonlinear functions can be approximated as linear, such that $\mathbf{f}(\boldsymbol{x}) \approx A\boldsymbol{x}$ and $\mathbf{g}(\boldsymbol{x})u \approx Bu$. Depending on the example being considered, $\mathbf{h}(\boldsymbol{x}) \approx \bar{C}\boldsymbol{x}$ may also be applicable. In this case the nonlinear system can be approximated by Eq. (1.32), usually only for some limited range of system parameters.

One of the simplest ways of dealing with nonlinear systems is to use the control signal to cancel the nonlinear part of the system and effectively turn it back into a linear system—see Slotine and Li (1991). Consider the Duffing oscillator (similar to that discussed in Sect. 1.3.2), only this time with a control input which can be written in first-order matrix form as

$$\begin{bmatrix} \dot{x}_1 \\ \dot{x}_2 \end{bmatrix} = \begin{bmatrix} 0 & 1 \\ -\omega_n^2 & -2\zeta\omega_n \end{bmatrix} \begin{bmatrix} x_1 \\ x_2 \end{bmatrix} + \begin{bmatrix} 0 \\ -\alpha x_1^3 \end{bmatrix} + \begin{bmatrix} 0 \\ p \end{bmatrix} u(t). \tag{1.33}$$

In vector form this becomes

$$\dot{\boldsymbol{x}} = A\boldsymbol{x} + \mathscr{N}(\boldsymbol{x}) + Bu, \tag{1.34}$$

where $A\boldsymbol{x} + \mathscr{N}(\boldsymbol{x}) = \mathbf{f}(\boldsymbol{x})$, $B = \mathbf{g}(\boldsymbol{x})$ and $\boldsymbol{x} = [x_1, x_2]^T$ in this case. Note that the non-linear function, $\mathbf{f}(\boldsymbol{x})$, has been split into a linear part, $A\boldsymbol{x}$, and a vector of nonlinear terms, denoted by $\mathscr{N}(\boldsymbol{x})$.

It can be seen by inspection of Eq. (1.33), that if one sets $u = \alpha x_1^3/p$ then, B can be chosen such that $\mathscr{N}(\boldsymbol{x}) + Bu = 0$, ignoring the possibility of transient effects (initial conditions etc.) destabilising the system. The system then reduces to $\dot{\boldsymbol{x}} = A\boldsymbol{x}$, which is linear and, providing A has stable eigenvalues, is also stable.

However, although this has removed the nonlinearity, it has applied no additional control to the linear part of the system. To apply proportional control, one could set $u = \alpha x_1^3/p - k_p x_1$, which would remove the nonlinearity and apply proportional control to the resulting linear system, as now $\mathscr{N}(\boldsymbol{x}) + Bu \to -k_p x_1$ as $t \to \infty$, providing in this case that the closed-loop proportional controller is stable. However, during the transient phase, there will be both linear feedback and nonlinear effects present, which will make the assessment of stability and robustness difficult—techniques for dealing with this will be discussed in Chap. 3.

An alternative approach is to try to establish a linear relationship between the output and input to the system. In general, the output from the nonlinear system is a vector $\mathbf{y} = \mathbf{h}(\boldsymbol{x})$, and in practice this will be one or more measurements of displacement, velocity or acceleration. For example, assume that the output from the Duffing system in Eq. (1.33) is the velocity, x_2, so that $y = x_2$. The input to the Duffing system is the control signal, u. If a linear relationship can be established between y and u, then the system will have been input-output linearized. Consider the Duffing system as an example.

Example 1.5 Nonlinear control

Problem Design a feedback linearisation controller for the Duffing system defined by Eq. (1.33), assuming that the output is the velocity, x_2.

Solution First take the output and differentiate with respect to time, t, to give $\dot{y} = \dot{x}_2$. From Eq. (1.33), the expression for $\dot{x}_2$ can be used to write

$$\dot{y} = \dot{x}_2 = -\omega_n^2 x_1 - 2\zeta\omega_n x_2 - \alpha x_1^3 + pu(t).$$

If the control signal is chosen as

$$u = \frac{1}{p}[\nu(t) + \omega_n^2 x_1 + 2\zeta\omega_n x_2 + \alpha x_1^3],$$

then the effect is to subtract off all the terms and replace them with a new control signal $\nu(t)$, giving an input-output relationship of the form $\dot{y} = \nu(t)$. Then a linear expression can be chosen for $\nu(t)$ to give the required control effect. ■

It should be noted in this example that although only x_2 is the measured output, both states are required to form the control signal. Compared with the more basic feedback linearisation discussed above, this approach has not just cancelled the nonlinear dynamics, the linear part of the $\dot{x}_2$ dynamics has been removed as well and replaced with $\nu(t)$. The whole process can be formalized by using *Lie derivatives*, and will be described in Chap. 3.

The usefulness of feedback linearisation is that, once the system has been linearized, additional linear control tasks can be included using well known techniques. Typically the control tasks of interest are to remove unwanted vibration and/or get the system to follow some predefined reference signal. This and other nonlinear control techniques are discussed in more detail in Chap. 3.

1.5 Continuous Structural Elements

Engineering structures are composed of *structural elements*. These elements can be thought of as the basic building blocks of all structures. In this book, structural

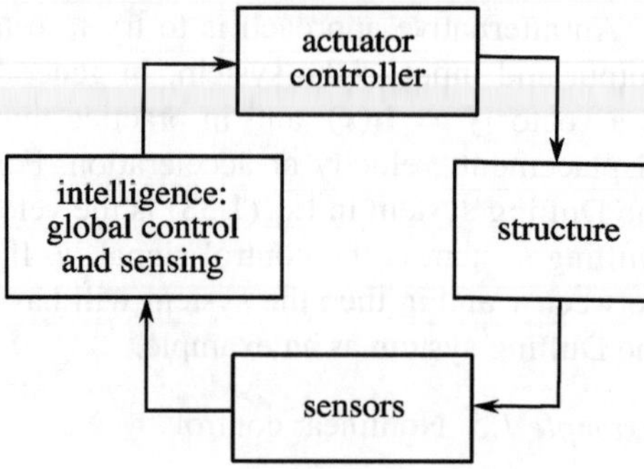

Fig. 1.17 Schematic diagram of a smart structure control system

elements with *continuous* mass distribution are considered. The simplest of such elements carry axial loads only. Examples are strings or *cables* which can carry tensile axial loads only, but have no lateral stiffness. Elements with lateral stiffness which carry only axial loads are called *rods* (or bars). Vertical rods which carry compressive loads are called *columns* or struts.

The next level of complexity is bars with flexural stiffness which can carry both axial and bending loads. These are *beams*. Beams which become increasingly wide eventually become *plates*, which can carry both axial and bending loads in two-directions. Other two-dimensional elements are *membranes*, which carry in-plane loads only, and *shells*, which carry both in-plane loads and bending moments.

To control the vibrations of continuous flexible structural elements such as beams, plates and cables, one first needs a mathematical model. Two main approaches are used to develop models of vibration problems. The first approach is to derive a set of equations that govern the behaviour of the system based on the physics of the system. Often a small section of the structural element is considered. Then the *governing equations of motion* are derived based on analytical techniques such as force and moment balance.[17] Models for these types of structural elements will be derived in detail in Chaps. 6–8. The second approach is to identify a model for the system based on experimentally recorded data. For vibration analysis this approach is normally referred to as *modal analysis*—see Ewins (2000). The modal aspects of nonlinear vibration will be discussed in more detail in Chap. 5 and a case study for a bi-stable shell structure is considered in Chap. 8.

1.6 Smart Structures

The ultimate objective of adding control capability to structures is typically to create smart (or adaptive or intelligent) structures. The key elements of a smart structure

[17] This is the method adopted in most derivations in this book. Other techniques such as Lagrange's equation for energy or Hamilton's principle can also be used in many cases.

are shown in Fig. 1.17. The structure needs to have some *awareness* of its condition and/or the environment it is in. This is achieved by having a series of measurement sensors mounted on (or integrated into) the structure. Information from the sensors is then used by the global control system. This is where the smart (or intelligent) behaviour is generated. The global control system will monitor the condition of the structure (via the sensors) and when required give command signals to a series of *actuators* which act on the structure.

An understanding of nonlinear structural mechanics and vibration is important for many smart structures, because they typically have one or more of the following characteristics: (i) the ability to have large deformations, (ii) non-homogeneous material properties, (iii) material parameters which vary (or can be varied), (iv) multiple stable states, (v) highly flexible elements, (vi) very light damping and (vii) the need to be operated in a dynamic environment. The most common application is to design some form of active or semi-active vibration control for the structure to reduce unwanted vibration.

Throughout this book, reference will be made to the relevance of smart structures. Particularly: Chap. 3 will discuss the control aspects of nonlinear structures; in Chap. 6 modal control to reduce unwanted vibration in beams will be discussed and in Chap. 8, shape change of a bi-stable shell is discussed.

1.7 Chapter Notes

The aim of this chapter is to give an introduction to nonlinear vibrations and control. Good introductions to nonlinear vibrations are given by Den Hartog (1934), McLachlan (1950), Moon (1987), Nayfeh and Mook (1995), Thompson and Stewart (2002), Cartmell (1990) and Worden and Tomlinson (2000). An excellent description of the history of mechanics of material, elasticity and structural mechanics is given by Timoshenko (1953). Additional information on the mechanics of materials in this chapter comes mainly from Nelkon (1969). A full description of the dynamics of the simple pendulum can be found in Jordan and Smith (1999). Classical results on large deformation of elastica can be found in Frish-Fay (1962) and are also discussed in Timoshenko (1953). See also Virgin (2007) for analysis of buckled beams and Thompson and Champneys (1996) for twisted rods and localized behaviour. References to the more recent applications, such as supercoiling and packing of DNA molecules are discussed in the overview by van der Heijden (2008).

There are a large numbers of texts on linear vibrations. Good overviews are given by Bishop and Johnson (1960), Weaver et al. (1990), Meirovitch (2001), Géradin (1997), Ewins (2000) and Inman (2007). A number of texts discuss linear vibration with control, such as Inman (2006), Beards (1981), Fuller et al. (1996) and Moheimani et al. (2003). There are also texts which discuss the vibration and control of smart structures such as Clark et al. (1998), Srinivasan and McFarland (2001), Preumont (2002), Worden et al. (2003) and Leo (2007). A good overview of linear control is given by Goodwin et al. (2000). A good introduction to nonlinear control can be found in Slotine and Li (1991).

References

Beards, C. F. (1981). *Vibration analysis and control system dynamics*. Chichester: Ellis Horwood.

Bishop, R. E. D., & Johnson, D. C. (1960). *The mechanics of vibration*. Cambridge: Cambridge University Press.

Brogliato, B. (1999). *Nonsmooth mechanics: Models, dynamics and control*. London: Springer-Verlag.

Cartmell, M. (1990). *Introduction to linear, parametric and nonlinear vibrations*. New York: Chapman and Hall.

Clark, R. L., Saunders, W. R., & Gibbs, G. P. (1998). *Adaptive structures; dynamics and control*. New York: John Wiley.

Clough, R. W., & Penzien, J. (1993). *Dynamics of structures* (2nd ed.). New York: McGraw-Hill.

Den Hartog, J. P. (1934). *Mechanical vibrations*. New York: McGraw-Hill.

di Bernardo, M., Budd, C. J., Champneys, A. R., & Kowalczyk, P. (2007). *Piecewise-smooth dynamical systems: Theory and applications*. London: Springer-Verlag.

Ewins, D. J. (2000). *Modal testing*. Philadelphia: Research Studies Press.

Frish-Fay, R. (1962). *Flexible bars*. London: Butterworths.

Fuller, C. R., Elliot, S. J., & Nelson, P. A. (1996). *Active control of vibration*. London: Academic Press.

Géradin, M., & Rixen, D. (1997). *Mechanical vibrations: Theory and application to structural dynamics*. Chichester: Wiley Blackwell.

Goodwin, G. C., Graebe, S. F., & Salgado, M. E. (2000). *Control system design*. Upper Saddle River: Pearson.

Inman, D. J. (2006). *Vibration with control*. New York: Wiley.

Inman, D. J. (2007). *Engineering vibration*. New York: Prentice Hall.

Jordan, D. W., & Smith, P. (1999). *Nonlinear ordinary differential equations; An introduction to dynamical systems* (3rd ed.). Oxford: Oxford University Press.

Krauskopf, B. (2005). Bifurcation analysis of lasers with delay. In *Unlocking dynamical diversity: Optical feedback effects on semiconductor lasers* (pp. 147–183). New York: Wiley.

Leo, D. J. (2007). *Smart material systems*. Hoboken: Wiley.

McLachlan, N. W. (1950). *Ordinary non-linear differential equations*. Oxford: Oxford University Press.

Meirovitch, L. (2001). *Fundamentals of vibration*. New York: McGraw-Hill.

Moheimani, S. O. R., Halim, D., & Fleming, A. J. (2003). *Spatial control of vibration*. Singapore: World Scientific.

Moon, F. C. (1987). *Chaotic vibrations: An introduction for applied scientists and engineers*. New York: John Wiley.

Moon, F. C., & Shaw, S. W. (1983). Chaotic vibrations of a beam with non-linear boundary conditions. *International Journal of Non-Linear Mechanics*, *18*(6), 465–477.

Nayfeh, A. H., & Mook, D. T. (1995). *Nonlinear oscillations*. New York: John Wiley.

Neill, D., Livelybrooks, D., & Donnelly, R. J. (2007). A pendulum experiment on added mass and the principle of equivalence. *American Journal of Physics*, *75*, 226–229.

Nelkon, M. (1969). *Mechanics and properties of matter*. Toronto: Heinemann.

Preumont, A. (2002). *Vibration control of active structures*. Dordrecht: Kluwer.

Rayleigh, J. W. S. (1894a). *Theory of sound* (Vol. 1). London: Macmillan and Co.

Rayleigh, J. W. S. (1894b). *Theory of sound* (Vol. 2). London: Macmillan and Co.

Slotine, J.-J. E., & Li, W. (1991). *Applied nonlinear control*. Englewood: Prentice Hall.

Srinivasan, A. V., & McFarland, D. M. (2001). *Smart structures*. Cambridge: Cambridge University Press.

Thompson, J. M. T., & Champneys, A. R. (1996). From helix to localized writhing in the torsional post-buckling of elastic rods. *Proceedings of the Royal Society A*, *452*, 117–138.

Thompson, J. M. T., & Stewart, H. B. (2002). *Nonlinear dynamics and chaos*. Chichester: John Wiley.

Timoshenko, S. P. (1953). *History of strength of materials*. New York: McGraw-Hill.

van der Heijden, G. H. M. (2008). *The nonlinear mechanics of slender structures undergoing large deformations*. Available for download from G. H. M. van der Heijden's website.

Virgin, L. N. (2007). *Vibration of axially-loaded structures*. Cambridge: Cambridge University Press.

Weaver Jr, W., Timoshenko, S. P., & Young, D. (1990). *Vibration problems in engineering*. New York: Wiley.

Worden, K., & Tomlinson, G. R. (2000). *Nonlinearity in structural dynamics*. Bristol: IOP.

Worden, K., Bullough, W. A., & Haywood, J. (2003). *Smart technologies*. River Edge: World Scientific.

Chapter 2
Nonlinear Vibration Phenomena

Abstract Nonlinear systems have a range of behaviour not seen in linear vibrating systems. In this chapter the phenomena associated with nonlinear vibrating systems are described in detail. In the absence of exact solutions, the analysis of nonlinear systems is usually undertaken using approximate analysis, numerical simulations and geometrical techniques. This form of analysis has become known as *dynamical systems theory* (or sometimes *chaos theory*) and is based on using a system *state space*. In this chapter the basic ideas of dynamical systems are applied to vibrating systems. Finally, the changes in system behaviour as one (or more) of the parameters is varied are discussed. Such changes are known as *bifurcations*, and they are highly significant for the understanding of nonlinear systems.

2.1 State Space Analysis of Dynamical Systems

In this chapter only continuous time *deterministic* dynamical systems are considered. This means that, even if the system output looks very complicated, it is entirely determined by the input with no random effects. The state of the system is measured by the state vector, $\boldsymbol{x} = \{x_1, x_2, \ldots, x_n\}^T$. The x_i are variable quantities such as positions and velocities, which describe what 'state' the system is in at any point in time, t. In a dynamical system, the states vary with time; so each state is a function of time, which for the ith state is written as $x_i(t)$. The states are sometimes referred to as the *dependent* variables, in that they depend on (are a function of) t. Time, t is then referred to as the *independent* variable, meaning that it is not a function of anything else.

The state vector has n states, often written as $\boldsymbol{x} \in \mathbb{R}^n$, which means that $\boldsymbol{x}$ can be plotted in an n-dimensional Euclidean space.[1] This is called the *state space* or sometimes the *phase space* of the system. When $n > 3$ the state space cannot be plotted directly, but instead *projections* of $\boldsymbol{x}$ are plotted, usually in either two or three dimensions. A plot of representative solutions (or trajectories) in phase space

[1] Think of $\mathbb{R}$ as representing the set of real numbers on an axis in an n-dimensional space.

D. Wagg and S. Neild, *Nonlinear Vibration with Control*,
Solid Mechanics and Its Applications 218, DOI 10.1007/978-3-319-10644-1_2

is called a *phase portrait*, and an important special case is a two-dimensional plot of displacement against velocity.

A dynamical system is one where the state of the system evolves over time. It is assumed that this evolution is governed by a differential equation which can be written in a general form as

$$\frac{\mathrm{d}\boldsymbol{x}}{\mathrm{d}t} = \mathbf{f}(\boldsymbol{x}, t). \tag{2.1}$$

Here $\boldsymbol{x}$ is the vector representing the state of the systems at any time, t, and $\mathbf{f}(\boldsymbol{x}, t)$ is a vector of nonlinear functions governing the time evolution of the system.[2] If $\mathbf{f}$ has no dependence on time, $\mathbf{f} = \mathbf{f}(\boldsymbol{x})$, the system is said to be autonomous, when $\mathbf{f} = \mathbf{f}(\boldsymbol{x}, t)$ is a function of time, the system is called non-autonomous.[3] Equation (2.1) is often written as

$$\dot{\boldsymbol{x}} = \mathbf{f}(\boldsymbol{x}, t),$$

where an overdot represents differentiation with respect to time, t, and the state vector $\boldsymbol{x} = \{x_1, x_2, \ldots, x_n\}^T$.

2.1.1 Harmonically Forced Linear Oscillator

First consider how this framework would work for the harmonically forced (and so non-autonomous) linear oscillator (Eq. (1.16) with $F_e = F_0 \cos(\Omega t)$)

$$\ddot{x} + 2\zeta\omega_n\dot{x} + \omega_n^2 x = \frac{F_0}{m}\cos(\Omega t), \tag{2.2}$$

where ζ is the damping ratio and ω_n is the natural frequency. Define the state vector $\boldsymbol{x} = \{x_1, x_2\}^T$, where $x_1 = x$ is the displacement and $x_2 = \dot{x}$ is the velocity. Using these definitions notice that $\dot{x}_1 = x_2 = \dot{x}$, and $\dot{x}_2 = \ddot{x}$ which enables the system to be written in first-order form. This means there are no terms differentiated more than once. In general, all linear systems can be reduced to this form, and in this case the first-order form gives

$$\begin{aligned} \dot{x}_1 &= x_2 \\ \dot{x}_2 &= -2\zeta\omega_n x_2 - \omega_n^2 x_1 + \frac{F_0}{m}\cos(\Omega t), \end{aligned}$$

or

[2] Throughout, it will be assumed that $\mathbf{f}$ is a smooth function, such that existence and uniqueness of solutions is always satisfied.

[3] For most vibration problems, non-autonomous means the system has time-dependent forcing, and autonomous means that the system is unforced. In fact, a non-autonomous system can usually be represented as autonomous by setting $t = x_3$ and adding an additional equation to the system $\dot{x}_3 = 1$.

$$\begin{bmatrix} \dot{x}_1 \\ \dot{x}_2 \end{bmatrix} = \begin{bmatrix} 0 & 1 \\ -\omega_n^2 & -2\zeta\omega_n \end{bmatrix} \begin{bmatrix} x_1 \\ x_2 \end{bmatrix} + \begin{bmatrix} 0 \\ \frac{F_0}{m}\cos(\Omega t) \end{bmatrix},$$

which can be written in matrix notation as

$$\dot{\boldsymbol{x}} = A\boldsymbol{x} + \mathbf{F}(t), \tag{2.3}$$

where A is a constant matrix and $\mathbf{F}(t)$ is a vector of forcing terms. If the system is unforced (autonomous), then $\mathbf{F} = 0$ so that $\dot{\boldsymbol{x}} = A\boldsymbol{x}$. This is the standard form for a linear dynamical system. It can be said that Eq. (2.3) is in 'state space form'.

Example 2.1 Steady-state solutions for a harmonic oscillator

Problem Find the steady-state behaviour of the example for the harmonic oscillator defined by Eq. (2.2), when the mass $m = 1$ kg, natural frequency $\omega_n = 1$ rad/s, forcing frequency $\Omega = 1$ rad/s, forcing amplitude $F_0 = 10$ N and damping ratios $\zeta = 0.0625, 0.125, 0.25$. Plot the steady-state solution results for the three different damping cases in a two-dimensional graph of displacement against velocity.

Solution For three different values of damping ratio, $\zeta = 0.0625, 0.125, 0.25$, the steady-state solutions can be computed using 4th order Runge-Kutta numerical integration,[4] to compute time series from initial values (see for example Fausett (1999)). The results are plotted in Fig. 2.1 which is a two-dimensional plot of displacement (x_1) against velocity (x_2). It can be seen that, in this case, the system behaviour is to form a circular orbit (if $\omega_n \neq 1$ the orbits will be elliptical) for each damping value. This steady-state behaviour is like simple harmonic motion, except with the addition of damping and forcing. The size of each ellipse is directly related to the balance of energy between the forcing input and the energy dissipated by damping. As the damping increases, the size of the ellipse reduces because more energy is dissipated by the viscous damper. This type of steady-state orbit is called a *limit cycle*, which is discussed in more detail in Sect. 2.5. ■

Now consider the more general case of a state space solution curve for a nonlinear system. An example of 20 s of data is shown in Fig. 2.2, as a solid black line with a start point in the plane $t = 0$ s and finishing in the plane $t = 20$ s. Three *projections* of this three-dimensional solution curve are also shown in Fig. 2.2. The displacement versus time plot is the projection onto the $(t,\ x_1)$ plane and the velocity versus time plot is the projection onto the $(t,\ x_2)$ plane. The projection onto the $(x_1,\ x_2)$ plane is called the phase portrait, and is used extensively as a way of analysing the dynamics of the second-order nonlinear oscillators—the example shown in Fig. 2.2 is from a forced Duffing oscillator.

Individual solutions curves, such as the one shown in Fig. 2.2 are called either *trajectories* or *orbits*. The time evolution of multiple nearby trajectories is called the *flow* of a dynamical system—see Guckenheimer and Holmes (1983) for a more detailed discussion.

---

[4] Although this linear system can be solved exactly, numerical integration is used as this will be required for the nonlinear examples.

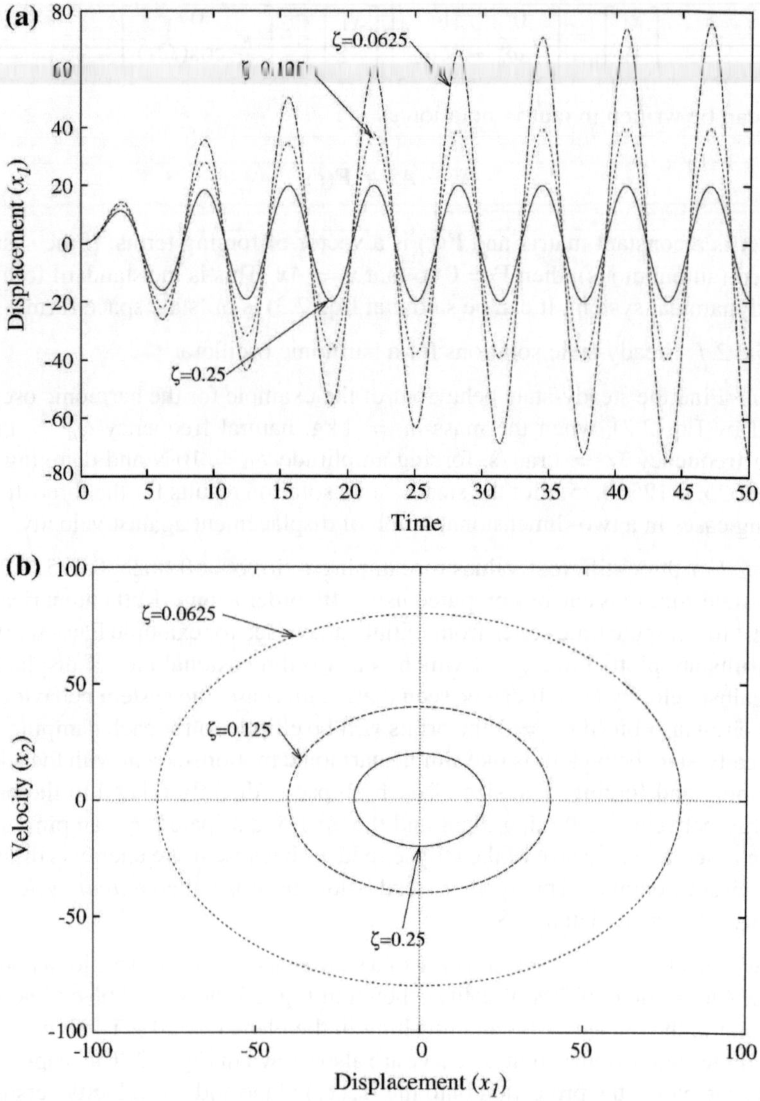

Fig. 2.1 Numerically computed solutions for a harmonic oscillator, **a** time series, and **b** steady-state periodic solutions

2.1.2 Equilibrium Points

When $\mathbf{f}(\boldsymbol{x}, t) = 0$ the system is said to have a *stationary point* or *equilibrium point*.[5] Equilibrium points will be denoted as $\boldsymbol{x}^*$, so that $\mathbf{f}(\boldsymbol{x}^*, t) = 0$.

[5] These are also sometimes called *fixed points* but here fixed point will only be used for maps.

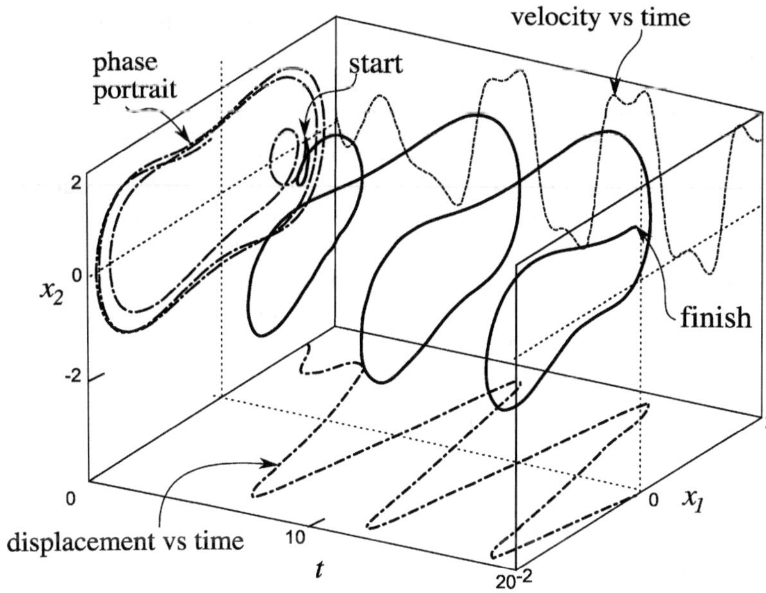

Fig. 2.2 State space for a forced Duffing oscillator

Equilibrium points play an important role in the analysis of nonlinear systems. This is because the dynamic behaviour close to an equilibrium point can normally be studied using a localized linear analysis. As a result, the study of a nonlinear system usually begins with identifying the equilibrium points. For simple systems, like the majority of those discussed in this book, the equilibrium points can be found by inspection of the state equations, Eqs. (2.1) and (2.2). In more complex systems, a Newton algorithm, or equivalent numerical technique, can be used to find solutions to $\mathbf{f}(\boldsymbol{x}, t) = 0$.

For harmonically forced nonlinear oscillators, the equation of motion can be rewritten as

$$\frac{\mathrm{d}\boldsymbol{x}}{\mathrm{d}t} = \mathbf{f}(\boldsymbol{x}) + \mathbf{F}(t), \tag{2.4}$$

where $\mathbf{F}(t)$ is the harmonic forcing vector, and the nonlinear function is autonomous. Then the analysis of the system equilibrium points can be simplified by first considering the unforced case when $\mathbf{F} = 0$. Here is an example.

Example 2.2 Equilibrium points and phase portrait for undamped pendulum oscillations

Problem Find the equilibrium points and draw the phase portrait for the undamped pendulum (introduced in Chap. 1) shown schematically in Fig. 1.5a, where the mass is assumed to act at a single point supported by a massless inextensible rod of length l.

Solution If no damping is assumed, the equation governing the motion of the pendulum is

$$\ddot{\theta} + \omega^2 \sin\theta = 0, \tag{2.5}$$

where θ is the angle from the downward equilibrium position, $\omega^2 = g/l$ rad/s, l is the length of the pendulum (of mass m) and g is the acceleration due to gravity (m/s^2) (see Sect. 1.2.2 for a physical explanation of how this equation is derived). The state of the pendulum at any time t is uniquely defined by its position (angle) and velocity (angular velocity). So, to define the state vector, let $x_1 = \theta$ and $x_2 = \dot{\theta}$ then $x_2 = \dot{x}_1$ and substitute these relationships into Eq. (2.5), $\dot{x}_2 = -\omega^2 \sin(x_1)$. So the system can be written in the first-order form

$$\begin{bmatrix} \dot{x}_1 \\ \dot{x}_2 \end{bmatrix} = \begin{bmatrix} x_2 \\ -\omega^2 \sin(x_1) \end{bmatrix},$$

which is an autonomous first-order system of the form $\dot{\boldsymbol{x}} = \mathbf{f}(\boldsymbol{x})$.

The equilibrium points occur when $\mathbf{f}(\boldsymbol{x}) = 0$. For $\mathbf{f}(\boldsymbol{x}) = 0$ to occur, x_2 must be zero and x_1 must be either zero or $\pm n\pi$, where $n = 1, 2, 3\ldots$ because the sine function is zero at integer multiples of π. The case where $x_1 = 0$ corresponds to $\theta = 0$ which is the downward resting position—see Fig. 1.5a. If the pendulum rotates by 360° (2π rad) then it is back at the downward resting position, which is the case for all even n. When $x_1 = \pi$ the pendulum is in the upward position, which is unstable (like trying to balance a pencil on your finger tip), so that just the smallest disturbance will destabilize the equilibrium position.[6]

For the case when $x_1 - x^*$ is small, it is possible to examine the dynamic behaviour of the system close to the equilibrium points. For the x_1 equilibrium point, when x_1 is small, the approximation $\sin(x_1) \approx x_1$ can be made so that

$$\begin{bmatrix} \dot{x}_1 \\ \dot{x}_2 \end{bmatrix} \approx \begin{bmatrix} x_2 \\ -\omega^2 x_1 \end{bmatrix}. \tag{2.6}$$

This equation approximates the dynamics of the system close to the origin, where there is an equilibrium point $(x_1, x_2) = (0, 0)$, and additional equilibrium points for even values of n. To eliminate time, and observe the behaviour in the displacement versus velocity plane, take the ratio of the two velocity expressions in Eq. (2.6) to give

$$\frac{\dot{x}_2}{\dot{x}_1} = \frac{\mathrm{d}x_2/\mathrm{d}t}{\mathrm{d}x_1/\mathrm{d}t} = \frac{\mathrm{d}x_2}{\mathrm{d}x_1} = \frac{-\omega^2 x_1}{x_2}. \tag{2.7}$$

Equation (2.7) can be rearranged to give $-\omega^2 x_1 \mathrm{d}x_1 = x_2 \mathrm{d}x_2$, which can then be integrated. By incorporating the 1/2 factors into the constant of integration, the following

[6] An interesting nonlinear control problem is to stabilize a pendulum in the upright (or inverted) position—see, for example the discussion in Sontag (1998) and Chap. 3.

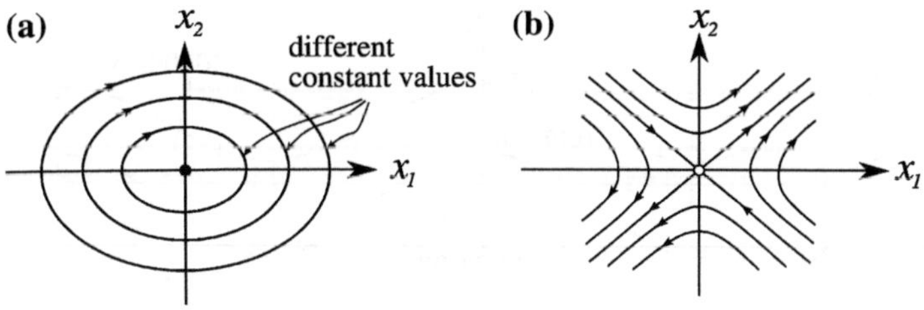

Fig. 2.3 Local equilibrium point dynamics: **a** centre, **b** saddle, *arrows* on the flow *lines* indicate increasing time

expression are obtained[7]

$$-\omega^2 x_1^2 = x_2^2 - \text{const} \rightsquigarrow x_2^2 + \omega^2 x_1^2 = \text{const}.$$

This is the equation of an ellipse that goes around the equilibrium point. The size of the ellipse depends on the constant (const), and as the constant is arbitrary, multiple ellipses exist close to the equilibrium point, as shown in Fig. 2.3a. This type of equilibrium point is known as a *centre*.

To study what happens close to the second equilibrium point, $(x_1, x_2) = (\pi, 0)$ when the pendulum is in the upward position, the x_1 coordinate needs to be shifted from zero to π, so that the sine term becomes

$$\begin{aligned}\sin(x_1 + \pi) &= \sin(x_1)\cos(\pi) + \cos(x_1)\sin(\pi),\\ &= \sin(x_1)(-1) + 0 = -\sin(x_1).\end{aligned}$$

So in this case the linearized equation becomes

$$\begin{bmatrix}\dot{x}_1\\ \dot{x}_2\end{bmatrix} \approx \begin{bmatrix}x_2\\ \omega^2 x_1\end{bmatrix},$$

and dividing $\dot{x}_2$ by $\dot{x}_1$ gives

$$\frac{\dot{x}_2}{\dot{x}_1} = \frac{\mathrm{d}x_2/\mathrm{d}t}{\mathrm{d}x_1/\mathrm{d}t} = \frac{\omega^2 x_1}{x_2} = \frac{\mathrm{d}x_2}{\mathrm{d}x_1}. \tag{2.8}$$

Equation (2.8) can be rearranged to give $\omega^2 x_1 \mathrm{d}x_1 = x_2 \mathrm{d}x_2$, which can then be integrated and, as before, by incorporating the 1/2 factors into the constant of integration, we can obtain the following expression

$$\omega^2 x_1^2 = x_2^2 - \text{const} \rightsquigarrow x_2^2 - \omega^2 x_1^2 = \text{const},$$

[7] Note that this is now the same as the equation of motion for an unforced, undamped, harmonic oscillator, where the constant is determined by the initial displacement and velocity.

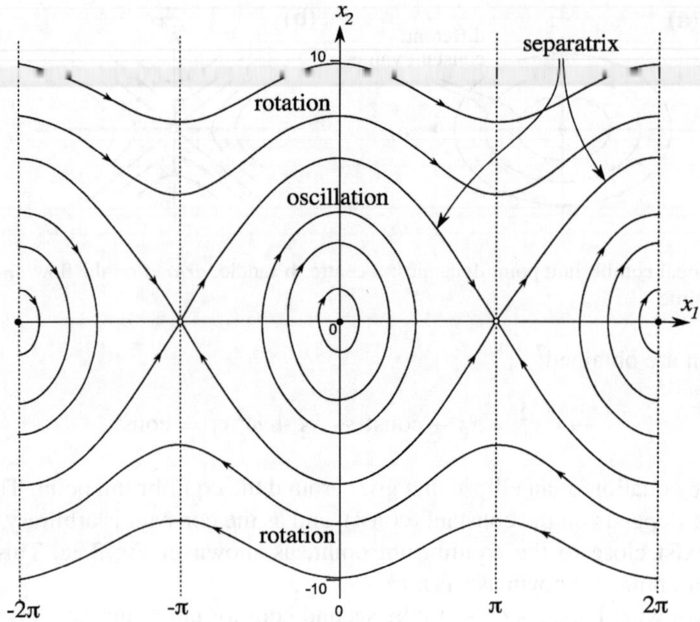

Fig. 2.4 Phase portrait for the undamped pendulum

which is the equation for a hyperbola, as shown in Fig. 2.3b. This type of equilibrium point is known as a *saddle*.[8]

Away from the equilibrium points the solution curves are governed by

$$\frac{\mathrm{d}x_2}{\mathrm{d}x_1} = \frac{-\omega^2 \sin(x_1)}{x_2},$$

from which

$$\int x_2 \mathrm{d}x_2 = \int \omega^2 \sin(x_1)\mathrm{d}x_1 \rightsquigarrow \frac{x_2^2}{2} = \omega^2 \cos(x_1) + \text{const},$$

so that the relation governing the phase portrait trajectories is

$$x_2 = \pm\sqrt{2(\omega^2 \cos(x_1) + \text{const})}.$$

Using this relationship we can draw the complete phase portrait. An example is shown in Fig. 2.4. ■

Notice that the equilibrium points are shown as solid black dots in the stable case (pendulum down) and as an unfilled circle to represent the unstable case (pendulum

[8] In three dimensions this point looks like a horse saddle. See Sect. 2.3.

balanced upright). In Fig. 2.4 only a section of the phase portrait is shown from $-2\pi \leq x_1 \leq 2\pi$. The space continues in both directions with exactly the same pattern of alternating saddles and centres. The orbit which goes through the saddle point is called the *separatrix*[9] because it separates two different types of behaviour. On the inside of the separatrix, oscillations occur around the stable centre point, which corresponds to constant values of $0 < \text{const} < \omega^2$. Outside the separatrix, the solution curves travel continuously to the right (above) or to the left (below), which corresponds to $\text{const} > \omega^2$. Physically, this corresponds to the pendulum rotating continuously with either positive angle (to the right) or negative angle (to the left). On the separatrix $\text{const} = \omega^2$.

2.1.3 Local Linear Approximation Near Equilibrium Points

Close to the equilibrium points the nonlinear system can be *linearized*. This is done by first changing coordinates to make the equilibrium point the origin of a new coordinate system. Defining a new coordinate vector $\boldsymbol{\xi} = \boldsymbol{x}(t) - \boldsymbol{x}^*$, means that $\boldsymbol{\xi} = 0$ corresponds to the equilibrium point of interest, If $\mathbf{f}$ is autonomous (i.e. a function of $\boldsymbol{x}$ and not t) then

$$\frac{\mathrm{d}\boldsymbol{\xi}}{\mathrm{d}t} = \frac{\mathrm{d}\boldsymbol{x}}{\mathrm{d}t} = \mathbf{f}(\boldsymbol{x}) = \mathbf{f}(\boldsymbol{x}^* + \boldsymbol{\xi}),$$

such that the dynamical system in terms of the new coordinates becomes

$$\frac{\mathrm{d}\boldsymbol{\xi}}{\mathrm{d}t} = \mathbf{f}(\boldsymbol{x}^* + \boldsymbol{\xi}). \tag{2.9}$$

This is because $\boldsymbol{x}^*$ is a constant so $\dot{\boldsymbol{\xi}} = \dot{\boldsymbol{x}}$. To find the linearized system, $\mathbf{f}(x^* + \boldsymbol{\xi})$ is expanded as a Taylor series expansion

$$\mathbf{f}(\boldsymbol{x}^* + \boldsymbol{\xi}) = \mathbf{f}(\boldsymbol{x}^*) + D_{\boldsymbol{x}^*}\mathbf{f}\ \boldsymbol{\xi} + \mathscr{O}(\|\boldsymbol{\xi}\|^2), \tag{2.10}$$

where $D_{\boldsymbol{x}^*}\mathbf{f}$ is the *Jacobian matrix* evaluated at $\boldsymbol{x}^*$ and $\mathscr{O}(\|\boldsymbol{\xi}\|^2)$ denotes terms of second-order and higher.[10] The Jacobian gives the linear gradients of the nonlinear function, $\mathbf{f}$, with respect to the states x_1 and x_2.

For an equilibrium point, $\mathbf{f}(\boldsymbol{x}^*) = 0$ by definition, so Eq. (2.10) becomes

$$\mathbf{f}(\boldsymbol{x}^* + \boldsymbol{\xi}) = D_{\boldsymbol{x}^*}\mathbf{f}\ \boldsymbol{\xi} + \mathscr{O}(\|\boldsymbol{\xi}\|^2). \tag{2.11}$$

---

[9] Also known as a heteroclinic orbit, which joins two separate saddle points. Not to be confused with a homoclinic orbit, an orbit which starts and finishes at the same saddle point.

[10] Note that the norm is used here because $\boldsymbol{\xi}$ is a vector.

Both Eqs. (2.10) and (2.11) indicate that terms of second-order and higher-orders are ignored. If $\boldsymbol{\xi}$ is small, i.e. $\|\boldsymbol{\xi}\| \ll 1$, then this type of assumption can be justified,[11] but it means that the analysis is valid only in a small region close to the equilibrium point.

In general, the Jacobian matrix is given by

$$D_{\boldsymbol{x}}\mathbf{f} = \begin{bmatrix} \frac{\partial f_1}{\partial x_1} & \frac{\partial f_1}{\partial x_2} & \cdots & \frac{\partial f_1}{\partial x_n} \\ \frac{\partial f_2}{\partial x_1} & \frac{\partial f_2}{\partial x_2} & \cdots & \frac{\partial f_2}{\partial x_n} \\ \vdots & \vdots & \ddots & \vdots \\ \frac{\partial f_n}{\partial x_1} & \frac{\partial f_n}{\partial x_2} & \cdots & \frac{\partial f_n}{\partial x_n} \end{bmatrix},$$

where f_i are the terms in the nonlinear vector $\mathbf{f}$ and x_i are the states. Then $D_{\boldsymbol{x}^*}\mathbf{f}$ is found by substituting the state values at $\boldsymbol{x}^*$ into the Jacobian expression. For autonomous systems, this matrix is usually just a constant matrix, in which case we define $D_{\boldsymbol{x}^*}\mathbf{f} = A$ where A is a $n \times n$ matrix of constant terms. Substituting this into Eq. (2.9) using Eq. (2.11) and ignoring the higher-order terms gives

$$\frac{\mathrm{d}\boldsymbol{\xi}}{\mathrm{d}t} = A\boldsymbol{\xi}, \tag{2.12}$$

which is a linear system, valid only as an approximation to the nonlinear system close to the equilibrium point, i.e. for $\boldsymbol{\xi}$ 'small'.

2.2 Systems with Two States

Now consider the case when the system has two states, $\boldsymbol{x} = \{x_1, x_2\}^T$ and it is required to solve Eq. (2.12) close to an equilibrium point $\boldsymbol{x}^* = \{x_1^*, x_2^*\}^T$. To solve Eq. (2.12) we assume a solution of the form $\boldsymbol{\xi}(t) = \mathbf{c}\mathrm{e}^{\lambda t}$, where $\mathbf{c} = \{c_1, c_2\}^T$ is a vector of arbitrary constants.[12] Substituting this into Eq. (2.12) gives

$$\mathbf{c}\lambda\mathrm{e}^{\lambda t} = A\mathbf{c}\mathrm{e}^{\lambda t} \quad \text{so that} \quad \mathbf{c}\,\lambda = A\mathbf{c} \quad \text{or} \quad (A - \lambda I)\mathbf{c} = 0. \tag{2.13}$$

This is a linear eigenvalue problem, where λ are the eigenvalues. For Eq. (2.13) to be satisfied either $\mathbf{c} = 0$ or $\det(A - \lambda I) = 0$. The $\mathbf{c} = 0$ case involves no dynamics, so attention is focused on solving $\det(A - \lambda I) = 0$. For a two-dimensional system, this can be written as

---

[11] In fact, the formal definition is that the equilibrium points are *hyperbolic*. The structure of the trajectories close to a hyperbolic equilibrium point are topologically equivalent to the trajectory structure of the linearized dynamical system, see Guckenheimer and Holmes (1983) for a detailed discussion.

[12] An alternative approach is to use the solution $\boldsymbol{\xi} = \boldsymbol{\xi}_0\mathrm{e}^{At} = \boldsymbol{\xi}_0 P\mathrm{e}^{Jt}P^{-1}$ where J is the Jordan normal form of A—see for example Glendinning (1994).

$$\det\begin{bmatrix} a_{11}-\lambda & a_{12} \\ a_{21} & a_{22}-\lambda \end{bmatrix} = 0,$$

which when multiplied out becomes

$$\lambda^2 - (a_{11} + a_{22})\,\lambda + (a_{11}a_{22} - a_{21}a_{12}) = 0. \tag{2.14}$$

This is the characteristic equation and $(a_{11} + a_{22}) = \mathrm{tr}(A)$ is called the *trace* of A and $(a_{11}a_{22} - a_{21}a_{12}) = \det(A)$ is the *determinant* of A. So now Eq. (2.14) can be rewritten as

$$\lambda^2 - \mathrm{tr}(A)\,\lambda + \det(A) = 0,$$

which has the solution

$$\lambda_{1,2} = \frac{1}{2}(\mathrm{tr}(A) \pm (\mathrm{tr}(A)^2 - 4\det(A))^{\frac{1}{2}}). \tag{2.15}$$

Let $\mathrm{tr}(A)^2 - 4\det(A) = \Delta$, where Δ is called the *discriminant*, then the eigenvalue solutions can be written as

$$\lambda_{1,2} = \frac{1}{2}(\mathrm{tr}(A) \pm (\Delta)^{\frac{1}{2}}).$$

The sign of Δ determines whether the eigenvalues are real $\Delta > 0$, complex $\Delta < 0$ or repeated $\Delta = 0$. The signs of $\lambda_{1,2}$ are significant in determining the type of equilibrium points.

2.2.1 Equilibrium Points for Linear Harmonic Oscillator

Now consider the linear harmonic oscillator defined by $m\ddot{x} + c\dot{x} + kx = F(t)$, Eq. (1.3). First set the forcing parameter, $F(t)$, to zero so that the state space representation, Eq. (2.4), becomes

$$\begin{bmatrix} \dot{x}_1 \\ \dot{x}_2 \end{bmatrix} = \begin{bmatrix} 0 & 1 \\ -\frac{k}{m} & -\frac{c}{m} \end{bmatrix}\begin{bmatrix} x_1 \\ x_2 \end{bmatrix}. \tag{2.16}$$

Provided that all the parameters are non-zero, this system has a single equilibrium point at $x_1 = x_2 = 0$. Physically this is because, for an unforced, but damped, linear oscillator, releasing the mass from any non-zero displacement and velocity values results in the system gradually losing energy until it reaches the 'at rest' (zero displacement and velocity) point. The at rest position corresponds to the equilibrium point $x_1 = x_2 = 0$ in Eq. (2.16). The starting values are called the *initial conditions* and are values of displacement $x_1(0) = x(t = 0)$ and velocity $x_2(0) = \dot{x}(t = 0)$ at time $t = 0$. For all initial $x_1(0), x_2(0) \neq 0$ values, the solution curves end up at the equilibrium point, and so the equilibrium point is said to be *attracting*.

From Eq. (2.16) the $\text{tr}(A) = -c/m$, $\det(A) = k/m$, $\Delta = (c^2 - 4\,km)/m^2$ and

$$\lambda_{1,2} = -\frac{c}{2m} \pm \frac{1}{2m}\sqrt{c^2 - 4\,km}.$$

Now consider the case when $\Delta < 0$. This means that the eigenvalues, $\lambda_{1,2}$ are complex, which physically corresponds to underdamped vibrations. Using the relationships $\omega_n = \sqrt{k/m}$ as the undamped natural frequency, $\zeta = c/2m\omega_n$ as the damping ratio and $\omega_d = \omega_n\sqrt{1-\zeta^2}$ as the damped natural frequency, we can write the eigenvalues as $\lambda_{1,2} = -\zeta\omega_n \mp i\omega_d$. Substituting this into the original assumed solution, $\boldsymbol{\xi}(t) = \mathbf{c}e^{\lambda t}$ results in a solution which can be expressed as

$$\boldsymbol{\xi}(t) = \mathbf{c}e^{-\zeta\omega_n t}\sin(\omega_d).$$

This is a sine wave oscillation multiplied by an exponential envelope. There are two cases, depending on whether the damping is positive ($c > 0$ therefore $\zeta > 0$) or negative ($c < 0$ therefore $\zeta < 0$). The two cases are shown in Fig. 2.5a, c in terms of time history plots. For $\zeta > 0$ (Fig. 2.5a), the exponential envelope causes the sinusoidal oscillation to decay, but when $\zeta < 0$ (Fig. 2.5c) the oscillations grow. These two types of behaviour can be plotted in the state space x_1 versus x_2, which is shown in Fig. 2.5b, d relative to the equilibrium point at the origin.

Figure 2.5b is known as a *stable spiral* (or a focus or sink), and Fig. 2.5d is an *unstable spiral* (or a repeller or source). The idea of stability can be related to whether the oscillations grow or decay. If they decay, such that solution curves of the governing equation of motion are *attracted* to the equilibrium point, then this is

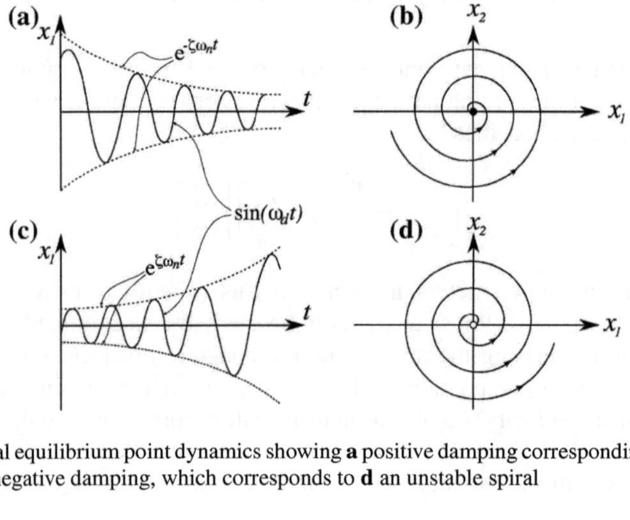

Fig. 2.5 Local equilibrium point dynamics showing **a** positive damping corresponding to **b** a stable spiral, and **c** negative damping, which corresponds to **d** an unstable spiral

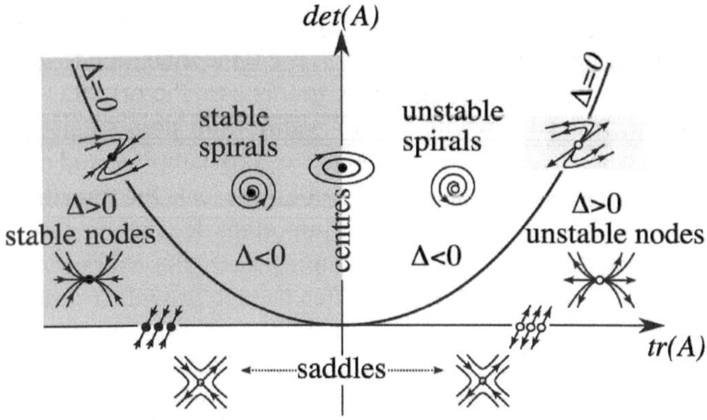

Fig. 2.6 Type and local stability of equilibrium points for a two state linear(ised) system

a stable behaviour. If the oscillations grow such that solution curves of the governing equation of motion are *repelled* from the equilibrium point, then this is an unstable behaviour.

Figure 2.6 shows the relationship between $\mathrm{tr}(A)$, $\det(A)$, Δ and the type and stability of all equilibrium points for the linear oscillator with two states, $x = \{x_1, x_2\}^T$. From a physical perspective (at least for this book) the mass, m, is always a positive constant. But both the stiffness and damping can be either positive or negative constants. In Fig. 2.6 the upper left-hand quadrant is shaded grey to indicate the region of *stable* solutions. The physical nature of the *unstable* regions corresponds to solutions with negative damping and negative stiffness. For constant, positive mass, the $\mathrm{tr}(A)$ is proportional to the damping, c, and $\det(A)$ is proportional to the stiffness, k. So $\det(A) < 0$ corresponds to negative stiffness $k < 0$, and from Fig. 2.6 it can be seen that this corresponds to saddle solutions. In addition $\mathrm{tr}(A) > 0$ corresponds to negative damping $c < 0$. The transition from the stable quadrant as damping changes sign can be thought of as a *dynamic instability*. Conversely, the transition that occurs as stiffness becomes negative is a form of *static instability*.

Notice that in the stable quadrant all solution curves have arrows pointing towards the equilibrium point. The equilibrium points are also shown as solid black dots, indicating that these equilibrium points are stable and *attract* solution curves toward them. The unstable quadrants have at least some of the solution curves pointing away from the equilibrium point, and the open circles indicate that the equilibrium points are unstable, and *repel* solution curves. The concept of stability will be discussed in further detail in Sect. 2.5.

The nodes shown in Fig. 2.6 correspond to the case when the eigenvalues are real and distinct. Physically this corresponds to the case when $\Delta = (c^2 - 4\,km)/m^2 > 0$ which is otherwise known as the overdamped case, i.e. when $\zeta > 1$, and will not be discussed in detail here—see Inman (2006) for a discussion on linear overdamped vibration.

Figure 2.6 also shows three special types of behaviour (i) when $\mathrm{tr}(A) = 0$, (ii) when $\det(A) = 0$, and (iii) when $\Delta = 0$. The last case, when $\Delta = 0$ corresponds to a critically damped system, $\zeta = 1$, so that $\lambda_1 = \lambda_2$ and the behaviour is known as a degenerate (or inflected) node.[13] When $\det(A) = 0$ the transition to static instability occurs as the stiffness passes through zero. In this case the origin is not an isolated equilibrium point, instead a whole line of equilibrium points exists, see Strogatz (2001). The case when $\mathrm{tr}(A) = 0$ corresponds to zero damping, and the positive half of the $\det(A)$ axis in Fig. 2.6 (where $\mathrm{tr}(A) = 0$) represents the transition from stable to unstable spirals. This is by far the most important degenerate case for vibration analysis, as it corresponds to the zero damping case. For vibration systems with small damping, the zero damping case is often used to provide a simplified analysis of the system behaviour.[14] An example of finding equilibrium point behaviour is considered next.

Example 2.3 Equilibrium points for an oscillator with quadratic stiffness nonlinearity (the escape equation)

Problem Find the type and stability of the equilibrium points for the following equation with a quadratic nonlinear term

$$m\ddot{x} + c\dot{x} + k_1 x + k_2 x^2 = 0,$$

with mass $m = 1\,\mathrm{kg}$, damping $c = 0.1\,\mathrm{Ns/m}$, linear stiffness $k_1 = 1\,\mathrm{N/m}$ and quadratic stiffness $k_2 = 1\,\mathrm{N/m^2}$. This equation is sometimes referred to as the *escape equation.*

Solution First, put the system into first-order form by defining $x_1 = x$ and $x_2 = \dot{x}$, such that $\ddot{x} = \dot{x}_2$. This gives

$$\begin{aligned} \dot{x}_1 &= x_2 = f_1, \\ \dot{x}_2 &= -(0.1)x_2 - x_1 - x_1^2 = f_2. \end{aligned} \tag{2.17}$$

The equilibrium points are values of (x_1, x_2) which make the right-hand side of Eq. (2.17) equal to zero ($\mathbf{f} = 0$, which means that $f_1 = 0,\ f_2 = 0$). So, by inspection it can be seen that $(x_1 = 0, x_2 = 0) = \boldsymbol{x}_a^*$ is an equilibrium point. Note that x_2 must always be zero to satisfy the first line of Eq. (2.17), $f_1 = 0$. When $x_2 = 0$, the second line of Eq. (2.17) gives $f_2 = -x_1 - x_1^2 = 0$ from which $x_1 = -1$ is a solution, so there is a second equilibrium point at $(x_1 = -1, x_2 = 0) = \boldsymbol{x}_b^*$.

---

13 In fact the behaviour depends on the *multiplicity* of the repeated eigenvalue. The degenerate node corresponds to the case where there is only a single eigenvector. For the case with two eigenvectors the degenerate equilibrium point becomes a *star* see Strogatz (2001). See Seyranian and Mailybaev (2003) for a more detailed discussion of multiplicity.

14 When $\mathrm{tr}(A) = 0$ and $\det(A) = 0$, there is a doubly-degenerate equilibrium point. This is not discussed further here.

To find the type and stability of the equilibrium points, the Jacobian, $D_{\boldsymbol{x}^*}\mathbf{f}$, must be determined for each equilibrium point. The Jacobian of Eq. (2.17) is

$$D_{\boldsymbol{x}}\mathbf{f} = \frac{\partial(f_1, f_2)}{\partial(x_1, x_2)} = \begin{bmatrix} \frac{\partial f_1}{\partial x_1} & \frac{\partial f_1}{\partial x_2} \\ \frac{\partial f_2}{\partial x_1} & \frac{\partial f_2}{\partial x_2} \end{bmatrix} = \begin{bmatrix} 0 & 1 \\ -1 - 2x_1 & -0.1 \end{bmatrix}. \tag{2.18}$$

Now substitute each of the equilibrium points into the Jacobian in turn. First for $\boldsymbol{x}_a^* = (x_1 = 0, x_2 = 0)$ the transformed localized coordinates are $\xi_1 = x_1 - 0 = x_1$ and $\xi_2 = x_2 - 0 = x_2$. The Jacobian becomes

$$D_{\boldsymbol{x}_a^*}\mathbf{f} = \begin{bmatrix} 0 & 1 \\ -1 & -0.1 \end{bmatrix}. \tag{2.19}$$

As $D_{\boldsymbol{x}_a^*}\mathbf{f}$ is a constant matrix, define $D_{\boldsymbol{x}_a^*}\mathbf{f} = A$. From Eq. (2.19), the trace of $D_{\boldsymbol{x}_a^*}\mathbf{f}$ is given by $\mathrm{tr}(D_{\boldsymbol{x}_a^*}\mathbf{f}) = -0.1$ and the determinant is $\det(D_{\boldsymbol{x}_a^*}\mathbf{f}) = 1$. Recall from Eq. (2.15) that the discriminant, Δ is given by $\Delta = \mathrm{tr}^2 - 4\det$. So for equilibrium point $\boldsymbol{x}_a^*$, the discriminant, $\Delta = -3.99$. Then using Fig. 2.6, it can be seen that this equilibrium point has negative $\mathrm{tr}(A)$, positive $\det(A)$ and $\Delta < 0$. This means that the equilibrium point is a stable spiral.

For the equilibrium point $\boldsymbol{x}_b^*$, the localized coordinates are $\xi_1 = x_1 - (-1) = x_1 + 1$ and $\xi_2 = x_2 - 0 = x_2$. Substituting $\boldsymbol{x}_b^* = (x_1 = -1, x_2 = 0)$ into Eq. (2.18), the Jacobian becomes

$$D_{\boldsymbol{x}_b^*}\mathbf{f} = \begin{bmatrix} 0 & 1 \\ 1 & -0.1 \end{bmatrix}.$$

So in this case $\mathrm{tr}(A) = -0.1$ and $\det(A) = -1$, which from Fig. 2.6 means that this equilibrium point is a saddle. ■

2.3 The Link Between State Space and Mechanical Energy

Consider an unforced linear oscillator with negligible damping such that the system is approximated as being undamped, in which case

$$m\ddot{x} + kx = 0,$$

where k is the spring stiffness and x is the displacement of the mass, m. Considering the work done over a small increment of distance $\mathrm{d}x$, as the mass moves from resting $x = 0$ to an arbitrary x value gives the integral

$$\int_0^x (m\ddot{x} + kx)\mathrm{d}x = m\int_0^x \ddot{x}\mathrm{d}x + k\int_0^x x\mathrm{d}x = E_t, \tag{2.20}$$

where E_t is the total energy. Note that velocity, $v = \frac{\mathrm{d}x}{\mathrm{d}t}$ and acceleration $\frac{\mathrm{d}^2x}{\mathrm{d}t^2} = \frac{\mathrm{d}v}{\mathrm{d}t}$ so that $\frac{\mathrm{d}v}{\mathrm{d}t}\mathrm{d}x = v\mathrm{d}v$, which can be substituted into Eq. (2.20) (with a change of integration limits) to give

$$m\int_0^v v\mathrm{d}v + k\int_0^x x\mathrm{d}x = \frac{1}{2}mv^2 + \frac{1}{2}kx^2 = E_t. \tag{2.21}$$

Equation (2.21) represents the kinetic plus potential energy of the mass-spring system, where $\frac{1}{2}mv^2$ is the kinetic and $\frac{1}{2}kx^2$ is the potential energy.

Equation (2.21) also represents the *Hamiltonian* for the system. For dynamical systems, Hamiltonians are typically used to model systems that are undamped,[15] also called *conservative* as energy is conserved. In vibration analysis of mechanical systems, applications are typically *non-conservative*, as damping is nearly always present. However, when damping is small, analysis is often developed for the undamped system, as the results are very close to (but not exactly the same) as for the system with damping.[16]

2.3.1 Potential Functions

Now a direct link can be made between the system state space and the energy in the system. To see this, first notice that in terms of state variables the velocity, $v = \dot{x} = x_2$ and the displacement $x = x_1$. Now consider the unforced, undamped nonlinear oscillator

$$m\ddot{x} + p(x) = 0 \rightsquigarrow mv\frac{\mathrm{d}v}{\mathrm{d}x} + p(x) = 0,$$

where $p(x)$ is the stiffness function. Integrating to find the energy gives

$$\frac{1}{2}mv^2 + \int_0^x p(x) = E_t \rightsquigarrow \frac{1}{2}mv^2 + V(x) = E_t, \tag{2.22}$$

where $V(x) = \int_0^x p(x)$ is called the *potential* function.[17]

[15] See, for example, Guckenheimer and Holmes (1983) and Strogatz (2001) for an introduction and further references.

[16] The fact that the undamped solutions *persist* with the addition of small damping, is an important underlying assumption in vibration analysis.

[17] Not to be confused with potential energy.

As an example, consider the generic version of the escape equation, from Example 2.3. The equation of motion is given by

$$m\ddot{x} + c\dot{x} + k_1 x + k_2 x^2 = 0,$$

where m is mass, c the damping, k_1 is linear stiffness and k_2 nonlinear stiffness. In this case $p(x) = k_1 x + k_2 x^2$, such that the potential function $V(x)$ is given by

$$V(x) = \int_0^x (k_1 x + k_2 x^2)\mathrm{d}x = \frac{1}{2}k_1 x^2 + \frac{1}{3}k_2 x^3. \tag{2.23}$$

Example 2.4 Phase portrait for the undamped escape equation

Problem Construct the phase portrait and potential function for the unforced, undamped escape equation given by

$$m\ddot{x} + kx + k_2 x^2 = 0, \tag{2.24}$$

where $m = 1\,\text{kg}$, $k = 1\,\text{N/m}$ and $k_2 = 1\,\text{N/m}^2$.

Solution First put the system into first-order form by defining $x_1 = x$ and $x_2 = \dot{x}$, such that $\ddot{x} = \dot{x}_2$. This gives

$$\begin{aligned} \dot{x}_1 &= x_2 = f_1, \\ \dot{x}_2 &= -x_1 - x_1^2 = f_2. \end{aligned}$$

The equilibrium points for this system are $\boldsymbol{x}_a^* = (x_1 = 0, x_2 = 0)$ and $\boldsymbol{x}_b^* = (x_1 = -1, x_2 = 0)$. The Jacobian is

$$D_{\boldsymbol{x}}\mathbf{f} = \frac{\partial(f_1, f_2)}{\partial(x_1, x_2)} = \begin{bmatrix} \frac{\partial f_1}{\partial x_1} & \frac{\partial f_1}{\partial x_2} \\ \frac{\partial f_2}{\partial x_1} & \frac{\partial f_2}{\partial x_2} \end{bmatrix} = \begin{bmatrix} 0 & 1 \\ -1 - 2x_1 & 0 \end{bmatrix}.$$

First, for $\boldsymbol{x}_a^* = (x_1 = 0, x_2 = 0)$, the Jacobian becomes

$$D_{\boldsymbol{x}_a^*}\mathbf{f} = \begin{bmatrix} 0 & 1 \\ -1 & 0 \end{bmatrix},$$

and localized coordinates $\xi_1 = x_1$ and $\xi_2 = x_2$. So for equilibrium point $\boldsymbol{x}_a^*$, $\mathrm{tr}(A) = 0$ and $\det(A) = 1$ which using Fig. 2.6 is a centre. For equilibrium point $\boldsymbol{x}_b^* = (x_1 = -1, x_2 = 0)$, the Jacobian becomes

$$D_{\boldsymbol{x}_a^*}\mathbf{f} = \begin{bmatrix} 0 & 1 \\ 1 & 0 \end{bmatrix},$$

and localized coordinates $\xi_1 = x_1$ and $\xi_2 = x_2$. So in this case $\mathrm{tr}(A) = 0$ and $\det(A) = -1$ which from Fig. 2.6 means that this equilibrium point is a saddle point.

The potential function can be found from Eq. (2.23). In this example $k_1 = k_2 = 1$ and $x = x_1$, so

$$V(x) = \frac{1}{2}x^2 + \frac{1}{3}x^3. \quad (2.25)$$

The phase space and potential function are shown in Fig. 2.7. ■

Figure 2.7 shows the link between the phase portrait (x_1, x_2) and the potential function $V(x)$. Notice that in terms of the state variables, the energy of any solution curve, from Eqs. (2.22) and (2.25), is defined by

$$E_t = \frac{1}{2}mx_2^2 + k\frac{1}{2}x_1^2 + k_2\frac{1}{3}x_2^3.$$

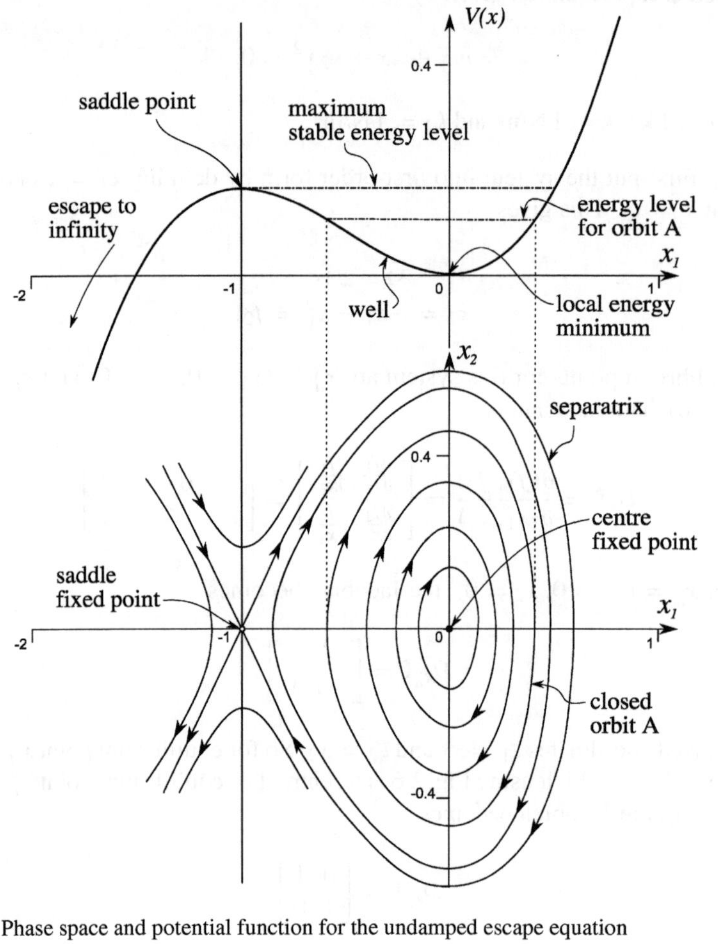

Fig. 2.7 Phase space and potential function for the undamped escape equation

So, on the x_2 axis, when $x_1 = 0$, the energy is purely kinetic $E_t = \frac{1}{2}mx_2^2$. This also corresponds to maximum and minimum x_2 values of any of the stable solution curves in the phase portrait, see for example, closed orbit A in Fig. 2.7. Conversely, when $x_2 = 0$, $E_t = V(x) = k\frac{1}{2}x_1^2 + k_2\frac{1}{3}x_1^3$, and maximum and minimum x_1 values for closed orbit A occur on the $x_2 = 0$ axis. The potential function, $V(x)$, is drawn in the top half of Fig. 2.7. This corresponds to the total energy along the x_1 axis. An example is shown for closed orbit A in the phase portrait (the lower part of Fig. 2.7), which has dashed lines from the intersections with the x_1 axis up to the energy plot. The constant energy level for closed orbit A is shown on the energy diagram.

As the size of the closed orbit increases, so does the energy level. At the point where the closed orbit touches the saddle point,[18] it also reaches a maximum energy level. Beyond this point the system becomes unstable, and solutions *escape* to infinity, which means they become infinitely large. The closed orbit which goes through the saddle point is the separatrix,[19] which in this case separates the stable area of solutions from the unstable area.

The stable part of the energy function containing the closed orbits is often called a *potential well*. Solutions which leave the potential well are sometimes said to have escaped from the potential well. From an engineering perspective, the escape phenomena are often related to catastrophic failure, for example, ship capsize or structural failure, Thompson and Stewart (2002).

Now consider an example of a system with a cubic nonlinear stiffness term.

Example 2.5 Phase portrait for oscillator with cubic stiffness nonlinearity (undamped Duffing oscillator)

Problem Construct the phase portrait and potential function for the unforced, undamped Duffing oscillator equation given by

$$m\ddot{x} - k_1x + k_3x^3 = 0, \tag{2.26}$$

with mass $m = 1$ kg, negative linear stiffness $k_1 = -1$ N/m and cubic stiffness $k_3 = 1$ N/m^3.

Solution First put the system into first-order form by defining $x_1 = x$ and $x_2 = \dot{x}$, such that $\ddot{x} = \dot{x}_2$. This gives

$$\dot{x}_1 = x_2 = f_1,$$
$$\dot{x}_2 = x_1 - x_1^3 = f_2.$$

By inspection, the equilibrium points for this system are $\boldsymbol{x}_a^* = (x_1 = 0, x_2 = 0)$, $\boldsymbol{x}_b^* = (x_1 = 1, x_2 = 0)$ and $\boldsymbol{x}_c^* = (x_1 = -1, x_2 = 0)$. The Jacobian is

---

[18] Also known as a homoclinic bifurcation.

[19] Also know as a homoclinic orbit, an orbit which starts and finishes at the same saddle point. Not to be confused with a heteroclinic orbit, which joins two separate saddle points.

$$D_x\mathbf{f} = \frac{\partial(f_1, f_2)}{\partial(x_1, x_2)} = \begin{bmatrix} \frac{\partial f_1}{\partial x_1} & \frac{\partial f_1}{\partial x_2} \\ \frac{\partial f_2}{\partial x_1} & \frac{\partial f_2}{\partial x_2} \end{bmatrix} = \begin{bmatrix} 0 & 1 \\ 1 - 3x_1^2 & 0 \end{bmatrix}.$$

For $x_a^* = (x_1 = 0, x_2 = 0)$, the Jacobian becomes

$$D_{x_a^*}\mathbf{f} = \begin{bmatrix} 0 & 1 \\ 1 & 0 \end{bmatrix}.$$

So for equilibrium point x_a^*, $\mathrm{tr}(A) = 0$ and $\det(A) = -1$ which using Fig. 2.6 is a saddle.

For equilibrium point $x_b^* = (x_1 = 1, x_2 = 0)$, the Jacobian becomes

$$D_{x_a^*}\mathbf{f} = \begin{bmatrix} 0 & 1 \\ -2 & 0 \end{bmatrix}, \tag{2.27}$$

so in this case $\mathrm{tr}(A) = 0$ and $\det(A) = 2$, which from Fig. 2.6 means that this equilibrium point is a centre. Equilibrium point $x_c^* = (x_1 = -1, x_2 = 0)$ has the same Jacobian as equilibrium point x_b^*, Eq. (2.27), and is also a centre.

The potential function can be found by integrating Eq. (2.26). In this example $p(x_1) = -x_1 + x_1^3$, so

$$V(x_1) = -\frac{1}{2}x_1^2 + \frac{1}{3}x_1^3 + \frac{1}{4},$$

where the $\frac{1}{4}$ constant ensures that the potential function is always positive, i.e. $V(x_1) \geq 0$ for all x_1. The phase space and potential functions are shown in Fig. 2.8. ■

Notice that the system plotted in Fig. 2.8 actually has a negative linear stiffness, $k_1 = -1$, which explains why there is a saddle point at the origin. This type of system may at first seem to have limited physical applications, but it can be used to model an interesting class of systems which have bi-stability. Or, in other words, they have two stable configurations (like the two equilibrium points at ± 1 in Fig. 2.8) separated by an unstable configuration (the saddle in Fig. 2.8). Classic examples are, buckled beams (discussed in Sect. 2.6), curved plates which can *snap-through* from one stable configuration to another (discussed in Chap. 8), and a cantilever beam with two magnetic fields acting on the tip—the so-called *Moon beam* (Moon 1987).

The form of $V(x)$ shown in Fig. 2.8 is often called a double potential well. The sides of the well continue to extend upwards, and energy levels for two different orbits are shown in Fig. 2.8. Orbit A is inside the potential well around the equilibrium point at $x_1 = 1, x_2 = 0$. Orbit B has a much higher energy level and is not confined to either of the centre equilibrium points. Here the separatrix marks the boundary between (i) the orbits confined to the potential wells around each of the centre equilibrium points and (ii) orbits which enclose both.

An analogy that is often used is to imagine that at any point in time the state of the system is represented by a ball rolling on the energy surface. As time evolves, the

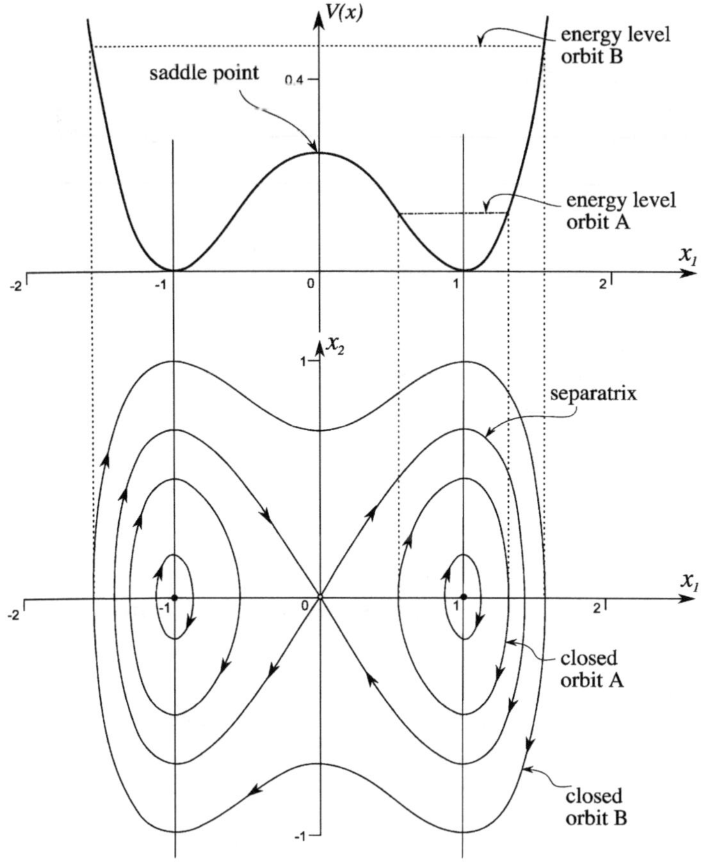

Fig. 2.8 Phase space and potential function for the undamped Duffing oscillator

ball will trace out a particular orbit in phase space, which in the case of an undamped, unforced system, would mean that the ball is constrained to be at a constant level on the energy surface—neither gaining or losing energy.

If the energy level can vary, then points where the ball crosses the separatrix correspond to the system escaping from one well into another, or—for the escape equation—to infinity. A further analogy is to imagine the phase space orbits as contours. These contours indicate lines of constant energy, in a similar way that contours on map indicate lines of constant height. By interpreting the phase portrait in this way an image of the energy surface, and therefore the system dynamics, can be obtained. This is shown in Fig. 2.9, which shows the complete energy surface for the Duffing oscillator example, Example 2.5. Extending the energy function in the x_2 direction gives a parabolic shape which can be seen from Eq. (2.22) where $E_t = V(x) + \frac{1}{2}mx_2^2$.

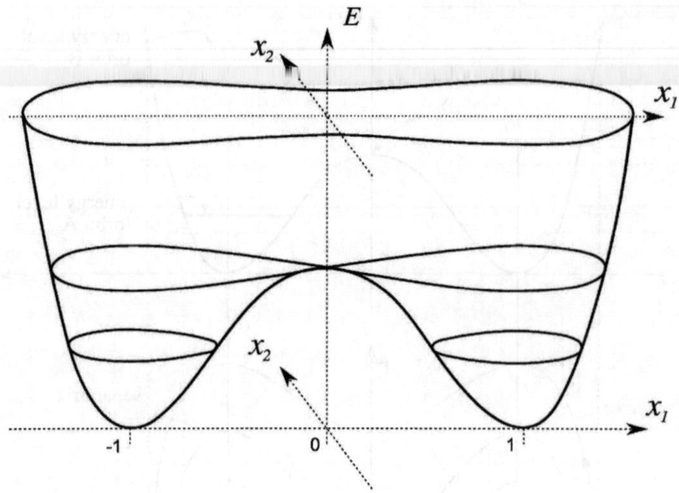

Fig. 2.9 Total energy surface for the Duffing oscillator

2.4 Multiple Solutions, Stability and Initial Conditions

A clear difference between linear oscillations and nonlinear oscillations is that nonlinear systems can have multiple solutions. For example, for the negative linear stiffness Duffing oscillator considered in Example 2.5, there are two stable equilibrium points, one at $x_1 = -1$ and the other at $x_1 = 1$. When damping is added to the system, these equilibrium points become attracting spirals, and the phase portrait changes to that shown in Fig. 2.10. This means that for a particular choice of initial conditions, $x_1(t_0), x_2(t_0)$, there are two potential finishing points. In fact, the regions of initial starting points for each stable equilibrium point are defined by the outer trajectories in Fig. 2.10. Starting points which are close to each other, but on either side of the saddle's *stable manifold*,[20] will diverge at the saddle point (due to the effect of the *unstable manifolds* coming out of the saddle point) and finish at different equilibrium points. An example is shown in Fig. 2.11, where two trajectories with initially close starting points[21] are attracted to different equilibrium points.

To find the initial condition values that are attracted to an equilibrium point for a whole region of initial values, the technique of *cell-to-cell mapping* can be used. This is a numerical technique which divides up the region of potential initial conditions into a grid. Each point in the grid is then used as an initial condition point, and a mapping from one cell to the next computed until an equilibrium point is reached. Each starting point is plotted (usually as a colour) based on which equilibrium point it is attracted to. Using this technique the *basins of attraction* can be seen. These are the regions of initial conditions which lead to trajectories which finish at the equilibrium

[20] This is the separatrix trajectory which goes directly to the saddle point.

[21] In fact they could be closer, but they have been slightly separated to make the figure clearer.

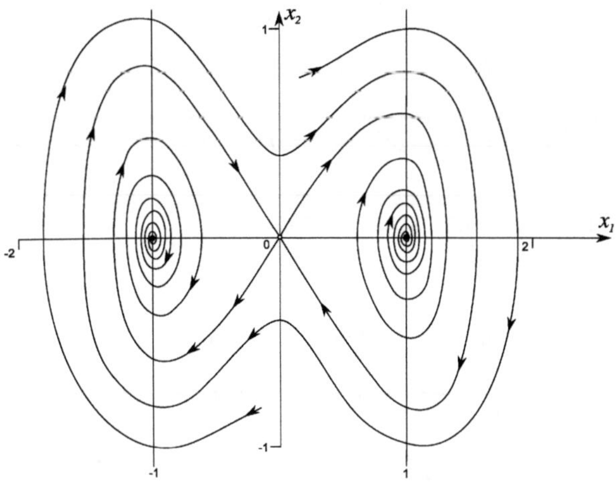

Fig. 2.10 Duffing oscillator phase portrait with damping

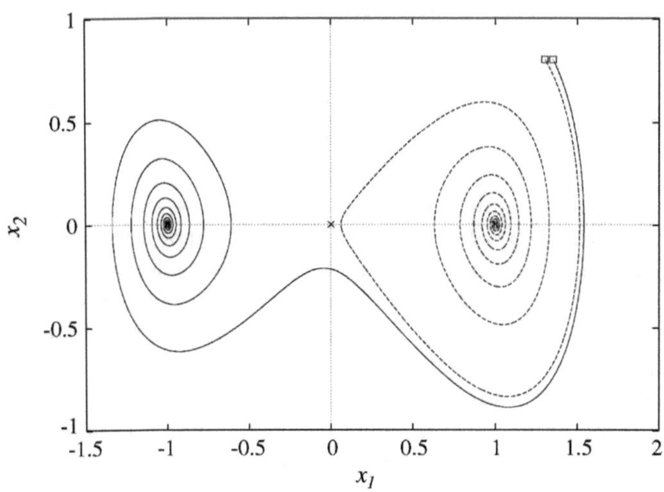

Fig. 2.11 Two close starting points finishing at different equilibrium points

point. These basins of attraction define the eventual, steady-state, behaviour of the oscillator. Using the same approach, but recording the *time taken* to reach a steady-state, can give information about the transient behaviour of the oscillator.

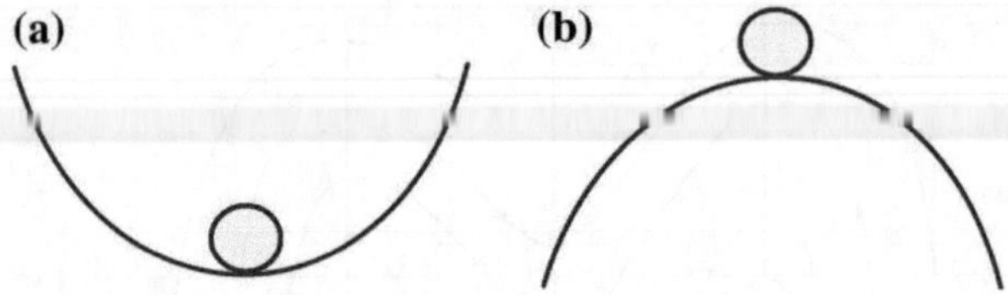

Fig. 2.12 Basic concept of stability. **a** Stable. **b** Unstable

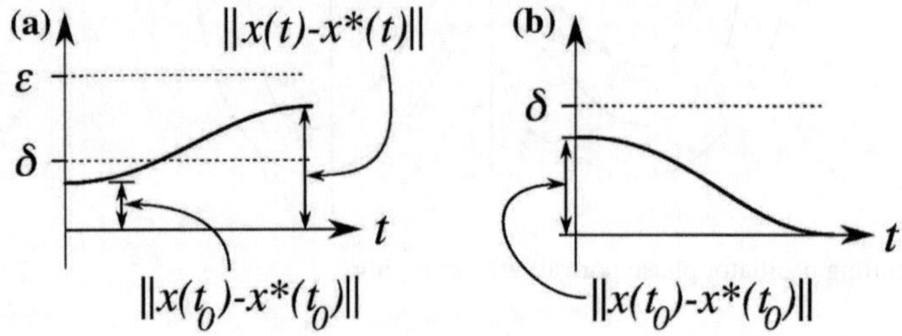

Fig. 2.13 Stability for a single state dynamical system, showing **a** Lyapunov stable, **b** asymptotically stable

2.4.1 Stability

In Sect. 2.1.3 attracting equilibrium points were said to be stable, but how is stability defined? The basic concept of stability can be visualized by considering a ball rolling over a surface. This is shown in Fig. 2.12 where an unstable position corresponds to the ball on a hill top, and a stable position corresponds to a ball resting in the bottom of a well. Any small disturbance will cause the ball on the hill top to roll down, whereas a small disturbance to the ball in the well would, under gravity, cause it to roll back to its original position.

In more precise terms it can be said that an equilibrium point, $\mathbf{x}^*$, is stable if a solution, $\mathbf{x}(t)$, close to it remains close for all time t. For example, for the system

$$\frac{\mathrm{d}\mathbf{x}}{\mathrm{d}t} = \mathbf{f}(\mathbf{x}, t), \qquad \mathbf{x}(t_0) = \mathbf{x}_0,$$

an equilibrium point, $\mathbf{x}^*$, is *Lyapunov stable* if $||\mathbf{x}(t_0) - \mathbf{x}^*(t_0)|| < \delta \Rightarrow ||\mathbf{x}(t) - \mathbf{x}^*(t)|| < \epsilon$, which is shown in Fig. 2.13a for a single state system. In other words, an equilibrium point is Lyapunov stable if trajectories that are initially close remain close. The idea of closeness is defined by δ and ϵ. For a choice of δ, the trajectory never goes further than ϵ if the system is to be Lyapunov stable. This type of stability is also called neutral stability. It includes centre equilibrium points where orbits stay close, but are neither pulled toward or pushed away from the equilibrium point.

An equilibrium point is *asymptotically* stable if nearby trajectories are 'attracted' to it as $t \to \infty$. This is written as $||\mathbf{x}(t_0) - \mathbf{x}^*(t_0)|| < \delta \Rightarrow \lim_{t\to\infty} ||\mathbf{x}(t) - \mathbf{x}^*(t)|| = 0$, which is shown in Fig. 2.13b. This type of stability applies to stable nodes and stable spirals, where nearby orbits are pulled strongly towards the equilibrium point.

2.5 Periodic and Non-periodic Oscillations

For unforced, undamped systems, steady-state periodic orbits can be observed which have amplitudes that are dependent on the initial conditions. However, the vast majority of mechanical engineering systems of interest are both forced and damped. For these systems, a common steady-state response is a periodic orbit with the same period as the harmonic forcing function. The response amplitudes will depend on the energy balance in the system. These types of periodic orbits are called *limit cycles* as noted in Example 2.1. Unlike the unforced, undamped case, where there are an infinite number of steady-state orbits which depend on the initial conditions, limit cycles are not entirely dependent on initial values. In fact they have the property of attracting nearby solution trajectories in state space. An example is shown in Fig. 2.14.

In addition to limit cycles it is possible to encounter a range of other types of behaviour. For example a closed orbit that takes two forcing periods to repeat the motion is called a *period-two* orbit. In fact multiple periodic responses can often be found and in general they are denoted as period n orbits.

Non-periodic responses can also occur, including *quasi-periodic* motion and *chaos*. Quasi-periodic motion (meaning almost, but not quite periodic) occurs when the response is composed of two or more signals with frequencies which are not-integer multiples of each other. In fact this type of motion can occur regularly in

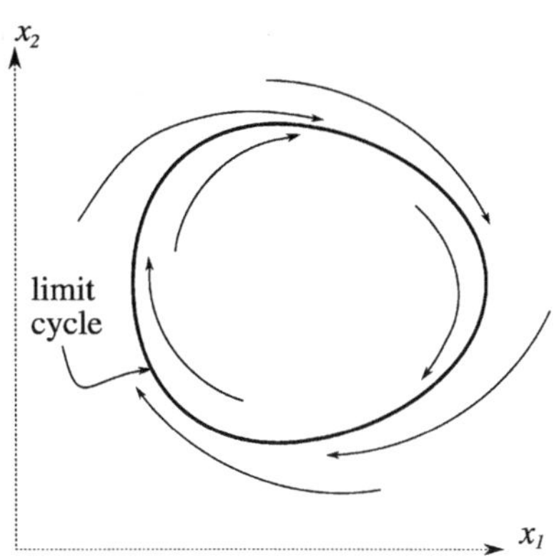

Fig. 2.14 A stable limit cycle oscillation which attracts nearby transient trajectories

vibration problems with multiple frequencies in the response. Linear natural frequencies are typically non-integer multiples, but they are spaced such that vibration modes at other frequencies are less significant to the response. However, closely spaced modes can result in a multi-frequency response which can appear similar to a quasi-periodic response for a nonlinear single degree-of-freedom system.

Viewed as a time series, chaos often appears to be non-repeatable and non-deterministic, but is in fact highly structured. However, how can chaos be identified as different from quasi-periodic or random time signals? Chaos is characterized by being highly sensitive to initial conditions, so that a very small change, leads to a very quick divergence of behaviour. This divergence can be quantified by calculating the rate of separation of initially close starting points as time increases. The rate of exponential divergence between nearby trajectories is measured by the *Lyapunov exponents* for the system. So if $\delta(t_0)$ is the initial distance between two nearby trajectories at the starting time t_0, and at some later time, t, the distance is modelled by $\delta(t) = e^{\lambda t}\delta(t_0)$, then λ is the Lyapunov exponent. For the general case where there would typically be more than one Lyapunov exponent, and one or more of the exponents is positive, then trajectories are separating with positive *exponential divergence*, and the system is chaotic. Exponential divergence is a key indication that the system is deterministic chaos, rather than a time signal from a stochastic system. This is because in a stochastic system trajectory separation is typically random rather than exponentially diverging.

A further test is to look at the data in the frequency domain. Chaos typically has a broad frequency content, as opposed to clear resonant peaks or harmonics from other non-chaotic signals. An example of this is discussed below.

In vibration analysis, chaos generally appears readily in low-dimensional problems, typically in forced nonlinear oscillators, such as the Duffing oscillator. In more flexible structures, which naturally have multiple modes, chaos is less readily apparent, and more difficult to distinguish as a specific response—especially when damping is very light. It should also be noted that chaos has primarily been studied and classified for low-dimensional problems with a single harmonic forcing input, which limits its relevance to higher-dimensional vibration problems. In some applications, chaos is seen as a desirable response—for example to distribute wear evenly across mechanical components—but in most vibration applications it is seen as undesirable, mainly because it typically has much larger amplitudes and is much less predictable than periodic motions. An example showing periodic and non-periodic responses is considered next.

Example 2.6 Periodic and non-periodic oscillations in a damped Duffing oscillator

Problem Consider the following forced, damped Duffing oscillator with negative linear stiffness

$$\ddot{x} + 0.4\dot{x} - x + x^3 = F\cos(1.8t), \tag{2.28}$$

where F is the forcing amplitude. Use time series and frequency spectra to investigate what type of periodic and non-periodic behaviour this oscillator has for F in the range $0.6 \leq F \leq 1.8$.

Solution Equation (2.28) can be written in first-order form

$$\dot{x}_1 = x_2,$$
$$\dot{x}_2 = -(0.4)x_2 + x_1 - x_1^3 + F\cos(1.8t).$$

The dynamics of this first-order system can now be simulated by first computing time series from initial values (this is typically done using 4th-order Runge-Kutta numerical integration as mentioned in Example 2.1). The simplest way to examine the behaviour for a range of F values is just to select some across the given range. Four cases of the dynamics of the forced Duffing oscillator with forcing values in the range $0.6 \le F \le 1.8$ have been computed and are shown in Fig. 2.15. The four cases selected are; $F = 0.6$ shown in (al)–(a3), $F = 0.73$ shown in (b1)–(b3), $F = 1.51$ shown in (c1)–(c3) and $F = 1.8$ shown in (d1)–(d3). In each case the left-hand picture shows the time series at the selected F value. Then in the centre the corresponding steady-state attractor in the $(x, \dot{x})$ plane is shown for each case. Finally on the right-hand side, the frequency spectrum is shown.

The frequency spectrum is obtained by taking the Fourier transform (FFT) of the time series and then using the absolute value of FFT amplitude, X. The log (to base 10) of $|X|$ is plotted against a linear scale of frequencies in Hz. Logs of the amplitudes are used so that all relevant frequency content can be viewed in the plot.[22] The angular forcing frequency in this example is $\omega = 1.8\,\mathrm{rad/s}$, which is related to the frequency, f, in Hz, by the relation $\omega = 2\pi f$, so that $f = 0.286$ in this case. The position of f in the frequency spectrum is marked on Fig. 2.15a3–d3 with an arrow. Frequency peaks at integer multiples of f from 1 to 5 are also marked on the frequency spectrum.

In the first case, $F = 0.6$ shown in (a1)–(a3), the motion repeats after one forcing period. The response is periodic but non-harmonic. As a result, in the frequency spectrum, as well as a response at f, the second and third harmonics are clearly evident. Notice also that the time series is not centred at zero displacement. This leads to a significant value in the frequency spectrum at zero, which is sometimes called the DC offset.[23]

In the second case, $F = 0.73$ shown in (b1)–(b3), the motion repeats after three forcing periods. As before this is periodic but non-harmonic and, in the frequency spectrum, multiple response peaks are evident. However, in this case only 1 and 3 are integer harmonics of f. The peak lower in the spectrum than f is a 1/3 *subharmonic*. Other peaks correspond to non-integer harmonics at 5/3, 7/3, 11/3 and 13/3.

In the third case, $F = 1.51$ shown in (c1)–(c3), the motion repeats after four forcing periods. Now, in the frequency spectrum all the integer harmonics from 1 to

---

22 There is a range of alternatives for plotting frequency spectra, the most common of which are various definitions of *power spectra*. Further discussion of the merits of these methods can be found in Newland (1993) and Press et al. (1994).

23 This term originates from electronics, where it refers to a direct current voltage, but the concept has been extended to any representation of a waveform.

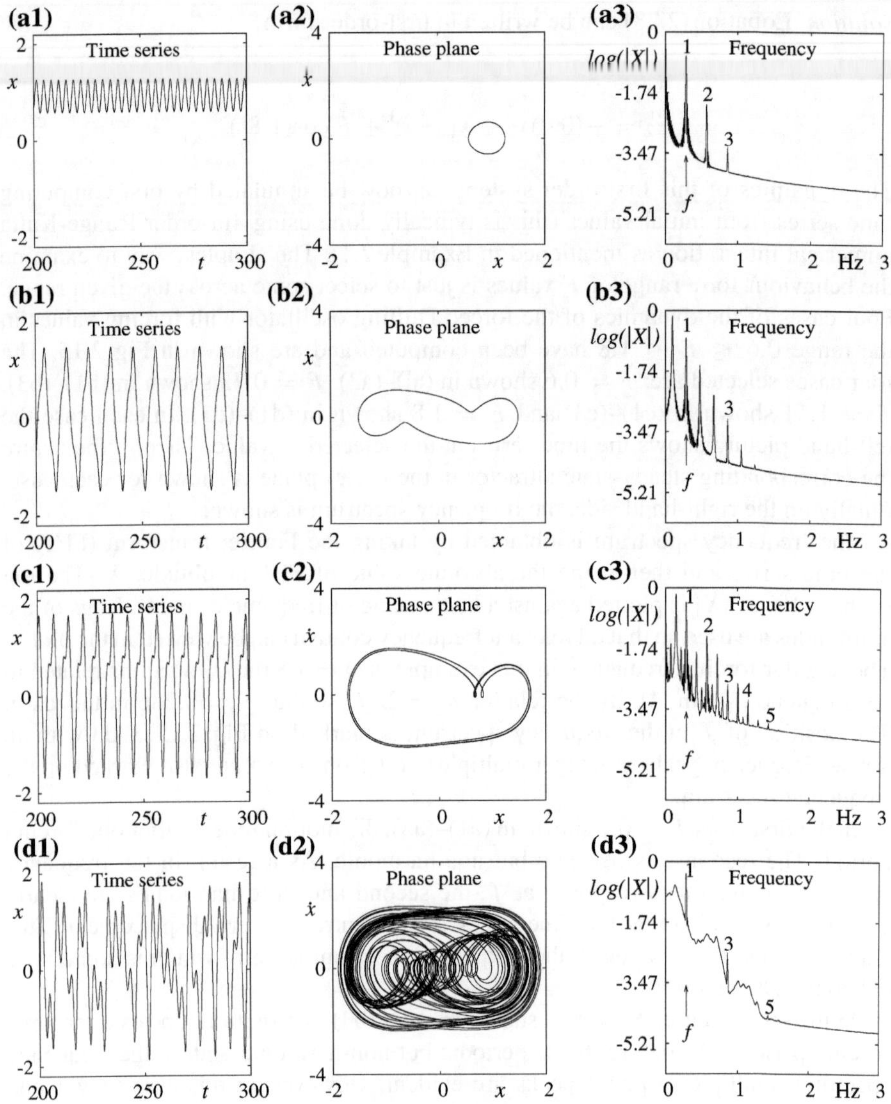

Fig. 2.15 Four different examples of Duffing oscillator dynamics, showing the time series on the *left*, the phase portrait in the *centre* and the frequency spectrum on the *right*

5 are present, as well as a 1/2 subharmonic and non-integer harmonics at 3/2, 5/2, 7/2 and 9/2.

Finally, for the case when $F = 1.8$ shown in (d1)–(d3), the response is non-periodic, and in fact in this case is chaotic. Clear evidence for chaotic motion can be seen from the frequency spectrum, which has noticeable peaks at 1, 3 and 5, but is generally much broader in its response compared to the periodic motions. ■

It should be clear from Example 2.6 that (i) in a nonlinear system a range of complex dynamic responses can occur over a relatively short parameter range, and (ii) as a parameter is varied, key changes take place between different dynamic responses. These changes are called *bifurcations* and they are discussed next.

2.6 Parameter Variation and Bifurcations

To investigate the steady-state behaviour of a particular system, one or more of the system parameters can be varied. In vibration engineering, the amplitude and frequency of the external forcing terms are often used to characterize the steady-state system response. As a result, these are natural parameters to vary, but other system parameters may also be used. For linear systems with harmonic forcing, the steady-state response will be made up of one or more resonance peaks, as discussed for example in Chap. 1, Sects. 1.3.1 and 1.3.3.

The stability criterion used so far for linear (or linearized) systems is that an equilibrium point in the upper left-hand quadrant of Fig. 2.6 indicates stability. An alternative way of representing this stability criterion is to plot the system eigenvalues in the complex plane.[24] Then for a linear system, if the real parts of the eigenvalues are in the left-hand plane the system is stable, as shown in Fig. 2.16. Conversely, if Re(λ) are in the right-hand plane, the system is unstable. This is because of the exponential form of the solution, given for example in Eq. (2.13). The behaviour follows that shown in Fig. 2.5, in that when the eigenvalue has negative real parts the solution shrinks (Fig. 2.5a), or if the eigenvalue has positive real parts the solution grows exponentially (Fig. 2.5c).

For nonlinear systems, we consider each equilibrium point individually. If for an equilibrium point, the eigenvalues of $D_{x^*}\mathbf{f}$ are in the left-hand plane, then the equilibrium point is locally stable. What happens to the eigenvalues of a stable equilibrium point if a system parameter, μ, is varied? The position of the eigenvalues

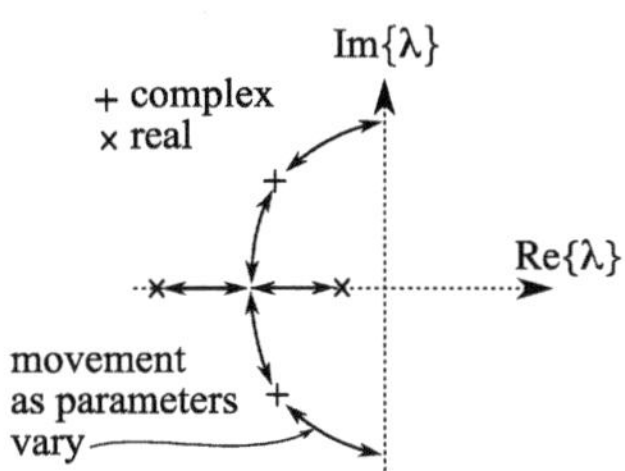

Fig. 2.16 Stable eigenvalues for a linear (or linearized) system showing real, and complex conjugate, cases

[24] This is the representation typically used in linear control theory.

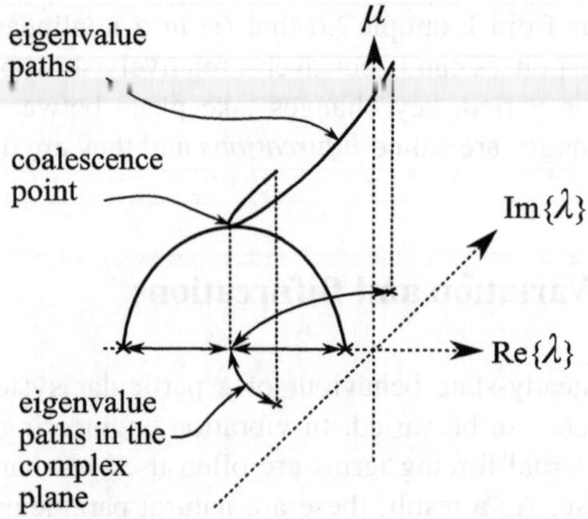

Fig. 2.17 Eigenvalue paths as a parameter is varied

of the linearized system will change,[25] as shown in Fig. 2.16 and for a particular value, μ_0, the real part of one or more of the eigenvalues will become zero. This is when a bifurcation occurs, meaning a substantive change in behaviour—originally a branching point.

In Fig. 2.16 there are two directions in which the stable eigenvalues can change along the paths shown. If the eigenvalues start as a complex conjugate pair and as the parameter changes they reduce in amplitude, they will eventually coalesce on the real axis before diverging as a pair of real eigenvalues. The second case is when the eigenvalues start as real and diverge to become a complex conjugate pair. Figure 2.16 can be replotted to include the parameter being varied, so that the paths of the eigenvalues can be viewed in a three-dimensional space, as shown in Fig. 2.17. A detailed treatment of this type of parameter dependent eigenvalue behaviour is given by Seyranian and Mailybaev (2003).

In terms of the vibration of mechanical, complex eigenvalues indicate underdamped vibration and real eigenvalues overdamped vibration. So in terms of bifurcations, real eigenvalues crossing the imaginary axis tend to relate to *static bifurcations* such as buckling of struts,[26] whereas complex eigenvalues crossing the imaginary axis relate to *dynamic bifurcations* such as the sudden appearance of oscillations like flutter.

In fact for equilibrium points in the linear unforced, undamped case, two types of *local* bifurcation have already been discussed. These correspond to the cases of static and dynamic stability loss, as solutions leave the stable upper left quadrant in Fig. 2.6. The dynamic instability corresponds to damping changing sign from positive to negative. In this case the system has complex conjugate eigenvalues, which simultaneously cross the imaginary axis at the point of instability. This type of behaviour is

25 Like a root-locus in linear control theory.

26 Aeroelastic divergence is another example.

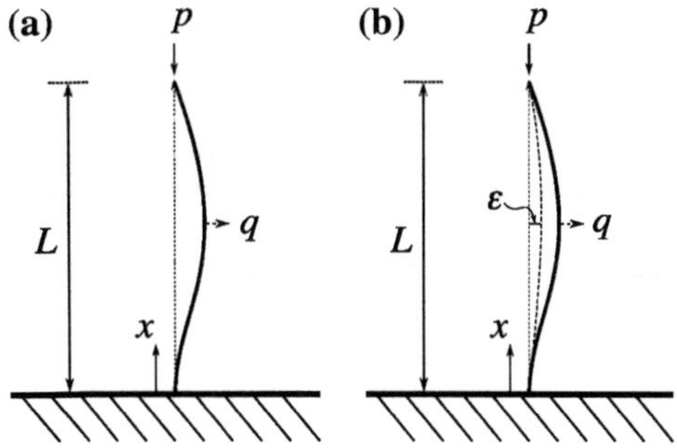

Fig. 2.18 Beam buckling: **a** perfect column, **b** column with eccentricity

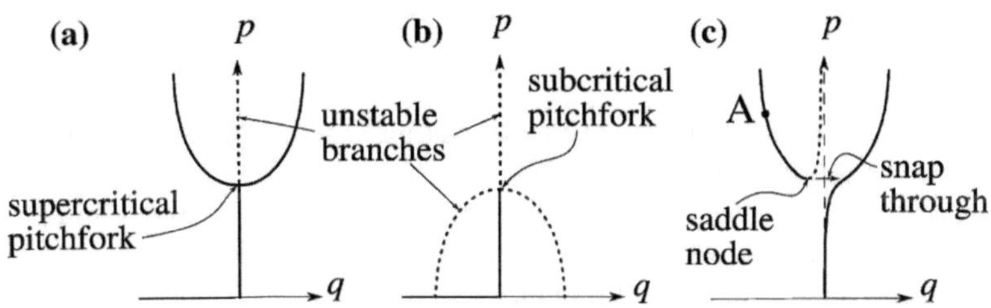

Fig. 2.19 Pitchfork bifurcation: **a** supercritical, **b** subcritical and **c** imperfect

a form of *Hopf bifurcation*[27] which will be discussed in greater detail in Sect. 2.6.1. Static instability corresponds to the case when the system has real eigenvalues, and one of the eigenvalues becomes zero at the point when linear stiffness changes sign from positive to negative. This type of bifurcation is characterized by a node changing into a saddle at the point of instability. Depending on the symmetry of the problem, this is either a *saddle-node* bifurcation or a *pitchfork bifurcation*.[28]

A classic engineering example where both saddle-node and pitchfork bifurcations can be observed is the buckling of an axially-loaded (planar) vertical column, as shown in Fig. 2.18. In Fig. 2.18a a perfectly straight, planar column is loaded with an axial load, p, and the mid-point transverse deflection is q. As the axial load reaches the critical Euler buckling load,[29] a pitchfork bifurcation occurs, which is shown in Fig. 2.19a, b. Figure 2.19a is a supercritical pitchfork bifurcation, which corresponds to the physical case when the column adopts a buckled shape but does not collapse. Figure 2.19b is a subcritical pitchfork bifurcation, which corresponds to the physical case when the column fails catastrophically at the point of bifurcation.

---

[27] In fact this is a special case, as there is no limit cycle close to the bifurcation point, see Strogatz (2001).

[28] There is a third variation called the *transcritical bifurcation* see Strogatz (2001).

[29] See for example Frish-Fay (1962) for details of Euler buckling. For discussions on more complex buckling problems in structural engineering, such as arches and shells, see Thompson and Hunt (1973), Thompson (1982) and Virgin (2007).

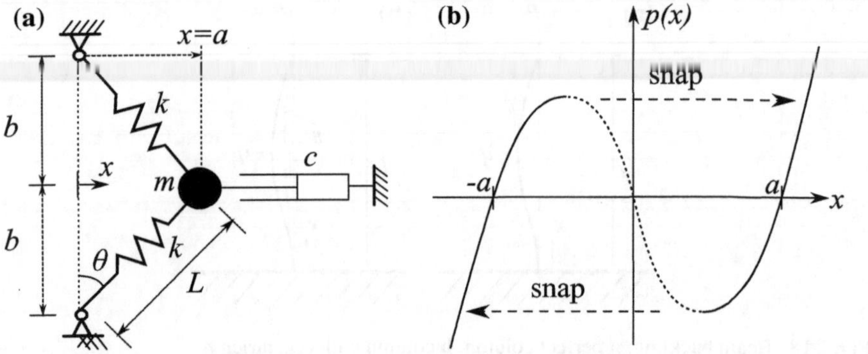

Fig. 2.20 Snap-through system showing, **a** schematic, and **b** nonlinear stiffness function

The two dashed curves linking the bifurcation point to zero correspond to the collapse solutions to the left or right.

In the supercritical case, Fig. 2.19a, after the bifurcation point, the original straight solution becomes unstable (shown as a dashed line) and two stable solutions emerge corresponding to the column buckling either to the left or the right. As the column is perfectly straight, there is an equal chance of the column buckling in either direction. Physically, it is unlikely that the column is perfectly straight, and so the case in Fig. 2.18b is for an imperfect column, where the initial imperfection is represented by the deflection ϵ. The initial imperfection means that the column will always buckle in the same direction. The case for positive ϵ is shown in Fig. 2.19c, where it can be seen that there are now two disconnected solution paths. Increasing p from zero always leads to a buckled shape to the right. If the beam is forced into the opposite (left-hand) buckled shape, and loaded above the Euler load, it can be held in this position, for example at point A in Fig. 2.19c. Physically, the axial load is holding the beam in the buckled state which is opposite to its initial imperfection. Then if the axial load is decreased, at the Euler load the beam will suddenly snap through to the other branch of solutions. The point of snap-through is a saddle-node bifurcation where the stable branch joins an unstable branch that corresponds to the original unbuckled solution.

Notice that in Fig. 2.19 the solid lines indicate the paths of the stable equilibrium points (node/spiral) as p is varied and the dashed lines indicate the unstable equilibrium points (saddles). The unstable and stable branches join at the bifurcation point.

Example 2.7 Bifurcation due to linear stiffness changing sign (Pitchfork)

Problem The physical system shown in Fig. 2.20a has a geometric nonlinearity due to the angle, θ, of the springs. This type of nonlinearity can be approximated by a Duffing-type oscillator with nonlinear stiffness shown in Fig. 2.20b. The equation of motion is given by

$$m\ddot{x} + c\dot{x} - \mu x + \alpha x^3 = 0,$$

where μ and α are coefficients which depend on k, θ and L, the natural length of the springs, and c is viscous damping.[30] Assuming $m = 1$ and $\alpha = 1$, find the change in behaviour which occurs as the linear stiffness parameter, μ, is varied and changes sign. What does this change in μ correspond to physically for the system in Fig. 2.20?

Solution First, put the system into first-order form

$$\dot{x}_1 = x_2 = f_1,$$
$$\dot{x}_2 = \mu x_1 - x_1^3 - cx_2 = f_2.$$

By inspection, the equilibrium points for this system are found by equating $f_1 = f_2 = 0$ which gives

$\mu < 0$	$x_1 = 0$	$x_2 = 0,$	one equilibrium point
$\mu = 0$	$x_1 = 0$	$x_2 = 0,$	one equilibrium point
$\mu > 0$	$x_1 = 0$	$x_2 = 0,$	
and	$x_1 = \pm\sqrt{\mu}$	$x_2 = 0,$	three equilibrium points.

To investigate the behaviour, the system is linearized locally close to the equilibrium points. For all μ values the equilibrium point $\boldsymbol{x}_a^* = (x_1 = 0, x_2 = 0)$ exists. For $\mu > 0$ values, two additional equilibrium points exist and are labelled as $\boldsymbol{x}_{b,c}^* = (x_1 = \pm\sqrt{\mu}, x_2 = 0)$. In general, the Jacobian for the system is

$$D_{\boldsymbol{x}}\mathbf{f} = \frac{\partial(f_1, f_2)}{\partial(x_1, x_2)} = \begin{bmatrix} \frac{\partial f_1}{\partial x_1} & \frac{\partial f_1}{\partial x_2} \\ \frac{\partial f_2}{\partial x_1} & \frac{\partial f_2}{\partial x_2} \end{bmatrix} = \begin{bmatrix} 0 & 1 \\ \mu - 3x_1^2 & -c \end{bmatrix}.$$

First for $\boldsymbol{x}_a^* = (x_1 = 0, x_2 = 0)$, the Jacobian becomes

$$D_{\boldsymbol{x}_a^*}\mathbf{f} = \begin{bmatrix} 0 & 1 \\ \mu & -c \end{bmatrix}.$$

So for equilibrium point $\boldsymbol{x}_a^*$, $\mathrm{tr}(A) = -c$ and $\det(A) = -\mu$.

For equilibrium points $\boldsymbol{x}_{b,c}^* = (x_1 = \pm\sqrt{\mu}, x_2 = 0)$, the Jacobian becomes

$$D_{\boldsymbol{x}_a^*}\mathbf{f} = \begin{bmatrix} 0 & 1 \\ -2\mu & -c \end{bmatrix}.$$

So in this case $\mathrm{tr}(A) = -c$ and $\det(A) = 2\mu$.

---

[30] The derivation of a Duffing oscillator from the snap-through system can be found as the solution to Problem 2.1. Note that the mass is at static equilibrium when $x = \pm a$, and the springs are at their natural length, L.

Note that the expression for $\mathrm{tr}(A)$ and $\det(A)$ are computed assuming that $\mu > 0$. In the case when $\mu < 0$ the sign of μ terms *will change*. So for equilibrium point $\boldsymbol{x}_a^*$ (using Fig. 2.6) when

$$\begin{array}{lll} \mu < 0, & \mathrm{tr}(A) = -c,\ \det(A) = -(-\mu) = \mu, & \text{stable node/spiral} \\ \mu = 0, & \mathrm{tr}(A) = -c,\ \det(A) = 0, & \text{degenerate case} \\ \mu > 0, & \mathrm{tr}(A) = -c,\ \det(A) = -\mu, & \text{saddle} \end{array}$$

so this equilibrium point changes from a stable node/spiral to a saddle point as μ passes through zero. In general, for $\mu > 0$ the discriminant is $\Delta = \mathrm{tr}^2 - 4\det = c^2 + 4\mu$. So the μ value at which $\Delta = 0$ is $\mu = -c^2/4$, marking the degenerate node case from Fig. 2.6. So for $-c^2/4 < \mu < 0$, $\boldsymbol{x}_a^*$ is a stable node and for $\mu < -c^2/4$ a stable spiral.

For equilibrium points $\boldsymbol{x}_{b,c}^*$ when

$$\begin{array}{lll} \mu < 0, & \text{n/a}, & \text{no equilibrium point} \\ \mu = 0, & \mathrm{tr}(A) = -c,\ \det(A) = 0, & \text{degenerate case} \\ \mu > 0, & \mathrm{tr}(A) = -c,\ \det(A) = 2\mu, & \text{stable node/spiral} \end{array}$$

So, for $\mu < 0$, there are no equilibrium points. For $\boldsymbol{x}_b^*, c$ the discriminant is $\Delta = \mathrm{tr}^2 - 4\det = c^2 - 8\mu$. So the μ value at which $\Delta = 0$ is $\mu = 1/8$, marking the degenerate node case from Fig. 2.6. So for $0 < \mu < c^2/8$, $\boldsymbol{x}_{b,c}^*$ are stable nodes and for $\mu > c^2/8$ they become a stable spirals.

Physically changing μ from negative to positive corresponds to the system in Fig. 2.20 having positive linear stiffness. Geometrically, this can be considered in terms of the length of the springs. So, in the case when $b < L$ the springs are in compression and $\mu < 0$. Conversely, when $b > L$ the springs are in tension and $\mu > 0$. ■

The physical interpretation of the snap-through can be seen from Fig. 2.21, where in (a1) and (a2) the linear stiffness is negative, (b1) and (b2) shows the $\mu = 0$ case and (c1) and (c2) shows the case where the linear stiffness is positive. This corresponds to moving the end supports apart from (a1)–(b1) and finally (c1). In (b1) there is no tension or compression in the springs, whereas in (c1) the springs are in tension.

This type of transition is known as a *cusp bifurcation*, because if plots (a2), (b2) and (c2) are combined into a surface plot with μ as the additional coordinate, then the surface has a cusp at $\mu = 0$. See Thompson (1982) for further details of this phenomenon.

From a mechanical vibration perspective, bifurcations of equilibria relate primarily to unforced systems and/or static stability. A more significant class of bifurcations for vibration analysis are those that lead to oscillations. These are considered next.

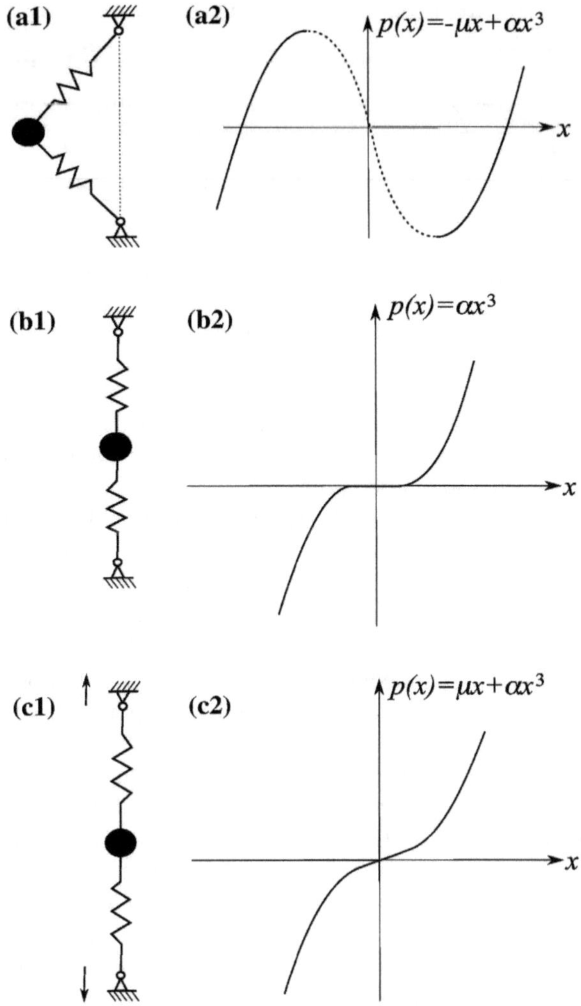

Fig. 2.21 Change in stiffness function as μ varies for Example 2.7

2.6.1 The Onset of Oscillations via a Hopf Bifurcation

This subsection describes an important phenomenon that occurs in nonlinear vibrations. It occurs when a stable equilibrium point becomes unstable and is replaced by a limit cycle as a parameter is varied. The point at which this happens is called a Hopf bifurcation. Numerous physical examples of this type of phenomenon exist. In engineering, Hopf bifurcations are often associated with systems where fluid flow is the external forcing, and the flow speed is the parameter which triggers the bifurcation. A simple example is fluid flowing through a hose pipe at slow speeds does not

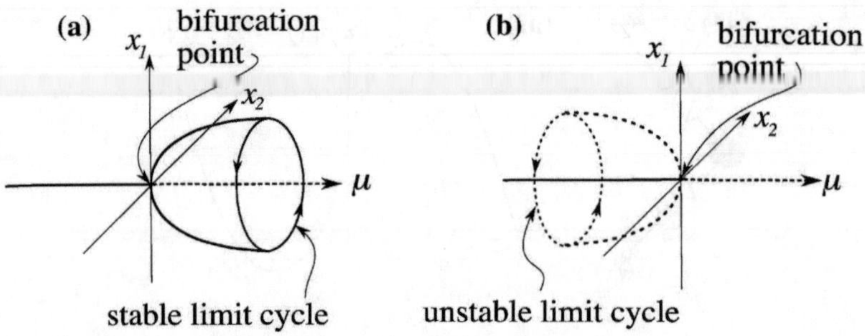

Fig. 2.22 Hopf bifurcation: **a** supercritical, **b** subcritical

induce any (large) oscillations in the pipe. As the flow speed is increased, there comes a critical point where large oscillations in the pipe occur due to a Hopf bifurcation (try an experiment with your garden hose).

A significant physical example is flutter in aeroelastic vibration problems. This is a major design consideration for aerospace structures. Other examples include oscillations in cables immersed in a fluid flow, for example on bridges, and the onset of lateral oscillations in train carriages at a critical speed.

A schematic representation of the two different types of Hopf bifurcation is shown in Fig. 2.22. As the system parameter, μ, is varied, a stable equilibrium point at the origin ($x_1 = 0,\ x_2 = 0$) goes unstable at the bifurcation point, $\mu = 0$. There are two cases (like the pitchfork), in the supercritical case (Fig. 2.22a) a stable limit cycle exists for $\mu > 0$. In the subcritical case (Fig. 2.22b) no stable limit cycle exists for $\mu > 0$, instead an unstable limit cycle exists for $\mu < 0$. As a result the subcritical case can be viewed as potentially catastrophic, because after the bifurcation the system has no (local) stable solution to stabilize onto. Instead, if there are no solutions nearby, the system could jump to a distant solution or escape to infinity. Determining whether a particular bifurcation is super- or sub-critical can be done by either numerical simulation or using *centre manifold theory*. This is a major topic of dynamical systems theory, and good introductions are given by Guckenheimer and Holmes (1983), Thompson and Stewart (2002), Glendinning (1994) and Strogatz (2001).

Another important class of systems in which Hopf bifurcations occur are those with delays. This is especially important when applying control to dynamic systems, as actuators can introduce delays into the overall system. An example is considered next.

Example 2.8 Hopf bifurcation due to actuator delay

Problem An experimental test is configured such that an actuator is attached to a linear spring, k_s, as shown in Fig. 2.23. The actuator is controlled to follow (track) the output, z, of a single-degree-of-freedom mass-spring oscillator ($m,\ c,\ k$) system and the force from the actuator, F, is fed into the single degree of freedom—see Chap. 7 of Bursi and Wagg (2008) for a complete description of this system. Assuming perfect control tracking, underdamped vibrations, and that the actuator dynamics

Fig. 2.23 Schematic representation of the experimental test system. See Chap. 7 of Bursi and Wagg (2008) for a complete description of this system

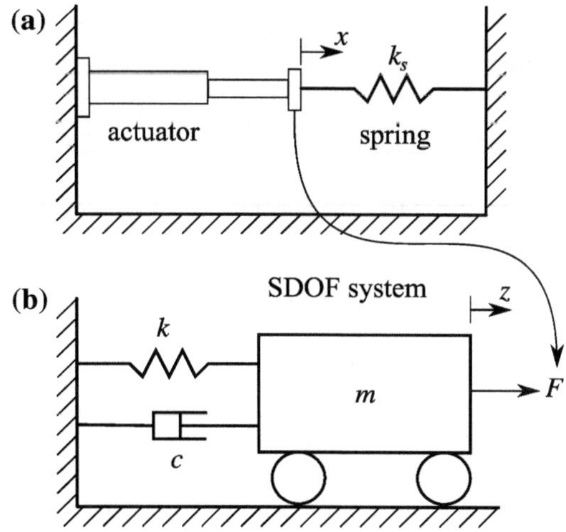

can be modelled as a small fixed delay, τ, find the eigenvalues of the system and use them to examine the stability of the equilibrium point at the origin in terms of the parameter τ.

Solution The governing equation of the single-degree-of-freedom system is

$$m\ddot{z} + c\dot{z} + kz = F,$$

where the feedback force $F = -k_s x$ and x is the actuator displacement (from Fig. 2.23a). For perfect control tracking $x = z$, *but* the key observation is that the actuator introduces a delay such that $x(t) = z(t - \tau)$, $\tau > 0$. The overall system is then governed by the delay differential equation[31]

$$m\ddot{z} + c\dot{z} + kz + k_s z(t - \tau) = 0. \tag{2.29}$$

The characteristic equation can be found by assuming solutions of the form $z(t) = Ce^{\lambda t}$, where C is an arbitrary constant, which for the delay term gives $z(t - \tau) = Ce^{\lambda(t-\tau)}$. Substituting these expressions into Eq. (2.29) gives

$$m\lambda^2 + c\lambda + k + k_s e^{-\lambda\tau} = 0,$$

because $e^{\lambda(t-\tau)} = e^{\lambda t}e^{-\lambda\tau}$ so that the $Ce^{\lambda t}$ factors can be divided out, leaving just the $e^{-\lambda\tau}$ exponential term.[32]

---

[31] The introduction of a fixed delay means that the delay differential equation actually has an infinite-dimensional state space, see Stépan (1989) or Diekmann et al. (1995) for an introduction to delay differential equations.

[32] In fact, this is an infinite-dimensional eigenvalue problem. However, only two are significant in this case. See Stépan (1989) or Diekmann et al. (1995) for further details.

If the delay, τ, is small, then the approximation $e^{-\lambda\tau} \approx (1 - \lambda\tau)$ can be made, which gives

$$m\lambda^2 + c\lambda + k + k_s(1 - \lambda\tau) \rightsquigarrow m\lambda^2 + \lambda(c - k_s\tau) + k + k_s \approx 0. \qquad (2.30)$$

The solution of Eq. (2.30) is

$$\lambda_{1,2} = \frac{1}{2m}\left[(k_s\tau - c) \pm \sqrt{(c - k_s\tau)^2 - 4m(k + k_s)}\right].$$

For underdamped vibrations and τ small, physically realistic parameters result in complex eigenvalues. Or, in other words, assume that $4m(k + k_s) > (c - k_s\tau)^2$ for all realistic choices of m, c, k, k_s and τ. Then the stability of the eigenvalues is governed by the real part, specifically by the sign of $(k_s\tau - c)$. If $(k_s\tau - c) < 0$ the eigenvalues are complex and stable. When $(k_s\tau - c) > 0$ the eigenvalues are unstable. The transition occurs when $(k_s\tau - c) = 0$ or $\tau = \dfrac{c}{k_s}$. The value of τ at which the transition occurs corresponds to complex eigenvalues crossing the imaginary axis from left to right, which is a Hopf bifurcation.

Physically, the delay can be interpreted as negative damping, with an equivalent negative damping term of $c_{neg} = -k_s\tau$. ■

2.6.2 *Bifurcations in Forced Nonlinear Oscillations*

When a (damped) nonlinear system is forced, one of the most likely steady-state responses is a limit cycle (also called periodic orbit) type of behaviour. Note that when the system has more than one degree-of-freedom, limit cycles typically exist for each degree-of-freedom (or mode of vibration, discussed further in Chap. 5). Examining the response behaviour as a parameter varies allows a comprehensive picture to be built up of the system dynamics. In the linear single- and multi-degree-of-freedom systems (for example, those discussed in Chap. 1, Sects. 1.3.1 and 1.3.3), varying the forcing frequency leads to changes in limit cycle amplitude, but no changes to the *structure*, of the limit cycle occurs. For nonlinear systems, the limit cycle structure can change and the points at which this happens are bifurcation points.

To analyse bifurcations of limit cycles, the cycle is typically linearized in some local region of state space. A way to linearize a limit cycle is to use the stroboscopic map. For a steady-state vibration in the form of a limit cycle, the stroboscopic map is formed by *sampling* the cycle once per forcing period. For example, if T is the forcing period, then at times $t = 0,\ T,\ 2T,\ 3T,\ \ldots$ the values of displacement d_n and velocity v_n are recorded, to get a series of data points $(d_0, v_0),\ (d_1, v_1),\ (d_2, v_2),\ (d_3, v_3),\ \ldots$, corresponding to the sampling times. For example in Fig. 2.24a two planes a distance T apart are shown intersecting with the trajectory, L. In the first plane at $t = 0$ the path of L intersects the plane at

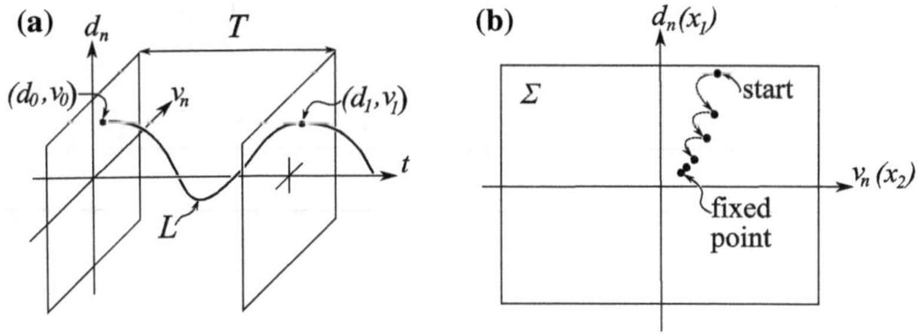

Fig. 2.24 The stroboscopic map, showing **a** the continuous orbit in d_n, v_n, t space, and **b** and example of a transient motion, showing the *points* that the orbit makes in the d_n, v_n plane. Notice that as the transient motion decays and the system becomes a steady state limit cycle, the *points* move towards a fixed point in the plane

point (d_0, v_0) (the initial conditions) and at time $t = T$ the intersection is at point (d_1, v_1). Plotting all these points on top of each other in a single displacement-velocity plane, denoted Σ, shows the evolution of the trajectory towards it's final steady state behaviour, which in this case is a limit cycle. An example of the Σ plane is shown in Fig. 2.24b. This shows a series of transient points in the plane, sampled at $t = 0,\ T,\ 2T,\ 3T,\ \ldots$, converging towards a *fixed point*[33] in the plane Σ.

Now, the fixed point in the map corresponds to the limit cycle in the *flow*, where flow is the evolution of multiple nearby, continuous time, trajectories in state space. So if the mapping can be linearized close to the fixed point, the eigenvalues[34] of the linearized system will indicate the type of fixed point behaviour and where bifurcations occur.

The state vector for the mapping is written in a discrete time formulation as $\boldsymbol{x}_n = [d_n, v_n]$, and the general nonlinear relationship is $\boldsymbol{x}_{n+1} = \boldsymbol{h}(\boldsymbol{x}_n)$, where each n in the map corresponds to a forcing period in the continuous time oscillator. A fixed point in the map[35] is denoted $\boldsymbol{x}^*$, and has the property that $\boldsymbol{x}^* = \boldsymbol{h}(\boldsymbol{x}^*)$. Linearising the mapping close to the fixed point means first defining a new coordinate with an origin at the fixed point $\boldsymbol{\xi}_n = \boldsymbol{x}_n - \boldsymbol{x}^*$. Then $\boldsymbol{\xi}_{n+1} = \boldsymbol{x}_{n+1} - \boldsymbol{x}^* = \boldsymbol{h}(\boldsymbol{x}_n) - \boldsymbol{x}^*$ or $\boldsymbol{\xi}_{n+1} = \boldsymbol{h}(\boldsymbol{x}^* + \boldsymbol{\xi}_n) - \boldsymbol{x}^*$. Taking a Taylor expansion of the nonlinear mapping function, $\boldsymbol{h}(\boldsymbol{x}^* + \boldsymbol{\xi}_n)$, gives the approximation

$$\boldsymbol{\xi}_{n+1} \approx \boldsymbol{h}(\boldsymbol{x}^*) + D_{\boldsymbol{x}^*}\boldsymbol{h}\,\boldsymbol{\xi}_n + \mathcal{O}(||\boldsymbol{\xi}_n||^2) - \boldsymbol{x}^*,$$

where the notation $D_{\boldsymbol{x}^*}$ is the same as that used for computing the Jacobian in Eq. (2.11). Now it can be seen that because $\boldsymbol{h}(\boldsymbol{x}^*) = \boldsymbol{x}^*$, the expression for $\boldsymbol{\xi}_{n+1}$ reduces to

---

33 A fixed point in a map can be thought of as analogous to equilibrium point in a continuous flow.

34 Also sometimes referred to as the Floquet multipliers of the periodic orbit.

35 This definition is for the lowest order return period i.e. one. Higher order periodicity maps can be defined, and the interested reader can find details in Thompson and Stewart (2002).

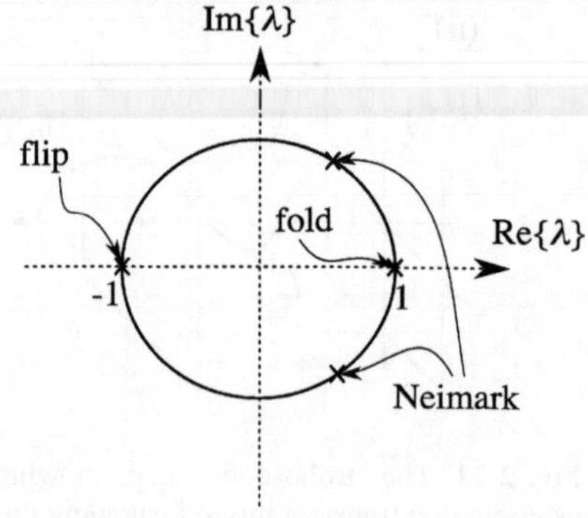

Fig. 2.25 Eigenvalues of a linearized mapping

$$\boldsymbol{\xi}_{n+1} \approx D_{\boldsymbol{x}^*}\boldsymbol{h}\,\boldsymbol{\xi}_n + \mathcal{O}(||\boldsymbol{\xi}_n||^2).$$

For mappings, the stability criterion for eigenvalues in the complex plane is different from that for equilibria. Now, any eigenvalue with a modulus greater than one[36] will lead to instability. This is shown in Fig. 2.25. There are three ways in which stability can be lost. For an eigenvalue of $\lambda = 1$ the system undergoes a cyclic saddle-node or *fold* bifurcation. For an eigenvalue of $\lambda = -1$ the system undergoes a *flip* bifurcation (also known as a *period-doubling* bifurcation). For a complex eigenvalue with $|\lambda| = 1$, the system undergoes a secondary Hopf or *Neimark-Sacker* bifurcation.

Example 2.9 Fixed points in the Henón map[37]

Problem The Henón map is a two state mapping which can be represented as

$$x_{n+1} = 1 - ax_n^2 + y_n,$$
$$y_{n+1} = bx_n,$$

where x_n, y_n are the system states, and a, b are parameters. Determine the condition for the fixed points of the system to be real. Then compute the stability of the fixed point which exists at values of $a = -\frac{1}{16}$ and $b = \frac{1}{2}$. Comment on the eigenvalues of this fixed point in terms of the expected bifurcation behaviour of the system.

Solution First we write the map as

$$x_{n+1} = 1 - ax_n^2 + y_n = h_1,$$
$$y_{n+1} = bx_n = h_2.$$

36 In discrete systems λ act as multipliers, so $|\lambda| > 1$, solution grows, unstable; $|\lambda| < 1$, solution shrinks, stable.

37 See Thompson and Stewart (2002) for details of the derivation of this and other maps.

For a fixed point of the map x_n^*, y_n^* we have by definition

$$\begin{aligned} x_n^* &= h_1(x_n^*, y_n^*) = 1 - ax_n^{*2} + y_n^*, \\ y_n^* &= h_2(x_n^*, y_n^*) = bx_n^* \end{aligned} \tag{2.31}$$

and by substituting the second of these expressions into the first we obtain

$$x_n^* = 1 - ax_n^{*2} + bx_n^*, \quad \rightsquigarrow \quad ax_n^{*2} + (1-b)x_n^* - 1 = 0$$

from which it can be determined that the solutions are given by

$$x_{n:1,2}^* = \frac{(1-b) \pm \sqrt{(1-b)^2 - 4a(-1)}}{2a}. \tag{2.32}$$

As a result, the condition for real solutions is the same as ensuring that the discriminant is greater than or equal to zero, such that

$$(1-b)^2 - 4a(-1) \geq 0 \quad \rightsquigarrow \quad (1-b)^2 + 4a \geq 0 \quad \rightsquigarrow \quad a \geq -\frac{1}{4}(1-b)^2.$$

When $a = -\frac{1}{16}$ and $b = \frac{1}{2}$ the discriminant is zero, so this will lead to a real fixed point(s). Substituting these values into Eq. 2.32 gives $x_n = 4$, and then from Eq. 2.31 it is found that $y_n = 2$.

The Jacobian of the map is given by

$$D_{\boldsymbol{x}}\boldsymbol{h} = \frac{\partial(h_1, h_2)}{\partial(x_n, y_n)} = \begin{bmatrix} \frac{\partial h_1}{\partial x_n} & \frac{\partial h_1}{\partial y_n} \\ \frac{\partial h_2}{\partial x_n} & \frac{\partial h_2}{\partial y_n} \end{bmatrix} = \begin{bmatrix} -2ax_n & 1 \\ b & 0 \end{bmatrix}.$$

Now to evaluate the stability of the fixed point at $x_n = 4$, $y_n = 2$, we substitute these values into the Jacobian to obtain

$$D_{\boldsymbol{x}^*}\boldsymbol{h} = \begin{bmatrix} -2(-\frac{1}{16})4 & 1 \\ \frac{1}{2} & 0 \end{bmatrix} = \begin{bmatrix} \frac{1}{2} & 1 \\ \frac{1}{2} & 0 \end{bmatrix}.$$

The eigenvalues of this Jacobian matrix are $\lambda_1 = 1$ and $\lambda_2 = -\frac{1}{2}$. An eigenvalue of $+1$ means that the system is on the boundary of stability (as shown in Fig. 2.25), and the system is undergoing a fold bifurcation. ■

A set of numerically computed data of the Henón map is shown in Fig. 2.26. First in Fig. 2.26a a *chaotic attractor*[38] which occurs for the parameter values $b = 0.3$ and

[38] We will not give further detailed discussion on chaos, but a good references are the books by Guckenheimer and Holmes (1983), Moon (1987), Glendinning (1994), Strogatz (2001) and Thompson and Stewart (2002).

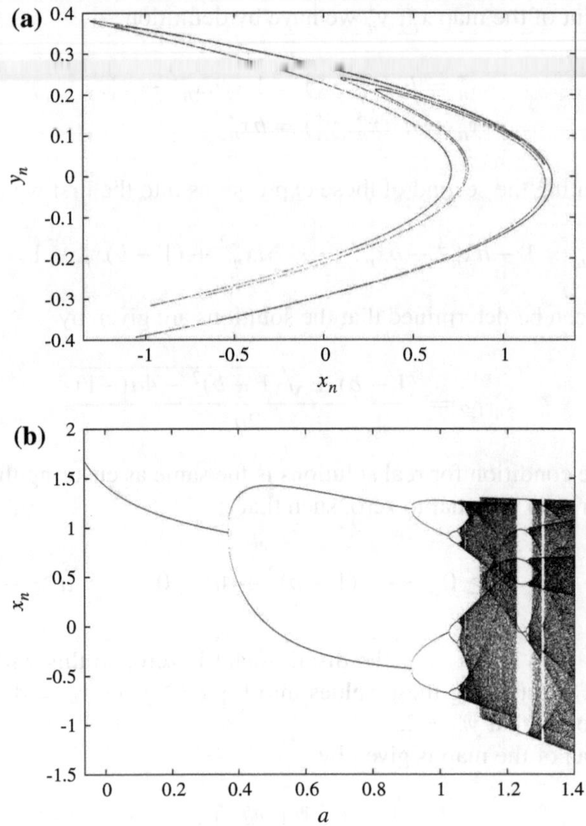

Fig. 2.26 Henón map, showing **a** the chaotic attractor at $b = 0.3$ and $a = 1.42$, **b** a bifurcation diagram as a is varied

$a = 1.42$ is shown. For maps, a stable limit cycle is a fixed point attractor because nearby transient points are attracted towards it (as shown in Fig. 2.24b). Chaotic attractors act in the same way, in that nearby transient points are drawn onto the attractor.

In Fig. 2.26b, the value of x_n is shown as the parameter a is varied. This is called a bifurcation diagram. Starting on the extreme left of the diagram the system is at the fold bifurcation discussed in Example 2.9. Then for $-0.0625 < a < 0.4$ (approx) there is just one x_n value for each a value, so this is a period one behaviour. At approximately $a = 0.4$ (for increasing a), a flip (period doubling) bifurcation occurs, so now there are two x_n values for each a. This happens again at approx $a = 0.93$ and so then 4 x_n values occur. After $a \approx 1.1$ there is a rapid transition into chaotic motion, via a period doubling cascade. Note that the chaotic region has some narrow windows of periodic behaviour within it.

Now consider what happens when the period of the limit cycle changes from one forcing period to two forcing periods. This scenario is shown in Fig. 2.27, where a

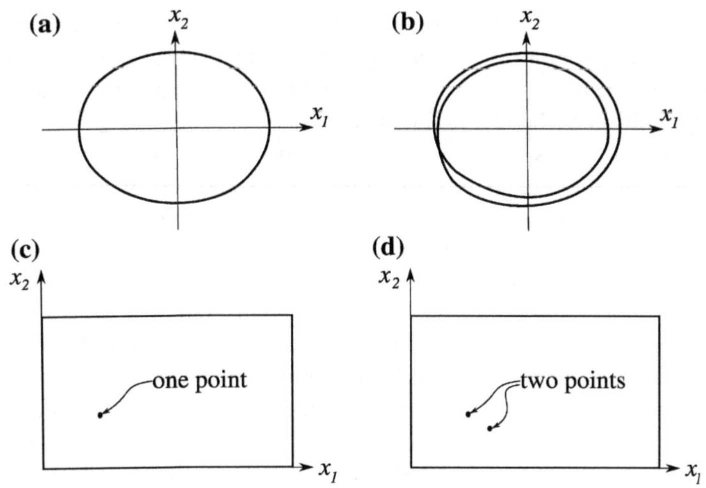

Fig. 2.27 Period-doubling (flip) bifurcation: **a** before, **b** after, **c** map before and **d** map after

period-1 limit cycle in (a) becomes a period-2 orbit in (b). Schematic representations of the stroboscopic maps[39] are shown in Fig. 2.27c, d, where there is a change from 1 to 2 points in the map.

In practice, it is very rarely possible to write down the mapping explicitly for the majority of nonlinear vibration problems. As a result, investigating a particular system is usually done by computing the map numerically. Numerical estimations of the Jacobian can be found, for example, by using finite differences—see Foale and Thompson (1991) for a description of numerical investigations of these types of systems. The numerical results are then usually plotted as a series of *bifurcation diagrams*. Typically, for vibration problems, the bifurcation parameters of interest will be the forcing amplitude and frequency, so the bifurcation diagrams will be closely linked to the frequency and amplitude response plots used for linear vibration studies.

One way to obtain a bifurcation diagram is to compute a time series of the system, allowing a large enough number of forcing periods to decay such that steady-state behaviour has been reached. Then plot the amplitude of one of the system states (usually displacement) for a number of steady state periods, before incrementing the parameter by a small amount and repeating. Note that it is important *not* to reset the initial conditions after each parameter increment. In other words, keep the last $\boldsymbol{x}_n$ and t values from the steady-state, to use as the initial conditions after the parameter has been incremented. This is how the bifurcation diagram was computed for the Henón map shown in Fig.2.26b. The idea is to observe how a particular solution changes its behaviour as a parameter is varied (for example the parameter a in Fig.2.26b). Once the maximum parameter value of interest is reached, the process should be repeated

[39] Note that in general these types of maps are called *Poincaré* maps, see Strogatz (2001).

for decreasing parameter values through the full range, back to the starting value. This will allow any regions of *hysteresis* to be captured. Also, multiple steady-state points (usually at least ten or more) are plotted for each parameter value, in order to capture any multi-periodic or chaotic behaviour.

The approach described here is one of the most basic, and is sometimes referred to as the *brute-force* approach. It is useful for a quick and approximate assessment of the system, but care is needed, as problems can arise. Of course bifurcation theory is a highly developed field in its own right (see for example Guckenheimer and Holmes 1983; Kuznetsov 2004) and there is a range of sophisticated associated numerical techniques (see for example Doedel et al. 1998; Krauskopf et al. 2007), which make it possible to start from a fixed point and then *continue* the path of the fixed point, in state space, as a parameter is varied. Where brute force will normally only capture stable steady-state solutions,[40] continuation methods can be used to capture both stable and unstable branches, as shown in Fig. 2.28 which is discussed next.

Bifurcations of limit cycles lead to structural changes in the resonance behaviour of nonlinear oscillators. One of the most common examples is shown in Fig. 2.28. This resonance peak has been simulated from a Duffing oscillator (similar to Example 2.6), using the brute-force method described above. The parameter varied is the ratio of the forcing frequency to the (linear) natural frequency of the oscillator, $\mu = \Omega/\omega_n$. The measurement taken, for each frequency value, is the maximum displacement per forcing period. The resulting bifurcation curve is then the envelope function defining the resonance amplitude of the oscillator. This is similar to the linear dynamic amplification function for a linear system plotted in Chap. 1, Fig. 1.10.

The resonance peak in Fig. 2.28 is distorted (or bent) to the right, and contains two fold bifurcations, on either side of a region of hysteresis. In this context, hysteresis means that a different behaviour is obtained for increasing or decreasing ω. As ω is increased, a stable solution path gradually increases in amplitude until it reaches fold A. Here, the stable path joins an unstable path of solutions. If ω is increased beyond the bifurcation point, there is a jump to the lower stable branch. When decreasing ω from above the resonance, the stable path continues until fold B, where there is a jump up to the upper stable branch. The region between fold A and B is the region of hysteresis.

This type of resonance is associated with a *hardening* spring nonlinearity, meaning a spring which becomes stiffer as it displaces further. The opposite case is a softening spring, meaning a spring that becomes *less* stiff as it displaces further. This leads to a resonance peak that bends to the left, as shown in Fig. 2.29a.

In Fig. 2.29b a double fold behaviour is shown. This is in fact also for the Duffing oscillator, but instead of using forcing frequency, Ω, as the bifurcation parameter, in this example the forcing amplitude, F, is varied, with Ω selected within the hysteresis region.

Finally, note that the discussion in this section has been for *local* bifurcations. Global bifurcations are significant changes in dynamics which happen when struc-

---

[40] Running time backwards reverses the stability of solution branches, so in this way brute force can be used in some cases to find unstable solutions like repellers, but not saddles.

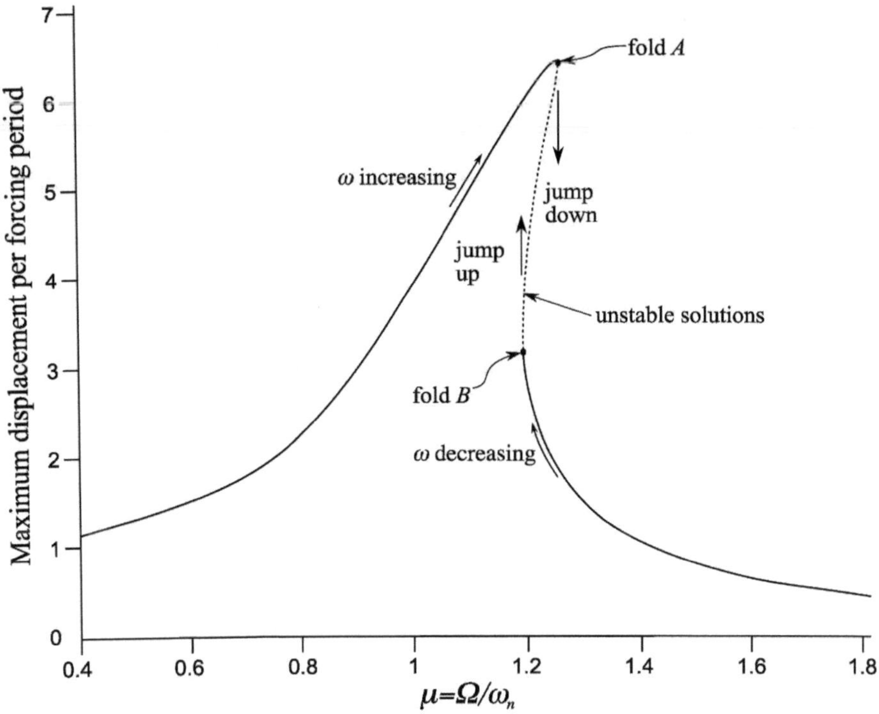

Fig. 2.28 Nonlinear resonance with hysteresis for the Duffing oscillator $\ddot{x} + \zeta\omega_n\dot{x} + \omega_n x + \alpha x^3 = F\cos(\Omega t)$ with a hardening spring $\alpha > 0$

tures in the phase plane, like equilibrium/fixed points, limit cycles come together as a parameter is varied. One such case has already been mentioned in the discussion of Example 2.4, where the limit cycle grows in size until it touches the saddle equilibrium point resulting in a homoclinic bifurcation. Discussions on global bifurcations can be found in Guckenheimer and Holmes (1983) and Glendinning (1994).

2.7 Systems with Harsh Nonlinearities

So far in this chapter little has been said about how *strong* or *severe* the nonlinearity in any particular vibrating system might be. In fact, most of the systems discussed in this chapter have relatively *weak* nonlinear terms. In Chap. 1, Sect. 1.2.4 some examples of what can collectively be called *harsh* nonlinearity were introduced. In this context, harsh means the most severe type of nonlinearity. This is characterised by sudden, and large changes in parameters. Common examples include; impact, friction, freeplay and backlash phenomena. In this section we will briefly discuss two important examples, the *friction oscillator* and the *impact oscillator*.

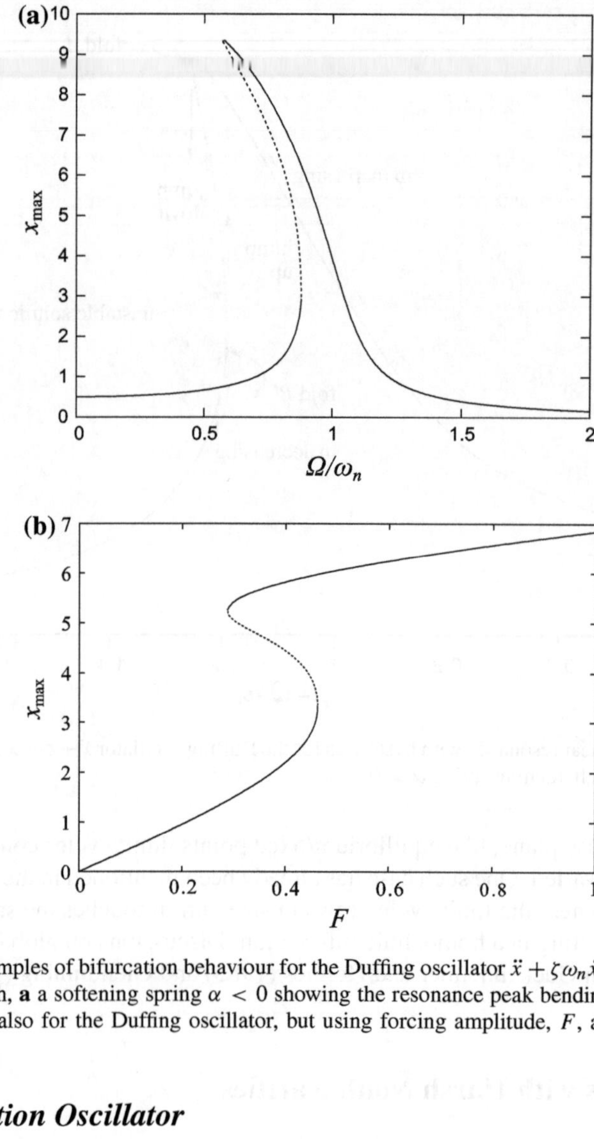

Fig. 2.29 Examples of bifurcation behaviour for the Duffing oscillator $\ddot{x} + \zeta\omega_n\dot{x} + \omega_n x + \alpha x^3 = F\cos(\Omega t)$ with, **a** a softening spring $\alpha < 0$ showing the resonance peak bending to the *left*, and **b** double fold also for the Duffing oscillator, but using forcing amplitude, F, as the bifurcation parameter

2.7.1 Friction Oscillator

One of the most difficult to model nonlinear vibration phenomena is when a dynamic system involves friction. It is also a phenomena that is widely exploited in engineering applications to damp out vibration and provide braking mechanisms (Guran et al. 1996; Sextro 2002). There are many models for friction, and three such models are shown schematically in Fig. 2.30. The models shown in Fig. 2.30 are defined in terms of the friction force, F, against a velocity, v, which is typically taken to be the velocity of a moving mass at the friction interface. Both Fig. 2.30a, b show very strong changes in friction force in the vicinity of $v = 0$, which represents the

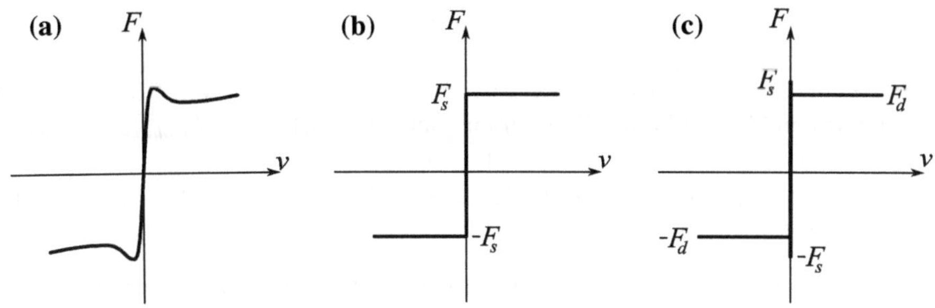

Fig. 2.30 Models for friction force, F, showing **a** continuous, **b** simple Coulomb, and **c** modified Coulomb

behaviour observed from physical systems. The main difference is that Fig. 2.28a is a continuous function curve while Fig. 2.30b is a *non-smooth* function. The continuous model is more representative of the physics involved,[41] but when implementing it as a dynamic model, the complex continuous function is normally approximated. The model in Fig. 2.30b is considerably easier to implement, but the key difference between static and dynamic friction is not captured. This limitation can be partly overcome by using the model in Fig. 2.30c which has been modified to include both the static friction, F_s, and dynamic (or sliding) friction F_d. Depending on the context, models of the type shown in Fig. 2.30b, c are referred to as the Coulomb friction model.

The Coulomb model can be represented mathematically as

$$F_{Coulomb} = \begin{cases} +F_d, & v > 0 \\ -F_s < F < F_s, & v = 0 \\ -F_d, & v < 0 \end{cases}$$

where F_s is the static friction force, often defined for simple friction models as $F_s = \mu N$ where μ is the coefficient of friction and N is the normal force at the frictional interface.[42]

Example 2.10 The friction oscillator

Problem A lumped mass oscillator system with friction is shown schematically in Fig. 2.31a. This is a single degree-of-freedom spring-mass-damper system where the mass, m, is resting on a taut belt rotating at a constant velocity, V. This oscillator system is known as a *friction oscillator*, and can be represented by a governing equation of the form

---

[41] For example it can capture the down then upward curving trend for $|v|$ increasing, known as the Stribeck effect.

[42] Note that this definition relies on Amonton's laws of friction. i.e. that the friction force is directly proportional to the normal load and independent of area. This will restrict the situations in which it could be applied in practice.

$$m\ddot{x} + c\dot{x} + kx + F_f = 0$$

where F_f is the friction force between the mass, m, and the belt. This oscillator equation is combined with the Coulomb model, such that $F_f = F_{Coulomb}$, to give a complete model of the system. Using numerical techniques find one characteristic behaviour of the system when $m = 2$ kg, $k = 10$ N/m, $F_s = 9.81$ N, $F_d = 5.89$ N, $V = 0.2$ m/s, and $c = 0$.

Solution The friction oscillator can exhibit a range of dynamic behaviour, and for the parameters given in this problem the resulting numerically computed result is shown in Fig. 2.31b. This is known as a *stick-slip* oscillation. In this type of oscillation the mass is stuck to the belt for part of the period, which is represented by the horizontal line of constant velocity $v = V = 0.2$. As x increases the restoring force due to the spring grows, until at the top right hand point of the cycle it overcomes the static friction force and the mass starts to move. This happens repeatedly in a periodic motion due to the difference between the spring force and the dynamic friction force at the onset of slip. Note that as we have used a non-smooth friction law, the periodic cycle is also non-smooth. Other motions are possible, such as constant slipping and chaotic motion.

2.7.2 Impact Oscillator

The combined effect of vibration with impacting components is a harsh nonlinearity encountered in a wide range of engineering applications. For example consider a ball experiencing an impact with a wall as shown schematically in Fig. 2.32a. As the ball impacts the wall there is a process of compression followed by restitution,[43] as shown in Fig. 2.32b. If no energy is lost (not possible in practice) the restitution phase will be exactly symmetric with the compression phase. However, in practice, some energy is always lost during impact and this is seen by the fact that the velocity of the ball after impact is both reversed and reduced, compared to the velocity before impact. This is shown in Fig. 2.32c.

This change and reduction in velocity can be captured by a simple model called the coefficient of restitution[44] law

$$\dot{x}(t_e) = -r\dot{x}(t_s) \tag{2.33}$$

[43] Note both the ball and the wall will deflect. How much depends on the material and geometric properties.

[44] Also referred to as the Newtonian coefficient of restitution, this type of impact model assumes that there is no tangential force during the impact. See Stronge (2000) for details of this and other more complex impact cases.

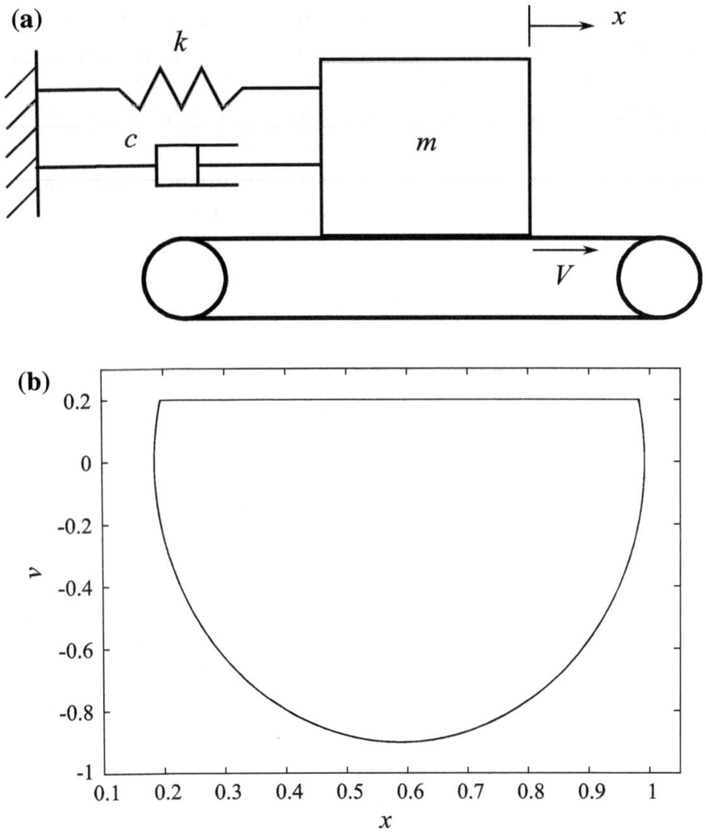

Fig. 2.31 Friction oscillator $m\ddot{x}+c\dot{x}+kx+F_{Coulomb}=0$, showing **a** a schematic representation of the system, and **b** a phase portrait of a steady-state stick slip oscillation with $m=2\,\text{kg}$, $k=10\,\text{N/m}$, $F_s=9.81\,\text{N}$, $F_d=5.89\,\text{N}$, $V=0.2\,\text{m/s}$, and $c=0$

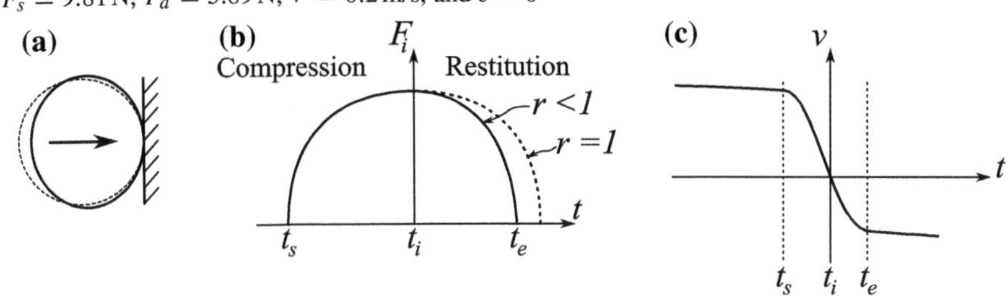

Fig. 2.32 Models for impact, showing **a** a compliant ball impacting a wall, **b** impact force, F_i, against time during impact, and **c** the velocity profile of the ball during impact

where r is the coefficient of restitution, which can be in the range $0 \le r \le 1$ with the idealised cases $r=1$ corresponding to a perfectly elastic impact and $r=0$ a perfectly plastic impact.

The time of impact starts at time t_s and ends at time t_e. The time at which the impact force, F_i, is a maximum is denoted t_i. For "hard" materials the total time of impact $t_e - t_s$ is relatively small, and r is close to one. In this case is may be justifiable to assume that the time of impact is so small compared to the overall timescales, that the impact can be considered to be instantaneous. This implies that $t_s = t_e = t_i$ and the function in Fig. 2.32b becomes an instantaneous impulse, that can be modelled by a force value multiplied by a Dirac delta function. This in turn leads to a form of the coefficient of restitution law which becomes

$$\dot{x}(t_{i+}) = -r\dot{x}(t_{i-}) \tag{2.34}$$

where t_{i-} is the time immediately before impact and t_{i+} is the time immediately after impact. This simplifies the implementation of the model, as impact time does not have to be computed. However, the model becomes non-smooth and, as was mentioned above for friction models, this can have significant limitations. Several authors have shown examples where this type of instantaneous coefficient model can capture the behaviour of engineering applications, and for an example where the limitations are discussed, see for example Melcher et al. (2013).

Example 2.11 The impact oscillator

Problem The example system shown in Fig. 2.33a consists of a linear, harmonically forced spring-mass-damper system with a motion limiting constraint so that the displacement of the mass, x, is limited to $x \leq d$ and when $x = d$ an impact takes place. A complete model for this system is obtained by combining the equation of motion for a linear harmonically forced oscillator with a coefficient of restitution rule when impact occurs. This system is called an impact oscillator. Using numerical computation find the nonlinear behaviour of this oscillator close to the first natural frequency of the linear oscillator, when $m = 1\,\text{kg}$, $k = 1\,\text{N/m}$, $F = 0.5\,\text{N}$, $c = 0.05\,\text{N s/m}$, $d = 1\,\text{m}$, and Ω is varied between 0 and 2 rad/s.

Solution We can understand the behaviour of this impact oscillator close to the first natural frequency of the linear oscillator by computing a bifurcation diagram. The resulting bifurcation diagram is shown in Fig. 2.33b as the forcing frequency, Ω, is varied. When the displacement amplitude measure $x_{max-min}$ is less than d the behaviour is that of a linear oscillator, and it can be seen that the bottom of a resonance curve is visible. However at $x_{max-min} = d$ impacts start to occur and complex behaviour such as chaotic motion and multi-period responses can be observed. Overall the resonant peak is hardening, and a region of hysteresis is evident for $\Omega > 1$.

Several alternatives to coefficient of restitution models exist, the most widely used for engineering applications being the Hertzian law of impact. This can be derived by considering the continuum mechanics close to the point of contact, see Stronge (2000) for details. This model is particularly useful for cases when the time of contact and impact force are important to the overall modelling process, but will not be considered further here.

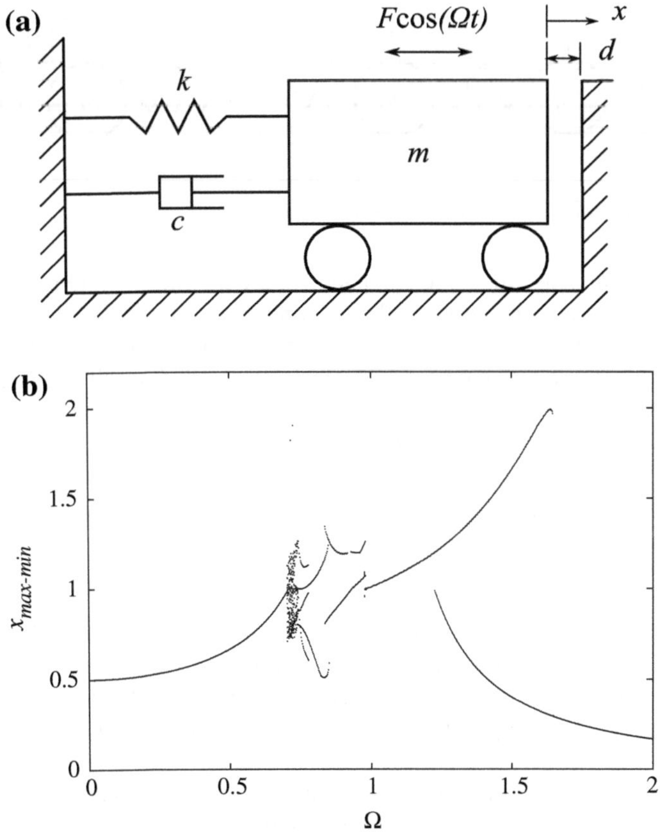

Fig. 2.33 Impact oscillator $m\ddot{x} + c\dot{x} + kx = F\cos(\Omega t)$, for $x < d$ and $\dot{x}(t_{i+}) = -r\dot{x}(t_{i-})$ at $x = d$, showing **a** a schematic representation of the system, and **b** a bifurcation diagram of stable response solutions with $m = 1\,\text{kg}$, $k = 1\,\text{N/m}$, $F = 0.5\,\text{N}$, $c = 0.05\,\text{N s/m}$, $d = 1\,\text{m}$, and Ω is varied between 0 and 2 rad/s. Note that $x_{max-min}$ is the maximum minus minimum (or peak to peak) x value per forcing period divided by 2

2.8 Nonlinear Phenomena in Higher Dimensions

So far in this chapter the nonlinear phenomena discussed are those which occur in second-order oscillators such as the escape equation or Duffing oscillator. From a vibrations perspective, second-order oscillators are used to model single-degree-of-freedom systems, so what happens when there are multiple degrees-of-freedom?

Some characteristics appear to be similar to linear systems. For example, in a multi-degree-of-freedom nonlinear system there can be multiple resonance peaks, just as in linear systems. However, the resonance peaks will typically be distorted in some way, as in Fig. 2.28. As one would expect, each of these resonance peaks can potentially be excited by an external forcing input. However, unlike for linear systems, in nonlinear multi-degree-of-freedom systems it is possible for *internal resonances*

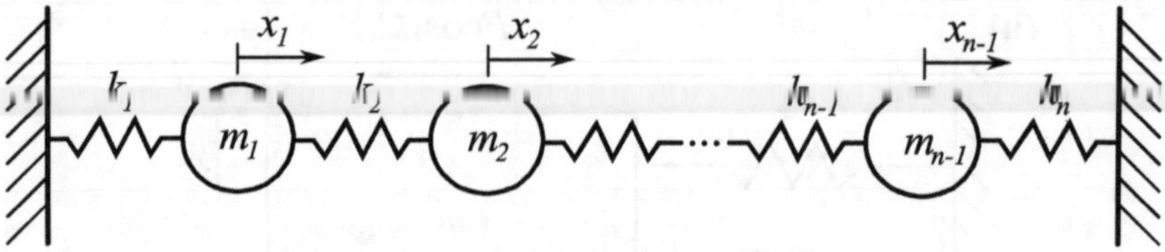

Fig. 2.34 Nonlinear resonance with hysteresis (Duffing)

to occur. This happens when one resonance in the nonlinear system excites another resonance due to coupling between them. Nonlinear systems can also be excited (or forced) by oscillations in the system parameters, so-called *parametric excitation*. When parametric excitation occurs in systems of coupled nonlinear oscillators, resonant behaviour can occur, that may also include internal resonances, sometimes referred to as *auto-parametric excitation*.

2.8.1 The Fermi-Pasta-Ulam Paradox

One of the earliest attempts to understand the dynamics of multi-degree-of freedom nonlinear systems took place in the early 1950s when a group of researchers working at Los Alamos led by Enrico Fermi decided to study the problem of molecular dynamics. The problem they studied looked schematically like that shown in Fig. 2.34, which consists of a chain of masses coupled by nonlinear springs used to represent the interaction of molecules (or atoms) in a solid material. It had already been well established that for a multi-degree-of freedom linear (undamped, unforced) system, energy put into a single mode would remain in that mode for all time. Fermi with his co-workers, Pasta, and Ulam thought that if they put energy into the lowest mode, the nonlinear coupling from the springs would cause the energy to gradually redistribute (or equipartition) into all the modes as time increased. This did happen, however, they also found that if the simulation was run for long enough, the energy flowed back out of all the other modes into the mode where it started. This phenomena was called the Fermi-Pasta-Ulam (FPU) paradox, and gave rise to a large field of research, particularly for Hamiltonian dynamical systems.

Attempts to solve the FPU paradox established important phenomena for nonlinear dynamical systems with multiple degrees-of-freedom. For example the existence of solitary waves or *solitons*, not quite periodic motion otherwise known as *quasi-periodic* motion, and chaos. Another important feature was the interaction of nonlinear resonances, which will be discussed in more depth for vibration problems in Chap. 5.

2.8.2 Localization

Another phenomena that was first established in solid state physics is that of localization. In particular the idea that disorder (or irregularities) in a lattice like structure leads to the confinement of vibrational energy to one localized part of the structure. This phenomena also occurs in structural dynamics and is sometimes referred to as *periodic structure theory*.[45] An example is shown in Fig. 2.35a, where two pendula are coupled by a linear spring. The length of the pendulum attached to mass m_2 can be varied by a small amount Δl. When the spring stiffness, k, is small and therefore provides only a *weak* coupling between the two masses, localization phenomena can occur. This can be observed by considering the effect of changing Δl on the *linearised* natural frequencies of the system (i.e. assuming small angles). The result of such an investigation is shown in Fig. 2.35b. This shows how the two natural frequency values appear to veer away from each other as they pass through $\Delta l = 0$. This phenomena is known as *mode veering*. It can also be seen that when $\Delta l = 0$ the two natural frequencies are almost identical in value. However, at $\Delta l = 0$ the usual normal modes (equal amplitude in-phase and out-of-phase responses) exist as would be expected for a symmetric two degree-of-freedom linear system. This changes significantly as Δl is increased or decreased, leading to nearly all the energy being confined, or localized, in the vibration of one or other of the pendula.

Note that the natural frequencies veer rather than cross. This is because there is not a multiple eigenvalue for this system at $\Delta l = 0$.

2.8.3 Modelling Approaches

Figure 2.36 shows what can be thought of as a structural dynamics "landscape". The vertical axis represents the severity of the nonlinearity, from linear at the origin to harsh nonlinearities such as impact and friction in the upper ranges. The horizontal axis represents geometrical complexity, from single-degree-of-freedom oscillators at the origin to millions of degrees-of-freedom (or discretisation points) at the upper end. There are other factors which are not included in this simplified "landscape", for example the type of external excitation which the system is subjected to. Despite this, the vast majority of nonlinear structural dynamics applications could be plotted at specific points, or regions, in the landscape.

The harmonically forced linear oscillator that was described in Sect. 2.1 is at the origin of Fig. 2.36. All the systems described in Sects. 2.1–2.7, lie on the vertical axis. In Chaps. 4–8 classical structural elements such as beams, cables and plates will be considered. These structures have many degrees-of-freedom and nonlinearity conditions from weak to harsh, and therefore lie away from the axes in Fig. 2.36. That said, analytical techniques such as those described in Chaps. 4–8 are only typically used for relatively small numbers of degrees-of-freedom, typically up to 20.

---

[45] See Hodges and Woodhouse (1983) for more details.

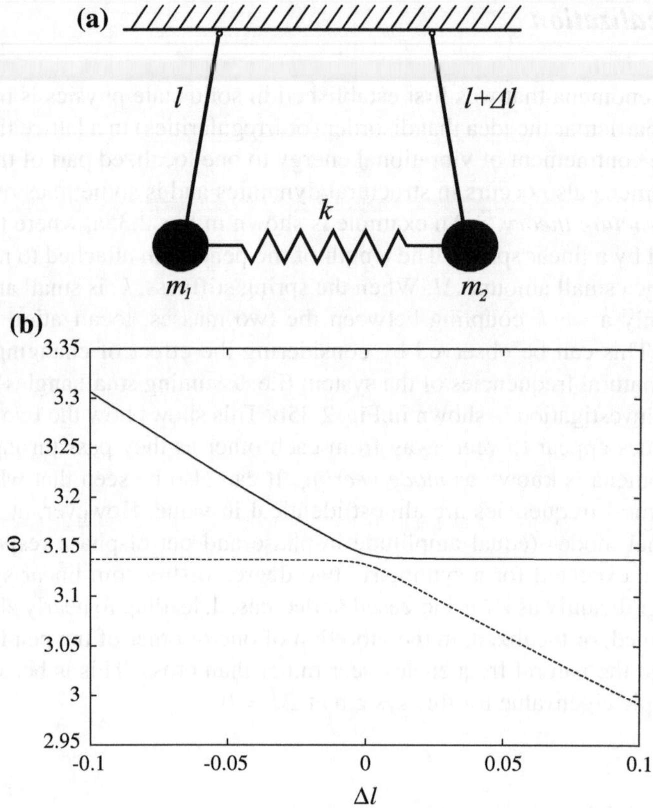

Fig. 2.35 Coupled pendulum system, showing **a** a schematic of the system, and **b** the two natural frequencies of the linearised system as Δl is varied. Parameter values, $m_1 = m_2 = 3$ kg, $l = 1$ m and $k = 0.1$ N/m

But many other systems have complex geometry and/or many degrees of freedom so how can they be modelled? In this case, for structural dynamics, finite element analysis (FEA) is a very powerful tool for creating a model of a system with complex geometry,[46] see for example Crisfield (1997). However, there is a trade off between representing the complexity of the geometry and understanding the effect of the nonlinearity. This is because creating a model that captures the complexities of the geometry is both computationally expensive and restricts the dynamic analysis which can subsequently be carried out. In fact, most nonlinear phenomena have only really been thoroughly studied and understood for systems with simple geometry, primarily single degree-of-freedom systems.

As a result the part of the landscape in Fig. 2.36 where there is a good understanding of the dynamic behaviour is on the axes and close to the origin. This region is

[46] This is assuming that the frequencies of interest are in the low range. For mid-frequency problem statistical energy analysis is often more appropriate, see Langley (1989).

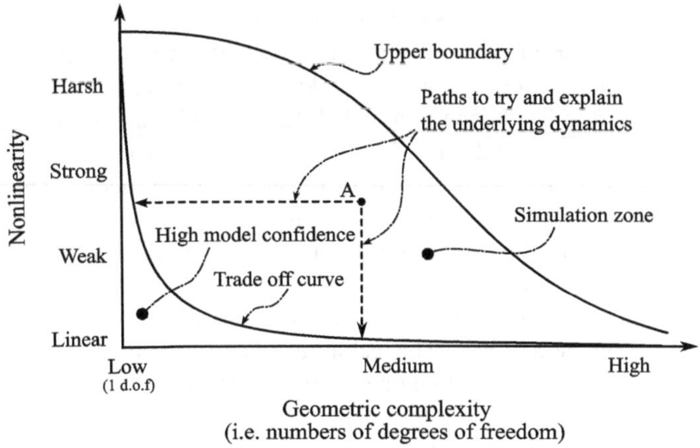

Fig. 2.36 A nonlinear structural dynamics landscape

where high confidence in a model can be achieved.[47] This is because (i) there is a thorough understanding of the underlying physical behaviour, so that a model can be constructed based on understanding the underlying physics and (ii) the model can typically be validated against experimental or in service data. Beyond this high confidence region, numerical simulations can be used to provide information on the potential behaviour of systems that have combined nonlinearity and complex geometry. However, moving further into this region will typically reduce confidence levels significantly. The upper boundary shown in Fig. 2.36 is set by the available computational power (or cost of computation) and also the confidence in physical models such as impact and friction.

One of the major challenges in structural dynamics is to improve the modelling capability of the simulation zone shown in Fig. 2.36. This is because the ability to carry out a simulation doesn't imply a clear understanding of the simulation results. For example, consider a structure at point A, that has strong nonlinearity and medium geometric complexity, say 1,000 degrees-of-freedom. To try and explain the results of a time-stepping simulation for this system, the choices are to either talk in terms of linear modal systems which exist on the horizontal axis, or strongly nonlinear single degree-of-freedom systems. In other words to project the problem to where understanding and model confidence are greatest. This topic will be discussed further in Chaps. 5 and 6.

2.9 Chapter Notes

This chapter gives an introduction to the nonlinear phenomena observed in vibrating systems. An in-depth treatment of dynamical systems theory is given by Guckenheimer and Holmes (1983), Moon (1987), Cartmell (1990), Glendinning

[47] Assuming that the modelling techniques employed are used with sufficient care.

(1994), Jordan and Smith (1999), Strogatz (2001) and Thompson and Stewart (2002). Note that Strogatz (2001) offers a particularly good introduction for those who are unfamiliar with the subject. More detailed mathematical treatments of bifurcations can be found in Guckenheimer and Holmes (1983) and Glendinning (1994). A very good treatment of stability via eigenvalue analysis is given by Seyranian and Mailybaev (2003). Relevant aspects of linear vibration theory can be found in Inman (2006). The numerical techniques required can be found in Fausett (1999) for general Matlab and time-integration information and see also Newland (1993), Press et al. (1994) for additional discussions on frequency domain transformations. For an overview of computing mappings and bifurcation diagrams, an excellent paper is by Foale and Thompson (1991). The technique of cell-to-cell mapping is described by Hsu (1987). The delay in actuator example is from Bursi and Wagg (2008), Chap. 7. Mathematical analysis of bifurcation theory is given by Kuznetsov (2004) and Guckenheimer and Holmes (1983), while numerical aspects are discussed by Krauskopf et al. (2007). Topics relating to buckling can be found in Frish-Fay (1962), Thompson and Hunt (1973), Thompson (1982) and Virgin (2000). For topics relating to harsh nonlinearities, friction is covered by Sextro (2002) and Guran et al. (1996), and impact is discussed in depth by Babitsky (1998) and Stronge (2000). Applications of impact oscillator systems can be found in Thompson and Stewart (2002) and a more recent example applied to a flexible beam application is described by Melcher et al. (2013). Detailed treatments of non smooth modelling techniques are given in Brogliato (1999) and di Bernardo et al. (2008). For discussions on the Fermi-Pasta-Ulam problem see Berman and Izrailev (2005) and for localisation see Hodges and Woodhouse (1983) and Pierre (1988) and references therein. Discussion of several multi-degree-of-freedom nonlinear systems are given in Thomsen (2003).

Problems

2.1 Derive the equation of motion for the system shown in Fig. 2.20a. Show that this equation can be approximated by the Duffing equation

$$m\ddot{x} + c\dot{x} - \mu x + \alpha x^3 = 0,$$

and estimate when this might be a valid assumption.

2.2 The normal form of the Hopf bifurcation is usually written as

$$\begin{aligned}\dot{x} &= \mu x + y - x(x^2 + y^2),\\ \dot{y} &= -x + \mu y - y(x^2 + y^2).\end{aligned} \tag{2.35}$$

Show that this system can also be represented as

$$\dot{r} = r(\mu - r^2),$$
$$\dot{\theta} = -1,$$

in polar coordinates. Examine the stability of the equilibrium point at the origin $(x = 0,\ y = 0)$ by finding the Jacobian of Eq. (2.35).

2.3 A nonlinear system is governed by the following set of first-order differential equations

$$\dot{x}_1 = x_2,$$
$$\dot{x}_2 = x_1 - x_1^2 - \mu x_2, \qquad (2.36)$$

where μ is a parameter which can be varied. Find the equilibrium points for the system when $0 < \mu < \sqrt{4}$ and find the type and stability of each equilibrium point. Sketch typical trajectories in the system state space.

2.4 Consider the potential function[48] given by

$$V = \frac{x_2^2}{2} - \frac{x_1^2}{2} + \frac{x_1^3}{3}.$$

Finding the time derivative of V and substituting for $\dot{x}_1$ and $\dot{x}_2$ gives an indication of the stability of equilibrium points at the origin. For the case when x_1 and x_2 are small, use this function to determine the stability of the origin for the system given in Eq. (2.36). How does the sign of $\dot{V}$ relate to the stability?

2.5 For the system given in Eq. (2.36), when μ passes through zero a bifurcation occurs. Use local analysis to explain what happens at the bifurcation point. What type of bifurcation occurs?

2.6 The dynamics of a damped unforced pendulum can be modelled using the nonlinear differential equation

$$\ddot{\theta} + \delta\dot{\theta} + \omega^2 \sin\theta = 0,$$

where θ is the angle of the pendulum from the downwards resting position, δ is the damping parameter and $\omega = \sqrt{\frac{g}{l}}$ is the natural frequency of the pendulum where g is the force due to gravity and l is the length of the pendulum.
Find the equilibrium points for the pendulum in the range $2\pi \leq \theta \leq 2\pi$ when $\delta^2 < 4\omega^2$. Indicate the type and stability of each equilibrium point and sketch the pendulum trajectories in the $\theta, \dot{\theta}$ plane.

2.7 For small angles the motion for a pendulum can be approximated by

$$\ddot{\theta} + \delta\dot{\theta} + \omega^2(\theta - \frac{\theta^3}{3!}) = 0.$$

---

[48] Used in this context, this is usually called a Lyapunov function, although limitations exist—see Chap. 3, Sect. 3.2.

Use the potential (i.e. Lyapunov) function

$$V = \frac{1}{2}\dot{\theta}^2 + \frac{\omega^2}{2}\theta^2 - \left(\frac{\omega^2}{3!}\right)\frac{\theta^4}{4}$$

to determine the stability of the point $\theta = 0, \dot{\theta} = 0$, by finding the sign of $\dot{V}$. Assume that $\omega = 1$.

2.8 The logistic map is a single state mapping which is used to model population dynamics represented as

$$x_{n+1} = \lambda x_n(1 - x_n),$$

Identify the period one fixed points for the system and their stability. Which bifurcations occur at $\lambda = 1$ and $\lambda = 3$?

References

Babitsky, V. I. (1998). *Theory of vibro-impact systems and applications*. Berlin: Springer.

Berman, G., & Izrailev, F. (2005). The Fermi-Pasta-Ulam problem: Fifty years of progress. *Chaos*, *15*(1), 15104.

Brogliato, B. (1999). *Nonsmooth mechanics: Models, dynamics and control*. London: Springer.

Bursi, O. S. & Wagg, D. J. (eds.). (2008). *Modern testing techniques for structural systems*. New York: Springer.

Cartmell, M. (1990). *Introduction to linear, parametric and nonlinear vibrations*. London: Chapman and Hall.

Crisfield, M. A. (1997). *Non-linear finite element analysis of solids and structures*. In: Advanced Topics (Vol. 2). Chichester: Wiley.

di Bernardo, M., Budd, C., Champneys, A. R., & Kowalczyk, P. (2008). *Piecewise-smooth dynamical systems: Theory and applications*. London: Springer.

Diekmann, O., van Gils, S., Verduyn Lunel, S., & Walther, H. (1995). Delay equations. In: Applied mathematical sciences (Vol. 110). New York: Springer.

Doedel, E. J., Champneys, A. R., Fairgrieve, T. F., Kuznetsov, Y. A., Sandstede, B., & Wang, X. (1998). *Auto97. Continuation and bifurcation software for ordinary differential equations*. Citesteer

Fausett, L. V. (1999). *Applied numerical analysis using Matlab*. Upper Saddle River, NJ: Prentice Hall.

Foale, S., & Thompson, J. M. T. (1991). Geometrical concepts and computational techniques of nonlinear dynamics. *Computer Methods for Applications in Mechanical Engineering*, *89*, 381–394.

Frish-Fay, R. (1962). *Flexible bars*. London: Butterworths.

Glendinning, P. (1994). *Stability, instability and chaos*. Cambridge: Cambridge University Press.

Guckenheimer, J., & Holmes, P. (1983). *Nonlinear oscillations, dynamical systems, and bifurcations of vector fields*. New York: Springer.

Guran, A., Pfeiffer, F., & Popp, K. (1996). *Dynamics with friction*. Singapore: World Scientific Publishing.

Hodges, C. H., & Woodhouse, J. (1983). Vibration isolation from irregularity in a nearly periodic structure—theory and measurements. *Journal of the Acoustical Society of America*, *74*(3), 894–905.

Hsu, C. S. (1987). *Cell-to-cell mapping*. New York: Springer.

Inman, D. J. (2006). *Vibration with control*. Chichester: Wiley.

Jordan, D. W., & Smith, P. (1999). *Nonlinear ordinary differential equations; an introduction to dynamical systems* (3rd ed). Oxford: Oxford University Press.

Krauskopf, B., Osinga, H. M., & Galan-Vioque, J. (eds.). (2007). *Numerical continuation methods for dynamical systems*. New York: Springer.

Kuznetsov, Y. A. (2004). *Elements of applied bifurcation theory*. New York: Springer.

Langley, R. S. (1989). A general derivation of the statistical energy analysis equations for coupled dynamic systems. *Journal of Sound and Vibration*, *135*(3), 499–508.

Melcher, J., Champneys, A. R., & Wagg, D. J. (2013). The impacting cantilever: Modal non-convergence and the importance of stiffness matching. *Philosophical Transactions of the Royal Society A: Mathematical, Physical and Engineering Sciences*, *371*(1993), 1471–2962.

McInnes, C. R., Gorman, D. G., & Cartmell, M. P. (2008). Enhanced vibrational energy harvesting using nonlinear stochastic resonance. *Journal of Sound and Vibration*, *318*, 655–662.

Moon, F. C. (1987). *Chaotic vibrations: An introduction for applied scientists and engineers*. New York: Wiley.

Newland, D. E. (1993). *An introduction to random vibrations and spectral analysis*. Pearson: Prentice Hall.

Pierre, C. (1988). Mode localization and eigenvalue loci veering phenomena in disordered structures. *Journal of Sound and Vibration*, *126*(3), 485–502.

Press, W. H., Teukolsky, S. A., Vettering, W. T., & Flannery, B. P. (1994). *Numerical recipes* (2nd ed). Cambridge: Cambridge University Press.

Sextro, W. (2002). *Dynamical contact problems with friction: Models, methods. Experiments and applications*. New York: Springer.

Seyranian, A. P., & Mailybaev, A. A. (2003). *Multiparameter stability theory with mechanical applications*. Singapore: World Scientific Publishing.

Sontag, E. D. (1998). *Mathematical control theory*. New York: Springer.

Stépan, G. (1989). *Retarded dynamical systems: Stability and characteristic functions*. Essex: Longman Scientific & Technical.

Strogatz, S. H. (2001). *Nonlinear dynamics and chaos*. Cambridge: Perseus Books Group.

Stronge, W. J. (2000). *Impact mechanics*. Cambridge: Cambridge University Press.

Thompson, J. M. T. (1982). *Instabilities and catastrophes in science and engineering*. Chichester: Wiley.

Thompson, J. M. T., & Hunt, G. W. (1973). *A general theory of elastic stability*. Chichester: Wiley.

Thompson, J. M. T., & Stewart, H. B. (2002). *Nonlinear dynamics and chaos*. Chichester: Wiley.

Thomsen, J. J. (2003). *Vibrations and stability: Advanced theory, analysis and tools*. New York: Springer.

Virgin, L. N. (2000). *An introduction to experimental nonlinear dynamics*. Cambridge: Cambridge University Press.

Virgin, L. N. (2007). *Vibration of axially-loaded structures*. Cambridge: Cambridge University Press.

Chapter 3
Control of Nonlinear Vibrations

Abstract In this chapter, methods which can be used to control nonlinear structural vibrations are discussed. Introductory examples showing the control of linear and nonlinear single-degree-of-freedom oscillators have already been discussed in Sect. 1.4 of Chap. 1. This chapter extends the ideas presented in these introductory examples to a range of controllers, which can be designed to control nonlinear vibrations. Control of structural vibrations is different from the majority of control problems, in that there are typically multiple lightly damped resonances in the system response. In addition, when an actuator is attached to the structure, its effect will be coupled to some resonances much more strongly than others. As a result, careful design is required to reduce particular resonant responses. Even with careful design, other resonances will exist which cannot be effectively controlled. Using feedback can induce instability in the system, and so ensuring the stability of any control design is of primary importance. The underlying ideas of stability for nonlinear systems have been introduced in Sect. 2.3, Chap. 2. In this chapter, these ideas are extended to include systems with feedback control, and the stability analysis is carried out using a particular type of potential function, called a *Lyapunov function*. The basic ideas of Lyapunov-based control design can be extended to a range of other approaches. The main control method described here is the effective linearisation of a system using feedback. Adaptive control, which can also be a useful method for nonlinear or uncertain systems is also discussed in the later part of the chapter.

3.1 Control Design for Nonlinear Vibrations

Here we will consider different methods for controlling nonlinear vibrations. The simplest way to reduce vibrations is to design the system with additional damping, by using special materials or adding physical damping devices. This approach is called *passive* vibration control (or redesign) and is a very well developed subject area for linear vibration problems, see Soong and Dargush (1997) and Inman (2006). Passive techniques, such as the classical *tuned mass damper* (see Hartog 1934 for a description relating to linear vibration) have been extended to nonlinear systems—see for example Ibrahim (2008) and references therein. Passive solutions are often

D. Wagg and S. Neild, *Nonlinear Vibration with Control*,
Solid Mechanics and Its Applications 218, DOI 10.1007/978-3-319-10644-1_3

preferred in practice as they can be built into the system and there is no control element, which eliminates any issues with stability or robustness. However, for a growing class of structures for which reduced weight and flexibility are important features, passive redesign is not an effective design solution.

Alternatives to passive design are to use either active control or semi-active vibration reduction techniques. This chapter introduces these concepts as applied to nonlinear vibration problems. Throughout this chapter, the nonlinear governing equations of interest are those which have quadratic and cubic nonlinearities in the restoring force. This is because these are the type of nonlinear terms which naturally arise in the nonlinear vibration problems discussed in the later chapters of this book, Chaps. 6–8.[1]

Active vibration-control methods are also increasingly being used in conjunction with *energy harvesting* techniques. In such techniques, the mechanical energy from the vibrating system is transformed into electrical energy which can then be used for a range of applications. Often this is done as part of the multifunctional behaviour, typical of a smart structure. Further information on energy harvesting can be found in Priya and Inman (2009).

3.1.1 Passive Vibration Control

Passive vibration control works either by *isolation* or *absorption*. A passive vibration isolator works by minimising the transmission of vibration from the support to the moving parts of the system being controlled. In contrast, a vibration absorber is a device attached to a structure to reduce vibration in other parts of the same structure. Examples of these two types of passive vibration control device are shown in Fig. 3.1.

Consider the case of mass subjected to a support excitation, $r(t)$, as shown in Fig. 3.1a. A vibration isolator consisting of a spring and viscous damper is placed between the mass and the excitation to try and minimise the transmission of $r(t)$ to the mass. Then in Fig. 3.1b the case of single-degree-of-freedom oscillator plus a vibration absorber is shown where m_a is the mass of the absorber, f_{sa} is the stiffness function of the absorber and f_{ds} is the damping function. Note that isolators can only be applied to systems with a support excitation, such as $r(t)$ shown in Fig. 3.1, whereas absorbers can be designed both for support and direct excitation of the mass. Notice also that in the case of Fig. 3.1a $r(t)$ is a support motion. i.e. a displacement, whereas in Fig. 3.1b it is used to denote a force. There are well developed linear theories for designing both isolators and absorbers for the systems shown in Fig. 3.1, and these are now considered in the following two examples.

[1] Note that some of the nonlinearities discussed later are *non-smooth* for example, the impacting beam in Chap. 6. These types of nonlinearities require special treatment in terms of control.

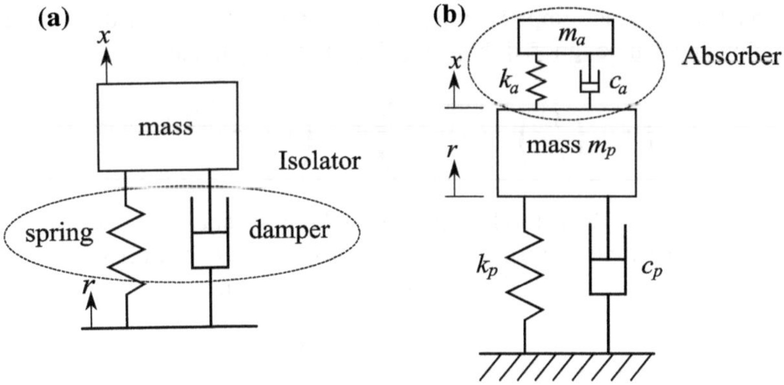

Fig. 3.1 Passive vibration control. **a** shows a mass with support excitation $r(t)$ and displacement $x(t)$ and a vibration isolator consisting of a spring and viscous damper, and **b** shows a single-degree-of-freedom oscillator plus vibration absorber

Example 3.1 Linear passive vibration isolator

Problem Derive the steady state motion of the system shown in Fig. 3.1a for the case of when there is a ground excitation $r(t) = R\cos(\omega t)$, and a linear spring and damper are present such that the governing equation is

$$m\ddot{x} + c(\dot{x} - \dot{r}) + k(x - r) = 0, \tag{3.1}$$

where m, c and k are the scalar parameters of the system.

Solution Assume that the excitation is $r = R\sin(\Omega t)$, which can be written as

$$r = \frac{R}{2\mathrm{i}}(\mathrm{e}^{\mathrm{i}\Omega t} - \mathrm{e}^{-\mathrm{i}\Omega t}), \tag{3.2}$$

where R is real. The response needs to include a phase lag, ϕ, but must be real, so it is assumed that

$$x = \frac{X}{2\mathrm{i}}\mathrm{e}^{\mathrm{i}\Omega t} - \frac{\overline{X}}{2\mathrm{i}}\mathrm{e}^{-\mathrm{i}\Omega t}, \tag{3.3}$$

where X is a complex constant and $\overline{X}$ is the complex conjugate of X. The use of complex function X allows both amplitude and phase information to be included in the response sine wave. The amplitude function, X_r, is the modulus of X and the phase, ϕ, is the argument of X.

Carrying out the substitution of Eqs. (3.2) and (3.3) into Eq. (3.1) and comparing coefficients of the $\mathrm{e}^{\mathrm{i}\Omega t}$ and $\mathrm{e}^{-\mathrm{i}\Omega t}$ terms leads to the relationships

$$(k - m\Omega^2 + \mathrm{i}c\Omega)X = (k + i\Omega c)R, \tag{3.4}$$
$$(k - m\Omega^2 - \mathrm{i}c\Omega)\overline{X} = (k - i\Omega c)R, \tag{3.5}$$

Notice that Eq. 3.5 is the complex conjugate of Eq. 3.4, and so it provides no additional information and can be ignored. Now Eq. 3.4 can be written as

$$\left(1-\left(\frac{\Omega}{\omega_n}\right)^2+i2\zeta\frac{\Omega}{\omega_n}\right)X=\left(1+i2\zeta\frac{\Omega}{\omega_n}\right)R, \tag{3.6}$$

where $\omega_n=\sqrt{k/m}$ is the natural frequency, and $\zeta=c/2m\omega_n$ is the damping ratio.

To find the amplitude of response, X_r, and the phase lag, ϕ, Eq. (3.6) is separated into real and imaginary parts. This is done by first writing

$$\frac{X}{R}=\frac{\left(1+i2\zeta\frac{\Omega}{\omega_n}\right)}{(1-(\frac{\Omega}{\omega_n})^2+i2\zeta\frac{\Omega}{\omega_n})}. \tag{3.7}$$

Multiplying the numerator and denominator of the right-hand sides of Eq. (3.7) by the complex conjugate of the denominator, gives

$$\frac{X}{R}=\frac{(1+i2\zeta\frac{\Omega}{\omega_n})((1-(\frac{\Omega}{\omega_n})^2)-i2\zeta\frac{\Omega}{\omega_n})}{(1-(\frac{\Omega}{\omega_n})^2)^2+4(\zeta\frac{\Omega}{\omega_n})^2},$$

from which it can be found that the real and imaginary parts are

$$\frac{X}{R}=\frac{(1-(\frac{\Omega}{\omega_n})^2)+4\zeta^2(\frac{\Omega}{\omega_n})^2)}{(1-(\frac{\Omega}{\omega_n})^2)^2+4(\zeta\frac{\Omega}{\omega_n})^2}+i\frac{-2\zeta(\frac{\Omega}{\omega_n})+2\zeta(\frac{\Omega}{\omega_n})(1-(\frac{\Omega}{\omega_n})^2)}{(1-(\frac{\Omega}{\omega_n})^2)^2+4(\zeta\frac{\Omega}{\omega_n})^2}.$$

So the magnitude of the complex vector is

$$X_r=\left|\frac{X}{R}\right|=\sqrt{\frac{(1+4\zeta^2(\frac{\Omega}{\omega_n})^2)}{(1-(\frac{\Omega}{\omega_n})^2)^2+4\zeta^2(\frac{\Omega}{\omega_n})^2}}, \tag{3.8}$$

and phase lag is

$$\arg(X)=\arctan\left(\frac{-2\zeta(\frac{\Omega}{\omega_n})^3}{(1-(\frac{\Omega}{\omega_n})^2)+4\zeta^2(\frac{\Omega}{\omega_n})^2}\right)=-\phi. \tag{3.9}$$

As described for in Sect. 1.3.1 of Chap. 1, these functions define the response of the linear system as

$$\begin{aligned} x &= -\mathrm{i}\frac{X_r}{2}\mathrm{e}^{-\mathrm{i}\phi}\mathrm{e}^{\mathrm{i}\Omega t}+\mathrm{i}\frac{X_r}{2}\mathrm{e}^{\mathrm{i}\phi}\mathrm{e}^{-\mathrm{i}\Omega t} \\ &= X_r\sin(\Omega t-\phi), \end{aligned}$$

which corresponds to two counter-rotating complex vectors in the complex plane ■

The solutions found in Example 3.1 can be used to plot the steady state response functions for the linear vibration isolator. These are shown in Fig. 3.2 where Eq. 3.8 has been used to compute the displacement amplitude divided by input amplitude, and Eq. 3.9 has been used to compute the phase shift between the response, $x(t)$ compared to input, $r(t)$. It can be seen that for $0 < \Omega/\omega_n < \sqrt{2}$ the ratio of $|X/R| > 1$ and for $\Omega/\omega_n > \sqrt{2}$ the ratio $|X/R| < 1$. This means that to design the isolator the relationship $\omega = \sqrt{2}\omega_n$ will define which parts of the frequency range amplify or attenuate the vibration.[2] The $|X/R|$ relationship is called the *transmissibility* of the isolator system. Before including nonlinearity into the isolator design, an absorber example is considered.

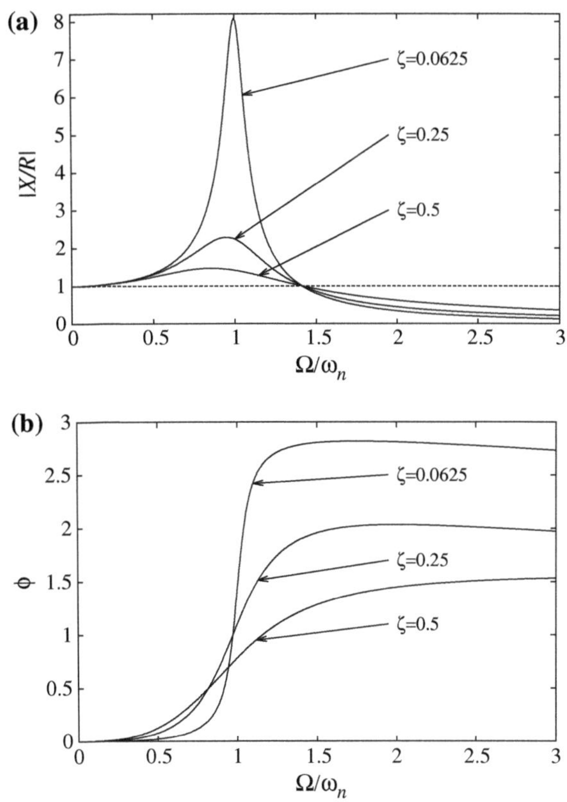

Fig. 3.2 Steady state response of linear vibration isolator showing, **a** displacement amplitude divided by input amplitude, and **b** phase shift of response compared to input

[2] This relationship can be obtained by setting $|X/R| = 1$ in Eq. 3.8.

Example 3.2 Linear passive vibration absorber

Problem Design the optimum undamped vibration absorber for the system shown in Fig. 3.1b for the case of when, $r(t) = F\sin(\Omega t)$, $c_a = c_p = 0$, such that the governing equations are[3]

$$
\begin{aligned}
m\ddot{x}_1 + kx_1 + k_a(x_1 - x_2) &= F\sin(\Omega t),\\
m_a\ddot{x}_2 + k_a(x_2 - x_1) &= 0.
\end{aligned}
\tag{3.10}
$$

where m and k are the scalar parameters of the primary system, and m_a and k_a are the absorber parameters. Note that there is zero phase lag[4] in the response of x_1 and x_2 relative to F because there is no damping in the system.

Solution In this example the excitation is a force $r = F\sin(\Omega t)$. As there is no damping present, a solution of the following form

$$
\mathbf{x} = \begin{bmatrix} x_1 \\ x_2 \end{bmatrix} = \begin{bmatrix} a_1 \\ a_2 \end{bmatrix} \sin(\Omega t)
$$

is assumed, where a_1 and a_2 are real scalar coefficients. Substituting $x_1, x_2, \ddot{x}_1$ and $\ddot{x}_2$ into Eq. (3.10) and dividing by $\sin(\Omega t)$ gives:

$$
\begin{bmatrix} k + k_a - \Omega^2 m & -k_a \\ -k_a & k_a - \Omega^2 m_a \end{bmatrix} \begin{bmatrix} a_1 \\ a_2 \end{bmatrix} = \begin{bmatrix} F \\ 0 \end{bmatrix} \tag{3.11}
$$

Using the second line in Eq. (3.11) gives

$$
a_2 = \frac{k_a a_1}{k_a - \Omega^2 m_a} \tag{3.12}
$$

Substituting (3.12) into the first line of Eq. (3.11) gives:

$$
a_1 = \frac{F(k_a - \Omega^2 m_a)}{(k + k_a - \Omega^2 m)(k_a - \Omega^2 m_a) - k_a^2} \tag{3.13}
$$

Then substituting (3.13) into (3.12) to eliminate a_1 leaves:

$$
a_2 = \frac{F k_a}{(k + k_a - \Omega^2 m)(k_a - \Omega^2 m_a) - k_a^2}
$$

From Eq. (3.13) it can be seen that $a_1 = 0$ when $k_a - \Omega^2 m_a = 0$. This is the optimal condition for vibration suppression referred to in the question. The frequency value

[3] For a full derivation of these equations see Hartog (1934).

[4] This assumption can be arguably justified based on the fact that the system is undamped. Energetically, however, this system is unrealistic as there is excitation being added to the system but no mechanism for dissipating energy.

of the combined system at which $a_1 = 0$ is

$$\Omega = \sqrt{\frac{k_a}{m_a}} = \omega_a, \tag{3.14}$$

where ω_a is the natural frequency of the absorber. The natural frequency for the primary system is $\omega_p = \sqrt{\frac{k}{m}}$, so the best vibration suppression will occur when

$$\omega_p = \sqrt{\frac{k}{m}} = \sqrt{\frac{k_a}{m_a}} = \omega_a. \tag{3.15}$$

■

The frequency at which $a_1 = 0$ in Example 3.2 is called the *anti-resonance* of the system. By setting the anti-resonance of the combined system equal to the resonance of the primary (i.e. SDOF) system, the maximum vibration reduction can be obtained. An example is shown in Fig. 3.3. It can be seen that this type of vibration absorber reduces the amplitude of the original resonance very significantly, but creates two other resonance peaks. As a result, this type of approach works well for systems where a single resonance is a problem, and the excitation is quite narrow band. A more detailed analysis of the (linear) damped case can be found in Hartog (1934) or Soong and Dargush (1997).

Many authors have considered extending the concept of a tuned mass damper (TMD)[5] to the nonlinear domain. Most commonly this extension to nonlinearity has been by adding nonlinear stiffness terms. Typically cubic terms are added, but sometimes other polynomial terms are used, see for example Ibrahim (2008), Alexander and Schilder (2009) and references therein for further details. Generally adding nonlinear stiffness to an otherwise linear TMD is of limited benefit,[6] as the main effect is to distort the resonance curves, which can be counter-productive, for example by producing isola, see Alexander and Schilder (2009). We note also, that several practical implementations of TMDs have taken the form of damped pendulum systems,[7] which have inherent geometric nonlinearities. In some cases, these are designed by being linearized such that linear TMD design rules can be applied. However, there are a class of more general torsional vibration absorbers which have a range of practical applications. For example, details of a highly sophisticated automotive application which exploits torsional vibration absorbers is given by Shaw (2012).

For TMDs an alternative approach to adding nonlinear stiffness, is to add nonlinear damping. For example, this approach was taken by Gattulli et al. (2004) when considering the application of damping vibrations in cables. It is reasonable to ask rather than have nonlinear dampers, why not just install larger linear dampers? The

[5] Also called a damped vibration absorber (DVA) or tuned vibration absorber (TVA).

[6] Note this is not the same as exploiting the properties of a geometrically nonlinear structure, which will be discussed in Chap. 5.

[7] For example in the Taipai 101 Tower.

answer is that (i) no damper in practice behaves exactly in a linear way, and (ii) the larger the damper the more the nonlinear effects are generally accentuated. For example, one drawback of scaling up fluid based dampers, is that the static friction forces in the seals become very large, so that the damper will only mobilise if this force is overcome. If the damper fails to mobilise, it just acts like a rigid link in the structure, and only modifies the stiffness (see for example the study by Londoño et al. 2013). This point is often ignored (or misunderstood) as viscous damping terms such as $c\dot{x}$ are used in models to approximate physical dampers, which take no account of the nonlinear effects. As a result studying nonlinear dampers is important to (i) understand physical damper behaviour, and (ii) to improve the performance of structural systems. Some cases are discussed later in this chapter.

Finally we mention a relatively recent alternative to the TMD called the *inerter*. This is a device which generates a passive inertial force that is proportional to the relative acceleration. The inerter can be used in combination with springs and dampers to form a *tuned inertial damper* (*TID*). For details of the inerter applied to vehicle dynamics see Smith and Wang (2004), and details of the TID are given in Lazar et al. (2014).

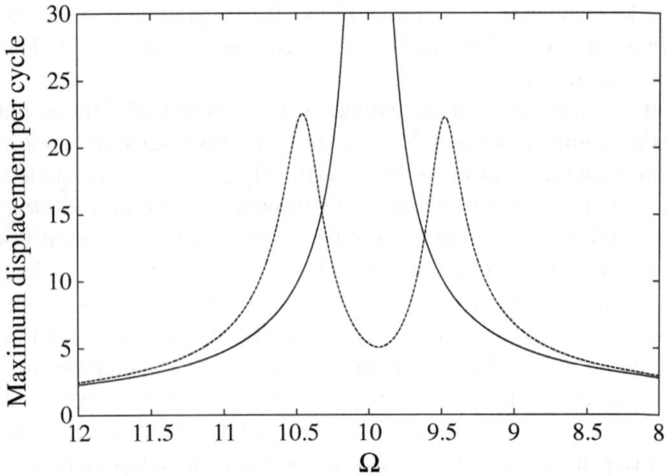

Fig. 3.3 Vibration absorber, showing maximum displacement amplitude per cycle of the primary system response (*solid line*) and the combined primary plus absorber system (*dashed line*). In this Figure $\omega_p = 10$ rad/s and a small amount of damping has been added to the absorber so that the peaks of the combined system are finite. Adding damping also has the effect that the anti-resonance does not go to zero, as predicted in Example 3.2. Furthermore, to get the peaks of the combined system to be approximately equal, the absorber frequency has been de-tuned slightly. Further details of this analysis can be found in Hartog (1934) or Soong and Dargush (1997)

3.1.2 Nonlinear Passive Vibration Isolators

In the case of vibration isolation, nonlinear spring mechanisms can be used to significantly improve the performance of the isolator. It was noted in Example 3.1, and shown in Fig. 3.2 that a linear vibration isolator only reduces vibrations when $\Omega/\omega_n > \sqrt{2}$. Therefore the isolator performance can be improved by reducing ω_n as much as practical. For a linear isolator this is problematic because the mass m is normally fixed and so reducing ω_n can only be achieved by reducing the stiffness, k. In many cases this can lead to excessive *static* deflections if the k value is too small.

Therefore, there is a case for designing an isolator with nonlinear stiffness to give a high static stiffness, but a low dynamic stiffness. This scenario is shown in Fig. 3.4, where a linear spring is combined with a nonlinear spring to give a combined stiffness which has very low (i.e. almost zero) tangent stiffness at zero displacement, and hardening stiffness at larger displacements. The effect is shown in Fig. 3.5 where the resonance curve is shifted to the lower frequency range and is hardening. This leads to a increased isolation region, although the maximum amplitude of displacement response is similar to the linear case. The parameters have been chosen such that the amplitude of the resonance peak is the same as the linear case, but the isolation region is significantly increased. In practice, this type of isolator can be realised by combining the linear spring with stiffness, k, with a snap-through structure similar to the buckled beam described in Chap. 2. It should be noted that to obtain a combined stiffness curve shown in Fig. 3.4 that is correctly zeroed, the linear spring will need to be preloaded to compensate for the static displacement from the mass. Further details of this type of nonlinear isolator can be found in Shaw et al. (2012).

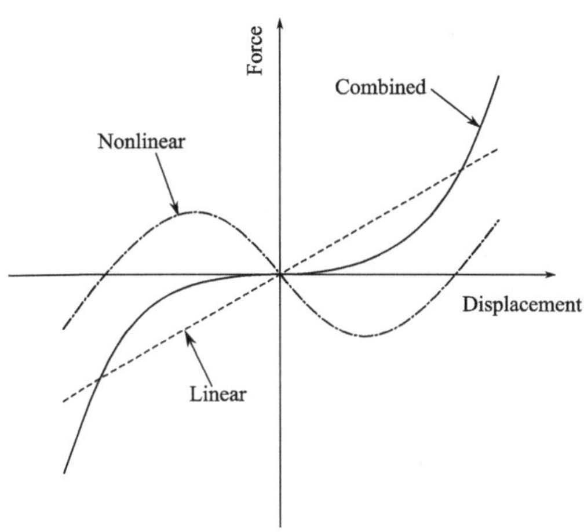

Fig. 3.4 Nonlinear vibration isolator stiffness function, showing a linear and nonlinear spring combined

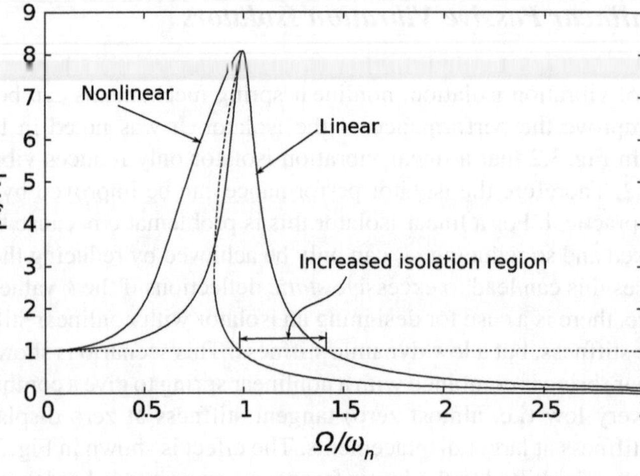

Fig. 3.5 Nonlinear passive vibration isolator using a high static, low dynamic stiffness function $N = -\mu(x - r) + \mu_3(x - r)^3$. The complete system is $m\ddot{x} + c(\dot{x} - \dot{r}) + k(x - r) + N = 0$ where the following parameters have been used: $m = 1\,\text{kg}$, $c = 0.375\,\text{kg/s}$, $k = 9\,\text{N/m}$, $\mu = 5\,\text{N/m}$ and $\mu_3 = 0.1\,\text{N/m}^3$. The linear isolator shown here is the same type as the passive linear isolator from Example 3.1. For the linear simulation all parameters are as above except $\mu = \mu_3 = 0$. The excitation signal is $r = \sin(\Omega t)$

The passive nonlinear isolation system shown in Fig. 3.5 can be analysed using the approximate techniques described in Chap. 4.

3.2 Semi-active Vibration Control

Semi-active control is a method for effecting change in a vibrating system without using control actuators. Instead, a semi-active element, usually a damper, is used. Within the semi-active element it is typically possible to vary one or sometimes more system parameters. An important difference between semi-active and active control is that semi-active control cannot add energy to the system (the device can only resist motion), and therefore is normally an unconditionally stable form of control.

An example of a single-degree-of-freedom oscillator with semi-active vibration control is shown in Fig. 3.6. In this example, the mass-spring-damper system is excited by a moving support input of $r(t)$. To decide how to select c_v, the damping value of the variable damper, information is needed about the relative velocity of the mass and the input. This can be achieved by using accelerometers to measure the acceleration of the mass and the support, $\ddot{x}$ and $\ddot{r}$ respectively, which can be integrated to give $\dot{x}$ and $\dot{r}$.

One of the most common semi-active control strategies is *sky-hook*. The idea is that the mass can be isolated from the support input by getting the semi-active damper to mimic a grounded[8] passive damper as closely as possible. If this can be achieved

[8] Grounded means one end of the damper is attached to a surface which does not move. Not to be confused with ground-hook control which is a variant of sky-hook.

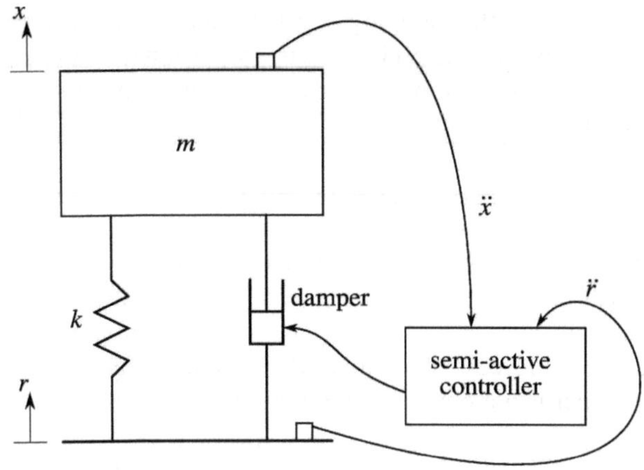

Fig. 3.6 Single-degree-of-freedom oscillator with semi-active vibration control

perfectly, a damping force which resists the absolute velocity of the mass will be provided. A common way to implement this in a semi-active element is to switch between a high and a low damping value. Usually, the high damping is selected when the damper force is resisting the direction of motion of the mass, and the low damping force is used when this is not the case. Practically, this can be achieved in various ways, for example one of the most common is by switching between high and low viscosity in a magneto-rheological damper. An example of how this can be done is considered next.

Example 3.3 Piecewise linear semi-active vibration control example

Problem Design a semi-active vibration control strategy, for the mass-spring-damper system shown in Fig. 3.6. The feedback to the controller is the acceleration of the mass, $\ddot{x}$, and input, $\ddot{r}$, measured by accelerometers. Assume that the semi-active damper can be switched between two constant values, c_{high} and 0.

Solution The acceleration signals are fed to a semi-active controller, which then needs to switch between c_{high} and 0 depending on the state of the system. The governing equation of motion for the oscillator is given by

$$m\ddot{x} + c_v(\dot{x} - \dot{r}) + k(x - r) = 0,$$

where the displacement of the mass, m, is given by x, k is the spring stiffness and c_v is a variable damping parameter which can be controlled by the semi-active controller. The control objective is to reduce vibration in the system as much as possible, or in other words, to *isolate* the mass (or minimize the absolute acceleration of the mass, $\ddot{x}$).

The sky-hook strategy is based on mimicking the situation where a grounded damper is attached to the mass. The assumed damping force for the grounded damper would be $c_g\dot{x}$, where c_g is the damping constant and $\dot{x}$ is the absolute velocity. Of course the actual damper force, $c_v(\dot{x} - \dot{r})$, is proportional to the relative velocity, $\dot{x} - \dot{r}$. So to get the system to mimic the grounded damper as closely as possible the ideal situation would be

$$c_v(\dot{x} - \dot{r}) = c_g\dot{x}. \tag{3.16}$$

If c_v can be continuously varied, then Eq. 3.16 can be satisfied subject to constraints on the amplitudes of $\dot{x}$ and $\dot{r}$. However, the question states that c_v can only be switched between two constant values, c_{high} and c_{low} where in this example $c_{low} = 0$. Therefore the best that can be achieved is to switch off the damper, by setting $c_v = 0$, at any time when the damper is not opposing the motion

$$\text{sign}(\dot{x} - \dot{r}) = \text{sign}(\dot{x}) \rightsquigarrow (\dot{x} - \dot{r})\dot{x} \geq 0. \tag{3.17}$$

This is achieved by noting that when the *relative* velocity has the same sign as the absolute velocity, then the damper is opposing the mass. So the semi-active control law applied to the single-degree-of-freedom system can be written as

$$c_v = \begin{cases} c_{high} & (\dot{x} - \dot{r})\dot{x} \geq 0, \\ 0 & \text{otherwise.} \end{cases}$$

The semi-active control will act like additional damping in the linear oscillator, which in turn will reduce the height of the resonance peak. ■

The switching strategy defined in Example 3.3 is usually called *on-off* sky hook control. It should be noted that switching strategies can introduce problems of their own such as *chatter*, when the control rapidly switches back and forth between the two switching states. Varying c_v to try and satisfy Eq. 3.16 is called *modified* or *continuous* skyhook.

A time simulation for a systems similar to Example 3.3 when $r = \sin(10t)$ is shown in Fig. 3.7 with $x(0) = 1.1$ and $\dot{x}(0) = 1.0$, $m = 1$, $k = 1$ and $c_v = 0.1$. Initially the sky hook control is switched off, and then at time $t = 15$ s the control is switched on, with $c_{high} = 0.7$ and $c_{low} = 0.1$. A significant reduction in vibration amplitude can be seen as soon as the control is switched on. However, a residual amount of vibration remains in the steady state. This is because there is some time when the damper is switched off, and so not all the vibration can be eliminated.

As a comparison between sky-hook and linear passive isolation, Fig. 3.8 shows a comparison of three isolator systems. Two are linear passive isolators, one with low damping, labelled $\zeta = 0.0625$ and another with higher damping, labelled $\zeta = 0.5$. The third isolator is a on-off sky-hook system which is switching between two damping values $\zeta = 0.0625$ and $\zeta = 0.5$. Here we see that sky-hook reduces the overall response compared to just damping values, although the reduction is not that significant compared to the higher damping value.

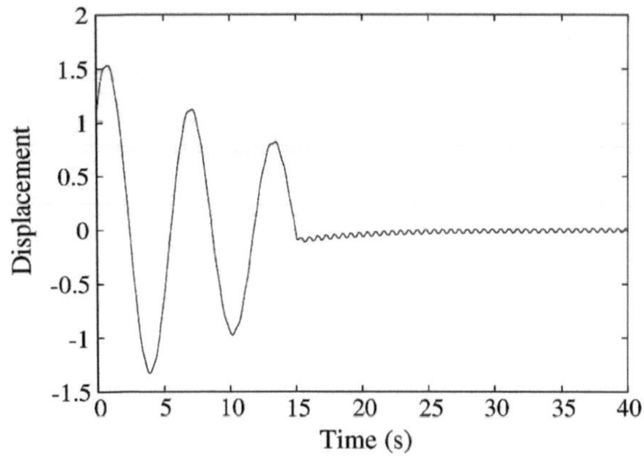

Fig. 3.7 Time simulation of on-off sky hook control for Example 3.3

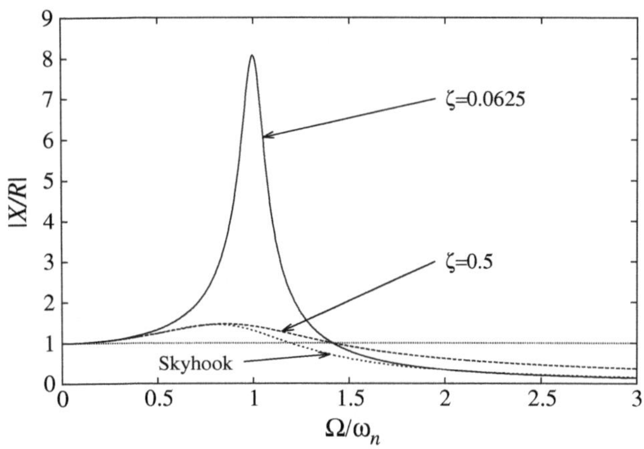

Fig. 3.8 Transmissibility curves for sky hook control compared to two passive linear isolators with the same form as Example 3.1. The linear passive isolators, are labelled $\zeta = 0.0625$ (*low damping*) and $\zeta = 0.5$ (*high damping*)

However, combining this type of semi-active control with the passive nonlinear isolator discussed in the previous Section does give a significant increase in the isolation region. An example of this is shown in Fig. 3.9, where nonlinear sky-hook is compared with lightly damped linear, lightly damped nonlinear and heavily damped nonlinear.

Sky-hook and other semi-active approaches can produce significantly improved vibration isolation compared to passively damped systems. The most common application is in automotive and other suspension systems. They can also be applied to other types of base isolation systems. For systems with multiple degrees of freedom, such as continuous structural elements it becomes increasingly difficult to maintain the benefits of semi-active control methods like sky-hook.

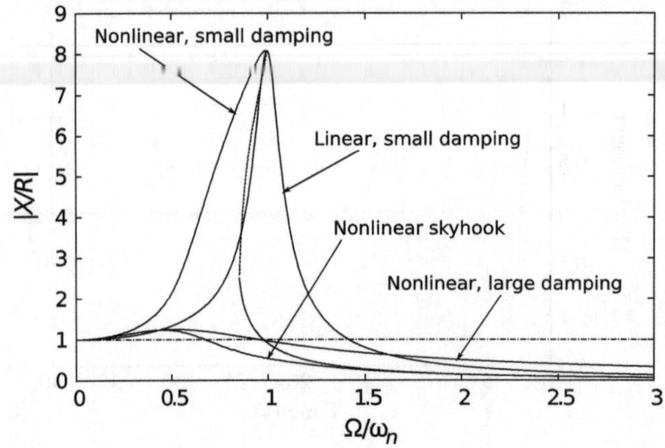

Fig. 3.9 Transmissibility curves for nonlinear sky hook control compared to passive linear and nonlinear isolators. All parameters as for Fig. 3.5, and the skyhook strategy is the same as Figs. 3.7 and 3.8

3.3 Active Vibration Control

An introduction to the basic ideas of feedback control has been given in Sect. 1.4. These ideas will now be used to develop a more systematic method for *designing* controllers for nonlinear vibrating systems.

The general governing equation for an autonomous nonlinear control system was given by Eq. (1.32). For a system with time-dependent nonlinear dynamics (usually from external forcing), Eq. (1.32) becomes

$$\dot{\boldsymbol{x}} = \mathbf{f}(\boldsymbol{x}, t) + \mathbf{g}(\boldsymbol{x})u, \tag{3.18}$$

where $\mathbf{f}$ is the nonlinear system function and $\mathbf{g}$ is the nonlinear controller function, $\boldsymbol{x}$ is the state vector $\boldsymbol{x} = [\mathbf{x}^T, \dot{\mathbf{x}}^T]^T$ and $\mathbf{x}$ is the displacement vector.[9] The output is defined as $\mathbf{y} = \mathbf{h}(\boldsymbol{x})$, where $\mathbf{h}$ is the nonlinear output function.

To specify the governing equations of the system, an approximate model of the vibration behaviour is required. For continuous structural elements, this is discussed in Chaps. 6–8. A typical vibration scenario is that the displacements and velocities in the state vector represent an approximate modal model of a continuous structural element (beam, cable or plate etc.). The idealized modal model is infinite (see for example Sect. 6.1.2, in Chap. 6), but in practice it must be truncated to the sum of N modal contributions.[10]

Two typical vibration-control configurations are shown schematically in Fig. 3.10. In both cases a cantilever beam is the vibrating element. In Fig. 3.10a the cantilever beam is being controlled by a linear actuator. The beam displacement at the point at

---

[9] Note that $\boldsymbol{x}$ is the $1 \times 2N$ *state* vector and $\mathbf{x}$ is the $1 \times N$ *displacement* vector.

[10] See Inman (2006) for a discussion of modal truncation.

Fig. 3.10 Two vibration control configurations: **a** beam with actuator and **b** beam with piezo patches

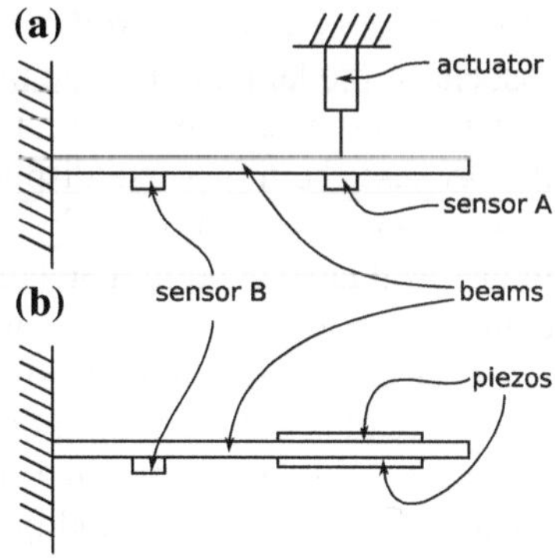

which the control force is applied is measured by sensor A, and the beam displacement at a second point along the cantilever is measured by sensor B. In Fig. 3.10b the cantilever beam is being controlled by a *collocated* pair of piezo actuator/sensors. Again the beam displacement at a second point along the cantilever is measured by sensor B. In both cases the underlying vibrating system (i.e. the cantilever beam) is infinite dimensional, but is acted on by only a small number of actuators, and measured with a small number of sensors. This leads to the question how can a small number of sensors and actuators be used to control the vibrations of the beam? This question is considered in the next subsection.

Note that in Fig. 3.10a the actuator is collocated with sensor A. However, in some situations, measurements from sensor B may be the only form of feedback available. Such a situation is called non-collocated control, see for example the discussion in Preumont (1997). Note also that in some situations the actuator-sensor positions are predetermined, and in other cases they can be positioned to give the best control effect.

3.3.1 Observability and Controllability

An important part of the control design is to determine to what degree the system can be *observed*, and to what degree it can be *controlled*. Sensors and actuators are located at a limited number of discrete points on the structure. When a modal decomposition is carried out,[11] the effect of discrete point forces (such as control actuators) appears on the right-hand side of the modal equations multiplied by a coefficient (as shown, for example, in the linear case by Eq. (1.24), Chap. 1). The coefficient is called the *modal participation factor* (formally derived in Sect. 6.1.3 in Chap. 6) and gives a measure of the effect of the applied force on the mode. Therefore, in modal models

[11] This is shown in detail in Chap. 5.

of vibrating systems controllability of a particular mode will depend directly on the associated modal participation factor.

Using a sensor to measure at a discrete point has a similar effect on observability, because the transverse displacement, w, at point A along the beam is typically approximated as $w(A, t) \approx \sum_{j=1}^{N} \phi_j(A)q_j(t)$, where the $\phi_j(A)$ values are the beam modes evaluated at point A and the $q_j(t)$ values are the modal coordinates. In fact the definition needs to be more precise for vibrating systems, because although N modes are taken in a truncated model of the continuous system, observability and controllability is related only to the *controlled* part of the system. For example, if the control objective is just to control the first mode of vibration of the beam, then only the controllability and observability of this mode is of interest. In addition, the uncontrolled modes may still have significant dynamics. For example, if the controlled modes[12] run from $1, 2, \ldots, N_c$ and the uncontrolled modes from $N_c + 1, \ldots, \infty$, then when making a measurement at point A, the response is actually given by $w(A, t) \approx \sum_{j=1}^{N_c} \phi_j(A)q_j(t) + \sum_{j=N_c+1}^{\infty} \phi_j(A)q_j(t)$. As a result if the response of the *uncontrolled modes* is significant, the measurement of $w(A, t)$ will be corrupted by their contribution to the measurement. This effect is called *observation spillover*. A similar effect occurs when the control force is applied at a single point, because in general the modal participation factors are non-zero for the uncontrolled modes, so the effect of the control force is to excite the uncontrolled modes. This is called *control spillover*. The scenario is shown as a control block diagram in Fig. 3.11.

The position of the sensors and actuators is important, because for many modes both the mode-shape, ϕ, and the modal participation factor can be zero (or close to zero) at some points along the beam. For linear modal systems, in-depth analysis of the effect of actuator and sensor placement has been developed, see for example Gawronski (2004).

If the sensor and actuator positions can be selected as part of the control design, the issue of finding whether the controlled modes are controllable and observable remains. Typically, the state vector, $\boldsymbol{x} = [\mathbf{x}^T, \dot{\mathbf{x}}^T]^T$, (from Eq. 3.18) will consist of

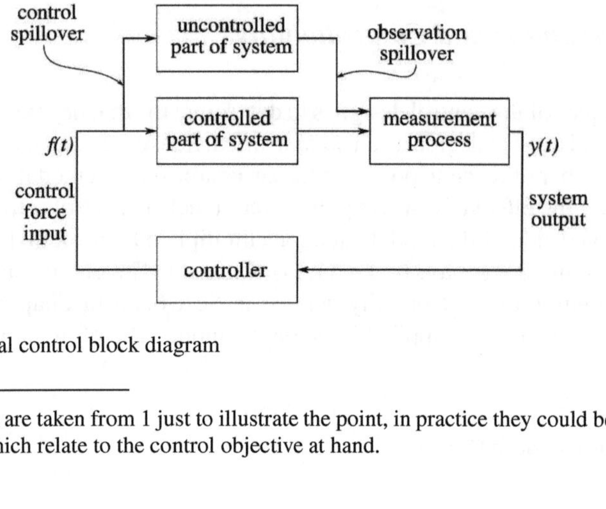

Fig. 3.11 Modal control block diagram

[12] These modes are taken from 1 just to illustrate the point, in practice they could be chosen as any set of modes which relate to the control objective at hand.

an equal number of system displacements and velocities, for the controlled modes, and so the vector length is $2N_c$. So to what extent can these modal displacements and velocities be measured and controlled? For the vibrating systems considered in this book it will be sufficient to consider the observability and controllability of the underlying linearized system. This can be derived from Eq. (3.18), by using the same approach described in Sect. 2.1.2 (Chap. 2) for linearizing about equilibrium points and will typically give a linearized system of the form

$$\dot{x} \approx Ax + Bu,$$

with an output $y = \bar{C}x$. Then the normal linear conditions for observability and controllability, that matrices $\mathscr{O}$ and $\mathscr{R}$ have full rank can be applied, where

$$\mathscr{O} = \begin{bmatrix} \bar{C} \\ \bar{C}A \\ \vdots \\ \bar{C}A^{2N-1} \end{bmatrix}, \quad \mathscr{R} = [B \ AB \ A^2B \ \ldots \ A^{2N-1}B].$$

In this chapter it will be assumed that the controlled part of the system is both observable and controllable. A more general discussion on controllability and observability in nonlinear systems, particularly those in which the underlying linear system is not necessarily controllable or observable can be found in Nijmeijer and van der Schaft (1990), Vidyasagar (1992) and Sastry (1999) .

3.3.2 Control Law Design

Once the model of the vibrating system has been selected and the questions of observability and controllability have been assessed, the *control law* needs to be designed. The control law effectively involves how to choose u. Once the control law is selected it is often referred to as the *controller*. However, before a control law for u can be chosen, the objective of the control process needs to be specified. In nonlinear control theory, control objectives fall broadly into the two main classes of *stabilization* and *tracking*. A stabilization problem is concerned with finding u such that the state vector tends to a stable equilibrium point (typically zero) for any initial conditions and parameter values in the required range. Tracking is concerned with getting the state, or output variables, to follow a predefined reference trajectory.

For active vibration control, the main objective is to reduce the amplitude of vibration in one or more points in the system. As a result this type of control problem can be thought of as being in the stabilization category. The other type of control task which could be required, for example in an adaptive structure, is some form of shape change or morphing. This type of control task would fall into the tracking category.

Other requirements which need to be considered, in order of importance, are (i) stability, (ii) robustness, (iii) performance, and (iv) cost. Stability and robustness are essential requirements for any control law and these will be discussed in detail in Sect. 3.4. There are a range of ways to measure the control performance. For tracking control, accuracy in replicating the required reference signal and the speed of response are of primary interest. For active vibration control, the amount of vibration reduction (or effective added damping) is the main performance measure. Cost is important both in terms of implementation cost (and practicality) and also how hard the control actuator has to work. Excessive actuation will quickly lead to wear and high maintenance costs. For example, *optimal control* can be used to take costs such as actuator effort into account by allowing actuator effort to be included into a system *cost function*.

To demonstrate a typical control design, active vibration control of a linear single-degree-of-freedom oscillator is considered as a first example.

Example 3.4 Linear active vibration control example

Problem Design an active vibration-control law, u, for the mass-spring-damper system shown in Fig. 3.12. The only feedback to the controller is the acceleration of the mass, measured by an accelerometer. The mass is also acted on by an external excitation force, F_e, which is an unknown force/disturbance signal. The control force produced by the actuator can be assumed to be $F_c = bu(t)$, where b is a scalar constant.

Solution The acceleration signal is fed to a controller that produces the control signal, u, which is used as the input into the control actuator. Then the actuator produces a control force that acts on the mass. The governing equation of motion for the oscillator is given by

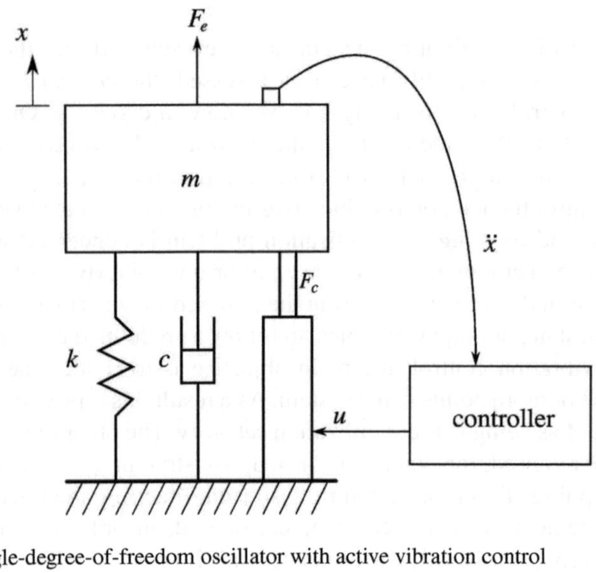

Fig. 3.12 Single-degree-of-freedom oscillator with active vibration control

$$m\ddot{x}(t) + c\dot{x}(t) + kx(t) = F_e(t) + bu(t), \tag{3.19}$$

where $F_c = bu(t)$.The control objective is to reduce vibration in the system as much as possible, or in other words, increase the effective damping of the system. To add damping, the control can be directly related to an extra velocity term by integrating the acceleration feedback signal

$$u(t) \; = -\kappa \int_0^t \ddot{x}\mathrm{d}t,$$

such that Eq. (3.19) becomes

$$m\ddot{x}(t) + (c + b\kappa)\dot{x}(t) + kx(t) = F_e(t), \tag{3.20}$$

which increases the damping by $b\kappa$ and providing both $b > 0$ and $\kappa > 0$ the control will act like additional damping in the linear oscillator, which in turn will reduce the height of the resonance peak. ■

This type of active vibration control is called velocity feedback or integral acceleration feedback[13] control, see Preumont (1997). In fact, for nonlinear oscillators with linear damping, such as the Duffing oscillator of the form,

$$m\ddot{x}(t) + c\dot{x}(t) + k_1x(t) + k_3x^3 = F_e(t) + F_c(t),$$

velocity feedback control can be applied in the same way to give,

$$m\ddot{x}(t) + (c + b\kappa)\dot{x}(t) + k_1x(t) + k_3x^3 = F_e(t), \tag{3.21}$$

which is a forced nonlinear oscillator with increased damping. The stability and robustness of the linear system in Example 3.4 can be analysed with standard control techniques. The stability of the controlled Duffing oscillator in Eq. (3.21), on the other hand, needs a different approach and this is considered next.

3.4 Stability Theory

The basic ideas of dynamic stability were introduced in Sect. 2.3 in Chap. 2. This section extends the stability concepts to the case where an oscillator, or vibration mode, is subject to a control signal.[14] The key idea from Sect. 2.3 is that an

---

[13] This type of control needs careful implementation in practice to avoid noise being amplified at higher frequencies.

[14] In fact, only a very limited discussion is presented here. For more detailed discussion of these topics consult, for example, Khalil (1992), Sastry (1999), Slotine and Li (1991), Isidori (1995), Fradkov et al. (1999) and Krstić et al. (1995).

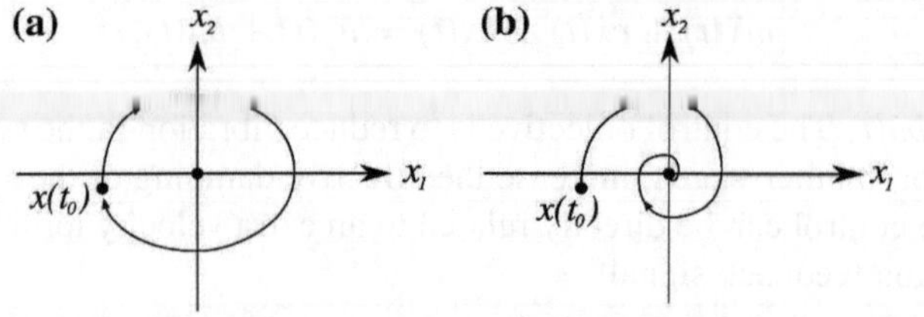

Fig. 3.13 Stability in the phase plane: **a** neutral stability and **b** asymptotic stability

equilibrium point, $\boldsymbol{x}^*$, is stable if a solution, $\boldsymbol{x}(t)$, starting close to the equilibrium point stays close for all time t, and there are two specific cases.[15] An equilibrium point, $\boldsymbol{x}^*$, is Lyapunov or neutrally stable if trajectories stay 'close' to it, and asymptotically stable if nearby trajectories are 'attracted' to it. For systems with two states, such that $\boldsymbol{x} = [x_1, x_2]^T$, where $x_1 = x$ is displacement and $x_2 = \dot{x}$ is velocity, the two types of stability are shown in the phase plane in Fig. 3.13. In both cases the equilibrium point is at the origin. In Fig. 3.13a, the trajectory starts at an initial point $\boldsymbol{x}(t_0)$ at time t_0, and orbits the equilibrium point, not getting much closer or further away, which corresponds to neutral or Lyapunov stability (also shown in Fig. 2.13a). In Fig. 3.13b, the trajectory starts at an initial point $\boldsymbol{x}(t_0)$ at time t_0, and is pulled towards the equilibrium point, which corresponds to asymptotic stability (also shown in Fig. 2.13b) .

3.4.1 Lyapunov Functions

When the nonlinear control system has only a single equilibrium point at the origin,[16] a powerful technique which can be used to analyse the stability of the system is to use a *Lyapunov function*. These are 'energy-like' functions which are similar to the potential functions discussed in Chap. 2. Using a Lyapunov function allows the system stability to be investigated without integrating the differential equation. The key idea is to take a Lyapunov function, $V(\boldsymbol{x}, t)$, and find the derivative of V with respect to t to get $\dot{V}$. If V is taken to represent the energy in the system, then $\dot{V}$ gives the rate of change of energy with time and there are three possible cases which relate directly to the stability of the equilibrium point at the origin:

- If $\dot{V}$ is positive → energy increasing → unstable
- If $\dot{V}$ zero → energy stays the same → neutrally stable
- If $\dot{V}$ negative → energy decreasing → asymptotically stable

---

[15] Exponential stability is a type of asymptotic stability.

[16] For systems with a single equilibrium point it is nearly always possible to change coordinates to move the equilibrium point to the origin.

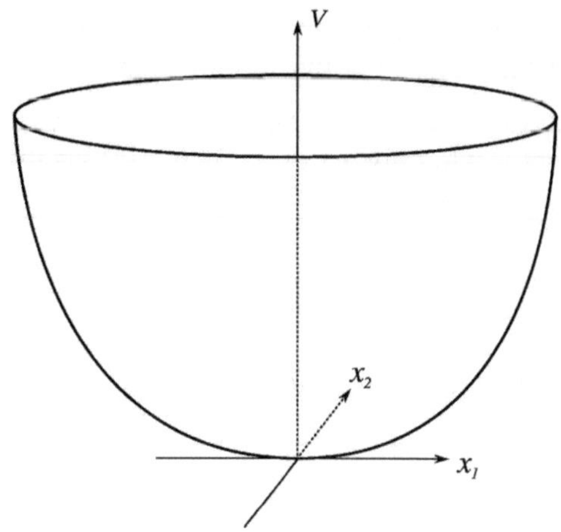

Fig. 3.14 Parabolic Lyapunov function

It is important to note that this is only true if V is a positive definite function,[17] such that for $\boldsymbol{x}^* = 0$, $V(0, t) = 0$ and then $V(\boldsymbol{x}, t) > 0$ must hold for all $\boldsymbol{x}$ other than $\boldsymbol{x} = 0$.

Note also that V is not unique for any particular system. In fact, the main difficulty with using Lyapunov function analysis is deciding which function to select. For oscillators with two states, $\boldsymbol{x} = [x_1, x_2]^T$, a parabolic potential function of the form $V(\boldsymbol{x}) = \frac{x_1^2}{2} + \frac{x_2^2}{2}$, can often be used as a first-guess Lyapunov function. The parabolic function is shown in Fig. 3.14 and, if the energy in the system is represented as a ball rolling on the inside surface of the function (as for the discussion in Sect. 2.2.1), it is easy to see that the ball can only come to rest at the origin. This is because V is positive definite, and therefore has a single unique minimum at the origin.

To carry out the Lyapunov function analysis, first differentiate V with respect to t to get $\dot{V}$, which for the parabolic function gives $\dot{V} = x_1\dot{x}_1 + x_2\dot{x}_2$ via the chain rule. Then substitute for $\dot{x}_1$ and $\dot{x}_2$ from the governing equations to derive a final expression for $\dot{V}$. This approach will be illustrated in the following example.

Example 3.5 Lyapunov function for controlled Duffing oscillator

Problem Use a Lyapunov function to assess the stability of the controlled Duffing equation, given by Eq. (3.21), for the case when only a single equilibrium point at the origin exists in the system and F_e is zero.

Solution First write the governing equations of motion for the oscillator in first-order form

$$\begin{aligned}\dot{x}_1 &= x_2\\ \dot{x}_2 &= -\left(\frac{c + b\kappa}{m}\right)x_2 - \frac{k_1}{m}x_1 - \frac{k_3}{m}x_1^3 + \frac{F_e}{m}.\end{aligned} \tag{3.22}$$

[17] More details on the strict definition of a Lyapunov function are given by Slotine and Li (1991).

The state vector is $\boldsymbol{x} = [x_1, x_2]^T$, and it will be assumed that $F_e = 0$. It is not clear which Lyapunov function to choose, so, as a first guess, assume that the parabolic function given by $V_1(\boldsymbol{x}) = \frac{x_1^2}{2} + \frac{x_2^2}{2}$ is the Lyapunov function. Differentiating V_1 with respect to t gives $\dot{V}_1 = x_1\dot{x}_1 + x_2\dot{x}_2$ via the chain rule. Then substituting for $\dot{x}_1$ and $\dot{x}_2$ from Eq. (3.22) gives

$$\dot{V}_1 = x_1 x_2 + x_2\left(-\left(\frac{c + b\kappa}{m}\right)x_2 - \frac{k_1}{m}x_1 - \frac{k_3}{m}x_1^3\right),$$

which can be expanded out to give

$$\dot{V}_1 = x_1 x_2 - \left(\frac{c + b\kappa}{m}\right)x_2^2 - \frac{k_1}{m}x_1 x_2 - \frac{k_3}{m}x_1^3 x_2. \tag{3.23}$$

The objective now is to get Eq. (3.23) into a form where the sign of $\dot{V}_1$ can be determined. To achieve this, terms of x_1^2 or x_2^2 (or higher even powers) are kept as these will always be positive, and any other terms, such as x_1x_2 and $x_1^3x_2$, are ideally eliminated. To do this, the initial guess of the Lyapunov function needs to be changed and the analysis repeated. However, the form of Eq. (3.23) often indicates what the new form of V should be.

In this example, it can be seen that Eq. (3.23) has positive and negative x_1x_2 terms, so just changing the coefficient of the assumed Lyapunov function to

$$V_2(\boldsymbol{x}) = \frac{k_1}{m}\frac{x_1^2}{2} + \frac{x_2^2}{2},$$

and repeating the analysis removes an additional term to give

$$\dot{V}_2 = -\left(\frac{c + b\kappa}{m}\right)x_2^2 - \frac{k_3}{m}x_1^3 x_2.$$

There is still a term of undetermined sign on the right-hand side and this can be eliminated by adding an extra term to the Lyapunov function so that

$$V_3(\boldsymbol{x}) = \frac{k_1}{m}\frac{x_1^2}{2} + \frac{x_2^2}{2} + \frac{k_3}{m}\frac{x_1^4}{4}.$$

This time, repeating the analytical process gives

$$\dot{V}_3 = -\left(\frac{c + b\kappa}{m}\right)x_2^2,$$

and providing $(c + b\kappa)/m$ is positive then $\dot{V}_3$ is always negative definite and the equilibrium point at the origin is asymptotically stable. Note this assumes that the new Lyapunov function is positive definite, which in this example is true. ■

Example 3.5 demonstrates how a Lyapunov function analysis can be carried out in an iterative way using an initial guess for V. However, notice that the form of the final Lyapunov function, V_3, is actually the sum of the potential energy function (as defined in Sect. 2.2.1) and the kinetic energy, all divided by the mass, m, i.e. $V_3 = E_t/m$ where E_t is the potential plus kinetic energy. It turns out that using energy as a guess for the Lyapunov function analysis is often a good choice for mechanical oscillators. As before, the process can be iterated to find a final Lyapunov function. It also possible to guess numerous functions and still not be able to determine the sign of $\dot{V}$, a problem that increases with the complexity of the system. It is also difficult to extend this technique to multi-degree-of-freedom systems.

3.4.2 Bounded Stability

In Example 3.5, the force/disturbance signal, F_e, was assumed to be zero, so how can stability be assessed in the more realistic situation when $F_e \neq 0$? One way this can be done is using *input-output stability* analysis for both forcing inputs such as F_e and also control inputs, u. In both cases the input must be *bounded.* Bounded is used to described a signal which is unknown or uncertain, but stays within prescribed limitations or *bounds.*

The first stage is to assess the stability of the system with the input set to zero, as was the case in the previous subsection. If the unforced system is stable and the input is bounded, then it is usually possible to determine the stability of the system using a Lyapunov function analysis with additional conditions in place for the bounded input. First, consider an example of *input to state* stability, which relates the input to the state via a Lyapunov function analysis.

Example 3.6 Input to state stability for controlled Duffing oscillator

Problem Assess the input to state stability of the controlled Duffing equation, given by Eq. (3.21), for the case when the forcing signal $F_e \neq 0$ but is bounded.

Solution First, write the governing equations of motion for the oscillator in first-order form

$$\begin{aligned}\dot{x}_1 &= x_2 \\ \dot{x}_2 &= -\left(\frac{c + b\kappa}{m}\right)x_2 - \frac{k_1}{m}x_1 - \frac{k_3}{m}x_1^3 + \frac{F_e}{m}.\end{aligned}$$

The state vector is $\boldsymbol{x} = [x_1, x_2]^T$, and $F_e \neq 0$, but bounded.

Now, using the Lyapunov function found in Example 3.5

$$V_3(\boldsymbol{x}) = \frac{k_1}{m}\frac{x_1^2}{2} + \frac{x_2^2}{2} + \frac{k_3}{m}\frac{x_1^4}{4},$$

and the time derivative of V_3 becomes

$$\dot{V}_3 = -\left(\frac{c + b\kappa}{m}\right) x_2^2 + \frac{F_e}{m} x_2.$$

The forcing term on the right-hand side is of unknown sign and, without knowledge of the size of this term compared to the first term on the right-hand side, the sign of $\dot{V}_3$ cannot be determined. However, it is possible to set a bound on the size of F_e which will make the system stable.

To do this, let $\alpha_1 = (c + b\kappa)/m$ and $u = F_e/m$. Then define a positive constant α_2 such that $0 < \alpha_2 < \alpha_1$, noting that c, b and κ are all positive so that $\alpha_1 > 0$ always. Then $\dot{V}_3$ can be rewritten as

$$\dot{V}_3 = -(\alpha_1 - \alpha_2)x_2^2 - \alpha_2 x_2^2 + u x_2,$$

where the first term on the right-hand side has been split into two. Now notice that if $u = \alpha_2 x_2$ then $\dot{V}_3 = -(\alpha_1 - \alpha_2)x_2^2$ which is always negative definite. So setting the bound for the input signal as $|u| \leq \alpha_2 |x_2|$ means that $-\alpha_2 x_2^2 + u x_2 \leq 0$, and

$$\dot{V}_3 = -(\alpha_1 - \alpha_2)x_2^2 - \alpha_2 x_2^2 + u x_2 \leq -(\alpha_1 - \alpha_2)x_2^2,$$

which ensures $\dot{V}_3$ is always negative definite and the system is asymptotically stable. ■

The condition limiting the size of the input in Example 3.6 ensures that the input term in $\dot{V}_3$ is always smaller than the x_2^2 term. But for most practical applications this is unnecessarily restrictive, and it is more reasonable to expect that a bounded input signal produces a bounded output signal. For linear control systems, this leads to the idea of bounded input, bounded output (BIBO) stability. In nonlinear systems, the same concept of bounded stability is usually defined in terms of *input-output* stability.

The key idea is that, providing the underlying unforced system is asymptotically stable, the output will be bounded if the input is bounded and all operators acting on the input are bounded as well. In terms of the governing equations for a nonlinear control system

$$\dot{\boldsymbol{x}} = \mathbf{f}(\boldsymbol{x}, t) + \mathbf{g}(\boldsymbol{x})u, \quad \mathbf{y} = \mathbf{h}(\boldsymbol{x}),$$

when $u = 0$ the underlying system $\dot{\boldsymbol{x}} = \mathbf{f}(\boldsymbol{x}, t)$ is asymptotically stable. For the stability analysis, the input is u and the output is $\mathbf{y}$. The input u is bounded which can be written as $|u| \leq C_1$, where C_1 is an arbitrary constant.

To have a bounded output, both $\mathbf{g}$ and $\mathbf{h}$ need to be bounded as well. This is usually done by imposing the conditions that $||\mathbf{g}|| \leq C_2$ and $||\mathbf{h}|| \leq C_3$, where C_2 and C_3 are constants. As a result $||\mathbf{y}||$ will be limited to some bounded range.[18]

[18] Formal derivations can be found in Khalil (1992), Sastry (1999) and Slotine and Li (1991).

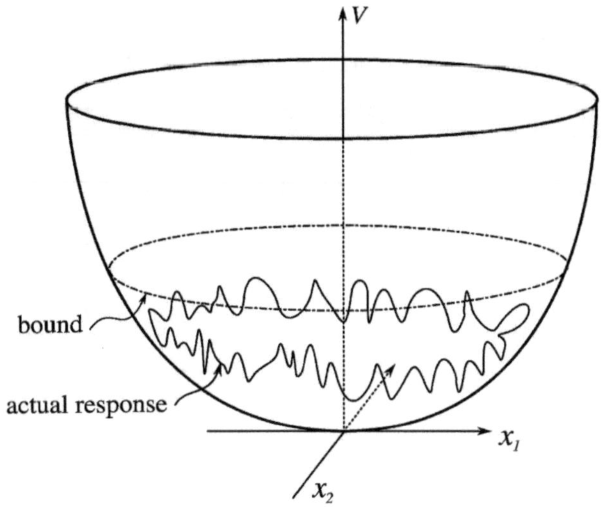

Fig. 3.15 Stability of a forced system using a Lyapunov function

This means that in terms of the Lyapunov function, another constant, C_4, can be defined such that $V_{max} < C_4$. The result is shown schematically in Fig. 3.15, where the actual V signal is unpredictable, because of the uncertain nature of the input, but bounded. This approach is now applied to the Duffing oscillator considered in Example 3.6.

Example 3.7 Bounded stability for controlled Duffing oscillator

Problem Assess the stability of the controlled Duffing equation, given by Eq. (3.21), for the case when the forcing signal F_e is bounded, such that $|F_e| \leq C_1$ where C_1 is a positive constant.

Solution First, write the governing equations of motion for the oscillator in the form $\dot{\boldsymbol{x}} = \mathbf{f}(\boldsymbol{x}, t) + \mathbf{g}(\boldsymbol{x})u$, where in this example the input u is the forcing signal F_e to give

$$\begin{bmatrix} \dot{x}_1 \\ \dot{x}_2 \end{bmatrix} = \begin{bmatrix} x_2 \\ -\left(\frac{c+b\kappa}{m}\right) x_2 - \frac{k_1}{m} x_1 - \frac{k_3}{m} x_1^3 \end{bmatrix} + \begin{bmatrix} 0 \\ \frac{1}{m} \end{bmatrix} F_e.$$

The state vector is $\boldsymbol{x} = [x_1, x_2]^T$, and F_e is bounded.

Now using the Lyapunov function found in Example 3.5

$$V_3(\boldsymbol{x}) = \frac{k_1}{m}\frac{x_1^2}{2} + \frac{x_2^2}{2} + \frac{k_3}{m}\frac{x_1^4}{4}, \tag{3.24}$$

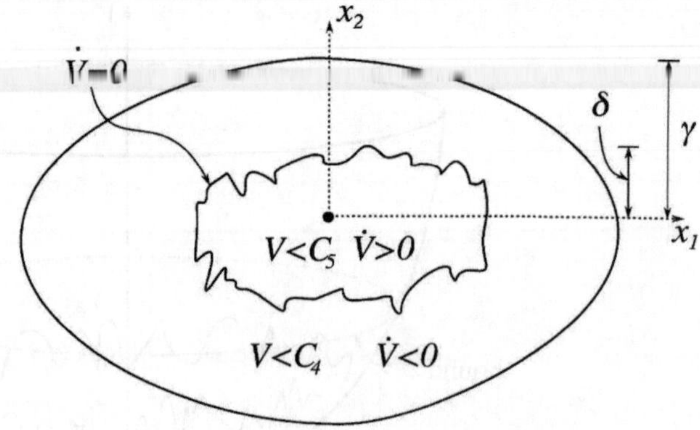

Fig. 3.16 Bounded response using a Lyapunov function

the time derivative of V_3 becomes

$$\dot{V}_3 = -\alpha_1 x_2^2 + \frac{F_e}{m} x_2 \leq -\alpha_1 x_2^2 + \frac{C_1}{m} x_2, \tag{3.25}$$

where $\alpha_1 = (c + b\kappa)/m$. The right-hand side $\dot{V}_3$ is not negative definite, but it is bounded, and $F_e \leq C_1$ is bounded. The sign of $\dot{V}_3$ actually depends on the relative magnitude of the two terms on the right-hand side of Eq. (3.25). The two terms can be thought of as *competing*, but as t becomes large, the system approaches the steady-state, such that $\dot{V}_3 \approx 0$, and V_3 becomes bounded. This can be seen by considering what occurs as x_2 grows; eventually the term with x_2^2 will always become large and dominate the x_2 term so the system is drawn back towards the origin. Setting Eq. (3.25) to zero gives the bounding value of x_2, which in this case occurs when $x_2 = C_1/(\alpha_1 m)$. Finally, the value of V_{3max} can be found by substituting this back into Eq. (3.24). In this example there is no bounding value for x_1, so only the max V_{3max} value along the $x_1 = 0$ axis can be specified exactly. This gives $V_{3max} = \frac{1}{2}\left(\frac{C_1}{\alpha_1 m}\right)^2$. ■

The bounded stability in Example 3.7 can be explained by Fig. 3.16. In Fig. 3.16 an outer set of points has $\dot{V} < 0$ and $C_5 < V < C_4$ for $\delta \leq |x_2| \leq \gamma$. The constant, γ, can be made as large as required to include any initial condition. In this outer set of points, the system behaves as if the origin were stable and, as $\dot{V} < 0$, any trajectories are pulled towards the origin.[19] Any trajectories starting close to the equilibrium point, inside the inner set, $\dot{V} > 0$ and $V < C_5$, behave as if the equilibrium point is unstable (as $\dot{V} > 0$), so the trajectories move away from the equilibrium point. As $\dot{V} \to 0$ the value of V becomes bounded, and this bound can be selected to be less than an arbitrary constant.

Note that, throughout this section, the examples used a Duffing system for which the control had been specified by the physics of the feedback process defined in

[19] For a proof that this happens in a finite time see Khalil (1992).

Fig. 3.12 and Example 3.4. The Lyapunov control design process is then concerned with finding a suitable Lyapunov function to demonstrate that the system is stable. It is possible, but less relevant to practical control problems, to do things the other way round. In other words, specify a Lyapunov function and then design a controller to stabilize the system. This approach will not be pursued in this section, but can be found in the references at the end of this chapter, and a related approach will be covered in Sect. 3.5.

This is also why the control signal does not appear explicitly in the Duffing system, Eq. (3.21), used for the examples. The main issue then becomes one of proving stability in the presence of an unknown forcing function. It is more typical in control theory to consider the effect of disturbances such as noise. For vibration applications this type of effect can be considered to be part of the forcing signal, and the input-output analysis then effectively includes a robustness element.

3.5 Linearisation Using Feedback

In Sect. 3.3, the stability of the active vibration-control system shown in Fig. 3.12, was studied by defining a linear velocity feedback control which led to the nonlinear governing equation, Eq. (3.21). This approach meant that the resulting nonlinear system, as given by Eq. (3.21), has to be proven stable. However, is it possible to achieve the same control objective and simultaneously remove the nonlinearity? In this section, *feedback linearisation* techniques are discussed, which are designed to linearize the nonlinear system using the feedback control signal.

To reduce vibration and linearize the system simultaneously, additional control effort will be required. This can be provided by either integrating the velocity signal to obtain displacement, or preferably, by adding a displacement transducer such as an LVDT.[20] The latter option is shown in Fig. 3.17 for a single-degree-of-freedom oscillator with a nonlinear spring.

The governing equations for the system in Fig. 3.17, with $F_e = 0$, can be written as

$$\begin{aligned}\dot{x}_1 &= x_2,\\ \dot{x}_2 &= -\frac{c}{m}x_2 - \frac{k_1}{m}x_1 - \frac{k_3}{m}x_1^3 + \frac{p}{m}u(t),\end{aligned} \tag{3.26}$$

where it is assumed that $F_c = pu(t)$. This can be written in the matrix form

$$\begin{bmatrix}\dot{x}_1\\ \dot{x}_2\end{bmatrix} = \begin{bmatrix}0 & 1\\ -\frac{k_1}{m} & -\frac{c}{m}\end{bmatrix}\begin{bmatrix}x_1\\ x_2\end{bmatrix} + \begin{bmatrix}0\\ -\frac{k_3}{m}x_1^3\end{bmatrix} + \begin{bmatrix}0\\ \frac{p}{m}\end{bmatrix}u(t), \tag{3.27}$$

or in general

$$\dot{\boldsymbol{x}} = A\boldsymbol{x} + \mathscr{N}(\boldsymbol{x}) + Bu, \tag{3.28}$$

[20] An LVDT is a linear variable differential transformer, which is a type of electrical transformer used for measuring linear displacement.

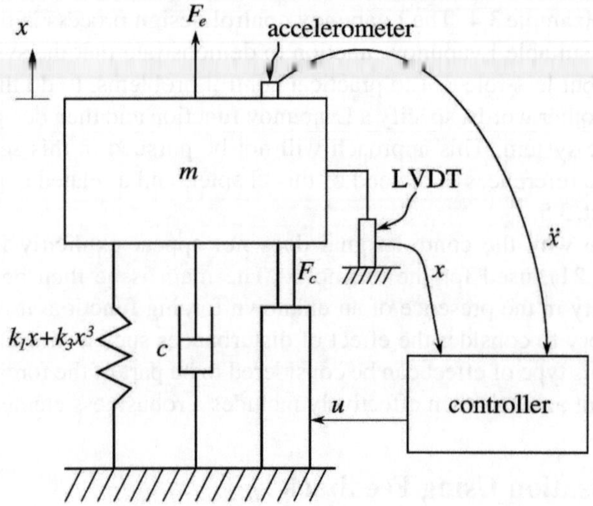

Fig. 3.17 Single-degree-of-freedom oscillator with active vibration control using an accelerometer and an LVDT

which is the same as Eq. (1.34) in Chap. 1.

By inspection of either Eq. (3.26) or (3.27), it can be seen that, by setting $u = \frac{k_3}{p}x_1^3$, gives $\mathcal{N}(\boldsymbol{x}) + Bu(t) = 0$ for all t. The system thereby reduces to $\dot{\boldsymbol{x}} = A\boldsymbol{x}$, which is linear and, providing A has stable eigenvalues, is also stable. For the general case, $\mathcal{N}(\boldsymbol{x})$ can be rewritten as $\mathcal{N}(\boldsymbol{x}) = B\mathcal{N}^*(\boldsymbol{x})$ and Eq. (3.28) becomes

$$\dot{\boldsymbol{x}} = A\boldsymbol{x} + B\mathcal{N}^*(\boldsymbol{x}) + Bu = A\boldsymbol{x} + B(\mathcal{N}^*(\boldsymbol{x}) + u). \tag{3.29}$$

Setting $u = -\mathcal{N}^*$ gives the feedback linearisation control signal for the system in the case where A is a stable matrix. We assume that the system states in the expression $-\mathcal{N}^*$ can be readily accessed for use in the control signal u. So systems which can be expressed in the form of Eq. (3.29) can be linearized using the feedback control signal.

The control signal can also include an additional control task, like adding damping. For example, $u = \frac{k_3}{p}x_1^3 - \frac{\kappa}{m}x_2$ means that $\mathcal{N}(x) + Bu(t) \rightarrow \left[0, -\frac{\kappa}{m}x_2\right]^T$ as $t \rightarrow \infty$, and the damping increases by $\frac{\kappa}{m}$. In general, the control input is chosen as $u = -\mathcal{N}^*(\boldsymbol{x}) + c(\boldsymbol{x})$, where $c(\bullet)$ is the desired control function.

If the underlying linear system happens to be *unstable*, then the control function needs to be designed to provide a stable linear control after the nonlinear terms have been removed. Now consider an example of an oscillator with nonlinear damping.

Example 3.8 Feedback linearisation for an oscillator with nonlinear damping

Problem Use feedback linearisation to remove the nonlinear damping terms in the following nonlinear oscillator

$$m\ddot{x} + c\dot{x}(1 + \delta x) + kx = pu(t),$$

where $u(t)$ is the control input. Is it possible to linearize the system and add more linear viscous damping simultaneously?

Solution First, write the governing equations of motion for the oscillator in the form $\dot{\boldsymbol{x}} = A\boldsymbol{x}(t) + \mathcal{N}(\boldsymbol{x}) + Bu$, to give

$$\begin{bmatrix} \dot{x}_1 \\ \dot{x}_2 \end{bmatrix} = \begin{bmatrix} 0 & 1 \\ -\frac{k}{m} & -\frac{c}{m} \end{bmatrix} \begin{bmatrix} x_1 \\ x_2 \end{bmatrix} + \begin{bmatrix} 0 \\ -\frac{c\delta}{m} x_1 x_2 \end{bmatrix} + \begin{bmatrix} 0 \\ \frac{p}{m} \end{bmatrix} u(t).$$

The state vector is $\boldsymbol{x} = [x_1, x_2]^T$, and u is the control input.

The control matrix is $B = \left[0, \frac{p}{m}\right]^T$, and the objective is to put the problem into the form of Eq. (3.29). This can be achieved by setting $\mathcal{N}^* = -\frac{c\delta}{p} x_1 x_2$, and to linearize the system by setting $u = \frac{c\delta}{p} x_1 x_2$. Additional damping can be obtained by adding an additional term to the control signal to give $u = \frac{c\delta}{p} x_1 x_2 - \frac{\kappa}{p} x_2$. ■

A numerical simulation of Example 3.8 is shown in Fig. 3.18, with numerical parameters $m = k = 1$, $c = 0.1$, $\delta = 30$ and $p = 10$. In each case the system is

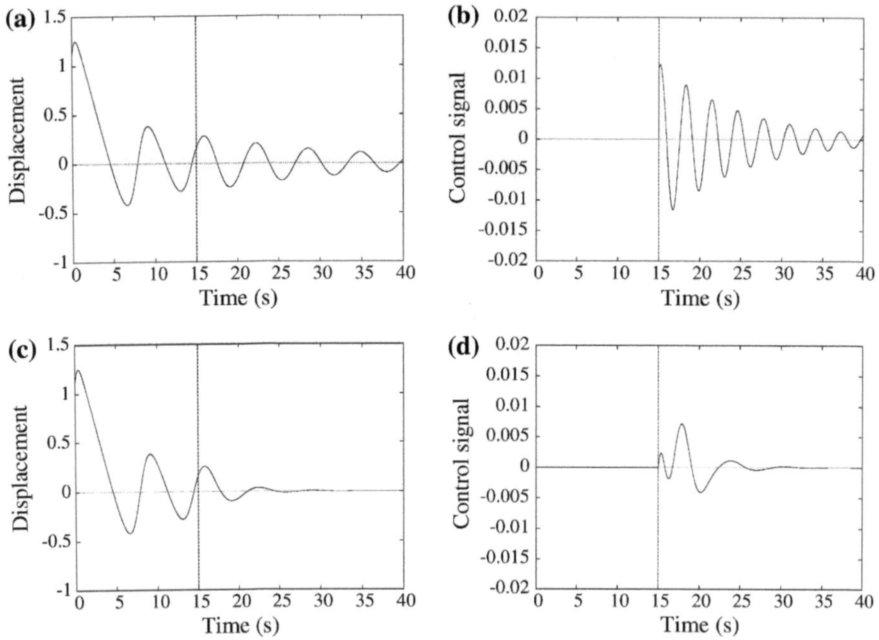

Fig. 3.18 Feedback linearisation for Example 3.8, with control starting at time $t = 15$ s; (**a**) and (**b**) show feedback linearisation response, (**c**) and (**d**) show feedback linearisation response with added damping

uncontrolled until time $t = 15$ s, when the feedback linearisation control is switched on. In Fig. 3.18a, b the case of feedback linearisation response on its own is shown. A clear change can be seen from the distorted non-harmonic response occurring before 15 s to a harmonic response afterwards. In Fig. 3.18c, d the case of feedback linearisation response with additional viscous damping is shown. The additional viscous damping parameter in this simulation is $\kappa = 0.5$. In this case, after 15 s the vibrations are significantly more damped than in the previous case.

In the next subsection, the idea of feedback linearisation is approached from an input-output perspective .

3.5.1 Input-Output Linearisation

The basic idea of input-output linearisation was introduced in Sect. 1.4.2 in Chap. 1. Consider how the method could be applied to the example system shown in Fig. 3.17. The main idea is to obtain a relationship between the output and the input by repeatedly differentiating the output until the input appears.

For example, in the case of the system shown in Fig. 3.17, which is governed by Eq. (3.26), assume the output is the displacement, x_1, so that $y = h(\boldsymbol{x}) = x_1$. Differentiating y with respect to time t gives $\dot{y} = \dot{x}_1 = x_2$ from Eq. (3.26). But the input, u, does not appear, so differentiate again to give

$$\ddot{y} = \dot{x}_2 = -\frac{c}{m}x_2 - \frac{k_1}{m}x_1 - \frac{k_3}{m}x_1^3 + \frac{p}{m}u(t),$$

which gives a relationship between the second derivative of the output (which is the acceleration $\ddot{y}$) and the control input u. It can be seen that choosing

$$u(t) = \frac{m}{p}\left(v(t) + \frac{c}{m}x_2 + \frac{k_1}{m}x_1 + \frac{k_3}{m}x_1^3\right),$$

leads to a linear input-output relationship

$$\ddot{y} = \dot{x}_2 = v(t),$$

where $v(t)$ is an input signal which can be chosen to achieve an additional control task.

A more systematic approach is to consider how the output from the nonlinear system $\mathbf{y} = \mathbf{h}(\boldsymbol{x})$ varies with time. Here, the analysis will be limited to the single-input-single-output case. These means that the output, $\mathbf{y}$, and output function, $\mathbf{h}$, are both scalar, written $y = h(\boldsymbol{x})$. Similarly, the control input, $\mathbf{u}$, is scalar, written as u. Then, differentiating the output with respect to time gives

$$\dot{y} = \frac{\partial h(\boldsymbol{x})}{\partial \boldsymbol{x}}\frac{\partial \boldsymbol{x}}{\partial t} = \frac{\partial h(\boldsymbol{x})}{\partial \boldsymbol{x}}\dot{\boldsymbol{x}}, \tag{3.30}$$

where $\frac{\partial h(\boldsymbol{x})}{\partial \boldsymbol{x}}$ is an $1 \times N$ row vector and $\frac{\partial \boldsymbol{x}}{\partial t}$ is a $N \times 1$ column vector. Equation 3.30 indicates that the rate of change of the output with time can be expressed as the rate of change of the output *with the state* multiplied by the system velocity vector. Substituting for $\dot{\boldsymbol{x}}$ from Eq. (3.18) ($\dot{\boldsymbol{x}} = \mathbf{f}(\boldsymbol{x}, t) + \mathbf{g}(\boldsymbol{x})u$ with $\mathbf{f}(\boldsymbol{x}, t) = \mathbf{f}(\boldsymbol{x})$) gives

$$\dot{y} = \frac{\partial h(\boldsymbol{x})}{\partial \boldsymbol{x}}(\mathbf{f}(\boldsymbol{x}) + \mathbf{g}(\boldsymbol{x})u) = \frac{\partial h(\boldsymbol{x})}{\partial \boldsymbol{x}}\mathbf{f}(\boldsymbol{x}) + \frac{\partial h(\boldsymbol{x})}{\partial \boldsymbol{x}}\mathbf{g}(\boldsymbol{x})u,$$

which can be rewritten as

$$\dot{y} = L_f h(\boldsymbol{x}) + L_g h(\boldsymbol{x})u, \qquad (3.31)$$

where $L_f h(\boldsymbol{x})$ and $L_g h(\boldsymbol{x})$ are the *Lie derivatives* of $h(\boldsymbol{x})$ with respect to $\mathbf{f}(\boldsymbol{x})$ and $\mathbf{g}(\boldsymbol{x})$. Effectively, the Lie derivatives are the directional derivatives of the output function, $h(\boldsymbol{x})$, along the vector fields $\mathbf{f}(\boldsymbol{x})$ and $\mathbf{g}(\boldsymbol{x})$, respectively.

Now, choosing the control to remove the system dynamics and replace them with a new control signal, $v(t)$, results in a control input of the form

$$u = \frac{1}{L_g h(\boldsymbol{x})}(v(t) - L_f h(\boldsymbol{x})), \qquad L_g h(\boldsymbol{x}) \neq 0, \qquad (3.32)$$

to give $\dot{y} = v(t)$. This gives a linear relationship between the new input $v(t)$ and the derivative of the output $\dot{y}$.

The number of times the equations need to be differentiated to get an input-output relation corresponds to the *relative degree* of the system. In other words, if the condition $L_g h(\boldsymbol{x}) \neq 0$ is true, the system is said to have relative degree one and no more differentiation is required. However, if the output does not appear directly in the expression $L_g h(\boldsymbol{x}) = 0$, the Lie derivative process needs to be iterated until it does, as was needed in the first example in this section, where $\ddot{y}$ was required. Although this can be done as many times as necessary in theory, here only the first derivative (velocity) and second derivative (acceleration) cases will be used, as these correspond to physically meaningful outputs for vibration-control problems.

So when $L_g h(\boldsymbol{x}) = 0$ differentiate Eq. (3.31) again to give

$$\ddot{y} = \frac{\partial}{\partial \boldsymbol{x}}(L_f h(\boldsymbol{x}))\frac{\partial \boldsymbol{x}}{\partial t} = \frac{\partial L_f h(\boldsymbol{x})}{\partial \boldsymbol{x}}\mathbf{f}(\boldsymbol{x}) + \frac{\partial L_f h(\boldsymbol{x})}{\partial \boldsymbol{x}}\mathbf{g}(\boldsymbol{x})u = L_f^2 h(\boldsymbol{x}) + L_g L_f h(\boldsymbol{x})u. \qquad (3.33)$$

In this case, if $L_g L_f h(\boldsymbol{x}) \neq 0$, the system is said to have relative degree two, and the control law is given by

$$u = \frac{1}{L_g L_f h(\boldsymbol{x})}(v(t) - L_f^2 h(\boldsymbol{x})), \qquad L_g L_f h(\boldsymbol{x}) \neq 0. \qquad (3.34)$$

This process will be demonstrated in the following example.

Example 3.9 Input-output linearisation for an oscillator with nonlinear damping

Problem Use input-output linearisation to linearize the following nonlinear oscillator

$$m\ddot{x} + c\dot{x}(1 + \delta x^2) + kx = pu(t),$$

where $u(t)$ is the control input. Assume that the output is the displacement so that $y = x$.

Solution First, write the governing equations of motion for the oscillator in the form $\dot{\boldsymbol{x}} = \mathbf{f}(\boldsymbol{x}) + \mathbf{g}(\boldsymbol{x})u$, to give

$$\begin{bmatrix} \dot{x}_1 \\ \dot{x}_2 \end{bmatrix} = \begin{bmatrix} x_2 \\ -\frac{c}{m}x_2(1+\delta x_1^2) - \frac{k}{m}x_1 \end{bmatrix} + \begin{bmatrix} 0 \\ \frac{p}{m} \end{bmatrix} u(t).$$

The state vector is $\boldsymbol{x} = [x_1, x_2]^T$, and u is the control input. The other vectors are

$$\mathbf{f}(\boldsymbol{x}) = \begin{bmatrix} x_2 \\ -\frac{c}{m}x_2(1+\delta x_1^2) - \frac{k}{m}x_1 \end{bmatrix}, \qquad \mathbf{g}(\boldsymbol{x}) = \begin{bmatrix} 0 \\ \frac{p}{m} \end{bmatrix}.$$

To compute the Lie derivative, the first step is to compute $\frac{\partial h(\boldsymbol{x})}{\partial \boldsymbol{x}}$, which in this case with $h(\boldsymbol{x}) = x_1$ gives

$$\frac{\partial h(\boldsymbol{x})}{\partial \boldsymbol{x}} = [1, 0].$$

Note that this is a row vector. Now the Lie derivatives can be computed

$$L_f h(\boldsymbol{x}) = [1, 0]\begin{bmatrix} x_2 \\ -\frac{c}{m}x_2(1+\delta x_1^2) - \frac{k}{m}x_1 \end{bmatrix} = x_2, \qquad L_g h(\boldsymbol{x}) = [1, 0]\begin{bmatrix} 0 \\ \frac{p}{m} \end{bmatrix} = 0.$$

As $L_g h(\boldsymbol{x}) = 0$, the Lie derivative process needs to be repeated.

To do this, first compute the derivative of $L_f h(\boldsymbol{x})$ giving

$$\frac{\partial L_f h(\boldsymbol{x})}{\partial \boldsymbol{x}} = [0, 1].$$

Then compute the second Lie derivative to give

$$L_f^2 h(\boldsymbol{x}) = \frac{\partial L_f h(\boldsymbol{x})}{\partial \boldsymbol{x}}\mathbf{f}(\boldsymbol{x}) = [0, 1]\begin{bmatrix} x_2 \\ -\frac{c}{m}x_2(1+\delta x_1^2) - \frac{k}{m}x_1 \end{bmatrix} = -\frac{c}{m}x_2(1+\delta x_1^2) - \frac{k}{m}x_1,$$

and

$$L_g L_f h(\boldsymbol{x}) = \frac{\partial L_f h(\boldsymbol{x})}{\partial \boldsymbol{x}}\mathbf{g}(\boldsymbol{x}) = [0, 1]\begin{bmatrix} 0 \\ \frac{p}{m} \end{bmatrix} = \frac{p}{m}.$$

So the system has a relative degree of two, and, as $L_g L_f h(\boldsymbol{x}) \neq 0$, a control input can be formed using Eq. (3.34) to give

$$u = \frac{m}{p}\left(v(t) + \frac{c}{m}x_2(1+\delta x_1^2) + \frac{k}{m}x_1\right). \tag{3.35}$$

When this control signal is used, the input-output relationship becomes $\ddot{y} = v(t)$, where $v(t)$ is a new control signal which can be defined to give the required linear behaviour. ■

A numerical simulation of Example 3.9 is shown in Fig. 3.19, with numerical parameters $m = k = 1$, $c = 0.1$, $\delta = 30$ and $p = 10$. The system is uncontrolled until time $t = 15$ s, when the input-output linearisation control is switched on. Two different cases are shown. In Fig. 3.19a, b the control objective is to eliminate vibration, so $v(t)$ is chosen to be $v(t) = -x_1 - x_2$, which is a stable linear oscillator with large damping. The result is a very sudden reduction in displacement response to zero after the control is turned on. In Fig. 3.19c, d the control objective is to get the nonlinear system to track a sine wave. One way to achieve this is to set $v(t) = -x_1 - x_2 + \sin(t)$, which is a strongly damped linear oscillator with sine wave forcing. The result is

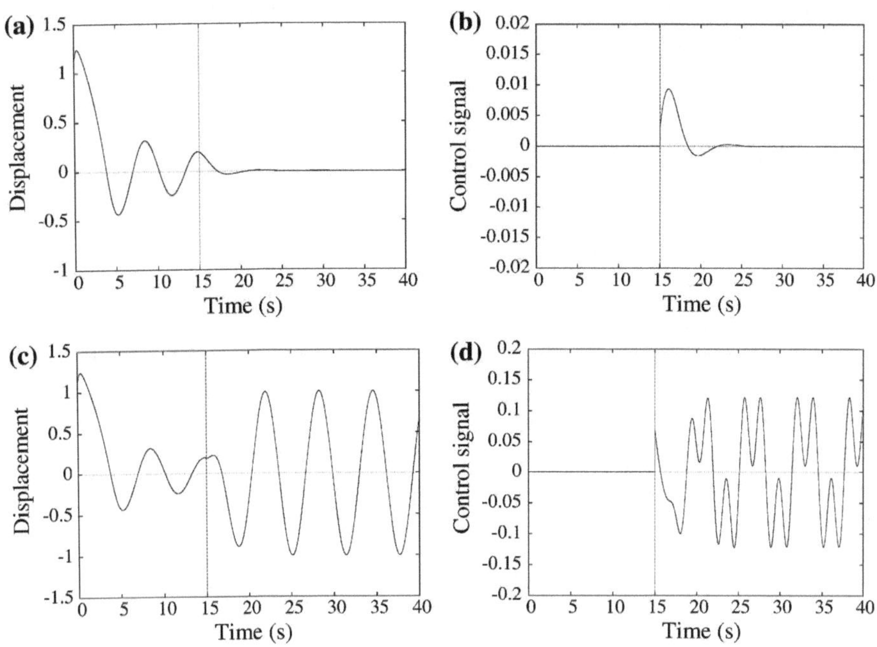

Fig. 3.19 Feedback linearisation for Example 3.9, with control starting at time $t = 15$ s; **a** and **b** show feedback linearisation response to eliminate vibration, **c** and **d** show feedback linearisation response to track a sine wave

that, when the control is switched on, the system quickly switches to the behaviour of the forced linear system, giving the required sine wave behaviour.

For systems with additional forcing and/or disturbance inputs, the input-output analysis can be followed by a bounded stability analysis of the forced system, either using a Lyapunov function analysis as in Sect. 3.4.1, or an appropriate linear robustness analysis Khalil (1992).

Note that significant problems can occur with input-output linearisation if the system has *internal dynamics* which are unstable. This leads to the concept of *zero dynamics* in the system, which must be stable, because they cannot be controlled otherwise. This is not a typical situation for vibration control, but the interested reader can find a detailed treatment of this and other related issues in Sastry (1999), Slotine and Li (1991) and Khalil (1992).

3.6 Control of Multi-Degree-of-Freedom Systems

In the previous sections, nonlinear control methods for single-degree-of-freedom oscillators have been discussed. In control terms, these systems are formulated as single-input-single-output systems. In this section, the case of multi-modal vibrations is introduced, which is the situation most commonly encountered in the practice of vibration engineering.

3.6.1 Modal Control

Transforming vibration problems into a modal space is a key modelling technique, and the basic concept for linear multi-degree-of-freedom systems was introduced in Sect. 1.3.3 of Chap. 1. A more detailed treatment of modal analysis is given in Chaps. 5–8, the modal approach will be applied to nonlinear vibration behaviour in beams, cables, plates and shells. Control forces can be included in the modal representation, and, if the control objective is formulated in terms of modal quantities as well, the approach can be called *modal control.*

For example, in Chap. 6, assuming proportional damping, it will be shown that nonlinear vibrations in beams result in modal equations which for the nth mode may be written as

$$\ddot{q}_n(t) + \zeta_n \omega_{nn} \dot{q}_n + \omega_{nn}^2 q_n(t) + \sum_{j=1}^{N} \sum_{k=1}^{N} \sum_{l=1}^{N} \mu_{jkln} q_j(t) q_k(t) q_l(t) = \alpha_n F_c(t) \quad (3.36)$$

where $q_n(t)$ is the modal displacement, ω_{nn} the modal natural frequency, ζ_n the modal damping ratio, μ_{jkln} the nonlinear coefficient, $F_c(t)$ is the control force and α_n is the modal participation factor of the nth mode. The summations over j, k and l and the coefficients μ_{jkln} represent the nonlinear cubic terms, which typically include coupling between the modes. In the first instance it is assumed that these modal equations can be reduced to a simpler form which for the nth mode may be written as

$$\ddot{q}_n + \zeta_n \omega_{nn} \dot{q}_n + \omega_{nn}^2 q_n + \mu_n q_n^3 = \alpha_j F_c(t), \tag{3.37}$$

in which the left-hand sides are nonlinear (Duffing type) oscillators, but now decoupled.

Consider the case when there is a single control force, F_c, acting on the beam, as shown in the schematic beam control scenario in Fig. 3.10a. This force acts like any other external force in a vibration problem, in that it has an influence on all modes via the α_j coefficients. To see this, take the number of controlled modes as $N_c = 2$ and form a state vector $\boldsymbol{x} = [q_1, q_2, \dot{q}_1, \dot{q}_2]^T$. This gives a state space form of $\dot{\boldsymbol{x}} = A\boldsymbol{x} + \mathscr{N}(\boldsymbol{x}) + Bu$ as

$$\frac{\mathrm{d}}{\mathrm{d}t}\begin{bmatrix} q_1 \\ q_2 \\ \dot{q}_1 \\ \dot{q}_2 \end{bmatrix} = \begin{bmatrix} 0 & 0 & 1 & 0 \\ 0 & 0 & 0 & 1 \\ -\omega_{n1}^2 & 0 & -\zeta_1\omega_{n1} & 0 \\ 0 & -\omega_{n2}^2 & 0 & -\zeta_2\omega_{n2} \end{bmatrix}\begin{bmatrix} q_1 \\ q_2 \\ \dot{q}_1 \\ \dot{q}_2 \end{bmatrix} - \begin{bmatrix} 0 \\ 0 \\ \mu_1 q_1^3 \\ \mu_2 q_2^3 \end{bmatrix} + \begin{bmatrix} 0 \\ 0 \\ \alpha_1 \\ \alpha_2 \end{bmatrix} F_c.$$

So, the control force F_c will act on both the modes. How much force affects each mode depends on the relative size of the modal participation factors, α_1 and α_2. If the aim of the control force F_c is primarily to reduce the modal displacement q_1, then the effect of $\alpha_2 F_c$ acting on modal displacement q_2 needs to be minimized. In some cases, it may be possible to position the actuator to minimize the size of α_2, such as placing the actuator at the node point of mode 2, where α_2 should be close to zero.[21] However, for modes $N_c + 1, \ldots, \infty$, F_c will act as unwanted control spillover. The spillover effect will also apply to the measurements from the sensors, leading to observation spillover (see Sect. 3.3 for further discussion on both control and observation spillover).

Consider an example when taking measurements from two sensors (like the schematic representation in Fig. 3.20). The transverse displacement of the beam is $w(x, t)$, where x is the length along the beam. So the transverse displacement at two points A and B is $w(A, t)$ and $w(B, t)$ respectively. These physical displacements are taken as the control outputs for the system, $y_A = w(A, t)$ and $y_B = w(B, t)$. The outputs are related to the modal displacements q_1 and q_2 by a modal matrix, as discussed in Sect. 1.3.3 of Chap. 1 (and also derived in Chaps. 5 and 6), so that

$$\begin{bmatrix} y_A(t) \\ y_B(t) \end{bmatrix} = \begin{bmatrix} w(A, t) \\ w(B, t) \end{bmatrix} = \begin{bmatrix} \phi_1(A) \; \phi_2(A) \\ \phi_1(B) \; \phi_2(B) \end{bmatrix}\begin{bmatrix} q_1 \\ q_2 \end{bmatrix},$$

[21] For a detailed discussion of how this applies to linear systems see, for example, Gawronski (2000).

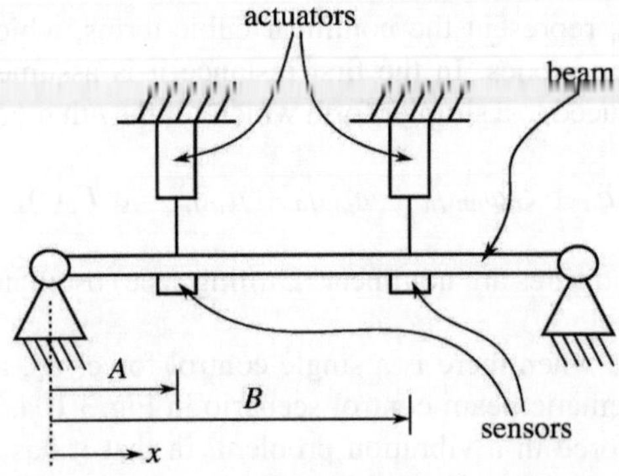

Fig. 3.20 Vibration control of a beam with two collocated actuators and sensors

where $\phi(x)$ is the beam mode-shape at point x along the beam. So, assuming a negligible contribution to the response from modes $3, 4, \ldots, \infty$ (i.e. no observation spillover) the output vector $\mathbf{y} = [\Phi]\mathbf{q}$, where $[\Phi]$ is the 2×2 modal matrix, and $\mathbf{q} = [q_1, q_2]^T$. As a result the modal displacement vector can be estimated directly from $\mathbf{q} = [\Phi]^{-1}\mathbf{y}$. If it is possible to place the sensors so that $\phi_2(A) \approx 0$ and $\phi_1(B) \approx 0$, then a direct relationship can be obtained between the outputs and the modal displacements.[22] The modal velocities, $\dot{q}_1$ and $\dot{q}_2$ also need to be estimated from the outputs, using the fact that $\dot{\mathbf{y}} = [\Phi]\dot{\mathbf{q}}$. The velocity of the output measurements can be estimated using a variety of numerical techniques, for example, the Savitsky-Golay filter—see Press et al. (1994) for more details.

Now consider what happens for a beam with two collocated actuators and sensors as shown in Fig. 3.20. Adding a second actuator to the beam results in an equation of the form

$$\frac{\mathrm{d}}{\mathrm{d}t}\begin{bmatrix} q_1 \\ q_2 \\ \dot{q}_1 \\ \dot{q}_2 \end{bmatrix} = \begin{bmatrix} 0 & 0 & 1 & 0 \\ 0 & 0 & 0 & 1 \\ -\omega_{n1}^2 & 0 & -\zeta_1\omega_{n1} & 0 \\ 0 & -\omega_{n2}^2 & 0 & -\zeta_2\omega_{n2} \end{bmatrix}\begin{bmatrix} q_1 \\ q_2 \\ \dot{q}_1 \\ \dot{q}_2 \end{bmatrix} - \begin{bmatrix} 0 \\ 0 \\ \mu_1 q_1^3 \\ \mu_2 q_2^3 \end{bmatrix} + \begin{bmatrix} 0 \\ 0 \\ \alpha_1 \\ \alpha_2 \end{bmatrix} F_{c1}$$
$$+ \begin{bmatrix} 0 \\ 0 \\ \beta_1 \\ \beta_2 \end{bmatrix} F_{c2},$$

where β_1 and β_2 are the modal participation factors for the second actuator. If the actuator locations can be chosen such that $\alpha_2 = 0$ and $\beta_1 = 0$, then the two system modes can be controlled independently (i.e. control spillover between the two modes

[22] However, the inverse, $[\Phi]^{-1}$ must remain defined.

is eliminated, but will still affect the uncontrolled modes).[23] In this case, taking $F_{c1} = p_1 u_1(t)$ and $F_{c2} = p_2 u_2(t)$, where p_1 and p_2 are constants and u_1 and u_2 are the control signals, then

$$\frac{\mathrm{d}}{\mathrm{d}t}\begin{bmatrix} q_1 \\ q_2 \\ \dot{q}_1 \\ \dot{q}_2 \end{bmatrix} = \begin{bmatrix} 0 & 0 & 1 & 0 \\ 0 & 0 & 0 & 1 \\ -\omega_{n1}^2 & 0 & -\zeta_1\omega_{n1} & 0 \\ 0 & -\omega_{n2}^2 & 0 & -\zeta_2\omega_{n2} \end{bmatrix}\begin{bmatrix} q_1 \\ q_2 \\ \dot{q}_1 \\ \dot{q}_2 \end{bmatrix} - \begin{bmatrix} 0 \\ 0 \\ \mu_1 q_1^3 \\ \mu_2 q_2^3 \end{bmatrix} + \begin{bmatrix} 0 \\ 0 \\ \alpha_1 p_1 \\ 0 \end{bmatrix} u_1 + \begin{bmatrix} 0 \\ 0 \\ 0 \\ \beta_2 p_2 \end{bmatrix} u_2. \tag{3.38}$$

This system can be controlled using the linearisation techniques developed in Sect. 3.5 as demonstrated by the next example.

Example 3.10 Feedback linearisation control for a decoupled two-mode nonlinear system

Problem Use feedback linearisation control techniques to linearize the two-mode nonlinear system defined by Eq. (3.38), where $u_1(t)$ and $u_2(t)$ are the control inputs. Assume that both observation and control spillover are negligible and that the outputs are the modal displacements $y_1 = q_1$ and $y_2 = q_2$.

Solution First, by inspection of Eq. (3.38), it can be seen that setting

$$u_1 = \frac{1}{\alpha_1 p_1}(\mu_1 q_1^3) \quad \text{and} \quad u_2 = \frac{1}{\beta_2 p_2}(\mu_2 q_2^3)$$

will linearize each mode directly. In fact, additional damping can also be included by using velocity feedback (assuming $\dot{q}_1$ and $\dot{q}_2$ can be measured). In this case the control signals become

$$u_1 = \frac{1}{\alpha_1 p_1}(\mu_1 q_1^3 - \kappa_1 \dot{q}_1) \quad \text{and} \quad u_2 = \frac{1}{\beta_2 p_2}(\mu_2 q_2^3 - \kappa_2 \dot{q}_2).$$

To obtain an input-output linearisation, first take the outputs $y_1 = q_1$ and $y_2 = q_2$, differentiate twice to get the relationship with the control inputs given by

$$\ddot{q}_1 = -\omega_{n1}^2 q_1 - \zeta_1\omega_{n1}\dot{q}_1 - \mu_1 q_1^3 + \alpha_1 p_1 u_1,$$
$$\ddot{q}_2 = -\omega_{n2}^2 q_2 - \zeta_2\omega_{n2}\dot{q}_2 - \mu_2 q_2^3 + \beta_2 p_2 u_2,$$

---

[23] This is often difficult (but not impossible) to do in practice. It becomes harder as the number of modes increases.

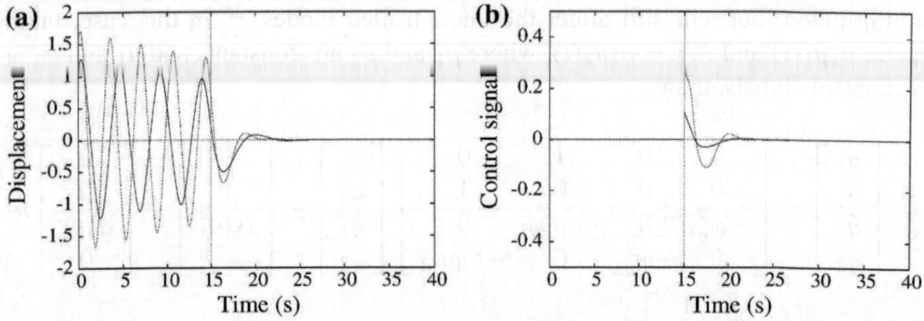

Fig. 3.21 Feedback linearisation for Example 3.10; **a** displacement response, **b** control signal, where *solid* and *dotted lines* refer to the modal displacements and modal control inputs of the first and second modes respectively

then choosing

$$u_1 = \frac{1}{\alpha_1 p_1}(v_1(t) + \omega_{n1}^2 q_1 + \zeta_1 \omega_{n1} \dot{q}_1 + \mu_1 q_1^3),$$

$$u_2 = \frac{1}{\beta_2 p_2}(v_2(t) + \omega_{n2}^2 q_2 + \zeta_2 \omega_{n2} \dot{q}_2 + \mu_2 q_2^3),$$

will give an input-output linearisation with the result that $\ddot{q}_1 = v_1(t)$ and $\ddot{q}_2 = v_2(t)$, where $v_1(t)$ and $v_2(t)$ are the new control signals, which can be chosen to give the desired linear system response. ■

A numerical simulation of Example 3.10 is shown in Fig. 3.21, with numerical parameters $\omega_1^2 = 1$, $\zeta_1 = 0.05$, $\omega_2^2 = 1.5$, $\zeta_2 = 0.04$, $\mu_1 = \mu_2 = 1$, $\alpha_1 p_1 = 10$ and $\beta_2 p_2 = 5$. The system is uncontrolled until time $t = 15$ s, when the input-output linearisation control is switched on. In this example $V_i(t) = -q_i - \dot{q}_i$ for $i = 1, 2$, which are stable modal linear oscillators with large damping. In Fig. 3.21a the displacement response is shown. A clear change can be seen from the large amplitude response occurring before 15 s to a reduced response afterwards. In Fig. 3.21b the control signal is shown. It can be seen that the control rapidly damps out the vibrations.

In fact using the same assumptions, feedback linearisation can be applied to the case when coupling exists between vibration modes. An example is shown next.

Example 3.11 Feedback linearisation control for a coupled two-mode nonlinear system

Problem Use feedback linearisation control techniques to linearize the two-mode nonlinear system defined by

$$\frac{\mathrm{d}}{\mathrm{d}t}\begin{bmatrix} q_1 \\ q_2 \\ \dot{q}_1 \\ \dot{q}_2 \end{bmatrix} = \begin{bmatrix} 0 & 0 & 1 & 0 \\ 0 & 0 & 0 & 1 \\ -\omega_{n1}^2 & 0 & -\zeta_1\omega_{n1} & 0 \\ 0 & -\omega_{n2}^2 & 0 & -\zeta_2\omega_{n2} \end{bmatrix} \begin{bmatrix} q_1 \\ q_2 \\ \dot{q}_1 \\ \dot{q}_2 \end{bmatrix} - \begin{bmatrix} 0 \\ 0 \\ \mu_1 q_1^3 + \delta_1 q_1^2 q_2 \\ \mu_2 q_2^3 + \delta_2 q_2^2 q_1 \end{bmatrix} + \begin{bmatrix} 0 \\ 0 \\ \alpha_1 p_1 \\ 0 \end{bmatrix} u_1 + \begin{bmatrix} 0 \\ 0 \\ 0 \\ \beta_2 p_2 \end{bmatrix} u_2, \tag{3.39}$$

where δ_1 and δ_2 are constant terms which determine the level of nonlinear cross-coupling between modes 1 and 2. Assume that both observation and control spillover are negligible and that the outputs are the modal displacements $y_1 = q_1$ and $y_2 = q_2$.

Solution First, by inspection of Eq. (3.39), it can be seen that setting

$$u_1 = \frac{1}{\alpha_1 p_1}(\mu_1 q_1^3 + \delta_1 q_1^2 q_2) \quad \text{and} \quad u_2 = \frac{1}{\beta_2 p_2}(\mu_2 q_2^3 + \delta_2 q_2^2 q_1)$$

will linearize each mode directly. In fact, additional damping can also be included by using velocity feedback (assuming $\dot{q}_1$ and $\dot{q}_2$ can be measured), in which case the control signals become

$$u_1 = \frac{1}{\alpha_1 p_1}(\mu_1 q_1^3 + \delta_1 q_1^2 q_2 - \kappa_1 \dot{q}_1) \quad \text{and} \quad u_2 = \frac{1}{\beta_2 p_2}(\mu_2 q_2^3 + \delta_2 q_2^2 q_1 - \kappa_2 \dot{q}_2).$$

To obtain an input-output linearisation, first take the outputs $y_1 = q_1$ and $y_2 = q_2$, differentiate twice to get the relationship with the control inputs given by

$$\begin{aligned} \ddot{q}_1 &= -\omega_{n1}^2 q_1 - \zeta_1\omega_{n1}\dot{q}_1 - \mu_1 q_1^3 - \delta_1 q_1^2 q_2 + \alpha_1 p_1 u_1, \\ \ddot{q}_2 &= -\omega_{n2}^2 q_2 - \zeta_2\omega_{n2}\dot{q}_2 - \mu_2 q_2^3 - \delta_2 q_2^2 q_1 + \beta_2 p_2 u_2, \end{aligned}$$

then choosing

$$\begin{aligned} u_1 &= \frac{1}{\alpha_1 p_1}(v_1(t) + \omega_{n1}^2 q_1 + \zeta_1\omega_{n1}\dot{q}_1 + \mu_1 q_1^3 + \delta_1 q_1^2 q_2), \\ u_2 &= \frac{1}{\beta_2 p_2}(v_2(t) + \omega_{n2}^2 q_2 + \zeta_2\omega_{n2}\dot{q}_2 + \mu_2 q_2^3 + \delta_2 q_2^2 q_1), \end{aligned}$$

will give an input-output linearisation with the result that $\ddot{q}_1 = v_1(t)$ and $\ddot{q}_2 = v_2(t)$, where $v_1(t)$ and $v_2(t)$ are the new control signals, which can be chosen to give the desired linear system response. ■

Feedback linearisation techniques can be applied to multi-modal systems when the modes are decoupled (or very weakly coupled) via the control forces. The modes themselves *can* be coupled, as has been shown in Example 3.11, although, to apply this control technique, detailed knowledge of the modal equations is required, and access to all modal states needs to be assumed. Note also that the nonlinear cross-coupling terms typically give rise to nonlinear resonance phenomena, which can dominate the vibration response. This will be discussed in more detail in Chap. 5. The effects of control and observer spillover are discussed further in Chap. 6, Sect. 6.3.1.

3.7 Adaptive Control

Vibration control when system parameters vary usually requires some form of adaption. The subject of adaptive control is an extensive one, and most linear and nonlinear adaptive control techniques have not been developed with vibration control in mind—the interested reader will find a good introduction to the subject in Åström and Wittenmark (1995), and a discussion of adaptive control for nonlinear systems in Slotine and Li (1991). The basic idea is that there is some form of uncertainty in the system, which means that system parameters change over time. Usually, the change is restricted to be slow over time. In this type of situation, a control design based on fixed parameter values will become increasing inaccurate as time increases. Allowing the control signal to adapt, based on some measurements of the changing parameters in the system, is one way of solving this problem. It should be noted that adaptive control can introduce additional problems, particularly those such as gain drift and lack of robustness.

To demonstrate how adaptive control techniques can be used to control nonlinear systems, the feedback linearisation approach discussed in Sect. 3.5 is extended to have an adaptive capability.

3.7.1 Adaptive Feedback Linearisation

It was shown in Sect. 3.5 that direct feedback linearisation can be applied to systems which can be expressed in the form of Eq. (3.29), giving

$$\dot{\boldsymbol{x}} = A\boldsymbol{x} + B(\mathscr{N}^*(\boldsymbol{x}) + u),$$

and setting $u = -\mathscr{N}^*$ gives the feedback linearisation control signal for the system in the case when A is a stable matrix. If the parameters in $\mathscr{N}^*$ are time varying or otherwise difficult to identify, an adaptive control law can be devised which allows variation over time, such that $\mathscr{N}^* + u = 0$ for all time, despite the uncertainty in $\mathscr{N}^*$. To do this, first redefine both $\mathscr{N}^*$ and u as

$$\mathscr{N}^* = \xi^T a^*, \quad u = -\xi^T a(t),$$

where ξ is a vector of nonlinear state terms, such as x_1^2, x_1x_2, etc. a^* is a vector of parameters and $a(t)$ is a vector of time-varying control gains. Substituting these expressions into the governing equation gives

$$\dot{\mathbf{x}} = A\mathbf{x} + B(\xi^T a^* - \xi^T a(t)) = B\xi^T \phi, \tag{3.40}$$

where $\phi = a^* - a(t)$ is the *parameter error*, meaning the difference between adaptive gain a_i and the uncertain parameter a_i^* which it is trying to match. Ideally $a \rightarrow a*$ and so $\phi \rightarrow 0$, which linearizes the system.

But how can a control law for the adaptive control gains a be defined? Typically, this is done by using a form of Lyapunov function analysis extended for multi-input, multi-output systems. Consider the following Lyapunov function for the system defined by Eq. 3.40

$$V = \mathbf{x}^T P\mathbf{x} + \phi^T \Gamma^{-1}\phi, \tag{3.41}$$

where P and Γ are to be defined as part of the control design. Differentiating with respect to time t gives the rate of change of V as

$$\dot{V} = \dot{\mathbf{x}}^T P\mathbf{x} + \mathbf{x}^T P\dot{\mathbf{x}} + \dot{\phi}^T \Gamma^{-1}\phi + \phi^T \Gamma^{-1}\dot{\phi}. \tag{3.42}$$

The expressions for $\dot{\mathbf{x}}$ and $\dot{\mathbf{x}}^T$ can be substituted from Eq. 3.40 to give

$$\dot{V} = (\mathbf{x}^T A^T + \phi^T \xi B^T)P\mathbf{x} + \mathbf{x}^T P(A\mathbf{x} + B\xi^T \phi) + \dot{\phi}^T \Gamma^{-1}\phi + \phi^T \Gamma^{-1}\dot{\phi}. \tag{3.43}$$

Expanding the brackets and gathering terms relating to the matrix A, gives

$$\dot{V} = \mathbf{x}^T (PA + A^T P)\mathbf{x} + \phi^T \xi B^T P\mathbf{x} + \mathbf{x}^T PB\xi^T \phi + \dot{\phi}^T \Gamma^{-1}\phi + \phi^T \Gamma^{-1}\dot{\phi}.$$

The matrix sum $(PA + A^T P)$ can be made negative definite by the appropriate choice of P and is usually written as $(PA + A^T P) = -Q$ so that

$$\dot{V} = -\mathbf{x}^T Q\mathbf{x} + \phi^T \xi B^T P\mathbf{x} + \mathbf{x}^T PB\xi^T \phi + \dot{\phi}^T \Gamma^{-1}\phi + \phi^T \Gamma^{-1}\dot{\phi}.$$

Now choosing $\dot{\phi} = -\Gamma \xi B^T P\mathbf{x}$ results in

$$\dot{V} = -\mathbf{x}^T Q\mathbf{x}, \tag{3.44}$$

which is negative definite with the appropriate choice of P. Note that this result also assumes that $P^T = P$ and $\Gamma^T = \Gamma$.

This shows what choice of $\dot{\phi}$ leads to a stable system, but the adaptive control law for $a(t)$ still needs to be defined. Integrating the expression for $\dot{\phi}$ gives

$$\phi = -\int_0^t \Gamma \xi B^T P\mathbf{x}\mathrm{d}\mathbf{x} = a^* - a(t),$$

so the time-varying adaptive gain is given by

$$a(t) = a^* + \int_0^t \Gamma \xi B^T P x dx. \tag{3.45}$$

In other words, the adaptive gain is the initial value a^* (or an initial estimation) plus the variation due to the changes in the parameters. The variation is an integral gain expression which involves the states, via ξ and x which are assumed to be accessible. The matrix Γ can be selected as part of the control design, as a control gain (sometimes also called adaptive weightings) matrix, in which the amount of adaptive effort can be selected by the control designer.

An example which demonstrates this approach is considered next.

Example 3.12 Adaptive feedback linearisation for an oscillator with nonlinear damping

Problem Use adaptive feedback linearisation to linearize the following nonlinear oscillator

$$m\ddot{x} + c\dot{x}(1 + \delta x) + kx + \mu x_1^3 = bu(t),$$

where $u(t)$ is the control input and both δ and μ are uncertain parameters. Assume that the mass, $m = 1$ kg, stiffness, $k = 1\,\text{N/m}^2$ and damping, $c = 0.1$ Ns/m. The control gain has the value $b = 10$. Both δ and μ have some uncertainty and initial estimated values can be assumed to be $\delta = 3$ and $\mu = 7$.

Solution First, write the governing equations of motion for the oscillator in the form $\dot{x} = A(x, t) + B(\mathcal{N}(x) + u)$, to give

$$\begin{bmatrix} \dot{x}_1 \\ \dot{x}_2 \end{bmatrix} = \begin{bmatrix} 0 & 1 \\ -\frac{k}{m} & -\frac{c}{m} \end{bmatrix} \begin{bmatrix} x_1 \\ x_2 \end{bmatrix} + \begin{bmatrix} 0 \\ \frac{b}{m} \end{bmatrix} \left(-\frac{c\delta}{b} x_1 x_2 - \frac{\mu}{b} x_1^3 + u(t) \right),$$

where $\mathcal{N}^* = -\frac{c\delta}{b} x_1 x_2 - \frac{\mu}{b} x_1^3$. The nonlinear state terms in $\mathcal{N}^*$ are $x_1 x_2$ and x_1^3, so form a vector $\xi = [x_1 x_2, x_1^3]^T$ so that $\mathcal{N}^*$ can be rewritten. To do this, first redefine both $\mathcal{N}^*$ and u as

$$\mathcal{N}^* = \xi^T a^* = [x_1 x_2,\ x_1^3] \begin{bmatrix} -\frac{c\delta}{b} \\ -\frac{\mu}{b} \end{bmatrix},$$

whilst defining $u = -\xi^T a(t)$ gives an equation in the form of Eq. (3.40)

$$\begin{bmatrix} \dot{x}_1 \\ \dot{x}_2 \end{bmatrix} = \begin{bmatrix} 0 & 1 \\ -\frac{k}{m} & -\frac{c}{m} \end{bmatrix} \begin{bmatrix} x_1 \\ x_2 \end{bmatrix} + \begin{bmatrix} 0 \\ \frac{b}{m} \end{bmatrix} [x_1 x_2,\ x_1^3] \begin{bmatrix} \phi_1 \\ \phi_2 \end{bmatrix}.$$

The A matrix is stable, since $m, c, k > 0$, and the Lyapunov stability analysis can be satisfied when matrices P and Q can be defined such that $PA + A^T P = -Q$ with

P and Q positive definite. In this example this leads to the relationship

$$\begin{bmatrix} p_{11} \, p_{12} \\ p_{21} \, p_{22} \end{bmatrix} \begin{bmatrix} 0 & 1 \\ -\frac{k}{m} & -\frac{c}{m} \end{bmatrix} + \begin{bmatrix} 0 & -\frac{k}{m} \\ 1 & -\frac{c}{m} \end{bmatrix} \begin{bmatrix} p_{11} \, p_{12} \\ p_{21} \, p_{22} \end{bmatrix} = - \begin{bmatrix} q_{11} \, q_{12} \\ q_{21} \, q_{22} \end{bmatrix},$$

which, by assuming that $p_{12} = p_{21}$ (a requirement of the Lyapunov analysis), can be written as

$$\begin{bmatrix} 2p_{21}\dfrac{k}{m} & -p_{11} + p_{21}\dfrac{c}{m} + p_{22}\dfrac{k}{m} \\ -p_{11} + p_{21}\dfrac{c}{m} + p_{22}\dfrac{k}{m} & -2p_{21} + 2p_{22}\dfrac{c}{m} \end{bmatrix} = \begin{bmatrix} q_{11} \, q_{12} \\ q_{21} \, q_{22} \end{bmatrix}.$$

For Q to be positive definite the conditions are that $q_{11} > 0$ and $q_{11}q_{22} - q_{21}q_{12} > 0$. Furthermore terms will be positive if $p_{11} < \left(p_{21}\frac{c}{m} + p_{22}\frac{k}{m}\right)$ and $p_{21} < p_{22}\frac{c}{m}$. For P to be positive definite, the conditions are that $p_{11} > 0$ and $p_{11}p_{22} - p_{21}p_{12} > 0$.

So selecting $p_{21} = 0.025$ gives $q_{11} = 0.05$, then choosing $p_{22} = 0.5$ gives $q_{22} = 0.05$. Finally, selecting $p_{11} = 0.5$ gives $q_{21} = q_{12} = 0.0025$, and the Lyapunov criteria, Eq. (3.44) is satisfied.

Now the adaptive control law can be found from Eq. (3.45). This gives

$$\begin{bmatrix} a_1 \\ a_2 \end{bmatrix} = \begin{bmatrix} a_1^* \\ a_2^* \end{bmatrix} + \int_0^t \begin{bmatrix} \gamma_{11} & 0 \\ 0 & \gamma_{22} \end{bmatrix} \begin{bmatrix} x_1 x_2 \\ x_1^3 \end{bmatrix} \begin{bmatrix} 0, & \frac{b}{m} \end{bmatrix} \begin{bmatrix} 0.5 & 0.025 \\ 0.025 & 0.5 \end{bmatrix} \begin{bmatrix} x_1 \\ x_2 \end{bmatrix} \mathrm{d}\boldsymbol{x}, \tag{3.46}$$

where the control gain matrix has been taken as diagonal, which satisfies the condition that $\Gamma^T = \Gamma$. This expression defines the adaptive gains, and the weightings γ_{11} and γ_{22} can be chosen by the control designer to give the required adaptive effort. ■

A numerical simulation of Example 3.12 is shown in Fig. 3.22, with numerical parameters $m = k = 1$, $c = 0.1$, $\delta = 3$, $\mu = 7$ and $b = 10$. The system is uncontrolled until time $t = 15$ s, when the input-output linearisation control is switched on. The uncertain parameters have 5 % error in the initial value plus a 30 % sinusoidal fluctuation. The adaptive weightings are selected as $\gamma_{11} = 0.05$ and $\gamma_{22} = 0.2$. This leads to a rapid adaption when the control is switched on at 15 s. As the parameter error is reduced, the adaptive gains become steady state and oscillatory to compensate for the fluctuating parameters. The system is also linearized after $t = 15$.

Note that the adaptive gain values are not unique, and they depend to some extent on the arbitrary choices of the P and Q matrices. More sophisticated adaptive control techniques (which are beyond the scope of this chapter) can be used, so that the gains can be utilized as part of a system identification process. Useful discussions on this are given in Slotine and Li (1991), Khalil (1992) and Fradkov et al. (1999).

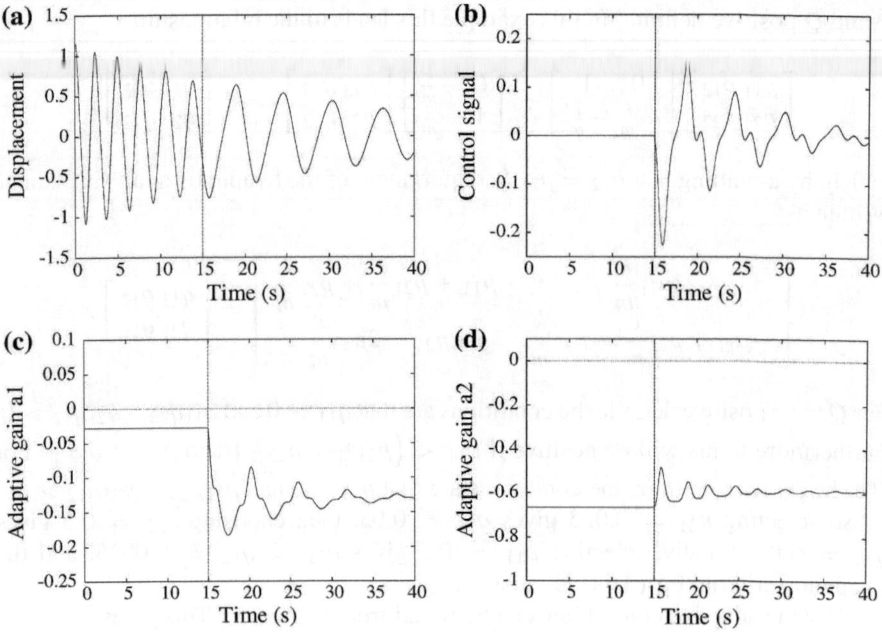

Fig. 3.22 Adaptive feedback linearisation for Example 3.12, with control starting at time $t = 15$ s; **a** shows the displacement response **b** shows the control signal, **c** and **d** show the adaptive gains

3.8 Chapter Notes

This chapter introduces the basic ideas of nonlinear control as applied to vibration engineering applications. The topic of nonlinear control is generally not discussed in terms of vibrating flexible bodies, but good introductions to the subject are given by Khalil (1992) and Slotine and Li (1991). More mathematical treatments of nonlinear control topics can be found in Sastry (1999), Isidori (1995), Fradkov et al. (1999) and Krstić et al. (1995). Semi-active control methods are discussed by Casciati et al. (2006) and Preumont and Seto (2008). A good discussion on velocity feedback and other linear techniques is given by Preumont (1997). The single-degree-of-freedom example was adapted from Malik (2009). The first reference to control and observation spillover appears to be that of Balas (1978). Controllability and observability concepts for nonlinear systems are discussed by Nijmeijer and van der Schaft (1990), Sastry (1999) and Slotine and Li (1991). The material on Lyapunov stability theory is widely available, with most of the texts mentioned above having an in-depth discussion. Example 3.6 is based on a similar example discussed in more depth by Khalil (1992). Input-output linearisation is commonly used in rigid body applications such as robotics, but can also be applied to multi-modal vibration problems where spillover is minimized by careful placement of sensors and actuators. A discussion of vibration-control techniques for linear systems, including spillover and modal participation, is given by Inman (2006). A good introduction to adaptive control techniques is given by Åström and Wittenmark (1995), and a discussion on adaptive control for nonlinear systems in Slotine and Li (1991). An application of adaptive feedback linearisation is given by Wagg (2002). Information on robustness of adaptive control is discussed by Yang (2008).

Problems

3.1 Draw the control block diagram and analyse the stability of the closed loop transfer function of the system given by Eq. (3.19), using a similar approach to that of Example 1.4 in Chap. 1. What effect do b and κ have on the stability of the system?

3.2 For the mass-spring-damper system shown in Fig. 3.12, design a control law which uses feedback terms proportional to the acceleration or displacement instead of the velocity term derived in Example 3.4. What effect will this have on the resonance characteristics of the mass-spring-damper system?

3.3 Construct a Lyapunov function to assess the stability of the controlled Duffing equation, given by Eq. (3.21), using the potential energy plus the kinetic energy of the system. Assume the case when only a single equilibrium point at the origin exists in the system and F_e is zero.

3.4 Choose a Lyapunov function to assess the stability of the equilibrium point at the origin for the system given by

$$\ddot{x} + \delta\dot{x} - x + \alpha x^3 = 0,$$

where δ and α are both positive constants. Can the energy equation associated with this oscillator be used as a Lyapunov function?

3.5 Design a feedback linearisation controller for the following Duffing oscillator. Note that the underlying linear system has an unstable equilibrium point at the origin, and the control objective is to stabilize this equilibrium point.

$$\begin{bmatrix} \dot{x}_1 \\ \dot{x}_2 \end{bmatrix} = \begin{bmatrix} 0 & 1 \\ \frac{k_1}{m} & -\frac{c}{m} \end{bmatrix} \begin{bmatrix} x_1 \\ x_2 \end{bmatrix} + \begin{bmatrix} 0 \\ \frac{p}{m}u(t) - \frac{k_3}{m}x_1^3 \end{bmatrix}.$$

3.6 Use feedback linearisation to remove the nonlinear damping terms in the following nonlinear oscillator

$$m\ddot{x} + c\dot{x}(1 + \delta x^2) + kx = pu(t),$$

where $u(t)$ is the control input. Is it possible to linearize the system and add more linear viscous damping simultaneously?

3.7 For the controlled Duffing oscillator given by

$$\dot{x}_1 = x_2,$$
$$\dot{x}_2 = -\frac{c}{m}x_2 - \frac{k_1}{m}x_1 - \frac{k_3}{m}x_1^3 + \frac{p}{m}u(t),$$

with an output $y = x_2$, use input-output linearisation to design a control input u. How does this compare with the case when $y = x_1$?

3.8 For the oscillator

$$m\ddot{x} + c\dot{x}x + kx = pu(t),$$

with an output $y = x_1 + x_2$, use input-output linearisation to design a control input u. How does this compare with the case when (i) $y = x_1$ and (ii) $y = x_2$?

3.9 Use feedback linearisation control techniques to linearize the two-mode nonlinear system defined by

$$\frac{\mathrm{d}}{\mathrm{d}t}\begin{bmatrix} q_1 \\ q_2 \\ \dot{q}_1 \\ \dot{q}_2 \end{bmatrix} = \begin{bmatrix} 0 & 0 & 1 & 0 \\ 0 & 0 & 0 & 1 \\ -\omega_{n1}^2 & 0 & -\zeta_1\omega_{n1} & 0 \\ 0 & -\omega_{n2}^2 & 0 & -\zeta_2\omega_{n2} \end{bmatrix}\begin{bmatrix} q_1 \\ q_2 \\ \dot{q}_1 \\ \dot{q}_2 \end{bmatrix} - \begin{bmatrix} 0 \\ 0 \\ \mu_1 q_1^2 + \delta_1 q_1 q_2 \\ \mu_2 q_2^2 + \delta_2 q_2 q_1 \end{bmatrix}$$
$$+ \begin{bmatrix} 0 \\ 0 \\ \alpha_1 p_1 \\ 0 \end{bmatrix} u_1 + \begin{bmatrix} 0 \\ 0 \\ 0 \\ \beta_2 p_2 \end{bmatrix} u_2,$$

where δ_1 and δ_2 are constant terms which determine the level of nonlinear cross-coupling between modes 1 and 2. Assume that both observation and control spillover are negligible and that the outputs are the modal displacements $y_1 = q_1$ and $y_2 = q_2$.

3.10 The vibration of a nonlinear beam can be approximated by the summation given in Eq. (3.37). If modes 2 and 3 need to be controlled, write down the equations of motion for the controlled modes in the system. Assuming the modes can be treated as effectively decoupled, suggest a feedback linearisation control scheme which would linearize the system.

3.11 Use adaptive feedback linearisation to linearize the following nonlinear oscillator

$$m\ddot{x} + c\dot{x}(1 + \delta x^2) + kx + \mu x_1^3 = bu(t),$$

where $u(t)$ is the control input and both δ and μ are uncertain parameters. Assume that the mass, $m = 2\,\mathrm{kg}$, stiffness, $k = 2\,\mathrm{N/m^2}$ and damping, $c = 0.2\,\mathrm{Ns/m}$. The control gain has the value $b = 1$. Both δ and μ have some uncertainty and initial estimated values can be assumed to be $\delta = 1$ and $\mu = 2$.

References

Alexander, N. A., & Schilder, F. (2009). Exploring the performance of a nonlinear tuned mass damper. *Journal of Sound and Vibration, 319*(1–2), 445–462.

Åström, K. J., & Wittenmark, B. (1995). *Adaptive control*. Boston: Addison Wesley.

Balas, M. J. (1978). Feedback control of flexible systems. *IEEE Transactions on Automatic Control, 23*(4), 673–679.

Casciati, F., Magonette, G., & Marazzi, F. (2006). *Semiactive devices and applications in vibration mitigation*. Chichester: Wiley.

Den Hartog, J. P. (1934). *Mechanical vibrations*. New York: McGraw-Hill.

Fradkov, A. L., Miroshnik, I. M., & Nikiforov, V. O. (1999). *Nonlinear and adaptive control of complex systems*. Dordrecht: Kluwer.

Gattulli, V., Di Fabio, F., & Luongo, A. (2004). Nonlinear tuned mass damper for self-excited oscillations. *Wind and Structures, 7*(4), 251–264.

Gawronski, W. (2000). Modal actuators and sensors. *Journal of Sound and Vibration, 229*(4), 1013–1022.

Gawronski, W. K. (2004). *Advanced structural dynamics and active control of structures*. New York: Springer.

Ibrahim, R. A. (2008). Recent advances in nonlinear passive vibration isolators. *Journal of Sound & Vibration, 314*(3–5), 371–452.

Inman, D. J. (2006). *Vibration with control*. New York: Wiley.

Isidori, A. (1995). *Nonlinear control systems*. New York: Springer.

Khalil, H. K. (1992). *Nonlinear systems*. New York: Macmillan.

Krstić, M., Kanellakopoulos, I., & Kokotović, P. (1995). *Nonlinear and adaptive control design*. New York: Wiley.

Lazar, I. F., Neild, S. A., & Wagg, D. J. (2014). Using an inerter-based device for structural vibration suppression. *Earthquake Engineering & Structural Dynamics, 43*(8), 1129–1147.

Londoño, J. M., Neild, S. A., & Wagg, D. J. (2013). A noniterative design procedure for supplemental brace-damper systems in single-degree-of-freedom systems. *Earthquake Engineering & Structural Dynamics, 42*(15), 2361–2367.

Malik, N. S. (2009). Adaptive vibration control of flexible structures using piezo-electric actuators. Ph.D. thesis, University of Bristol.

Nijmeijer, H., & van der Schaft, A. (1990). *Nonlinear dynamical control systems*. New York: Springer.

Press, W. H., Teukolsky, S. A., Vettering, W. T., & Flannery, B. P. (1994). *Numerical recipes in C* (2nd ed.). Cambridge: Cambridge University Press.

Preumont, A. (1997). *Vibration control of active structures*. Dordrecht: Kluwer Academic Publishers.

Preumont, A., & Seto, K. (2008). *Active control of structures*. Dordrecht: WileyBlackwell.

Priya, S., & Inman, D. J. (Ed.). (2009). *Energy harvesting technologies*. New York: Springer.

Sastry, S. (1999). *Nonlinear systems: Analysis, stability and control*. New York: Springer.

Shaw, A., Neild, S., & Wagg, D. (2012). Dynamic analysis of high static low dynamic stiffness vibration isolation mounts. *Journal of Sound and Vibration, 332*, 1437–1455.

Shaw, S. W. (2012). Designing nonlinear torsional vibration absorbers. In D. Wagg & L. Virgin (Eds.), *Exploiting nonlinear behaviour in structural dynamics* (pp. 134–168). New York: Springer.

Slotine, J.-J. E., & Li, W. (1991). *Applied nonlinear control*. Englewood Cliffs: Prentice Hall.

Smith, M. C., & Wang, F.-C. (2004). Performance benefits in passive vehicle suspensions employing inerters. *Vehicle System Dynamics, 42*(4), 235–257.

Soong, T. T., & Dargush, G. F. (1997). *Passive energy dissipation systems in structural engineering*. Chichester: Wiley.

Vidyasagar, M. (1992). *Nonlinear systems analysis*. Englewood Cliffs: Prentice Hall.

Wagg, D. J. (2002). Partial synchronization of non-identical chaotic systems via adaptive control, with applications to modelling coupled nonlinear systems. *International Journal of Bifurcation and Chaos*, *12*(3), 561–570.

Yang, L. (2008). Robustness compensator for adaptive control of mechanical systems. Ph.D. thesis, University of Bristol.

Chapter 4
Approximate Methods for Analysing Nonlinear Vibrations

Abstract Most nonlinear systems have no easily obtainable exact analytical solution. As a result, one has to use either graphical methods (via numerical solution) or approximate analytical methods. In this chapter a range of the most common approximation methods will be discussed. In particular, the effect of nonlinearity on resonance peaks is often analysed using approximate methods. For example, vibrating systems where the natural frequency changes as the amplitude of oscillation increases will be examined in detail. The discussion starts with the methods of harmonic balance and averaging. Then perturbation approximation methods are considered, the most important of which is the multiple scales method. Finally the method of normal forms is discussed when applied to vibration problems.

4.1 Backbone Curves

In Chap. 1, Sect. 1.3.1, the response of a linear second-order oscillator was derived assuming the response to be sinusoidal (i.e. harmonic). It was shown that for the linear oscillator

$$m\ddot{x} + c\dot{x} + kx = F_0 \sin(\Omega t),$$

the steady-state vibration response is given by $x = x_0 \sin(\Omega t - \phi)$ where the response amplitude function is

$$x_0 = \left(\frac{F_0}{k}\right) \frac{1}{\sqrt{(1-(\frac{\Omega}{\omega_n})^2)^2 + 4\zeta^2(\frac{\Omega}{\omega_n})^2}}, \tag{4.1}$$

(which is from Eq. (1.12) in Chap. 1) and the phase difference between the forcing sine wave and the response sine wave is given by

$$\phi = \arctan\left(\frac{2\zeta\frac{\Omega}{\omega_n}}{(1-(\frac{\Omega}{\omega_n})^2)}\right), \tag{4.2}$$

D. Wagg and S. Neild, *Nonlinear Vibration with Control*,
Solid Mechanics and Its Applications 218, DOI 10.1007/978-3-319-10644-1_4

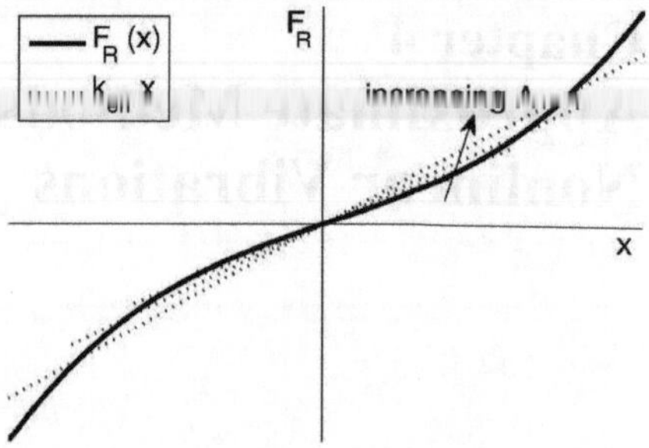

Fig. 4.1 Restoring force-displacement relationship for the Duffing oscillator along with the effective linearized stiffnesses, k_{eff}, for a range of oscillation amplitudes

(from Eq. (1.13) in Chap. 1), $\omega_n = \sqrt{k/m}$ and $\zeta = c/2m\omega_n$. The resonance behaviour of the linear oscillator is captured by Eqs. (4.1) and (4.2), which can be used to plot the resonance peak and corresponding phase lag, as demonstrated in Example 1.1, Fig. 1.10.

For lightly damped systems, resonance peaks are essential to modelling the vibration response because they represent significant amplifications of the input forcing signal. As a result, it is important to have a model for the frequency–amplitude behaviour of the system. For the linear oscillator this is given explicitly by Eq. (4.1) and is plotted in Fig. 1.10a. For nonlinear systems, resonant behaviour is much more complex and the natural frequency is often a function of oscillation amplitude. This can be seen by considering the unforced, undamped Duffing oscillator

$$m\ddot{x} + kx + \alpha x^3 = 0 \rightsquigarrow m\ddot{x} + F_R(x) = 0,$$

where $F_R(x)$ is the nonlinear restoring force (or the force due to the stiffness terms). One method of analysing this system is to linearize the restoring force $F_R(x) = kx + \alpha x^3$ by approximating it to a linear spring force $k_{eff}x$, where k_{eff} is the linearized stiffness. The natural frequency of the resulting linearized system is $\sqrt{k_{eff}/m}$. However, the effective stiffness is dependent on the amplitude of oscillation A. In addition, calculating k_{eff} to give the correct frequency is difficult as the system spends a proportionally longer time in lower stiffness regions than in higher stiffness regions. This can be seen graphically in Fig. 4.1, which shows the cubic restoring force and the corresponding approximate restoring force based on the effective stiffness for a range of oscillation amplitudes. Expressions governing the behaviour of the resonance peak for nonlinear systems normally need to be approximated. This can be done by numerical simulation, but additional insight can usually be obtained by considering approximate analysis.

For nonlinear systems which have relatively small nonlinear effects (often called weakly nonlinear systems[1]) the behaviour of a resonance curve can be measured relative to the purely linear case. An example resonance curve for a nonlinear system

[1] It is also assumed in this discussion that the nonlinear effects are smooth in nature.

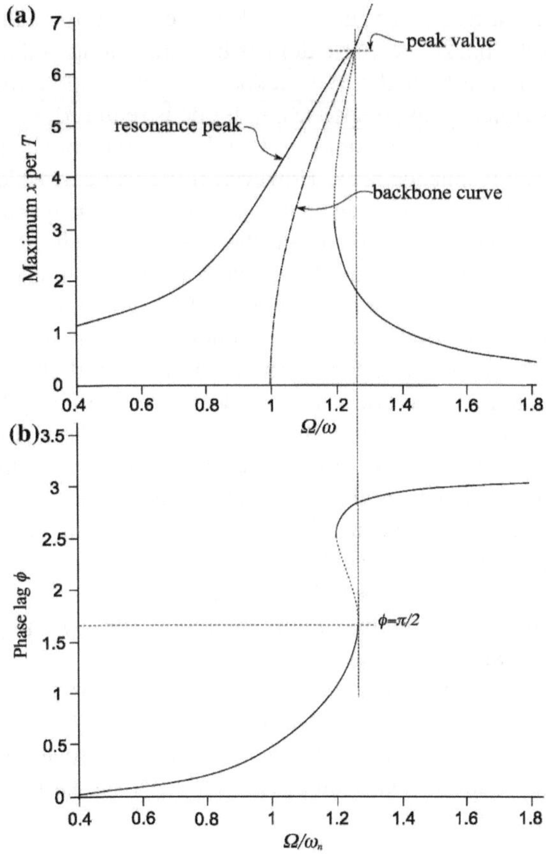

Fig. 4.2 Frequency response for a nonlinear oscillator showing **a** the backbone curve for the unforced, undamped system and the resonance peak for the forced and damped system in which the maximum displacement over a period of forcing, T, is plotted, and **b** the corresponding phase lag for the forced and damped system

is shown in Fig. 4.2a. One important tool in the understanding of nonlinear resonance curves is the *backbone curve*. The backbone curve defines the natural frequency as a function of the amplitude of response of the system when no damping or forcing is present. For a linear oscillator the backbone curve is a vertical line at $\Omega = \omega_n$, where $\omega_n = \sqrt{k/m}$ is the natural frequency, in the frequency–amplitude diagram. Taking this as a benchmark, finding an approximate backbone curve for a nonlinear oscillator gives an indication of the distortion from linear resonance. In addition, as the backbone curve represents the system without damping and forcing, calculating an approximate curve is considerably simpler than calculating the resonance curve for the full system. As the systems being considered are lightly damped (as well as weakly nonlinear), the backbone curve is normally a reasonable approximation to the system response. However, in common with linear resonance peaks, the width and extent of the nonlinear resonance curves are dependent mainly on the damping in the system. Note that the backbone curve defines how the undamped natural frequency increases with amplitude. So to define a backbone curve, an approximate relationship between natural frequency and response amplitude is required.

In addition to this, it is clear from Fig. 4.2a that the backbone curve alone is not enough to define the resonance peak. However, if the value at which the resonance curve crosses the backbone curve can be found, this can be taken as an approximation to the peak value of amplitude. To locate the point at which the resonance curve crosses the backbone curve, consider first what happens in a linear resonance peak. In this case, as the resonance peak passes the backbone curve, there is a rapid change in phase lag from values close to zero up to values close to π, with all cases passing through $\phi = \pi/2$, as shown in Fig. 1.10b. A similar rapid shift in phase occurs in nonlinear oscillators of the type shown in Fig. 4.2a, the phase lag being shown in Fig. 4.2b. The solid lines in Fig. 4.2b correspond to stable solution branches, so that there are jumps in the phase close to resonance, corresponding to the jumps in amplitude of the hysteresis region. Despite the jumps, an approximation to the phase at resonance as $\phi = \pi/2$ can often be used to approximate the point at which the resonance curve crosses the backbone curve[2] as shown for example in Fig. 4.2. Notice also that, for a hardening peak (leaning to the right), it is increasing Ω which gives a phase jump close to the maximum amplitude. For softening, decreasing Ω would give the equivalent response.

So to define a backbone curve, two things are required: (i) a frequency–amplitude relationship, and (ii) an estimate of the peak value. Next, a technique is considered that can be used to estimate the frequency–amplitude relationship. At the end of this section the backbone curve and peak response, shown in Fig. 4.2a, will be reconsidered.

4.2 Harmonic Balance

In Chap. 1, Sect. 1.3.2 the response of a nonlinear system, based on the approximation that the response was a single-frequency sine wave, was discussed. The example considered initially was the damped, forced Duffing oscillator

$$\ddot{x} + 2\zeta\omega_n\dot{x} + \omega_n^2 x + \alpha x^3 = F\sin(\Omega t) \tag{4.3}$$

from Eq. (1.17), where $F = F_0/m$ and F_0 is the forcing amplitude. To solve this equation it was assumed that the response was in the form $x = X_r\sin(\Omega t - \phi)$. Two key observations can be made. The first is that the assumption of a sine wave solution generates harmonic terms in the response (given by Eq. (1.19) in Chap. 1). The second is that the response frequency, ω_r, is a function of the response amplitude,

[2] See Eq. (4.10) for an approximate expression which can be used to compute the phase at resonance for the forced, damped Duffing oscillator.

X_r, and is given by the approximate relationship

$$\omega_r \approx \omega_n \sqrt{1 + \frac{3\alpha X_r^2}{4\omega_n^2}},$$

which is from Eq. (1.22) in Chap. 1. Note that $\omega_r = \omega_n$ when $X_r = 0$.

In fact, the process of assuming a harmonic (sine and cosine waves) solution is the first part of a technique known as *harmonic balance*. After the assumed solution is substituted into the governing equation, in this case Eq. (4.3), the second part of the process is to *balance* the coefficients of the harmonic terms.

Harmonic balancing is a technique that allows the calculation of the approximate steady-state system response. For a linear oscillator, an exact balance can be obtained, with the resulting expressions given by Eqs. (4.1) and (4.2). Unfortunately, finding a solution to the system response for nonlinear systems is more difficult. This is because nonlinear systems can have a response which is composed of multiple frequencies, as was demonstrated in Example 2.6, Fig. 2.15, in Chap. 2. So, in order to fully balance all the harmonic terms, the trial solution for x must be extended to include a summation of all the relevant harmonics and subharmonics. In general, the trial solution for a nonlinear system with a primary response at frequency ω_r would take the form

$$x = a_0 + \sum_{n=1}^{\infty} a_n \cos(n\omega_r t) + b_n \sin(n\omega_r t) \quad n = 1, 2, 3, \ldots,$$

where a_n are coefficients. For weakly nonlinear systems, the first harmonic frequency in this expression, ω_r, corresponds to the forcing frequency, Ω, if the system is forced, or to a natural frequency of an unforced system.

In summary, the harmonic balance approach is first to apply a trial solution with a reduced number of terms (often limited to just terms at frequency ω_r) and then to balance just the terms present in the trial solution whilst ignoring any higher frequency terms generated by the nonlinearity. This will be demonstrated in the following example, in which an approximation to the backbone curve is found for the Duffing oscillator.

Example 4.1 Harmonic balance and the backbone curve for the undamped Duffing oscillator

Problem Use the harmonic balance technique to find the nonlinear frequency–amplitude relationship needed to describe the backbone curve for the following undamped, unforced Duffing oscillator

$$\ddot{x} + \omega_n^2 x + \alpha x^3 = 0,$$

where ω_n is the natural frequency if no nonlinearity is present.

Solution The simplest trial solution is $x \approx X_r \cos(\omega_r t)$, where ω_r represents the nonlinear natural frequency which is amplitude dependent.[3] Making the substitution results in the equation

$$(\omega_n^2 - \omega_r^2)X_r \cos(\omega_r t) + \frac{\alpha}{4}X_r^3[3\cos(\omega_r t) + \cos(3\omega_r t)] \approx 0.$$

Applying the harmonic balance to the $\cos(\omega_r t)$ terms (and ignoring the $\cos(3\omega_r t)$ term) gives the amplitude frequency relationship

$$\omega_r^2 \approx \omega_n^2 + \frac{3\alpha}{4}X_r^2. \tag{4.4}$$

This is a first approximation to the backbone curve relating nonlinear natural frequency ω_r to the amplitude of oscillation X_r.

The number of terms considered in the harmonic balance can be extended to two, which results in the trial solution $x = X_r(\cos(\omega_r t) + X_{3r}\cos(3\omega_r t))$, where X_{3r} is the amplitude of the third harmonic relative to the fundamental amplitude.[4] This trial solution results in the equation

$$\begin{aligned} &X_r\left[(\omega_n^2 - \omega_r^2)c_1 + (\omega_n^2 - 9\omega_r^2)X_{3r}c_3\right] \\ &+ \frac{\alpha}{4}X_r^3[(3c_1 + c_3) + 3X_{3r}(c_1 + 2c_3 + c_5) + 3X_{3r}^2(2c_1 + c_5 + c_7) \\ &+ X_{3r}^3(3c_3 + c_9)] \approx 0, \end{aligned}$$

where the shorthand $c_k = \cos(k\omega_r t)$ is used. Applying the harmonic balance to the $\cos(\omega_r t)$ and $\cos(3\omega_r t)$ terms gives the two equations

$$(\omega_n^2 - \omega_r^2) + \frac{\alpha}{4}X_r^2(3 + 3X_{3r} + 6X_{3r}^2) \approx 0, \tag{4.5}$$

$$X_{3r}(\omega_n^2 - 9\omega_r^2) + \frac{\alpha}{4}X_r^2(1 + 6X_{3r} + 3X_{3r}^3) \approx 0. \tag{4.6}$$

Assuming that X_{3r} is small, such that X_{3r}^2 and X_{3r}^3 terms can be ignored, from Eq. (4.5) the backbone curve may be written as

$$\omega_r^2 \approx \omega_n^2 + \frac{3\alpha}{4}X_r^2(1 + X_{3r}), \tag{4.7}$$

[3] When there is no forcing function it makes no difference whether cos or sin is taken as the assumed solution.

[4] The $3\omega_r$ harmonic is added rather than the $2\omega_r$, because the cubic term gives harmonics at odd multiples of the fundamental frequency. See, Sect. 1.3.2 in Chap. 1.

where X_{3r} must be found. This can be done by firstly combining Eqs. (4.5) and (4.6) to eliminate X_{3r}, giving

$$144\omega_r^4 - (169\omega_n^2 + 132X_r^2\alpha)\omega_r^2 + (16\omega_n^4 + 36X_r^2\alpha\omega_n^2 + 15X_r^4\alpha^2) \approx 0.$$

Substituting in the backbone expression, Eq. (4.7), gives

$$(81\alpha X_r^2)X_{3r}^2 + (63\alpha X_r^2 + 96\omega_n^2)X_{3r} - (3\alpha X_r^2) \approx 0,$$

which has the small solution (using a Taylor series expansion on the expression for the roots)

$$X_{3r} = \frac{\alpha X_r^2}{21\alpha X_r^2 + 32\omega_n^2} \quad \rightsquigarrow \quad \omega_r^2 \approx \omega_n^2 + \frac{3\alpha}{4}X_r^2\left(1 + \frac{\alpha X_r^2}{21\alpha X_r^2 + 32\omega_n^2}\right).$$

So, by adding the third harmonic into the harmonic balance, an estimate of the amplitude of the response of the third harmonic has been calculated. ■

Now reconsider the resonance curve discussed at the start of this chapter, Fig. 4.2a. The previous example has generated approximate backbone curve (or frequency–amplitude relationship) expressions for the Duffing oscillator, Eq. (4.4), based on including only the first harmonic term in the harmonic balance. As already observed, this curve alone does not give enough information to understand the resonance behaviour of the full (damped and forced) system, as the peak amplitude of response is dominated by the damping. The response amplitude of the full system can be crudely estimated by assuming two things. The first is that the response is dominated by a response at the forcing frequency—this is similar to the assumption made when using the harmonic balance including just the first harmonic term. The second is that, as with linear systems, the resonance peak occurs when the phase lag of the response relative to the forcing is $\pi/2$. For the Duffing example, the forced equation can be written as

$$\ddot{x} + 2\zeta\omega_n\dot{x} + \omega_n^2 x + \alpha x^3 = F\sin(\Omega t),$$

where the response frequency is assumed to be equal to the forcing frequency $\omega_r = \Omega$ and F is the forcing amplitude divided by the mass. Using the assumption, that the response at resonance has exactly $\pi/2$ phase lag compared to the forcing, then the response can be written as $x = -X_r\cos(\omega_r t)$. Now the energy input from the forcing can be balanced with the energy dissipation through the damper over a cycle of oscillation

$$\int_{t-\pi/\omega_r}^{t+\pi/\omega_r} F\sin(\omega_r t)\dot{x}\,\mathrm{d}t = \int_{t-\pi/\omega_r}^{t+\pi/\omega_r} (c\dot{x})\dot{x}\,\mathrm{d}t,$$

to give $X_r = F/(2\zeta\omega_n\omega_r)$.

Alternatively, an approximation to the full response curve can be calculated using the harmonic balance technique, which will be discussed next.

4.2.1 Forced Vibration

For harmonically forced systems, with forcing $F_e = F_0 \sin(\Omega t)$, the simplest assumed response would be a single harmonic at $\omega_r = \Omega$ with a phase shift such that $x \approx X_r \sin(\omega_r t - \phi)$. A method which can be used to simplify the analysis is to shift the time origin by writing $\tau = t - \phi/\omega_r$, so that $F_e = F_0 \sin(\omega_r \tau + \phi)$ and $x = X_r \sin(\omega_r \tau)$. This time-shift allows the response to be purely sinusoidal, at the expense of introducing a cosine component to the forcing, which is useful when nonlinearities include terms such as αx^3. Now a forced vibration example will be considered.

Example 4.2 Harmonic balance for the forced Duffing oscillator

Problem Use the harmonic balance technique to find an approximate transfer function for the forced Duffing oscillator

$$\ddot{x} + 2\zeta\omega_n \dot{x} + \omega_n^2 x + \alpha x^3 = F \sin(\Omega t).$$

Solution Substituting $\omega_r = \Omega$, and $\tau = t - \phi/\omega_r$ gives

$$x'' + 2\zeta\omega_n x' + \omega_n^2 x + \alpha x^3 = F \sin(\omega_r \tau + \phi),$$

where $\bullet'$ is the derivative with respect to τ. Then taking $x \approx X_r \sin(\omega_r \tau)$ leads to

$$(\omega_n^2 - \omega_r^2) X_r \sin(\omega_r \tau) + 2\zeta\omega_n\omega_r X_r \cos(\omega_r \tau) + \frac{\alpha}{4} X_r^3 [3 \sin(\omega_r \tau) - \sin(3\omega_r \tau)]$$
$$\approx F \sin(\omega_r \tau + \phi),$$

Expanding the right-hand side using trigonometric identities and equating coefficients of $\sin \omega_r \tau$ and $\cos \omega_r \tau$ (and ignoring the $\cos(3\omega_r \tau)$ term) gives

$$\begin{aligned} (\omega_n^2 - \omega_r^2) X_r + \frac{3\alpha}{4} X_r^3 &\approx F \cos(\phi), \\ 2\zeta\omega_n\omega_r X_r &\approx F \sin(\phi). \end{aligned} \tag{4.8}$$

The phase, ϕ, can be eliminated by squaring and adding the expressions in Eq. (4.8), so that

$$X_r^2 [(\omega_n^2 - \omega_r^2 + \frac{3\alpha}{4} X_r^2)^2 + 4\zeta^2\omega_n^2\omega_r^2] \approx F^2,$$

which finally leads to

$$\frac{X_r}{F} \approx \frac{1}{[(\omega_n^2 - \omega_r^2 + \frac{3\alpha}{4}X_r^2)^2 + 4\zeta^2\omega_n^2\omega_r^2]^{\frac{1}{2}}}, \tag{4.9}$$

which is a form of amplitude-dependent transfer function.

The phase can be found from the ratio of the two expressions in Eq. (4.8), which gives

$$\phi = \arctan\left(\frac{2\zeta\omega_n\omega_r}{\omega_n^2 - \omega_r^2 + \frac{3\alpha}{4}X_r^2}\right). \tag{4.10}$$

Note that the phase is also dependent on the amplitude of response. ■

This approximate harmonic balance approach can be used to develop the concept of nonlinear frequency response functions. See for example Worden and Tomlinson (2000) for further discussion of Example 4.2 in this context.

For the resonance curve shown in Fig. 4.2, Eq. (4.9) is an approximation to the resonance curve. It is interesting to consider the point at which the resonance curve crosses the backbone curve, defined by Eq. (4.4). At this point, the resonance curve relationship reduces to $X_r \approx F/(2\zeta\omega_n\omega_r)$, which matches the resonance peak predicted using the energy balance.

In general, the harmonic balance method is considered to be inconsistent, as higher frequency terms are generally not balanced using a single-frequency substitution.[5] However, it is still a very useful technique for obtaining a quick and approximate idea of the underlying behaviour. Apart from its ease of application, it is also particularly useful because it can (in principle) be applied to systems with "large" nonlinear terms. As will be seen later in this chapter, most other approximate methods rely on the fact that the nonlinearity is "small", in order for the approximation to be valid. The precise definition of what constitutes a large or small nonlinearity, depends on the particular problem, and some examples are discussed later in this chapter. As far as the harmonic balance is concerned, the analysis often leads to unbalanced terms being neglected, and there is some implicit assumption that these terms are "small", in the sense that they are assumed to be insignificant. Because of these limitations, harmonic balance should ideally be verified by other analytical methods or numerical simulations. Note that in the control literature *describing functions* are sometimes used to capture the response of nonlinear plants, these are often derived using techniques similar to harmonic balance. Note also that the harmonic balance method is limited to calculating the steady-state response without providing information about the transient behaviour.

[5] An example of how the method can give inconsistent answers can be found in Nayfeh and Mook (1995).

4.3 Averaging

Averaging is a technique that can be applied to systems with dynamics of the following form

$$\ddot{x} + \omega_n^2 x = \varepsilon \mathcal{N}(x, \dot{x}), \tag{4.11}$$

which is sometimes referred to as the standard or normal form.

Here ε is used to indicate a small parameter, i.e. if an equation has the term εv the ε indicates that v is small compared to other terms in the equation. Hence ε may be thought of as a book-keeping aid—it allows the tracking of the significance of each term. Furthermore, ε can be set to unity when it is no longer necessary to track the significance of terms.

It can be seen that, for linear damping or forcing to be included in the system dynamics, they must be incorporated in $\mathcal{N}(x, \dot{x})$ and therefore must be small (i.e. lightly damped vibrating systems with $\zeta \ll 1$). In addition $\mathcal{N}(x, \dot{x})$ contains any nonlinear terms which therefore must also be small, i.e. the system is weakly nonlinear.

The averaging technique involves using a trial solution to Eq. (4.11) with the form

$$x = x_c(t)\cos(\omega_n t) + x_s(t)\sin(\omega_n t), \tag{4.12}$$

where $x_c(t)$ and $x_s(t)$ are unknown time functions. First-order differential equations are found for $x_c(t)$ and $x_s(t)$. These are then *averaged* over a cycle of oscillation to remove higher frequency oscillations and, as a result, find the underlying amplitude envelopes for $x_c(t)$ and $x_s(t)$. In contrast to the harmonic balance technique which provides information about just the steady-state response of a system, Averaging is capable of providing information on the transient behaviour of the system and, as a result, it can also be used to assess the stability of steady-state solutions.[6]

4.3.1 Free Vibration

Using the trial solutions for the displacement, Eq. (4.12), along with Eq. (4.11) the velocity can be written as

$$\dot{x} = -\omega_n x_c(t)\sin(\omega_n t) + \omega_n x_s(t)\cos(\omega_n t), \tag{4.13}$$

and the dynamics of $x_c(t)$ and $x_s(t)$ can be expressed as

$$\dot{x}_c = -\frac{\varepsilon}{\omega_n}\sin(\omega_n t)\mathcal{N}, \quad \dot{x}_s = \frac{\varepsilon}{\omega_n}\cos(\omega_n t)\mathcal{N}, \tag{4.14}$$

[6] Note that is also possible to use a trial solution with time-dependent amplitude and phase parameters: $x = X(t)\cos(\omega_n t + \phi(t))$, see for example Verhulst (1996). Discussion of this representation is beyond the scope of this book.

noting that now $\mathcal{N}(x,\dot{x}) = \mathcal{N}(x_c(t)\cos(\omega_n t)+x_s(t)\sin(\omega_n t), -\omega_n x_c(t)\sin(\omega_n t)+\omega_n x_s(t)\cos(\omega_n t))$. The derivation of Eq. (4.13) and the justification for writing the derivative of x in the form shown in Eq. (4.13) (i.e. without the $\dot{x}_c(t)\cos(\omega_n t) + \dot{x}_s(t)\sin(\omega_n t)$ terms) is shown in Box 4.1.

Note that while here the response is assumed to be at frequency ω_n, as indicated by the trial solution Eq. (4.12), it is possible to calculate the steady-state response frequency by introducing a small frequency detuning term, μ such that $\omega_r = \omega_n (1+\mu)$ along with a time-scaling $\tau = t(1+\mu)$. This approach is considered when the vibrations of a forced system are analysed in Sect. 4.3.2 and in the worked solution to Problem 4.3.

Although Eq. (4.14) are first-order differential equations, they are still potentially difficult to solve due to the sinusoidal term multiplied by $\mathcal{N}$ which also contains sinusoidal terms. To proceed, *averaging* is applied, the average value of $\dot{x}_c$ and $\dot{x}_s$ over a cycle of response is calculated from the integrals

$$\dot{x}_{ca} = \frac{\omega_n}{2\pi}\int_{t-\pi/\omega_n}^{t+\pi/\omega_n} \dot{x}_c \,\mathrm{d}t = -\frac{\omega_n}{2\pi}\int_{t-\pi/\omega_n}^{t+\pi/\omega_n} \frac{\varepsilon}{\omega_n}\sin(\omega_n t)\mathcal{N}\,\mathrm{d}t,$$

$$\dot{x}_{sa} = \frac{\omega_n}{2\pi}\int_{t-\pi/\omega_n}^{t+\pi/\omega_n} \dot{x}_s \,\mathrm{d}t = \frac{\omega_n}{2\pi}\int_{t-\pi/\omega_n}^{t+\pi/\omega_n} \frac{\varepsilon}{\omega_n}\cos(\omega_n t)\mathcal{N}\,\mathrm{d}t,$$

Box 4.1 *Derivation of Eq.* (4.14)
Using the trial solution, Eq. (4.12), the derivatives with respect to time can be written as

$$\begin{aligned}
x &= x_c(t)c_1(t) + x_s(t)s_1(t),\\
\dot{x} &= \dot{x}_c(t)c_1(t) + \dot{x}_s(t)s_1(t) - \omega_n x_c(t)s_1(t) + \omega_n x_s(t)c_1(t),\\
\ddot{x} &= \frac{\mathrm{d}}{\mathrm{d}t}\{\dot{x}_c(t)c_1(t) + \dot{x}_s(t)s_1(t)\} - \omega_n \dot{x}_c(t)s_1(t) + \omega_n \dot{x}_s(t)c_1(t)\\
&\quad - \omega_n^2(x_c(t)c_1(t) + x_s(t)s_1(t)),
\end{aligned}$$

where the shorthand $c_1(t) = \cos(\omega_n t)$ and $s_1(t) = \sin(\omega_n t)$ is used. Substituting these expressions into the equation for the system dynamics, Eq. (4.11), gives

$$\ddot{x} + \omega_n^2 x = \varepsilon\mathcal{N} \rightsquigarrow \frac{\mathrm{d}}{\mathrm{d}t}\{\dot{x}_c(t)c_1(t) + \dot{x}_s(t)s_1(t)\} - \omega_n \dot{x}_c(t)s_1(t) + \omega_n \dot{x}_s(t)c_1(t) = \varepsilon\mathcal{N}.$$

By inspection it can be seen that the differential term with respect to t, i.e. the first term on the left-hand side of the equation, is zero if $\dot{x}_s(t) = -\dot{x}_c(t)c_1(t)/s_1(t)$. Substituting this expression into the equation gives

$$-\omega_n \dot{x}_c(t)[s_1(t) + c_1^2(t)/s_1(t)] = \varepsilon \mathcal{N}.$$

and hence

$$\dot{x}_c = -\frac{\varepsilon}{\omega_n} \sin(\omega_n t) \mathcal{N},$$
$$\dot{x}_s = \frac{\varepsilon}{\omega_n} \cos(\omega_n t) \mathcal{N}.$$

In addition using these relationships it can be seen that the velocity equation can be simplified

$$\dot{x} = \dot{x}_c(t)c_1(t) + \dot{x}_s(t)s_1(t) - \omega_n x_c(t)s_1(t) + \omega_n x_s(t)c_1(t)$$
$$\rightsquigarrow \dot{x} = -\omega_n x_c(t)s_1(t) + \omega_n x_s(t)c_1(t).$$

where the subscript a is used to indicate that the variable has been averaged over a cycle. Now a significant simplification can be made. The function $\mathcal{N}(x_c(t)\cos(\omega_n t) + x_s(t)\sin(\omega_n t), -\omega_n x_c(t)\sin(\omega_n t) + \omega_n x_s(t)\cos(\omega_n t))$ contains various expressions containing $x_c(t)$ and $x_s(t)$ terms. These $x_c(t)$ and $x_s(t)$ terms are treated as constant over the cycle of oscillation. The justification for this approximation is that $x_c(t)$ and $x_s(t)$ are slowly varying, Eq. (4.14) shows that $\dot{x}_c(t)$ and $\dot{x}_s(t)$ are both of order ε and therefore considered small. The result of this approximation is that the integrals typically consist of sinusoidal terms which can be solved easily.

To understand how averaging works, consider the lightly-damped linear oscillator, with damping but no forcing

$$\ddot{x} + 2\varepsilon\zeta\omega_n\dot{x} + \omega_n^2 x = 0 \rightsquigarrow \ddot{x} + \omega_n^2 x = \varepsilon\mathcal{N}(x, \dot{x}),$$

where, to indicate that it is small, ε^1 has been added to the damping term and $\mathcal{N}(x, \dot{x}) = -2\zeta\omega_n\dot{x}$ is simply the linear viscous damping term. Making the substitutions

$$x = x_c(t)\cos(\omega_n t) + x_s(t)\sin(\omega_n t), \tag{4.15}$$
$$\dot{x} = -\omega_n x_c(t)\sin(\omega_n t) + \omega_n x_s(t)\cos(\omega_n t), \tag{4.16}$$

from Eqs. (4.12) and (4.13) into the expressions for $\dot{x}_c$ and $\dot{x}_s$, Eq. (4.14), gives

$$\dot{x}_c = -\frac{\varepsilon}{\omega_n}\sin(\omega_n t)\mathcal{N}$$
$$\rightsquigarrow \dot{x}_c = 2\varepsilon\zeta\omega_n \sin(\omega_n t)(-x_c(t)\sin(\omega_n t) + x_s(t)\cos(\omega_n t))$$

and

$$\dot{x}_s = \frac{\varepsilon}{\omega_n} \cos(\omega_n t)\mathcal{N}$$

$$\rightsquigarrow \dot{x}_s = -2\varepsilon\zeta\omega_n \cos(\omega_n t)(-x_c(t)\sin(\omega_n t) + x_s(t)\cos(\omega_n t))$$

respectively.

These equations can now be averaged over a cycle of oscillation at frequency ω_n

$$\dot{x}_{ca} = 2\varepsilon\zeta\omega_n \frac{\omega_n}{2\pi} \int_{t-\pi/\omega_n}^{t+\pi/\omega_n} (-x_c(t)s_1(t)^2 + x_s(t)c_1(t)s_1(t))\,\mathrm{d}t,$$

$$\dot{x}_{sa} = -2\varepsilon\zeta\omega_n \frac{\omega_n}{2\pi} \int_{t-\pi/\omega_n}^{t+\pi/\omega_n} (-x_c(t)c_1(t)s_1(t) + x_s(t)c_1(t)^2)\,\mathrm{d}t,$$

where the shorthand $c_1(t) = \cos(\omega_n t)$ and $s_1(t) = \sin(\omega_n t)$ is used. Approximating $x_c(t)$ and $x_s(t)$ to be constant over the period of the integral (taking the values $x_{ca}(t)$ and $x_{sa}(t)$ respectively), which allows them to be taken outside the integral, gives

$$\dot{x}_{ca} = 2\varepsilon\zeta\omega_n(-x_{ca}(t)\langle s_1^2\rangle + x_{sa}(t)\langle s_1 c_1\rangle) = -\varepsilon\zeta\omega_n x_{ca}(t), \tag{4.17}$$

$$\dot{x}_{sa} = -2\varepsilon\zeta\omega_n(-x_{ca}(t)\langle s_1 c_1\rangle + x_{sa}(t)\langle c_1^2\rangle) = -\varepsilon\zeta\omega_n x_{sa}(t), \tag{4.18}$$

where the subscript a indicates an averaged variable and the following notation is used

$$\langle\bullet\rangle = \frac{\omega_n}{2\pi} \int_{t-\pi/\omega_n}^{t+\pi/\omega_n} \bullet\,\mathrm{d}t.$$

Equations (4.17) and (4.18) can now be solved to give

$$\dot{x}_{ca} = -\varepsilon\zeta\omega_n x_{ca}(t) \rightsquigarrow x_{ca}(t) = x_{ca}(0)\mathrm{e}^{-\varepsilon\zeta\omega_n t}, \tag{4.19}$$

$$\dot{x}_{sa} = -\varepsilon\zeta\omega_n x_{sa}(t) \rightsquigarrow x_{sa}(t) = x_{sa}(0)\mathrm{e}^{-\varepsilon\zeta\omega_n t}, \tag{4.20}$$

where, in deriving these expressions, the following have been used

$$\langle s_1^2\rangle = \frac{1}{2}, \quad \langle s_1 c_1\rangle = 0, \quad \langle c_1^2\rangle = \frac{1}{2}.$$

Substituting these solutions into Eq. (4.15), assuming that x_{ca} and x_{sa} capture the key dynamics of x_c and x_s respectively and so the approximations $x_c \approx x_{ca}$ and

$x_s \approx x_{sa}$ can be made, to give

$$x = \mathrm{e}^{-\varepsilon\zeta\omega_n t}[x_c(0)\cos(\omega_n t) + x_s(0)\sin(\omega_n t)].$$

If the initial conditions are that the system is displaced a distance x_0 and is at rest, the system response is

$$x = x_0 \mathrm{e}^{-\varepsilon\zeta\omega_n t}\cos(\omega_n t),$$

which can be compared to the exact solution using linear analysis of the system (assuming light damping), namely

$$x = x_0 \mathrm{e}^{-\varepsilon\zeta\omega_n t}\cos(\omega_n\sqrt{1-\varepsilon^2\zeta^2}t).$$

For small ε it can be seen that the solution using averaging is very accurate.

Note that the expressions for x_{ca} and x_{sa}, given by Eqs. (4.19) and (4.20) respectively, define the amplitude envelope of the cosine and sine components of the response. This is shown in Fig. 4.3 for the case that has just been considered.

4.3.2 Forced Vibration

Before considering a nonlinear example, it is important to discuss how to deal with forcing. Forcing is often omitted from examples of averaging, but for vibration analysis, understanding how forcing affects the system is crucial. One of the

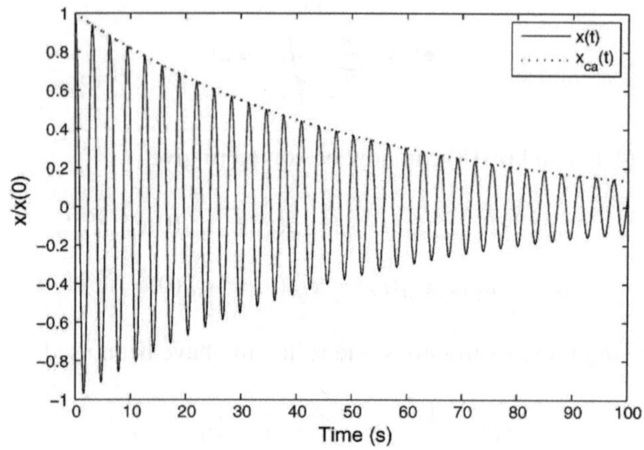

Fig. 4.3 Response of an unforced linear oscillator starting with an initial displacement of x_0 and zero velocity

key aspects of the averaging technique is that the frequency of the trial solution, $x = x_c(t)\cos(\omega_n t) + x_s(t)\sin(\omega_n t)$, must match the frequency of the left-hand side of the equation of motion, $\ddot{x} + \omega_n^2 x = \varepsilon\mathscr{N}$. So when forcing is present in $\mathscr{N}$, the frequency of the trial solution must match the forcing frequency, Ω. This is possible if the forcing frequency corresponds to the linear natural frequency $\Omega = \omega_n$. However, the case $\Omega \neq \omega_n$ must also be considered.

First, note that the forcing must be included within $\varepsilon\mathscr{N}$ to maintain the form of the equation of motion, $\ddot{x} + \omega_n^2 x = \varepsilon\mathscr{N}$. Due to the ε term, the excitation F (which represents the applied force divided by the mass) must be small in comparison with the response, and this usually occurs only if the excitation is *near a resonance*. Close to resonance, the forcing frequency Ω can be written as $\Omega = \omega_n(1 + \varepsilon\mu)$ where μ is a frequency detuning parameter and is small and so treated as of order ε^1. The excitation term, which is the force divided by the mass, is assumed to be in the form $F\cos(\Omega t) = \varepsilon F\cos(\omega_n(1 + \varepsilon\mu)t)$ and is also small, order ε as close to resonance the response is large compared with the excitation. To simplify the analysis, the right-hand side of the equation of motion can be written

$$\ddot{x} + \omega_n^2 x = \varepsilon\mathscr{N} \rightsquigarrow \ddot{x} + \omega_n^2 x = \varepsilon\bar{\mathscr{N}}(x, \dot{x}) + \varepsilon F\cos(\omega_n(1 + \varepsilon\mu)t), \quad (4.21)$$

where the forcing has been removed from $\mathscr{N}$, leaving $\bar{\mathscr{N}}$ which contains the nonlinear and damping terms.

Now a time-scaling is introduced such that the forcing has frequency ω_n in the new time-scale τ, i.e. $F\cos(\omega_n(1+\varepsilon\mu)t) = F\cos(\omega_n\tau)$ which means that $\tau = (1+\varepsilon\mu)t$. Before the time-scaling can be introduced into Eq. (4.21), the derivatives of x must be considered

$$\dot{x} = (1 + \varepsilon\mu)x', \quad \ddot{x} = (1 + \varepsilon\mu)^2 x'',$$

where $\bullet'$ is the derivative with respect to scaled time τ. Rewriting the equation of motion in scaled time gives

$$x'' + \omega_n^2 x = \varepsilon\bar{\mathscr{N}}(x, (1 + \varepsilon\mu)x') + \varepsilon F\cos(\omega_n\tau) - \varepsilon(2\mu + \varepsilon\mu^2)x''.$$

From inspection of this equation, it is noted that x'' can be written as $x'' = -\omega_n^2 x + \mathscr{O}(\varepsilon^1)$ such that $\varepsilon x'' = -\varepsilon\omega_n^2 x + \mathscr{O}(\varepsilon^2)$, where $\mathscr{O}(\varepsilon^2)$ denotes terms of order ε^2 or higher. Using this expression the $\varepsilon x''$ term on the right-hand side of the equation can be eliminated giving

$$x'' + \omega_n^2 x = \varepsilon\bar{\mathscr{N}}(x,\ (1 + \varepsilon\mu)x') + \varepsilon F\cos(\omega_n\tau) - (2\mu + \varepsilon\mu^2)(-\varepsilon\omega_n^2 x + \mathscr{O}(\varepsilon^2)).$$

Using a Taylor series to expand $\bar{\mathscr{N}}(x,\ (1 + \varepsilon\mu)x')$ and then ignoring ε^2 and higher terms gives

$$x'' + \omega_n^2 x = \varepsilon\left(\bar{\mathscr{N}}(x, x') + 2\mu\omega_n^2 x\right) + \varepsilon F\cos(\omega_n\tau).$$

This equation is in the standard form

$$x'' + \omega_n^2 x = \varepsilon \tilde{\mathcal{N}}, \quad \tilde{\mathcal{N}} = \mathcal{N}(x, x') + 2\mu\omega_n^2 x + F\cos(\omega_n \tau).$$

Now the trial solution

$$x = x_c(\tau)\cos(\omega_n \tau) + x_s(\tau)\sin(\omega_n \tau), \tag{4.22}$$

$$x' = -\omega_n x_c(\tau)\sin(\omega_n \tau) + \omega_n x_s(\tau)\cos(\omega_n \tau), \tag{4.23}$$

(the same as Eqs. (4.12) and (4.13) but now written in terms of scaled time τ) can be applied. Equations for x_c' and x_s' can be found using (4.14) (also expressed in terms of τ rather than t) to give

$$x_c' = -\frac{\varepsilon}{\omega_n}\sin(\omega_n \tau)\tilde{\mathcal{N}}, \quad x_s' = \frac{\varepsilon}{\omega_n}\cos(\omega_n \tau)\tilde{\mathcal{N}}. \tag{4.24}$$

Note that averaging is applied in the scaled-time τ over the region $\tau - \pi/\omega_n$ to $\tau + \pi/\omega_n$, i.e. over a scaled time $2\pi/\omega_n$, which is equivalent to averaging over time $2\pi/[\omega_n(1+\varepsilon\mu)] = 2\pi/\Omega$ corresponding to averaging over one cycle of forcing excitation.

Example 4.3 Forced vibration of a linear oscillator

Problem Calculate the system response as a function of excitation frequency for the lightly-damped linear oscillator with near resonant forcing

$$\ddot{x} + \omega_n^2 x = \varepsilon\mathcal{N}, \quad \mathcal{N} = -2\zeta\omega_n \dot{x} + F\cos(\Omega t),$$

where both the damping ratio and the forcing amplitude are small and hence labelled order ε^1.

Solution Making the substitution $\Omega = \omega_n(1+\varepsilon\mu)$ to reflect the fact that the forcing is close to resonance and then applying the time-scaling $\tau = (1+\varepsilon\mu)t$ gives

$$x'' + \omega_n^2 x = \varepsilon\tilde{\mathcal{N}}, \quad \tilde{\mathcal{N}} = -2\zeta\omega_n x' + 2\mu\omega_n^2 x + F\cos(\omega_n \tau),$$

where terms of order ε^2 and above have been ignored and where $\bullet'$ is the derivative with respect to scaled time τ. Substituting the trial solutions for x and x', Eqs. (4.22) and (4.23), into the expressions for x_c' and x_s', Eq. (4.24), gives

$$x_c' = -\frac{\varepsilon}{\omega_n}\sin(\omega_n \tau)\tilde{\mathcal{N}}$$

$$\rightsquigarrow x_c' = -\frac{\varepsilon}{\omega_n}s_1\left[-2\zeta\omega_n^2(-x_c(\tau)s_1 + x_s(\tau)c_1) + 2\mu\omega_n^2(x_c(\tau)c_1 + x_s(\tau)s_1) + Fc_1\right],$$

and

$$x_s' = \frac{\varepsilon}{\omega_n} \cos(\omega_n \tau)\tilde{\mathcal{N}}$$
$$\rightsquigarrow x_s' = \frac{\varepsilon}{\omega_n} c_1 \left[-2\zeta\omega_n^2(-x_c(\tau)s_1 + x_s(\tau)c_1) + 2\mu\omega_n^2(x_c(\tau)c_1 + x_s(\tau)s_1) + Fc_1 \right],$$

respectively, where $c_1(\tau) = \cos(\omega_n \tau)$ and $s_1(\tau) = \sin(\omega_n \tau)$. Applying averaging over the range $\tau - \pi/\omega_n$ to $\tau + \pi/\omega_n$, while making the assumption that $x_c(\tau)$ and $x_s(\tau)$ are constant over this time period (taking values $x_{ca}(\tau)$ and $x_{sa}(\tau)$ respectively, which can be taken outside the averaging integrals), gives

$$x_{ca}' = -\frac{\varepsilon}{\omega_n} \left[-2\zeta\omega_n^2 \left(-x_{ca}(\tau)\langle s_1^2 \rangle + x_{sa}(\tau)\langle c_1 s_1 \rangle \right) + 2\mu\omega_n^2 (x_{ca}(\tau)\langle c_1 s_1 \rangle \right.$$
$$\left. + x_{sa}(\tau)\langle s_1^2 \rangle) + F\langle c_1 s_1 \rangle \right],$$

and

$$x_{sa}' = \frac{\varepsilon}{\omega_n} \left[-2\zeta\omega_n^2(-x_{ca}(\tau)\langle c_1 s_1 \rangle + x_{sa}(\tau)\langle c_1^2 \rangle) + 2\mu\omega_n^2(x_{ca}(\tau)\langle c_1^2 \rangle \right.$$
$$\left. + x_{sa}(\tau)\langle c_1 s_1 \rangle) + F\langle c_1^2 \rangle \right],$$

where the following shorthand notation is used

$$\langle \bullet \rangle = \frac{\omega_n}{2\pi} \int_{\tau-\pi/\omega_n}^{\tau+\pi/\omega_n} \bullet \, d\tau,$$

as stated previously. Calculating the integrals simplifies the equations to give

$$x_{ca}' = -\frac{\varepsilon}{\omega_n}(\zeta\omega_n^2 x_{ca} + \mu\omega_n^2 x_{sa}),$$
$$x_{sa}' = \frac{\varepsilon}{\omega_n}\left(-\zeta\omega_n^2 x_{sa} + \mu\omega_n^2 x_{ca} + \frac{F}{2} \right).$$

Finally the steady-state amplitude response can be found by setting $x_{ca}' = x_{sa}' = 0$ to give

$$X_r = \frac{F}{2\omega_n^2\sqrt{\zeta^2 + \mu^2}},$$

where X_r is the response amplitude, $X_r^2 = x_{ca}^2 + x_{sa}^2$. This can be compared to the conventional linear analysis which predicts

$$X_r = \frac{F}{\sqrt{(\omega_n^2 - \Omega^2)^2 + (2\zeta\omega_n\Omega)^2}}.$$

Making the substitution $\Omega = \omega_n(1+\mu)$ gives

$$X_r = \frac{F}{2\omega_n^2\sqrt{\zeta^2 + \mu^2 + (\mu^2 + 2\zeta^2)\mu + (\mu^2/4 + \zeta^2)\mu^2}}.$$

This is the same as the averaged solution to order ε^1. This can be seen by recalling that both μ and ζ have been treated as of order ε^1 and hence $(\mu^2 + 2\zeta^2)\mu$ and $(\mu^2/4 + \zeta^2)\mu^2$ are of order ε^3 and ε^4 respectively and so can be dropped. ■

Additional information relating the stability of the steady-state response solutions can be found using this technique. This makes use of the fact that Eq. (4.24) can be written in the form

$$\begin{pmatrix} x_c \\ x_s \end{pmatrix}' = f\left(\begin{pmatrix} x_c \\ x_s \end{pmatrix}, \tau\right)$$

or $\mathbf{z}' = f(\mathbf{z}, \tau)$. The steady-state solutions to this equation, $\bar{\mathbf{z}}$, must satisfy $f(\bar{\mathbf{z}}, \tau) = 0$. In Example 4.3 these solutions are found for a linear system. For a stable solution, if a small perturbation is applied to the steady-state system, the perturbation will decay away with time and the system will return to the stable steady-state response. Consider a small perturbation $\mathbf{z}_p$ to the equilibrium solution $\bar{\mathbf{z}}$, such that the system is at point $\tilde{\mathbf{z}} = \bar{\mathbf{z}} + \mathbf{z}_p$. The system response may be written as

$$\tilde{\mathbf{z}}' = f(\tilde{\mathbf{z}}, \tau)$$

Substituting in $\tilde{\mathbf{z}} = \bar{\mathbf{z}} + \mathbf{z}_p$ and applying a Taylor series expansion to the nonlinear function, noting that the perturbation is small, gives

$$\bar{\mathbf{z}}' + {\mathbf{z}_p}' = f(\bar{\mathbf{z}}, \tau) + Df_z(\bar{\mathbf{z}}, \tau)\mathbf{z}_p,$$

where $Df_z(\mathbf{z}, \tau)$ is the Jacobian of $f(\mathbf{z}, \tau)$. Noting that, by definition, at a steady-state solution the system response equation can be written as $\bar{\mathbf{z}}' = f(\bar{\mathbf{z}}, \tau) = 0$, the equation can be simplified to

$${\mathbf{z}_p}' = Df_z(\bar{\mathbf{z}}, \tau)\mathbf{z}_p.$$

This equation represents the dynamics of the perturbation. For the perturbation to die away, indicating a stable steady-state solution, the real parts of the eigenvalues of $Df_z(\bar{\mathbf{z}}, \tau)$ must be negative.

The linear examples have shown that, using the averaging technique, both steady-state and transient information regarding the fundamental frequency response may be found. An example of the averaging technique applied to a nonlinear system, the dynamics of a cable, is considered in Sect. 7.4. The nonlinear cable example also includes a study of the stability of the solutions. Although the averaging method provides information regarding the amplitude of the response at the fundamental frequency, it is, however, difficult to extend the technique to include information regarding higher harmonics.

4.4 Perturbation Methods

The perturbation method[7] for analysing weakly nonlinear systems is based on the idea that the response can be made up of a power series in ε

$$x = x_0 + \varepsilon x_1 + \varepsilon^2 x_2 \ldots . \tag{4.25}$$

Often the first term in the series x_0 is the linear response (the response to the system if all nonlinear terms were removed). In these cases, the additional x_i terms can be thought of as perturbations away from the linear response, with reducing significance as i increases (reflected by the ε^i multiplier—a discussion on the use of ε is given at the start of Sect. 4.3).[8]

Two approaches will be discussed. The first is *regular perturbation theory*, in which Eq. (4.25) is substituted into the equation of motion. It will be seen that this technique can produce erroneous results (it can also work well). To overcome this difficulty a second approach, the *multiple scales* (or two-timing) technique will be discussed. This technique uses the observation that often the response consists of terms that change rapidly with time and others that change slowly.

4.4.1 Regular Perturbation Theory

Regular perturbation theory involves making the substitution given in Eq. (4.25) into the equation of motion. Consider the general unforced nonlinear system

$$\ddot{x} + 2\varepsilon\zeta\omega_n\dot{x} + \omega_n^2 x + \varepsilon\mathscr{N}(\dot{x}, x) = 0,$$

where the nonlinear term $\mathscr{N}(\dot{x}, x, t)$ and the damping term are assumed to be small and have been indicated as such by labelling them order ε^1. The full power series expansion for x can now be substituted into the equation and balanced for all powers

[7] Note that the term *perturbation* is also used in analysing the stability of equilibrium points. Perturbation analysis involves studying the effect of a small disturbance (or perturbation) applied to a system that was at an equilibrium position. See, for instance, the cable vibration example in Chap. 7.

[8] Note that this power series approximation should not be confused with a Taylor series expansion of x about an operating point.

of ε. However, for a technique to be useful, it is desirable to get a reasonably exact expression for the system response using a small number of terms in the power series expansion. Therefore, the power series expansion is approximated by the first two terms $x = x_0 + \varepsilon x_1$ and the ε^0 and ε^1 terms are balanced (while ignoring the higher order ε terms); this gives

$$\begin{aligned}\varepsilon^0&: \ddot{x}_0 + \omega_n^2 x_0 = 0,\\ \varepsilon^1&: \ddot{x}_1 + \omega_n^2 x_1 = -2\zeta\omega_n \dot{x}_0 - \mathcal{N}(\dot{x}_0, x_0),\end{aligned}$$

where a Taylor series expansion has been applied to the nonlinear term $\mathcal{N}(\dot{x}_0 + \varepsilon\dot{x}_1, x_0 + \varepsilon x_1)$. The first equation represents the linear undamped response (if the damping was assumed to be order ε^0 then it would represent the linear damped response). The second equation is also a linear differential equation for x_1, where the x_0 terms on the right-hand side act as the forcing terms. As x_0 represents the linear undamped response, x_1 can be thought of as the perturbation away from the linear undamped response.

The potential difficulty with this technique is that it can generate *secular* terms. These are terms that are unbounded with time, such as $t\sin(\omega_n t)$. This will be demonstrated in the following example.

Example 4.4 Vibration response of the damped Duffing oscillator

Problem Find the response of a lightly damped, but unforced, Duffing oscillator. Initially, the system is at rest with a displacement of $x(0)$.

Solution Consider the damped Duffing oscillator

$$\ddot{x} + 2\varepsilon\zeta\omega_n\dot{x} + \omega_n^2 x + \varepsilon\alpha x^3 = 0,$$

noting that in this case both the nonlinearity and the damping are assumed to be small and so labelled as order ε^1. Applying the substitution $x = x_0 + \varepsilon x_1$ and comparing the ε^0 and ε^1 terms (assuming the higher order ε terms are negligible) gives

$$\ddot{x}_0 + \omega_n^2 x_0 = 0, \tag{4.26}$$
$$\ddot{x}_1 + \omega_n^2 x_1 = -2\zeta\omega_n\dot{x}_0 - \alpha x_0^3. \tag{4.27}$$

These two linear equations can now be solved in turn. Firstly, the linearized response can be written as $x_l = x(0)\cos(\omega_n t)$, noting that the initial displacement $x(0) = x_0(0) + \varepsilon x_1(0)$ is valid for all ε such that $x_0(0) = x(0)$ and $x_1(0) = 0$. Equation (4.27) can now be solved. Substituting the solution of Eq. (4.26) for x_0 into Eq. (4.27) gives

$$\ddot{x}_1 + \omega_n^2 x_1 = 2\zeta\omega_n^2 x(0)\sin(\omega_n t) - \frac{\alpha}{4}x(0)^3(3\cos(\omega_n t) + \cos(3\omega_n t)).$$

On inspecting the right-hand side, the trial solution $x_1 = A\cos(\omega_n t) + B\sin(\omega_n t) + C\cos(3\omega_n t)$ might be selected. However, when this trial solution is substituted into the differential equation, the first two terms cancel out. Therefore, to ensure that the left-hand side has $\cos(\omega_n t)$ and $\sin(\omega_n t)$ terms to balance with those on the right-hand side, the substitution $x_1 = At\cos(\omega_n t) + Bt\sin(\omega_n t) + C\cos(3\omega_n t)$ must be made. This results in the full response

$$x = x(0)(1 - \zeta\omega_n t)\cos(\omega_n t) - \frac{3\alpha}{8\omega_n}x(0)^3 t\sin(\omega_n t) + \frac{\alpha}{32\omega_n^2}x(0)^3\cos(3\omega_n t),$$

where the ε labels have been removed by setting $\varepsilon = 1$.

Two interesting cases can now be examined. First, consider the case where there is no nonlinearity, $\alpha = 0$. The predicted response becomes

$$x = x(0)(1 - \zeta\omega_n t)\cos(\omega_n t),$$

which can be compared to the solution using standard linear analysis

$$x = x(0)\mathrm{e}^{-\zeta\omega_n t}\cos(\omega_n\sqrt{1-\zeta^2}t). \tag{4.28}$$

It can be seen that there is a slight error in the natural frequency, but, more significantly, it can be seen that the exponential decay term in the accurate solution has been replaced with the first two terms of its Taylor series expansion in the perturbation solution. This is acceptable at low values of $\zeta\omega_n t$ where the higher order terms are negligible. However, given that this term includes time, it suggests that, however small ζ is, at some time this approximation will break down. This is shown in Fig. 4.4a, where the perturbation solution is plotted against a time-stepping numerical simulation (using a Matlab-based integration method) for the case where $\omega_n = 4\pi,\ \zeta = 0.01$ and $x(0) = 0.01$.

Secondly, the undamped nonlinear response can be considered. The predicted system response is compared to a numerical simulation in Fig. 4.4b for the case where $\omega_n = 4\pi,\ \alpha = 2 \times 10^4$ and $x(0) = 0.01$. It can be seen that again the accuracy deteriorates with time, both in amplitude and (less clearly from the plot) in phase, due to inaccuracy in the estimation of the natural frequency. Note that, assuming the response is similar in magnitude to the linear response, the linear stiffness term has a maximum value $\omega^2 x = 1.58$ compared to the nonlinear term $\alpha x^3 = 0.02$. Therefore the nonlinear term is small and can be said to be order ε^1. ■

In Example 4.4, the response had secular terms in the form $t\cos(\omega_n t)$ and $t\sin(\omega_n t)$. For small t these accurately represent the system response. However, as time becomes larger, higher order terms are needed to counteract their effects. For the case where a linear damped system is considered, the response is $x = x(0)\mathrm{e}^{-\zeta\omega_n t}\cos(\omega_n\sqrt{1-\zeta^2}t)$. It can be seen that there are two time-scales occurring in this solution, a fast time-scale due to the oscillations at frequency $\omega_n\sqrt{1-\zeta^2} \approx \omega_n$ and a slow time-scale due to the exponential decay $\mathrm{e}^{-\zeta\omega_n t}$. Using the regular perturbation technique, the slow time-scale is poorly represented, the exponential decay $\mathrm{e}^{-\zeta\omega_n t}$ being represented as $1 - \zeta\omega_n t$, the first two terms of the Taylor series

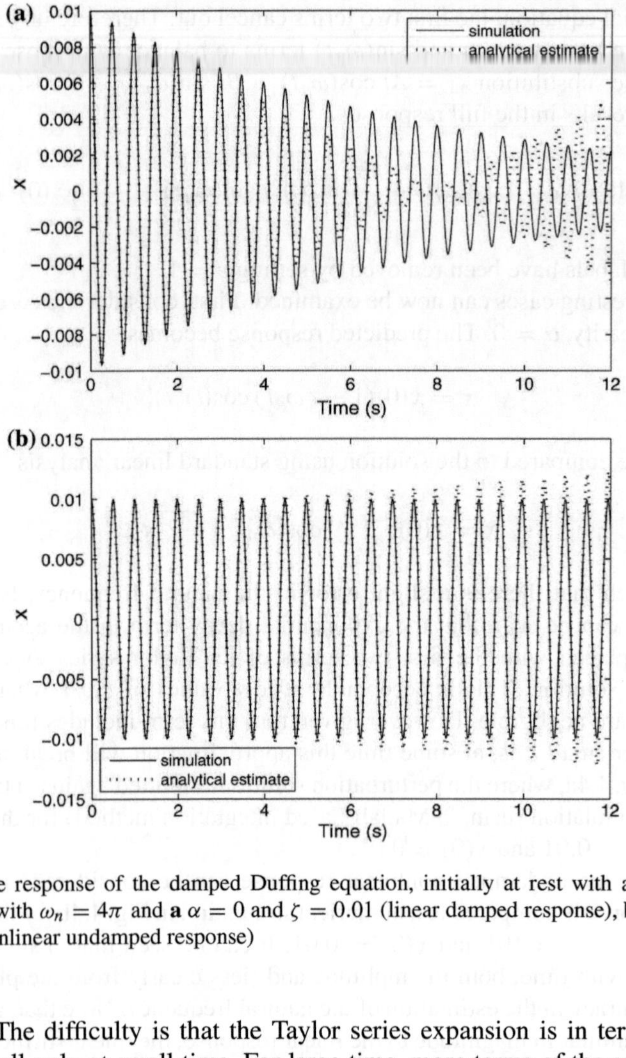

Fig. 4.4 Time response of the damped Duffing equation, initially at rest with a displacement $x(0) = 0.01$, with $\omega_n = 4\pi$ and **a** $\alpha = 0$ and $\zeta = 0.01$ (linear damped response), **b** $\alpha = 2 \times 10^4$ and $\zeta = 0$ (nonlinear undamped response)

expansion. The difficulty is that the Taylor series expansion is in terms of $\zeta\omega_n t$, which is small only at small time. For large time, more terms of the expansion are needed, along with balancing higher order ε terms in the subsequent equation of motion. This results in more terms in the power series expansion of x, Eq. (4.25).

4.4.2 Multiple Scales Method

From the discussions on the averaging technique and the regular perturbation technique, it is clear that dynamic responses often consist of terms that are functions of different time-scales. An example is the response of a linear unforced mass-spring-damper, $\ddot{x} + 2\zeta\omega_n\dot{x} + \omega_n^2 x = 0$, which if displaced and released will oscillate at frequency $\omega_n\sqrt{1+\zeta^2}$ with a decay due to the exponential term $\exp(-\zeta\omega_n t)$. If the damping is small, then over any one oscillation it could be thought of as almost constant (the approach taken when applying Averaging, Sect. 4.3). Therefore, the decay can be said to be occurring at a slower time-scale than the oscillations. In multiple scales a solution in the form

$$x = X_c(\varepsilon t)\cos(\omega t) + X_s(\varepsilon t)\sin(\omega t) \tag{4.29}$$

is sought.[9] Here the ε label is used to indicate that the amplitude terms X_c and X_s vary slowly over time giving $X_c(\varepsilon t)$ and $X_s(\varepsilon t)$—this does not necessarily mean that t is small but that the time-dependent terms within X_c and X_s are small (and hence slow).

The two time-scales can be labelled separately, fast-time over which oscillations occur $\tau = \omega t$ and slow-time over which the amplitudes evolve $T = \varepsilon t$. These times τ and T are treated as independent variables, such that derivatives with respect to t can be expressed, using the chain rule, as

$$\begin{aligned} \frac{\mathrm{d}x}{\mathrm{d}t} &= \frac{\partial x}{\partial \tau}\frac{\mathrm{d}\tau}{\mathrm{d}t} + \frac{\partial x}{\partial T}\frac{\mathrm{d}T}{\mathrm{d}t} = \omega\frac{\partial x}{\partial \tau} + \varepsilon\frac{\partial x}{\partial T}, \\ \frac{\mathrm{d}^2x}{\mathrm{d}t^2} &= \omega^2\frac{\partial^2 x}{\partial \tau^2} + 2\omega\varepsilon\frac{\partial^2 x}{\partial T\partial \tau} + \varepsilon^2\frac{\partial x}{\partial T}. \end{aligned} \tag{4.30}$$

Unforced Vibration: Firstly consider an unforced nonlinear system

$$\ddot{x} + 2\varepsilon\zeta\omega_n\dot{x} + \omega_n^2 x + \varepsilon\mathcal{N}(\dot{x}, x) = 0,$$

where both the nonlinear term, $\mathcal{N}(\dot{x}, x, t)$, and the damping term are taken to be small and hence order ε^1. Since the damping is small, the linear response has a natural frequency of ω_n and therefore the fast-time is set to $\tau = \omega t = \omega_n t$. Substituting the expressions for the derivatives with respect to t into this equation gives

$$\begin{aligned} &\left(\omega_n^2\frac{\partial^2 x}{\partial \tau^2} + 2\omega_n\varepsilon\frac{\partial^2 x}{\partial T\partial \tau} + \varepsilon^2\frac{\partial x}{\partial T}\right) + 2\varepsilon\zeta\omega_n\left(\omega_n\frac{\partial x}{\partial \tau} + \varepsilon\frac{\partial x}{\partial T}\right) \\ &+ \omega_n^2 x + \varepsilon\mathcal{N}(\omega_n\frac{\partial x}{\partial \tau} + \varepsilon\frac{\partial x}{\partial T}, x) = 0. \end{aligned} \tag{4.31}$$

---

[9] This type of solution could also be expressed in the amplitude-phase form $x = X(\varepsilon t)\sin(\omega t + \phi(\varepsilon t))$, but this is beyond the scope of this book. In addition, it is possible to consider very slow time-scale terms separately using a further time-scale. However this is not discussed here, instead see for example Glendinning (1994).

To proceed, the power series solution for the system response Eq. (4.25)

$$x(t) = x_0(\tau,\ T) + \varepsilon x_1(\tau,\ T) + \cdots$$

is substituted into Eq. (4.31) and the order ε^0 and ε^1 terms are balanced to give

$$\varepsilon^0\colon \omega_n^2 x_0'' + \omega_n^2 x_0 = 0, \tag{4.32}$$
$$\varepsilon^1\colon \omega_n^2 x_1'' + \omega_n^2 x_1 = -2\omega_n x_0'^* - 2\zeta\omega_n^2 x_0' - \mathscr{N}(\omega_n x_0', x_0), \tag{4.33}$$

where $\bullet'$ is the partial derivative with respect to fast-time τ, $\bullet^*$ is the partial derivative with respect to slow time T, and a Taylor series expansion has been applied to $\mathscr{N}$.

As with the regular perturbation technique, both these equations are linear in terms of x_0 and x_1 respectively, and in the second equation the x_0 terms can be viewed as forcing terms acting on x_1. Since the derivatives are partial with respect to τ, a solution to Eq. (4.32) is

$$x_0 = X_{0c}(T)\cos(\tau) + X_{0s}(T)\sin(\tau),$$

where the amplitudes X_{0c} and X_{0s} are functions of slow time T and the equation is in the form of Eq. (4.29). To proceed, this expression is substituted into the ε^1 equation, Eq. (4.33). In making this substitution there may be $A(T)\sin(\tau)$ or $B(T)\cos(\tau)$ terms on the right-hand side, where typically $A(T)$ and $B(T)$ are differential expressions in terms of X_{0c} and X_{0s}. These terms must be set to zero, resulting in conditions on X_{0c} and X_{0s}, to avoid secular terms. Imposing these conditions on X_{0c} and X_{0s} can be viewed as ensuring that the response at the resonant frequency (or the forcing frequency for the forced vibration case) is captured by x_0, since X_{0c} and X_{0s} have been selected such that the differential equation for x_1 (the ε^1 equation) has no $\sin(\tau)$ or $\cos(\tau)$ forcing terms. This will be seen by considering an example.

Example 4.5 Vibration response of a damped linear oscillator

Problem Find the response of a lightly damped, but unforced, linear oscillator. Initially the system is at rest with an initial displacement of $x(0)$.

Solution Consider the damped oscillator

$$\ddot{x} + 2\varepsilon\zeta\omega_n\dot{x} + \omega_n^2 x = 0,$$

noting that the damping is assumed to be small and hence of order ε^1.

Using the equations that have just been derived for the general unforced nonlinear system, the response is in the form $x(t) = x_0(\tau, T) + \varepsilon x_1(\tau, T)$ where x_0 and x_1 can be solved from Eqs. (4.32) and (4.33). For this case these are

$$\varepsilon^0\colon \omega_n^2 x_0'' + \omega_n^2 x_0 = 0, \tag{4.34}$$
$$\varepsilon^1\colon \omega_n^2 x_1'' + \omega_n^2 x_1 = -2\omega_n x_0'^* - 2\zeta\omega_n^2 x_0'. \tag{4.35}$$

The solution to the first of these equations is

$$x_0 = X_{0c}(T)\cos(\tau) + X_{0s}(T)\sin(\tau). \tag{4.36}$$

This solution can be substituted into the second equation to give

$$\begin{aligned}\omega_n^2 x_1'' + \omega_n^2 x_1 = 2\omega_n\left(X_{0c}(T)^* + \omega_n\zeta X_{0c}(T)\right)\sin(\tau)\\ - 2\omega_n\left(X_{0s}(T)^* + \omega_n\zeta X_{0s}(T)\right)\cos(\tau).\end{aligned} \tag{4.37}$$

If either the $\cos(\tau)$ or the $\sin(\tau)$ terms remain on the right-hand side then the solution to x_1 will be secular. This is because the substitution $x_1 = X_{1c}(T)\cos(\tau) + X_{1s}(T)\sin(\tau)$ would disappear, and as a result the trial solution would have to be of the form $x_1 = X_{1c}(T)\tau\cos(\tau) + X_{1s}(T)\tau\sin(\tau)$ to balance the right-hand side. Therefore to avoid secular terms the following conditions must be applied to X_{0c} and X_{0s}

$$X_{0c}(T)^* + \omega_n\zeta X_{0c}(T) = 0, \quad X_{0s}(T)^* + \omega_n\zeta X_{0s}(T) = 0.$$

These can be solved and substituted into Eq. (4.36) to give

$$x_0 = \mathrm{e}^{-\zeta\omega_n T}(\bar{X}_{0c}\cos(\tau) + \bar{X}_{0s}(T)\sin(\tau)),$$

where $\bar{X}_{0c}$ and $\bar{X}_{0s}$ are constants.

Finally, the initial conditions are considered. The initial displacement $x(0) = x_0(0) + \varepsilon x_1(0)$ is valid for all ε so that $\bar{X}_{0c} = x(0)$. Similarly, the initial velocity gives $\bar{X}_{0s} = \dot{x}(0) = 0$. Recalling that $\tau = \omega_n t$ and $T = \varepsilon t$, the approximate solution predicted by the multiple-scales technique is therefore

$$x = x_0 + \varepsilon x_1 = x(0)\mathrm{e}^{-\zeta\omega_n t}\cos(\omega_n t) + \mathcal{O}(\varepsilon^1). \tag{4.38}$$

Note that this analysis could proceed to solve for x_1 using Eq. (4.37), but this is often not necessary. By comparing Eq. (4.38) with the exact solution to the linear equation of motion given by Eq. (4.28), it can be seen that the exponential decay term has been captured accurately. There is, however, a slight error in the resonant frequency. This could be improved by including very slow time-scale terms, see for example Strogatz (2001). ■

Forced Vibration: When forcing is present, the general dynamic equation can be written as

$$\ddot{x} + 2\varepsilon\zeta\omega_n\dot{x} + \omega_n^2 x + \varepsilon\mathcal{N}(\dot{x}, x) = \varepsilon F\cos(\Omega t), \tag{4.39}$$

where the forcing is assumed to be small, of order ε^1. Using the fast and slow time scales $\tau = \Omega t$ and $T = \varepsilon t$ and substituting the expressions for the derivatives with respect to t gives

$$\left(\Omega^2\frac{\partial^2 x}{\partial\tau^2}+2\Omega\varepsilon\frac{\partial^2 x}{\partial T\partial\tau}+\varepsilon^2\frac{\partial x}{\partial T}\right)+2\varepsilon\zeta\omega_n\left(\Omega\frac{\partial x}{\partial\tau}+\varepsilon\frac{\partial x}{\partial T}\right)$$
$$+\omega_n^2 x+\varepsilon\mathcal{N}\left(\Omega\frac{\partial x}{\partial\tau}+\varepsilon\frac{\partial x}{\partial T},x\right)=\varepsilon F\cos(\tau).$$

The forcing is assumed to be small in comparison with the response, which is consistent with forcing close to resonance. As with the averaging technique, the forcing frequency Ω can be written as $\Omega=\omega_n(1+\varepsilon\mu)$, where μ is the frequency detuning parameter (and labelled ε to indicate its small since the forcing is near resonance). Introducing this relationship, and making the truncated power series substitution $x(t)=x_0(\tau, T)+\varepsilon x_1(\tau, T)$, Eq. (4.25) gives

$$\begin{aligned}&\omega_n^2(1+2\varepsilon\mu+\varepsilon^2\mu^2)(x_0''+\varepsilon x_1'')+2\varepsilon\omega_n(1+\varepsilon\mu)(x_0'^*+\varepsilon x_1'^*)\\&\quad+\varepsilon^2(x_0^{**}+\varepsilon x_1^{**})+2\varepsilon\zeta\omega_n\left(\omega_n(1+\varepsilon\mu)(x_0'+\varepsilon x_1')+\varepsilon(x_0^*+\varepsilon x_1^*)\right)\\&\quad+\omega_n^2(x_0+\varepsilon x_1)+\varepsilon\mathcal{N}(\omega_n(1+\varepsilon\mu)(x_0'+\varepsilon x_1')+\varepsilon(x_0^*+\varepsilon x_1^*),x_0+\varepsilon x_1)\\&\quad=\varepsilon F\cos(\tau),\end{aligned}\tag{4.40}$$

where $\bullet'$ and $\bullet^*$ are the partial derivatives with respect to fast and slow time respectively. Applying a Taylor series expansion to $\mathcal{N}$ and balancing the ε^0 and ε^1 terms, ignoring higher order terms, gives

$$\begin{aligned}&\varepsilon^0\!: \omega_n^2 x_0''+\omega_n^2 x_0=0,\\&\varepsilon^1\!: \omega_n^2 x_1''+\omega_n^2 x_1=-\omega_n^2 2\mu x_0''-2\omega_n x_0'^*-2\zeta\omega_n^2 x_0'-\mathcal{N}(\omega_n x_0',x_0)+F\cos(\tau).\end{aligned}$$

As with the unforced system, the first equation can be solved to give

$$x_0=X_{0c}(T)\cos(\tau)+X_{0s}(T)\sin(\tau).$$

This expression is then substituted into the ε^1 equation. The resulting $\cos(\tau)$ and $\sin(\tau)$ terms on the right-hand side of the equation (which can be thought of as forcing terms when considering the dynamics of x_1) are then set to zero to ensure that the solution for x_1 is not secular. Balancing the $\cos(\tau)$ and $\sin(\tau)$ terms respectively gives the two dynamic equations in $X_{0c}(T)$ and $X_{0s}(T)$

$$\begin{aligned}\omega_n^2 2\mu X_{0c}-2\omega_n X_{0s}^*-2\zeta\omega_n^2 X_{0s}-|\mathcal{N}(-\omega_n X_{0c}\mathrm{s}+\omega_n X_{0s}\mathrm{c},X_{0c}\mathrm{c}+X_{0s}\mathrm{s})|_{\mathrm{c}}+F=0\\ \omega_n^2 2\mu X_{0s}+2\omega_n X_{0c}^*+2\zeta\omega_n^2 X_{0c}-|\mathcal{N}(-\omega_n X_{0c}\mathrm{s}+\omega_n X_{0s}\mathrm{c},X_{0c}\mathrm{c}+X_{0s}\mathrm{s})|_{\mathrm{s}}=0\end{aligned}\tag{4.41}$$

where s and c have been used as shorthand for $\sin(\tau)$ and $\cos(\tau)$ respectively and $|\bullet|_{\mathrm{c}}$ indicates just the coefficients of the terms in $\bullet$ that contain $\cos(\tau)$.

The steady-state response solution may then be found by solving these equations for the case where the amplitude is non-varying and hence $X_{0c}^*=X_{0s}^*=0$. As an example, the application of the multiple-scales technique to the dynamics of an inclined cable is considered in Sect. 7.4.3. The results are compared to those obtained using the harmonic balance technique (Sect. 7.4.1) and the averaging technique (Sect. 7.4.2). This comparison shows that the multiple-scales technique produced the same result as the averaging technique. However, with the multiple scales it is also possible to extract information about the response at sub- and super-harmonics for the forcing frequency.

In addition, as with the averaging approach, the stability of the steady-state solutions can be identified using Eq. (4.41), as the equation can be written in matrix form of the pair of first-order dynamic equation in X_{0c} and X_{0s} with respect to T

$$\begin{pmatrix} X_{0c} \\ X_{0s} \end{pmatrix}^{*} = f\left(\begin{pmatrix} X_{0c} \\ X_{0s} \end{pmatrix}, T\right)$$

or $\mathbf{z}' = f(\mathbf{z}, T)$. The same approach as that presented for the averaging technique, see Sect. 4.3.2, can then be used to find the stability of the solutions. This is discussed further when the dynamics of an inclined cable are considered, see Sect. 7.4.3.

4.5 Normal Form Transformations

One of the difficulties with solving the equation of motion of a weakly nonlinear system is the harmonics that are present in the response. Using the harmonic balance with a simple guess such as the trial solution used in Example 4.2 results in the harmonic terms being ignored (and the balance not being perfect—the $\cos(3\omega_r\tau)$ term is ignored in the example). With the averaging technique the higher harmonic terms are removed by taking the average of the response over a cycle of oscillation.

The normal form transformation adopts a different approach that allows information regarding the harmonics to be retained while, at the same time, allowing a modified equation of motion to be solved by considering just the primary response (i.e. without needing to consider any harmonics). This is achieved by applying a nonlinear transformation to the equation of motion. This transformation is called a *near-identity transform*, reflecting the fact that in the derivation of the technique it is assumed that the transform is only a small perturbation away from a *linear unity transform*. The need for the transform to be near-identity places the same condition on the system as that placed by averaging and multiple scales, namely that the system is weakly nonlinear. The transform removes terms in the equation of motion that do not respond at the response frequency while retaining those that are "resonant", i.e. those that do respond at the response frequency and, as will be seen, would result in large transform terms. Once the resonant response is found by solving the resulting equation of motion the "non-resonant" terms can then be found by considering the near-identity transform.

The key difference with this method compared to those that have already been considered is that the method lends itself to a matrix formulation and hence can be easily extended to systems with multiple degrees-of-freedom, see Chap. 5. It also means that the technique lends itself to being used in symbolic programming packages.

It is usual to apply the normal form technique to first-order differential equations. Therefore the first step in applying the technique is to convert the equations of motion into their state-space representation. However Neild and Wagg (2011) have demonstrated that the technique can be applied, with slight modifications, directly to second-order differential equations representing nonlinear oscillator equations of motion. It is this second-order variant of the normal forms that will be discussed here. For information on the first-order (or state-space based) normal forms see, for example, Nayfeh (1993), Jezequel and Lamarque (1991) and Wagg and Neild (2009).

Considering free vibration the technique can be applied to a system through a series of steps:

Step 1: Apply the linear modal transformation to decouple the linear terms.
Step 2: Derive an equation of possible nonlinear transformations.
Step 3: Select and apply the near-identity transformation.
Step 4: Solve the normal form equations.

The distinction between the second and third steps is that the second step results in an equation defining a choice of possible nonlinear transformations. The choice is based on whether each nonlinear term in the dynamic equation is either included in the transformation or left in the dynamic equation. Ideally, all the terms would be included in the transformation, resulting in a set of linear governing equations. However, this is not normally possible, as some of the resulting transformation terms would be large and so would not represent a near-identity transform. The third step selects the valid near-identity transform and applies it.

For a system that is externally forced, an additional step is sometimes necessary. The step involves applying a further transformation to remove forcing terms that are are non-resonant.

In this section the application of the normal form method to unforced and undamped systems is considered first. As well as discussing the method, the technique is applied to an example that can be transformed to a purely linear dynamic equation, the escape equation, and to an example which cannot be fully transformed into a linear dynamic equation, the Duffing oscillator. Finally the discussion is extended to include external forcing and damping. In Chap. 5 the method is applied to systems with multiple degrees-of-freedom.

4.5.1 Free Vibration

Consider the nonlinear vibration problem taking the form

$$M\ddot{\mathbf{x}} + K\mathbf{x} + \mathscr{N}(\mathbf{x}) = 0, \tag{4.42}$$

where $\mathbf{x}$ is the displacement vector and M and K are the mass and stiffness matrices respectively. Note that the discussions in this chapter concentrate on single-degree-of-freedom systems in which $\mathbf{x}$, M and K are scalars. However, it is convenient to keep the discussion here more general, so that the technique can be used for multi-degree-of-freedom systems in Chap. 5

Step 1: Firstly the linear terms are decoupled by applying a linear modal decomposition to the dynamic equation. Here this decomposition is applied with minimal discussion as the chapter is primarily focussed on single degree-of-freedom systems, where the decomposition reduces to a trivial scaling. When systems with multiple degrees-of-freedom are considered in the next chapter the linear modal decomposition is discussed in detail, see Sect. 5.2.

The decomposition is achieved by writing $\mathbf{x} = \Phi\mathbf{q}$, where $\mathbf{q}$ are the modal coordinates for the linear system. Here Φ is a matrix of column vectors corresponding to the eigenvectors of $M^{-1}K$ and Λ is a diagonal matrix of the corresponding eigenvalues. Note that the kth eigenvalue corresponds to the square of kth natural frequency of the linearised system, ω_{nk}^2.

Eliminating $\mathbf{x}$ in Eq. (4.42) using $\mathbf{x} = \Phi\mathbf{q}$ and premultiplying by Φ^T gives

$$\Phi^T M\Phi\ddot{\mathbf{q}} + \Phi^T K\Phi\mathbf{q} + \Phi^T \mathscr{N}(\Phi\mathbf{q}) = 0.$$

Multiplying by $(\Phi^T M\Phi)^{-1}$ gives

$$\ddot{\mathbf{q}} + \Lambda\mathbf{q} + N_q(\mathbf{q}) = 0, \tag{4.43}$$

where

$$N_q(\mathbf{q}) = (\Phi^T M\Phi)^{-1}\Phi^T \mathscr{N}(\Phi\mathbf{q}).$$

Here the relationship $(\Phi^T M\Phi)^{-1}\Phi^T K\Phi = \Lambda$, is based on the observation that Λ and Φ are matrix forms of the eigenvalues and vectors of $M^{-1}K$ (see discussion around Eq. (5.6)). Note that for the case of a single degree-of-freedom system $\mathbf{q} = \mathbf{x}$, $\Lambda = k/m$ (where k and m are the linear stiffness and mass respectively) and $N_q(\mathbf{q}) = \mathscr{N}(\mathbf{q})/m$. This representation of the equations of motion is sometimes called the *linear modal normal form*.

Step 2: Following this linear transform a further nonlinear coordinate transform, $\mathbf{q} \rightarrow \mathbf{u}$, is sought. The aim is to remove non-resonant terms. Consider the kth equation of motion, resonant terms are those that respond at the response frequency ω_{rk} which, for an unforced system, corresponds to the nonlinear natural frequency. Non-resonant terms are those nonlinear terms that respond at harmonics of ω_{rk} in the case of the kth equation of motion. Note that as the nonlinearity is weak $\omega_{rk} \approx \omega_{nk}$.

First, the nonlinear term in the linear normal form representation of the equation of motion, Eq. (4.43), is partitioned into a series of functions with reduced levels of significance

$$\ddot{\mathbf{q}} + \Lambda\mathbf{q} + N_q(\mathbf{q}) = 0 \quad \text{where } N_q(\mathbf{q}) = \varepsilon\mathbf{n}_1(\mathbf{q}) + \varepsilon^2\mathbf{n}_2(\mathbf{q}) + \cdots. \tag{4.44}$$

As discussed at the start of Sect. 4.3, ε is used to label small terms. Hence N_q may be thought of as consisting of a series of terms of reducing significance—note that it is not a Taylor series expansion. Note also that this expansion assumes that the nonlinear term is small in comparison to the linear stiffness terms, which are represented in $\Lambda\mathbf{q}$.

Second, the as-yet unknown nonlinear transform is also decomposed into a series of terms of reducing significance

$$\mathbf{q} = \mathbf{u} + \mathbf{h}(\mathbf{u}) \quad \text{where } \mathbf{h}(\mathbf{u}) = \varepsilon\mathbf{h}_1(\mathbf{u}) + \varepsilon^2\mathbf{h}_2(\mathbf{u}) + \cdots. \tag{4.45}$$

Note again that the nonlinear transform terms are assumed to be small in comparison to the linear term $\mathbf{u}$, hence terming it a *near-identity transform*.

To find this transform, rather than seeking the ideal linear equation $\ddot{\mathbf{u}} + \Lambda\mathbf{u} = 0$, the following equation is sought

$$\ddot{\mathbf{u}} + \Lambda\mathbf{u} + N_u(\mathbf{u}) = 0 \quad \text{where} \quad N_u(\mathbf{u}) = \varepsilon\mathbf{n}_{u1}(\mathbf{u}) + \varepsilon^2\mathbf{n}_{u2}(\mathbf{u}) + \cdots. \tag{4.46}$$

The aim of the transform is to force the elements within $\mathbf{n}_{ui}$ vectors to zero wherever possible without invalidating the assumption that the transform, $\mathbf{h}(\mathbf{u})$, is small (order ε^1). This process results in the removal of all terms in $\mathbf{n}_{ui}$ except those that are resonant (i.e considering the kth equation of motion, terms that contribute to the response at ω_{rk}) from the equations of motion. These resonant terms are larger that the non-resonant ones and hence if removed would result in transform terms that cannot be considered as near-unity.

The process of finding the appropriate transform starts with combining Eqs. (4.44), (4.45) and (4.46). In doing so it is common to just include the terms up to order ε^1, as is done here, however later in the chapter the technique will be extended to handle ε^2 terms as well. Combining these equations it can be shown that for the ε^1 terms to be balanced, the following condition must be met

$$\mathbf{n}_1(\mathbf{u}) + \Gamma\mathbf{h}_1(\mathbf{u}) + \frac{\mathbf{d}^2}{\mathbf{d}t^2}\mathbf{h}_1(\mathbf{u}) = \mathbf{n}_{u1}(\mathbf{u}), \tag{4.47}$$

where Γ is a diagonal matrix with the kth diagonal element being ω_{rk}^2. See Box 4.2 for the derivation of Eq. (4.47).

Box 4.2 *Derivation of Eq.* (4.47)
Substituting the transform, Eq. (4.45), into the equation of motion, Eq. (4.44) to eliminate **q** gives

$$\ddot{\mathbf{u}} + \varepsilon\frac{\mathrm{d}^2}{\mathrm{d}t^2}\mathbf{h}_1(\mathbf{u}) + \varepsilon^2\frac{\mathrm{d}^2}{\mathrm{d}t^2}\mathbf{h}_2(\mathbf{u}) + \cdots + \Lambda(\mathbf{u} + \varepsilon\mathbf{h}_1(\mathbf{u}) + \varepsilon^2\mathbf{h}_2(\mathbf{u}) + \cdots)$$
$$+ \varepsilon\mathbf{n}_1(\mathbf{u} + \varepsilon\mathbf{h}_1(\mathbf{u}) + \cdots) + \varepsilon^2\mathbf{n}_2(\mathbf{u} + \cdots) = 0.$$

The Taylor series expansion can be applied to the nonlinear terms giving

$$\mathbf{n}_i(\mathbf{u} + \varepsilon\mathbf{h}_1(\mathbf{u}) + \cdots) = \mathbf{n}_i(\mathbf{u}) + \varepsilon D\{\mathbf{n}_i(\mathbf{u})\}\mathbf{h}_1(\mathbf{u}) + \cdots,$$

where $D\{\mathbf{n}_i\}$ is the Jacobian of $\mathbf{n}_i$. Using this expansion and Eq. (4.46), the equation for the dynamics in terms of **u**, to eliminate $\ddot{\mathbf{u}}$, giving

$$\varepsilon\frac{\mathrm{d}^2}{\mathrm{d}t^2}\mathbf{h}_1(\mathbf{u}) + \varepsilon^2\frac{\mathrm{d}^2}{\mathrm{d}t^2}\mathbf{h}_2(\mathbf{u}) + \Lambda(\varepsilon\mathbf{h}_1(\mathbf{u}) + \varepsilon^2\mathbf{h}_2(\mathbf{u})) + \varepsilon\mathbf{n}_1(\mathbf{u})$$
$$+ \varepsilon^2 D\{\mathbf{n}_1(\mathbf{u})\}\mathbf{h}_1(\mathbf{u}) + \varepsilon^2\mathbf{n}_2(\mathbf{u}) + \cdots$$
$$= \varepsilon\mathbf{n}_{u1}(\mathbf{u}) + \varepsilon^2\mathbf{n}_{u2}(\mathbf{u}) + \cdots. \qquad (4.48)$$

Since the nonlinearity is small it can be assumed that $\omega_{rk} \approx \omega_{nk}$. Using this a small, order ε^1, diagonal frequency detuning matrix, Δ, is introduced. Here the kth diagonal element in Δ is $\omega_{nk}^2 - \omega_{rk}^2$. By definition $\Lambda = \Gamma + \varepsilon\Delta$, and hence Eq. (4.48) may be rewritten as

$$\varepsilon\frac{\mathrm{d}^2}{\mathrm{d}t^2}\mathbf{h}_1(\mathbf{u}) + \varepsilon^2\frac{\mathrm{d}^2}{\mathrm{d}t^2}\mathbf{h}_2(\mathbf{u}) + \Gamma(\varepsilon\mathbf{h}_1(\mathbf{u}) + \varepsilon^2\mathbf{h}_2(\mathbf{u}))$$
$$+ \varepsilon^2\Delta\mathbf{h}_1(\mathbf{u}) + \varepsilon\mathbf{n}_1(\mathbf{u}) + \varepsilon^2 D\{\mathbf{n}_1(\mathbf{u})\}\mathbf{h}_1(\mathbf{u}) + \varepsilon^2\mathbf{n}_2(\mathbf{u}) + \cdots$$
$$= \varepsilon\mathbf{n}_{u1}(\mathbf{u}) + \varepsilon^2\mathbf{n}_{u2}(\mathbf{u}) + \cdots \qquad (4.49)$$

This detuning step is discussed further in Xin et al. (2013), where it is shown to have no effect of the resulting equation of motion in terms of **u** and improves the prediction of the harmonic terms (which are captured in the transform). The reason for this is that by using the response frequencies as the linear 'stiffness' term the size of the nonlinear terms is reduced and hence the approximation that the nonlinear terms is small is better satisfied.

Finally equating powers of ε in Eq. (4.49) gives:

$$\varepsilon^1\colon \mathbf{n}_1(\mathbf{u}) + \frac{d^2}{dt^2}\mathbf{h}_1(\mathbf{u}) + \Gamma\mathbf{h}_1(\mathbf{u}) = \mathbf{n}_{u1}(\mathbf{u}), \tag{4.50}$$

$$\varepsilon^2\colon \mathbf{n}_2(\mathbf{u}) + \frac{d^2}{dt^2}\mathbf{h}_2(\mathbf{u}) + \Gamma\mathbf{h}_2(\mathbf{u}) + \Delta\mathbf{h}_1(\mathbf{u}) + D\{\mathbf{n}_1(\mathbf{u})\}\,\mathbf{h}_1(\mathbf{u}) = \mathbf{n}_{u2}(\mathbf{u})$$

Note that the ε^0 terms are naturally balanced and that ε^3 and higher terms are not considered.

The next step in finding the transform requires consideration of the form of the response in $\mathbf{u}$ and exploits the fact that for each state in $\mathbf{u}$ the response is just at the response frequency. This is because the transform will have removed the non-resonant terms in the response. Using the complex exponential representation for the response, the state vector are split into two components $\mathbf{u} = \mathbf{u}_p + \mathbf{u}_m$ which capture the positive and negative complex exponential terms (subscripts 'plus' and 'minus'). So for the kth state, u_k, may be written as

$$u_k = u_{pk} + u_{mk} = \left(\frac{U_k}{2}e^{-i\phi_k}\right)e^{i\omega_{rk}t} + \left(\frac{U_k}{2}e^{i\phi_k}\right)e^{-i\omega_{rk}t}. \tag{4.51}$$

The ε^1-order terms $\mathbf{n}_1(\mathbf{u})$, $\mathbf{h}_1(\mathbf{u})$ and $\mathbf{n}_{u1}(\mathbf{u})$ can be expressed in matrix form such that

$$\mathbf{n}_1(\mathbf{u}) = \mathbf{n}^*\mathbf{u}^*(\mathbf{u}_p, \mathbf{u}_m), \quad \mathbf{h}_1(\mathbf{u}) = \mathbf{h}^*\mathbf{u}^*(\mathbf{u}_p, \mathbf{u}_m), \quad \mathbf{n}_{u1}(\mathbf{u}) = \mathbf{n}_u{}^*\mathbf{u}^*(\mathbf{u}_p, \mathbf{u}_m), \tag{4.52}$$

where $\mathbf{u}^*$ is a column vector containing all the combinations of $\mathbf{u}_p$ and $\mathbf{u}_m$ terms that appear within $\mathbf{n}_1(\mathbf{u})$ and $\mathbf{n}^*$, $\mathbf{h}^*$ and $\mathbf{n}_u^*$ are matrices of coefficient terms. If $\mathbf{u}$ is size $N \times 1$, corresponding to being an N degree-of-freedom system, and $\mathbf{u}^*$ is of size $L \times 1$ then the size of $\mathbf{n}^*$ is $N \times L$. Equation (4.47) leads to the selection of a similar form of matrix expression for $\mathbf{h}_1(\mathbf{u})$ and $\mathbf{n}_{u1}(\mathbf{u})$ (with $\mathbf{h}^*$ and $\mathbf{n}_u^*$ also being $N \times L$ matrices).

Matrices $\mathbf{h}^*$ and $\mathbf{n}_u^*$ are currently unknown whereas $\mathbf{n}^*$ is defined by the nonlinear vector in the dynamic equation for $\mathbf{q}$, $\mathbf{n}_1$. Substituting the relationships defined in Eq. (4.52) into Eq. (4.47) results in

$$\mathbf{n}^*\mathbf{u}^* + \mathbf{h}^*\frac{d^2\mathbf{u}^*}{dt^2} + \Gamma\mathbf{h}^*\mathbf{u}^* = \mathbf{n}_u^*\mathbf{u}^*. \tag{4.53}$$

To simplify this equation it is necessary to consider the general form for a term in $\mathbf{u}^*$. Recalling that $\mathbf{u}^*$ is a function of $\mathbf{u}_p$ and $\mathbf{u}_m$, the ℓth term $(1 \le \ell \le L)$ in $\mathbf{u}^*$ can be written as

$$u_\ell^* = \prod_{k=1}^{N}\left(u_{pk}^{s_{pk\ell}}u_{mk}^{s_{mk\ell}}\right), \tag{4.54}$$

where $s_{pk\ell}$ and $s_{mk\ell}$ are power indices.

Using this representation and with some algebraic manipulation Eq. (4.53) can be rewritten as

$$\mathbf{n}_u^* = \mathbf{n}^* - \tilde{\mathbf{h}}^*. \tag{4.55}$$

Here the element in the kth row and ℓth column of $\tilde{\mathbf{h}}^*$ is related to the corresponding element in $\mathbf{h}^*$ via

$$\tilde{h}_{k,\ell}^* = \beta_{k,\ell}^* h_{k,\ell}^* \quad \text{with:} \quad \beta_{k,\ell}^* = \left[\sum_{n=1}^{N} \{(s_{pn\ell} - s_{mn\ell})\omega_{rn}\}\right]^2 - \omega_{rk}^2, \tag{4.56}$$

where $\beta_{k,\ell}$ is the (k, ℓ)th element in matrix $\boldsymbol{\beta}^*$. The full derivation of Eqs. (4.55) and (4.56) is given in Box 4.3.

Since the calculation of $\boldsymbol{\beta}^*$ is critical to the selection of the near-identity transform in the next step of the process, it is worth summarising how it can be calculated starting from the equation of motion in terms of $\mathbf{q}$, $\ddot{\mathbf{q}} + \Lambda\mathbf{q} + N_q(\mathbf{q}) = 0$, i.e. Eq. (4.44). Using Eq. (4.44), $N_q(\mathbf{q})$ is written as $N_q(\mathbf{q}) = \varepsilon\mathbf{n}_1(\mathbf{q})$ where order ε^2 and higher terms are removed as here the derivation to order ε^1 accuracy is being discussed. Next $\mathbf{n}_1(\mathbf{q})$ is rewritten as $\mathbf{n}_1(\mathbf{u})$, $\mathbf{u}$ is replaced using $\mathbf{u} = \mathbf{u}_p + \mathbf{u}_m$ and then represented in matrix form using Eq. (4.52), $\mathbf{n}_1(\mathbf{u}) = \mathbf{n}^*\mathbf{u}^*(\mathbf{u}_p, \mathbf{u}_m)$ where $\mathbf{n}^*$ is a matrix of coefficients. Finally the terms in $\mathbf{u}^*$ may be expressed in the general form given by Eq. (4.54), $u_\ell^* = \prod_{k=1}^{N} \left(u_{pk}^{s_{pk\ell}} u_{mk}^{s_{mk\ell}}\right)$ by assigning values to the powers $s_{pk\ell}$ and $s_{mk\ell}$ for each term ℓ in $\mathbf{u}^*$ in term. These can then be used in Eq. (4.56) to calculate each term in $\boldsymbol{\beta}^*$.

Box 4.3 *Derivation of Eqs.* (4.55) *and* (4.56)
Equation (4.53) may be simplified by considering the double differential of $\mathbf{u}^*$ with respect to time. Taking the general form for the ℓth term in $\mathbf{u}^*$, Eq. (4.54), and differentiating gives

$$\begin{aligned}
\frac{\mathrm{d}u_\ell^*}{\mathrm{d}t} &= \sum_{n=1}^{N} \left(\frac{\partial u_\ell^*}{\partial u_{pn}}\frac{\mathrm{d}u_{pn}}{\mathrm{d}t} + \frac{\partial u_\ell^*}{\partial u_{mn}}\frac{\mathrm{d}u_{mn}}{\mathrm{d}t}\right) \\
&= \sum_{n=1}^{N} i\omega_{rn} \left(\frac{\partial u_\ell^*}{\partial u_{pn}} u_{pn} - \frac{\partial u_\ell^*}{\partial u_{mn}} u_{mn}\right) \\
&= i \sum_{n=1}^{N} \left[(s_{pn\ell} - s_{mn\ell})\omega_{rn}\right] u_\ell^*,
\end{aligned}$$

where the trial solution, Eq. (4.51), has been used to go from the first to the second line.

The key feature here is that the differential of u_ℓ^* contains only the ℓth term in $\mathbf{u}^*$. Therefore

$$\frac{\mathrm{d}^2\mathbf{u}^*}{\mathrm{d}t^2} = -\tilde{\Gamma}\mathbf{u}^*,$$

where $\tilde{\Gamma}$ is an $L \times L$ diagonal matrix with the ℓth diagonal element being

$$\tilde{\Gamma}_{\ell,\ell} = \left(\sum_{n=1}^{N}\left[(s_{pn\ell} - s_{mn\ell})\omega_{rn}\right]\right)^2. \tag{4.57}$$

Using this, Eq. (4.53) may be written as

$$\mathbf{n}^*\mathbf{u}^* - \mathbf{h}^*\tilde{\Gamma}\mathbf{u}^* + \Gamma\mathbf{h}^*\mathbf{u}^* = \mathbf{n}_u{}^*\mathbf{u}^*.$$

Hence

$$\mathbf{n}^* - \tilde{\mathbf{h}}^* = \mathbf{n}_u{}^*$$

where $\tilde{\mathbf{h}}^* = \mathbf{h}^*\tilde{\Gamma} - \Gamma\mathbf{h}^*$ is a Lie bracket. Using the fact that Γ and $\tilde{\Gamma}$ are both diagonal and using Eq. (4.57), the (k, ℓ)th element in $\tilde{\mathbf{h}}^*$ may be linked to the (k, ℓ)th element in $\mathbf{h}^*$ via matrix $\boldsymbol{\beta}^*$ using the relationship $\tilde{h}_{k,\ell}^* = \beta_{k,\ell}^* h_{k,\ell}^*$ where

$$\beta_{k,\ell}^* = \left[\sum_{n=1}^{N}\{(s_{pn\ell} - s_{mn\ell})\omega_{rn}\}\right]^2 - \omega_{rk}^2$$

Note this is an element by element relationship and so $\tilde{\mathbf{h}}^* \neq \boldsymbol{\beta}^*\mathbf{h}^*$.

Step 3: The near-identity transform to be applied to the system can now be found. Recall that the coefficients in $\mathbf{n}^*$ are known, whereas the transform coefficients in $\mathbf{h}^*$ and the coefficients for the nonlinear terms left in the transformed equation of motion, $\mathbf{n}_u{}^*$ are currently unknown. These unknown coefficients can be found using Eqs. (4.55) and (4.56) for each element in turn. Where possible it is desirable for the nonlinear terms in the equation of motion to be removed by setting elements in $\mathbf{n}_u{}^*$ to zero. In this case for the (k, ℓ) element

$$\text{non-resonant terms:} \quad n_{u,k,\ell}^* = 0, \quad h_{k,\ell}^* = n_{k,\ell}^*/\beta_{k,\ell}^*. \tag{4.58}$$

However if the term is near-resonant then $\beta_{k,\ell}^* \approx 0$. Setting $n_{u,k,\ell}^* = 0$ in these cases would result in a large $h_{k,\ell}^*$ term which would invalidate the assumption that the transform is near-identity. Therefore the near-resonant terms are kept in the equation of motion by writing

$$\text{near-resonant terms:} \quad n_{u,k,\ell}^* = n_{k,\ell}^*, \quad h_{k,\ell}^* = 0. \tag{4.59}$$

Step 4: Now that $\mathbf{n}_u$ and $\mathbf{h}$ have been found, the equation of motion in $\mathbf{u}$ needs to be solved. To order ε^1 this equation can be written as $\ddot{\mathbf{u}} + \Lambda\mathbf{u} + \mathbf{n}_u{}^*\mathbf{u}^* = 0$, from Eqs. (4.46) and (4.52). Then using the transform equation to convert to $\mathbf{q}$ using $\mathbf{q} = \mathbf{u} + \mathbf{h}^*\mathbf{u}^*$ from Eqs. (4.45) and (4.52) to order ε^1. Finally the response in $\mathbf{x}$ can be reconstructed using $\mathbf{x} = \Phi\mathbf{q}$.

This technique is now applied to two single-degree-of-freedom examples, the first of which transforms ideally and the second results in resonant terms.

Example 4.6 One DOF oscillator with a quadratic nonlinearity—the escape equation.

Problem Find the normal form for the unforced escape equation, discussed in Chap. 2,

$$\ddot{x} + \omega_n^2 x + \alpha x^2 = 0,$$

which may be thought of as a mass-spring-damper system with a weakly nonlinear spring, such that αx^2 is of order ε^1 relative to the linear terms.

Solution The solution follows the four step process described above.

Step 1: Since this is a single degree of freedom system, the linear transform is a unity transform $\mathbf{x} = \mathbf{q}$ (where $\mathbf{x} = x$). Rewriting the equation of motion in the form of Eq. (4.44) gives

$$\ddot{\mathbf{q}} + \Lambda\mathbf{q} + N_q(\mathbf{q}) = 0 \quad \text{where } \Lambda = \omega_n^2,\ N_q(\mathbf{q}) = \alpha q_1^2, \tag{4.60}$$

and where q_1 is the first (and, in this example, only) element in $\mathbf{q}$.

Step 2: Now consider the near-identity transformation. Firstly, the nonlinear term is written in terms of $\mathbf{u}$ giving $N_q(\mathbf{u}) = \alpha u_1^2$ and then, to order ε^1, $\mathbf{n}_1(\mathbf{u}) = \alpha u_1^2$ using Eq. (4.44). Making the substitution $\mathbf{u} = \mathbf{u}_p + \mathbf{u}_m$ gives $\mathbf{n}_1(\mathbf{u}) = \mathbf{n}_1(\mathbf{u}_p, \mathbf{u}_m) = \alpha(u_{p1}^2 + 2u_{p1}u_{m1} + u_{m1}^2)$. Rewriting this in matrix form using Eq. (4.52) gives

$$\mathbf{n}_1(\mathbf{u}) = \mathbf{n}^*\mathbf{u}^*(\mathbf{u}_p, \mathbf{u}_m) \quad \text{where:} \quad \mathbf{n}^* = \alpha\begin{bmatrix} 1 & 2 & 1 \end{bmatrix}, \quad \mathbf{u}^* = \begin{bmatrix} u_{p1}^2 \\ u_{p1}u_{m1} \\ u_{m1}^2 \end{bmatrix}.$$

Comparing each term in $\mathbf{u}^*$ to the general form, Eq. (4.54), the terms in $\boldsymbol{\beta}^*$ can be calculated as

$$\boldsymbol{\beta}^* = \begin{bmatrix} 3\omega_{r1}^2 & -\omega_{r1}^2 & 3\omega_{r1}^2 \end{bmatrix}$$

used Eq. (4.56).

Step 3: Using $\boldsymbol{\beta}^*$, the coefficient matrix for the near-identity transform and the nonlinear terms in $\mathbf{u}$ can now be identified using Eqs. (4.58) and (4.59) giving

$$\mathbf{n}^* = \alpha \begin{bmatrix} 1 \\ 2 \\ 1 \end{bmatrix}^T, \quad \boldsymbol{\beta}^* = \begin{bmatrix} 3\omega_{r1}^2 \\ -\omega_{r1}^2 \\ 3\omega_{r1}^2 \end{bmatrix}^T, \quad \mathbf{n}_u^* = \alpha \begin{bmatrix} 0 \\ 0 \\ 0 \end{bmatrix}^T, \quad \mathbf{h}^* = \frac{\alpha}{\omega_{r1}^2} \begin{bmatrix} 1/3 \\ 2 \\ 1/3 \end{bmatrix}^T.$$

Note that in this example there are no near-resonant terms (no values in $\boldsymbol{\beta}^*$ are close to zero) so all the nonlinear terms can be moved into the transform resulting in a linear equation of motion to order ε^1.

Step 4: To complete this example, now consider how the normal form can be used to find the nonlinear response of the escape equation when its initial conditions are $x(t = 0) = x(0)$ and $\dot{x}(t = 0) = 0$.

Through the use of transforms, the dynamics have been reduced to the linear equation $\ddot{\mathbf{u}} + \Lambda \mathbf{u} = 0$, and using the trial solution given in Eq. (4.51) gives

$$u_1(t) = u_{p1}(t) + u_{m1}(t) = \frac{u_1(0)}{2}\left(\mathrm{e}^{\mathrm{i}\omega_{r1}t} + \mathrm{e}^{-\mathrm{i}\omega_{r1}t}\right) = u_1(0)\cos(\omega_{r1}t). \quad (4.61)$$

Note that, in examples where the equation for $\mathbf{u}$ is nonlinear, the trial solutions will be in terms of frequency ω_{r1}, the response frequency which may be amplitude dependent. In this example $\omega_{r1} = \omega_n$ as the transformed system is linear.

Now the transform equation can be used to find $\mathbf{q}$

$$\mathbf{q} = \mathbf{u} + \mathbf{h}^*\mathbf{u}^* = u_1 + \frac{\alpha}{3\omega_{r1}^2}\left(u_{p1}^2 - 6u_{p1}u_{m1} + u_{m1}^2\right)$$

which is the same as $\mathbf{x}$ for this example. Using the solution for $\mathbf{u}$, Eq. (4.51), gives

$$\mathbf{q} = q_1 = x = u_1(0)\cos(\omega_{r1}t) + \frac{\alpha}{6\omega_{r1}^2}u_1(0)^2\cos(2\omega_{r1}t) - \frac{\alpha}{2\omega_{r1}^2}u_1(0)^2, \quad (4.62)$$

where $\omega_{r1} = \omega_n$. Finally the initial conditions can be related by setting $t = 0$ to give

$$u_1(0) = \frac{3\omega_{r1}^2}{2\alpha}\left(1 - \sqrt{1 - \frac{4\alpha}{3\omega_{r1}^2}x(0)}\right). \quad (4.63)$$

■

To assess the accuracy of the normal form technique, consider the transform derived in the previous example. For the values $x(0) = 0.01$, $\omega_n = 4\pi$ and $\alpha = 800$, Fig. 4.5a shows the numerically computed system response (using Matlab ode45 solver). The Fourier transform of the response, for 100s of data zero-padded to 1000s, is shown in the three subplots contained in Fig. 4.5b. The Fourier transform components predicted using the normal form analysis, from Eq. (4.62), are plotted as dots (for an infinitely long signal). It can be seen that the predicted frequency components closely match the simulation data in both amplitude and frequency. Assuming the linear behaviour is dominant, the maximum size of the nonlinear term is $\alpha x^2 = 0.08$ whereas the maximum size of the linear stiffness term is $\omega_n^2 x = 1.56$.

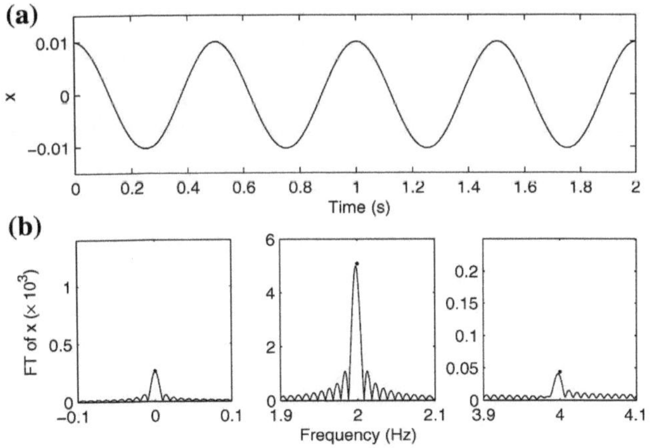

Fig. 4.5 Response of the escape equation to the initial conditions $x(0) = 0.01$ and $\dot{x}(0) = 0$, with $\omega_n = 4\pi$, $\zeta = 0$ and $\alpha = 800$, **a** in the time domain, and **b** in the Fourier domain using the ode45 Matlab time-stepping solver. The *dots* indicate the predicted Fourier content using the normal form analysis

Therefore the assumption that αx^2 is of order ϵ^1 relative to $\omega_n^2 x$ appears reasonable as the maximum value of the nonlinear term is 5 % that of the linear one.

To examine the effect of the assumption that the nonlinear term is small, Fig. 4.6 shows the response for the case where $\alpha = 3{,}200$ (such that the nonlinear term is 20 % of the linear one). In this case, it can be seen that the predicted Fourier content of the response is worse than that for $\alpha = 800$. The amplitudes are approximately correct, but the normal form method does not predict the shift in natural frequency. Moreover, at larger α values (over approximately 7,500), the system is unstable, because the nonlinear stiffness becomes negative and "escapes", as discussed in Chap. 2, Example 2.4. However, the normal form method fails to predict this feature since it is based on the assumption that the nonlinearity is small. This highlights how the normal form method is an accurate analysis tool provided that the assumption that the nonlinearity is small remains true—which in fact applies to all methods in perturbation theory where similar assumptions are made.

In the last example the escape equation, a system in which the dynamic equation can be linearized through a normal form transformation, was considered. It was seen that by considering the ε^1 terms, the transformation allowed the calculation of accurate frequency content of the free vibration response if the nonlinear terms were small. However, with larger nonlinearity the frequency content was inaccurately predicted; most noticeably the natural frequency was incorrect—the natural frequency is predicted to match ω_n regardless of the nonlinear behaviour. The normal form analysis can be extended to consider the ε^2 terms (the equation for which is given in Box 4.2 where $f_2(u)$ is zero). If the ε^2 terms are included, this results in a different response frequency which partially accounts for the error. This will be discussed in Sect. 4.5.2.

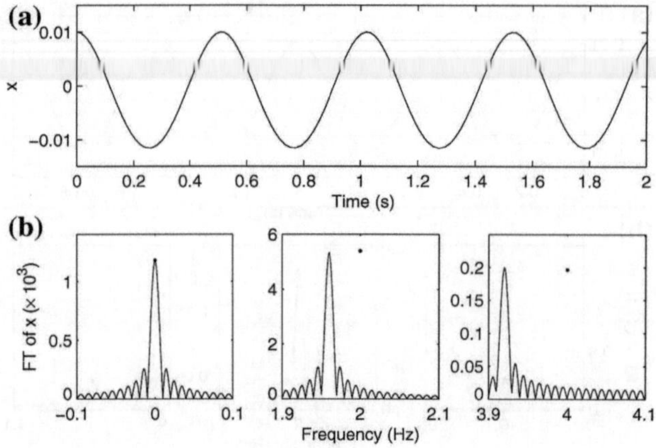

Fig. 4.6 Response of the escape equation to the initial conditions $x(0) = 0.01$ and $\dot{x}(0) = 0$, with $\omega_n = 4\pi$, $\zeta = 0$ and $\alpha = 3{,}200$, **a** in the time domain and **b** in the Fourier domain using the ode45 Matlab time-stepping solver. The *dots* indicate the predicted Fourier content using the normal form analysis

However, first, an example in which the dynamic equation cannot be totally transformed to a linear form is considered. This highlights the decision that must be made in Step 3 regarding which nonlinear terms should be included in the transform and which must remain in the dynamic equation.

Example 4.7 One DOF oscillator with a cubic nonlinearity—the Duffing equation.
Problem Consider the Duffing equation, an example of a system in which an ideal transform $\mathbf{q} = \mathbf{u} + \mathbf{h}^*\mathbf{u}^*$ such that $\dot{\mathbf{u}} = \Lambda\mathbf{u}$ does not exist. The equation of motion may be written as

$$\ddot{x} + \omega_n^2 x + \alpha x^3 = 0,$$

which, as with the escape equation, may be thought of as a mass-spring-damper system with a nonlinear spring. The nonlinear term is small and may be treated as of order ε^1.

Solution Again to find an expression for the response the four step process described above will be used.

Step 1: The linear transform is a unity transform, $\mathbf{q} = \mathbf{x} = x$ as the system only has one degree of freedom. Hence the system can be written in the form of Eq. (4.44) giving

$$\ddot{\mathbf{q}} + \Lambda\mathbf{q} + N_q(\mathbf{q}) = 0 \quad \text{where } \Lambda = \omega_n^2,\ N_q(\mathbf{q}) = \alpha q_1^3, \tag{4.64}$$

and where q_1 is the first and, as this is a one degree-of-freedom system, only element in $\mathbf{q}$.

Step 2: To find the near-identity transform, firstly the nonlinear vector is written in terms of $\mathbf{u}$ and then represented in orders of ε—see Eq. (4.44). Taking the nonlinearity to be order ε^1 gives $\mathbf{n}_1(\mathbf{u}) = \alpha u_1^3$. Now representing $\mathbf{u}$ as $\mathbf{u} = \mathbf{u}_p + \mathbf{u}_m$ gives $\mathbf{n}_1(\mathbf{u}) = \mathbf{n}_1(\mathbf{u}_p, \mathbf{u}_m) = \alpha(u_{p1} + u_{m1})^3$. Rewriting this in the matrix form given in Eq. (4.52) results in

$$\mathbf{n}_1(\mathbf{u}) = \mathbf{n}^*\mathbf{u}^*(\mathbf{u}_p, \mathbf{u}_m) \quad \text{where:} \quad \mathbf{n}^* = \alpha \begin{bmatrix} 1 \; 3 \; 3 \; 1 \end{bmatrix}, \quad \mathbf{u}^* = \begin{bmatrix} u_{p1}^3 \\ u_{p1}^2 u_{m1} \\ u_{p1} u_{m1}^2 \\ u_{m1}^3 \end{bmatrix}.$$

Using the general form for each term in $\mathbf{u}^*$ given by Eqs. (4.54), (4.56) can be used to calculate each of the terms in $\boldsymbol{\beta}^*$

$$\boldsymbol{\beta}^* = \begin{bmatrix} 8\omega_{r1}^2 \; 0 \; 0 \; 8\omega_{r1}^2 \end{bmatrix}.$$

where ω_{r1} is the response frequency, which for an unforced system is equal to the nonlinear natural frequency.

Step 3: The matrix $\boldsymbol{\beta}^*$ can now be used in conjunction with Eqs. (4.58) and (4.59) to identify the near-identity transform and the nonlinear resonant terms, $\mathbf{n}_u$, that remain in the dynamic equation for $\mathbf{u}$. In contrast to the previous example, here there are two zeros, and hence resonant, terms in $\boldsymbol{\beta}^*$. These require the use of Eq. (4.59) whereas the other terms can be included in the transform using Eqs. (4.58). The resulting transform and nonlinear matrices are

$$\mathbf{n}^* = \alpha \begin{bmatrix} 1 \\ 3 \\ 3 \\ 1 \end{bmatrix}^T, \quad \boldsymbol{\beta}^* = \begin{bmatrix} 8\omega_{r1}^2 \\ 0 \\ 0 \\ 8\omega_{r1}^2 \end{bmatrix}^T \rightsquigarrow \mathbf{n}_u{}^* = \alpha \begin{bmatrix} 0 \\ 3 \\ 3 \\ 0 \end{bmatrix}^T, \quad \mathbf{h}^* = \frac{\alpha}{\omega_{r1}^2} \begin{bmatrix} 1/8 \\ 0 \\ 0 \\ 1/8 \end{bmatrix}^T.$$

Step 4: Finally, using Eqs. (4.46) and (4.52), the equation of motion in the transformed co-ordinate system can be written as

$$\ddot{\mathbf{u}} + \Lambda\mathbf{u} + \mathbf{n}^*\mathbf{u}^* = 0 \rightsquigarrow \ddot{\mathbf{u}} + \Lambda\mathbf{u} + 3\alpha\left(u_{p1}^2 u_{m1} + u_{p1} u_{m1}^2\right) = 0 \qquad (4.65)$$

to order ε^1. Using Eqs. (4.45) and (4.52), to order ε^1 the corresponding transformation may be written as

$$\mathbf{q} = \mathbf{u} + \mathbf{n}^*\mathbf{u}^* \rightsquigarrow \mathbf{q} = \mathbf{u} + \frac{\alpha}{8\omega_{r1}^2}\left(u_{1p}^3 + u_{1m}^3\right)$$

Note that for both these equations $\varepsilon = 1$ has been used as it is no longer necessary to track the significance of the terms. ■

The key feature of the transformed equation is that when the trial solution for $\mathbf{u}$ is applied, an exact harmonic balance can be performed for each equation of motion. From Eq. (4.51), the trial solution may be written as

$$u_{p1} = \frac{U_1}{2} e^{i(\omega_{r1}t - \phi_1)}, \quad u_{m1} = \frac{U_1}{2} e^{-i(\omega_{r1}t - \phi_1)}$$

and $\mathbf{u} = u_1 = u_{p1} + u_{m1}$. Using this in Eq. (4.65) gives

$$\left(\omega_n^2 - \omega_{r1}^2\right) \frac{U_1}{2} \left(e^{i(\omega_{r1}t-\phi_1)} + e^{-i(\omega_{r1}t-\phi_1)}\right)$$
$$+ 3\alpha \frac{U_1^3}{8} \left(e^{2i(\omega_{r1}t-\phi_1)} e^{-i(\omega_{r1}t-\phi_1)} + e^{2i(\omega_{r1}t-\phi_1)} e^{-i(\omega_{r1}t-\phi_1)}\right) = 0,$$

where it has been recalled that in this example $\Lambda = \omega_n^2$. Hence balancing *exactly* either the positive or the negative complex exponential terms gives

$$\omega_{r1}^2 = \omega_n^2 + \frac{3\alpha U_1^2}{4}$$

which represents the relationship between the response frequency and amplitude. This exact balance eliminates the need for introducing trial solutions containing multiple frequency terms in order to balance the equation as is the case when the harmonic balance is applied directly to the dynamics in $\mathbf{x}$. This equation represents the backbone equation for the system, as discussed in Sect. 4.1.

4.5.2 Higher Order Accuracy

Before extending the normal form technique to include forced, damped systems, improvement in the accuracy, via including ε^2 terms in the near-identity transform and resulting nonlinear terms in the dynamic equation for $\mathbf{u}$, is considered.

When deriving the near-identity transform, the ε^2 terms can be balanced by finding $\mathbf{h}_2(\mathbf{u})$ and $\mathbf{n}_{u2}(\mathbf{u})$ vectors that satisfy Eq. (4.50). This equation can be rewritten to look very similar to that for ε^1, Eq. (4.47)

$$\tilde{\mathbf{n}}_2(\mathbf{u}) + \Gamma \mathbf{h}_2(\mathbf{u}) + \frac{d^2}{dt^2}\mathbf{h}_2(\mathbf{u}) = \mathbf{n}_{u2}(\mathbf{u}), \tag{4.66}$$
$$\text{where:} \quad \tilde{\mathbf{n}}_2(\mathbf{u}) = \mathbf{n}_2(\mathbf{u}) + \Delta \mathbf{h}_1(\mathbf{u}) + D\{\mathbf{n}_1(\mathbf{u})\}\mathbf{h}_1(\mathbf{u})$$

Comparing this to Eq. (4.47) it can be seen that the nonlinear term prior to the application of the transform $\tilde{\mathbf{n}}_1(\mathbf{u})$ is now more complex, rather than being simply $\mathbf{n}_1(\mathbf{u})$, now it contains both $\mathbf{n}_2(\mathbf{u})$ and terms resulting from the ε^1 transform. The rest of the equation is identical (except it relates to the ε^2 terms indicated by the subscript 2 rather than ε^1 terms). It therefore follows that the approach to solving this equation to find $\mathbf{h}_2(\mathbf{u})$ and $\mathbf{n}_{u2}(\mathbf{u})$ is almost the same as before.

Firstly the terms are expressed in matrix form, as was done in Eq. (4.47) for the ε^1 terms, giving

$$\tilde{\mathbf{n}}_2(\mathbf{u}) = \mathbf{n}^+\mathbf{u}^+(\mathbf{u}_p, \mathbf{u}_m), \quad \mathbf{h}_2(\mathbf{u}) = \mathbf{h}^+\mathbf{u}^+(\mathbf{u}_p, \mathbf{u}_m), \quad \mathbf{n}_{u2}(\mathbf{u}) = \mathbf{n}_u{}^+\mathbf{u}^+(\mathbf{u}_p, \mathbf{u}_m), \tag{4.67}$$

Note now however $\tilde{\mathbf{n}}_2(\mathbf{u})$ is considered when deciding what terms to include in $\mathbf{u}^+$ rather than $\mathbf{n}_1(\mathbf{u})$, which was used to define $\mathbf{u}^*$.

Following the same approach as before, terms in $\mathbf{u}^+$ can be expressed in a general form as in Eq. (4.54). Then matrix $\boldsymbol{\beta}^+$ can be calculated in exactly the same way as $\boldsymbol{\beta}^*$ was calculated using Eq (4.56). This matrix allows the non- and near-resonant terms to be identified for each element within the matrix in turn. As with Eqs. (4.58) and (4.59), for the (k, ℓ)th this gives

$$n^+_{u,k,\ell} = 0, \quad h^+_{k,\ell} = n^+_{k,\ell}/\beta^+_{k,\ell} \quad \text{except if } \beta^+_{k,\ell} \approx 0 \text{ when: } n^+_{u,k,\ell} = n^+_{k,\ell}, \, h^+_{k,\ell} = 0. \tag{4.68}$$

Finally the near-identity transform equation and the resulting dynamic equations can be expressed to order ε^2 as

$$\begin{aligned} \mathbf{q} &= \mathbf{u} + \mathbf{h}^*\mathbf{u}^* + \mathbf{h}^+\mathbf{u}^+, \\ \ddot{\mathbf{u}} + \Lambda\mathbf{u} + \mathbf{n}^*\mathbf{u}^* + \mathbf{n}_u{}^+\mathbf{u}^+ &= 0, \end{aligned} \tag{4.69}$$

respectively. Note that here ε has been set to unity as tracking the relevance of terms is no longer required.

The ε^2 accurate transform will now be derived for an example system, the escape equation considered in Example 4.6.

Example 4.8 One DOF oscillator with a quadratic nonlinearity—the Escape equation.

Problem Find the order ε^2 accurate normal form for the unforced escape equation

$$\ddot{x} + \omega_n^2 x + \alpha x^2 = 0,$$

where the nonlinear term αx^2 can be treated as an order ε^1 term.

Solution The solution to this problem builds of the ε^1 solution calculated in Example 4.6, where it was found that

$$\mathbf{u}_{u1} = \mathbf{n}_u{}^*\mathbf{u}^* = 0, \quad \mathbf{h}_1 = \mathbf{h}^*\mathbf{u}^* = \frac{\alpha}{3\omega_{r1}^2}\left(u_{p1}^2 - 6u_{p1}u_{m1} + u_{m1}^2\right).$$

In addition $\mathbf{n}_1(\mathbf{u}) = \alpha u_1^2$ such that the Jacobian of $\mathbf{n}_1(\mathbf{u})$ is $D\{\mathbf{n}_1(\mathbf{u})\} = 2\alpha u_1$. Now using Eq. (4.66), the pre-transform nonlinear term, $\tilde{\mathbf{n}}_2$, can be identified as

$$\begin{aligned}\tilde{\mathbf{n}}_2(\mathbf{u}) &= \Delta \mathbf{h}_1(\mathbf{u}) + D\left\{\mathbf{n}_1(\mathbf{u})\right\}\mathbf{h}_1(\mathbf{u})\\ &= \frac{\alpha}{3\omega_{r1}^2}\left(\delta + 2\alpha(u_{p1} + u_{m1})\right)\left(u_{p1}^2 - 6u_{p1}u_{m1} + u_{m1}^2\right)\end{aligned}$$

Note that there is no ε^2 nonlinear term in the original equation of motion, hence $\mathbf{n}_2(\mathbf{u}) = 0$ and that for this one degree of freedom system $\Delta = \Lambda - \Gamma = \omega_n^2 - \omega_{r1}^2 = \delta$.

Expressing this in matrix form $\tilde{\mathbf{n}}_2(\mathbf{u}) = \mathbf{n}^+\mathbf{u}^+(\mathbf{u}_p, \mathbf{u}_m)$, Eq. (4.67), results in

$$\tilde{\mathbf{n}}_2(\mathbf{u}) = \mathbf{n}^+\mathbf{u}^+(\mathbf{u}_p, \mathbf{u}_m) \quad \text{where: } \mathbf{n}^+ = \frac{\alpha}{3\omega_{r1}^2}\begin{bmatrix} 2\alpha \\ -10\alpha \\ -10\alpha \\ 2\alpha \\ \delta \\ -6\delta \\ \delta \end{bmatrix}^T, \mathbf{u}^+ = \begin{bmatrix} u_{p1}^3 \\ u_{p1}^2 u_{m1} \\ u_{p1} u_{m1}^2 \\ u_{m1}^3 \\ u_{p1}^2 \\ u_{p1}u_{m1} \\ u_{m1}^2 \end{bmatrix}.$$

Now that $\mathbf{u}^+$ has been identified the corresponding $\boldsymbol{\beta}^+$ matrix can be found to be

$$\boldsymbol{\beta}^+ = \left[\, 8\omega_{r1}^2 \; 0 \; 0 \; 8\omega_{r1}^2 \; 3\omega_{r1}^2 \; -\omega_{r1}^2 \; 3\omega_{r1}^2 \,\right].$$

using the same equation as that used to find $\boldsymbol{\beta}^*$, Eq. (4.56). Using $\boldsymbol{\beta}^+$ and Eq. (4.68) the transform matrix $\mathbf{h}^+$ and the nonlinear matrix in the transformed equation of motion $\mathbf{n}_u{}^+$ can be written as

$$\mathbf{n}^+ = \frac{\alpha}{3\omega_{r1}^2}\begin{bmatrix} 2\alpha \\ -10\alpha \\ -10\alpha \\ 2\alpha \\ \delta \\ -6\delta \\ \delta \end{bmatrix}^T, \; \boldsymbol{\beta}^+ = \begin{bmatrix} 8\omega_{r1}^2 \\ 0 \\ 0 \\ 8\omega_{r1}^2 \\ 3\omega_{r1}^2 \\ -\omega_{r1}^2 \\ 3\omega_{r1}^2 \end{bmatrix}^T \rightsquigarrow \mathbf{n}_u{}^+ = \frac{10\alpha^2}{3\omega_{r1}^2}\begin{bmatrix} 0 \\ -1 \\ -1 \\ 0 \\ 0 \\ 0 \\ 0 \end{bmatrix}^T, \; \mathbf{h}^+ = \frac{\alpha}{3\omega_{r1}^4}\begin{bmatrix} \alpha/4 \\ 0 \\ 0 \\ \alpha/4 \\ \delta/3 \\ 6\delta \\ \delta/3 \end{bmatrix}^T.$$

The order ε^2 equation for the near-identity transform and for the dynamics following the near-identity transform, Eq. (4.69), may therefore be written as

$$\ddot{\mathbf{u}} + \Lambda\mathbf{u} - \frac{10\alpha^2}{3\omega_{r1}^2}\left(u_{p1}^2 u_{m1} + u_{p1}u_{m1}^2\right) = 0.$$

$$\begin{aligned}\mathbf{q} = u_1 &+ \frac{\alpha}{3\omega_{r1}^2}\left(u_{p1}^2 - 6u_{p1}u_{m1} + u_{m1}^2\right)\\ &+ \frac{\alpha}{36\omega_{r1}^4}\left[3\alpha\left(u_{p1}^3 + u_{m1}^3\right) + 4\delta\left(u_{p1}^2 + u_{m1}^2\right) + 72\delta u_{p1}u_{m1}\right]\end{aligned}$$

respectively, where the expressions for the ε^1 terms have been used from Example 4.6.

With the ε^2 terms it can be seen that the dynamic equation now has nonlinear resonant terms and hence the response frequency will by amplitude dependent. Using the trial solution for $\mathbf{u}$, see Eq. (4.61) from Example 4.6, to solve the dynamic equation gives the frequency relationship

$$\omega_{r1}^2 = \omega_n^2 - \frac{10\alpha^2}{3\omega_{r1}^2}\frac{u_1(0)^2}{4}. \tag{4.70}$$

■

In the numerical simulations following Example 4.6 it was seen that the ε^1 accurate normal form prediction for free vibration response of the escape equation agreed well for $x(0) = 0.01$, $\omega_n = 4\pi$ and $\alpha = 800$. However the agreement was less strong, particularly in terms of the response frequency ω_{r1}, when α was increased to 3,200. Using the ε^1 accurate normal form solution the response amplitude and frequency are $(u_1(0), \omega_{r1}) = (0.0102, 4\pi)$ and $(u_1(0), \omega_{r1}) = (0.0108, 4\pi)$ for $\alpha = 800$ and $\alpha = 3{,}200$ respectively, where Eq. (4.63) has been used to find $u_1(0)$. Using these values an improved prediction for the response frequency can be found using Eq. (4.70) once rearranged as a quadratic in ω_{r1}^2. This gives $\omega_{r1} = 1.998 \times 2\pi$ and $\omega_{r1} = 1.958 \times 2\pi$ for $\alpha = 800$ and $\alpha = 3{,}200$ respectively. Comparing these values to the frequency content of the time-simulation data shown in Figs. 4.5 and 4.6, it can be seen that the ε^2 solution provides a good estimate of the response frequency.

4.5.3 Forced Vibration

The normal form analysis will now be extended to allow the study of systems that are excited by external harmonic forcing and have viscous damping. Following this, Example 4.7 will be extended to consider the forced, damped Duffing oscillator.

With forcing, the equation of motion, Eq. (4.42), becomes:

$$M\ddot{\mathbf{x}} + C\dot{\mathbf{x}} + K\mathbf{x} + \mathscr{N}(\mathbf{x}, \dot{\mathbf{x}}, \mathbf{r}) = P_x\mathbf{r}, \tag{4.71}$$

where P_x is a forcing amplitude matrix and $\mathbf{r} = \{r_p, r_m\}^T$ is a forcing vector with $r_p = \mathrm{e}^{\mathrm{i}\Omega t}$ and $r_m = \mathrm{e}^{-\mathrm{i}\Omega t}$, where Ω is the forcing frequency. Note that now the nonlinear terms are written as a function of velocity as well as displacement states and as a function of the forcing to permit parametric excitation terms—such terms will be discussed when cable dynamics are considered in Chap. 7.

The damping term is assumed to be small relative to the linear stiffness terms and hence or order ε^1. For algebraic convenience it is combined into the (also small) nonlinear term to give

$$M\ddot{\mathbf{x}} + K\mathbf{x} + \tilde{\mathscr{N}}(\mathbf{x}, \dot{\mathbf{x}}, \mathbf{r}) = P_x\mathbf{r}, \quad \text{where: } \tilde{\mathscr{N}}(\mathbf{x}, \dot{\mathbf{x}}, \mathbf{r}) = \mathscr{N}(\mathbf{x}, \dot{\mathbf{x}}, \mathbf{r}) + C\dot{\mathbf{x}}. \tag{4.72}$$

As with the unforced analysis, the application of the normal form technique will be presented as a series of steps. With the introduction of forcing it is necessary to introduce a new step, namely a forcing transformation, in which forcing terms that are non-resonant are removed from the equation of motion and their effects captured by the transform. This transformation ensures that an exact harmonic balance can still be performed on the equation for the dynamics in $\mathbf{u}$ using the trial solution given in Eq. (4.51). The additional step will be labelled **Step 1f** as it comes after the linear modal transform but before the near-identity transform. With the new step comes a new set of coordinates $\mathbf{v}$. The technique therefore involves the linear modal transform from $\mathbf{x} \to \mathbf{q}$, the forcing transformation from $\mathbf{q} \to \mathbf{v}$ and the near-identity transform from $\mathbf{v} \to \mathbf{u}$.

Before these steps are discussed, the response frequency for the kth mode, ω_{rk}, is considered. For the unforced system ω_{rk} was the nonlinear natural frequency of the kth mode. With forcing, the frequency of the response is taken to be near-resonant, such that $\omega_{rk} \approx \omega_{nk}$, but is also dependent on the forcing frequency Ω. If the forcing frequency is close to linear natural frequency for the kth mode, ω_{nk}, then the forcing is near-resonant and the relationship is straightforward: $\omega_{rk} = \Omega$. However for modes where the forcing is not close to resonance, the response frequency is taken to be a frequency close to the linear natural frequency that is an integer multiplier or divider of the forcing frequency. For example, for a one degree of freedom system with forcing at approximately three times the natural frequency the response frequency would be $\omega_{r1} = \Omega/3$ such that it captured the near-resonant response of the system.

Step 1: Firstly the equation of motion is written in *linear modal normal form*. As a linear modal decomposition it is unaffected by the presence of damping and forcing. As with the unforced case, the linear modal decomposition is achieved by writing $\mathbf{x} = \Phi\mathbf{q}$, where $\mathbf{q}$ are the modal coordinates for the linear system. Here Φ is a matrix of column vectors corresponding to the eigenvectors of $M^{-1}K$ and Λ is a diagonal matrix of the corresponding eigenvalues (with the kth corresponding to the square of kth natural frequency of the linearised system, ω_{nk}^2).

Eliminating $\mathbf{x}$ in Eq. (4.42) using $\mathbf{x} = \Phi\mathbf{q}$, premultiplying by Φ^T and then premultiplying by $(\Phi^T M \Phi)^{-1}$ gives

$$\ddot{\mathbf{q}} + \Lambda\mathbf{q} + N_q(\mathbf{q}, \dot{\mathbf{q}}, \mathbf{r}) = P_q\mathbf{r}, \tag{4.73}$$

where

$$N_q(\mathbf{q}, \dot{\mathbf{q}}, \mathbf{r}) = (\Phi^T M \Phi)^{-1}\Phi^T \tilde{\mathcal{N}}(\Phi\mathbf{q}, \Phi\dot{\mathbf{q}}, \mathbf{r}), \quad P_q = (\Phi^T M \Phi)^{-1}\Phi^T P_x$$

and the relationship $(\Phi^T M \Phi)^{-1}\Phi^T K \Phi = \Lambda$ (since by definition of the eigenvectors/values $M^{-1}K\Phi = \Phi\Lambda$) has been used.

Step 1f: A forcing transformation is now applied. The purpose of this transformation is to remove forcing terms that are non-resonant, where for the kth mode the resonant forcing terms are taken to be ones where the forcing frequency Ω is close to the linear natural frequency ω_{nk}. This is achieved using the transform $\mathbf{q} = \mathbf{v} + \mathbf{e}\mathbf{r}$ where for a system with N degrees of freedom, $\mathbf{e}$ is a matrix of size $N \times 2$. By substituting this transform into the equation for the dynamics in $\mathbf{q}$, Eq. (4.73), and rearranging the dynamic equation in transformed state $\mathbf{v}$ may be written as

$$\ddot{\mathbf{v}} + \Lambda\mathbf{v} + N_v(\mathbf{v}, \dot{\mathbf{v}}, \mathbf{r}) = P_v\mathbf{r}, \tag{4.74}$$

where the nonlinear vector has been transformed using

$$N_v(\mathbf{v}, \dot{\mathbf{v}}, \mathbf{r}) = N_q(\mathbf{v} + \mathbf{e}\mathbf{r}, \dot{\mathbf{v}} + \mathbf{e}W\mathbf{r}, \mathbf{r}), \tag{4.75}$$

and W is a 2×2 diagonal matrix with diagonal elements $i\Omega$ and $-i\Omega$. The transform matrix $\mathbf{e}$ and the post-transformed forcing term P_v (which, as with P_q, is size $N \times 2$) may be found element by element. Considering the kth row in these matrices, which corresponds to the kth mode with linear natural frequency ω_{nk}, if $\Omega \approx \omega_{nk}$ then

$$\text{near-resonant forcing:} \quad \begin{array}{ll} e_{k,1} = 0, & e_{k,2} = 0, \\ P_{v,k,1} = P_{q,k,1}, & P_{v,k,2} = P_{q,k,2}, \end{array} \tag{4.76}$$

otherwise

$$\text{non-resonant forcing:} \quad \begin{array}{ll} e_{k,1} = P_{q,k,1}/(\omega_{nk}^2 - \Omega^2) & e_{k,2} = P_{q,k,2}/(\omega_{nk}^2 - \Omega^2), \\ P_{v,k,1} = 0 & P_{v,k,2} = 0. \end{array} \tag{4.77}$$

The derivation of these equations is shown in Box 4.3.

Note that the forcing transformation step can be included for the unforced system, but results in the trivial unity transform $\mathbf{v} = \mathbf{q}$.

Box 4.4 *Derivation of Eqs.* (4.74), (4.76) *and* (4.77)
Substituting the, as yet unknown, forcing transform, $\mathbf{q} = \mathbf{v} + \mathbf{e}\mathbf{r}$, into the equation of motion in terms of $\mathbf{q}$, Eq. (4.43), gives

$$\ddot{\mathbf{v}} + \mathbf{e}WW\mathbf{r} + \Lambda\mathbf{v} + \Lambda\mathbf{e}\mathbf{r} + N_q(\mathbf{v} + \mathbf{e}\mathbf{r}, \dot{\mathbf{v}} + \mathbf{e}W\mathbf{r}, \mathbf{r}) = P_q\mathbf{r},$$

where $\dot{\mathbf{r}} = W\mathbf{r}$, such that W is a diagonal 2×2 matrix with elements $W_{1,1} = \mathrm{i}\Omega$ and $W_{2,2} = -\mathrm{i}\Omega$.

From this equation the relationship

$$\ddot{\mathbf{v}} + \Lambda\mathbf{v} + N_v(\mathbf{v}, \dot{\mathbf{v}}, \mathbf{r}) = P_v\mathbf{r},$$

can be written provided the nonlinear matrices are related by

$$N_v(\mathbf{v}, \dot{\mathbf{v}}, \mathbf{r}) = N_q(\mathbf{v} + \mathbf{e}\mathbf{r}, \dot{\mathbf{v}} + \mathbf{e}W\mathbf{r}, \mathbf{r}),$$

and the forcing terms are related by

$$\tilde{\mathbf{e}} + P_v = P_q \quad \text{with: } \tilde{\mathbf{e}} = \mathbf{e}WW + \Lambda\mathbf{e}.$$

Since Λ is a diagonal matrix with the kth diagonal element being ω_{nk}^2 the elements in the kth row of $\tilde{\mathbf{e}}$ can be related to those in $\mathbf{e}$ using

$$\tilde{e}_{k,1} = e_{k,1}(\omega_{nk}^2 - \Omega^2), \quad \tilde{e}_{k,2} = e_{k,2}(\omega_{nk}^2 - \Omega^2).$$

To satisfy the forcing term equation, while removing the non-resonant forcing terms (such that the exact harmonic balance can still be applied to the equation of motion in $\mathbf{u}$), the matrices are considered row by row. For the kth row of the matrices, if $\Omega \approx \omega_{nk}$ the forcing is resonant and therefore the forcing terms are kept in the equation for the dynamics, hence

$$\tilde{e}_{k,1} = 0, \quad \tilde{e}_{k,2} = 0, \quad P_{v,k,1} = P_{q,k,1}, \quad P_{v,k,2} = P_{q,k,2}.$$

Otherwise, the forcing is non-resonant and the forcing terms are moved to the transform matrix $\mathbf{e}$ resulting in

$$\tilde{e}_{k,1} = P_{q,k,1}, \quad \tilde{e}_{k,2} = P_{q,k,2}, \quad P_{v,k,1} = 0, \quad P_{v,k,2} = 0.$$

Step 2: Now the nonlinear near-identity transform is sought with the aim of removing the non-resonant nonlinear terms.

The process is very similar to that for the unforced system, albeit the transform is now from $\mathbf{v} \to \mathbf{u}$ whereas when the unforced system was discussed the transform was from $\mathbf{q} \to \mathbf{u}$.

Writing the nonlinear term in the equation of motion, Eq. (4.74), in terms of increasing powers of ε gives

$$\ddot{\mathbf{v}} + \Lambda\mathbf{v} + N_v(\mathbf{v}, \dot{\mathbf{v}}, \mathbf{r}) = P_v\mathbf{r},$$
$$\text{where } N_v(\mathbf{v}, \dot{\mathbf{v}}, \mathbf{r}) = \varepsilon\mathbf{n}_1(\mathbf{v}, \dot{\mathbf{v}}, \mathbf{r}) + \varepsilon^2\mathbf{n}_2(\mathbf{v}, \dot{\mathbf{v}}, \mathbf{r}) + \cdots. \quad (4.78)$$

Note here $\mathbf{n}_1$ is the ε^1 significant term in N_v, whereas in the unforced case it was used as the ε^1 significant term in N_q—this is consistent given that the forcing transform is a unity transform in the unforced case such that $N_v = N_q$ when there is no forcing. The near-identity transform, which is now a function of the forcing $\mathbf{r}$ as well as the states, may be expressed as

$$\mathbf{v} = \mathbf{u} + \mathbf{h}(\mathbf{u}, \dot{\mathbf{u}}, \mathbf{r}) \quad \text{where } \mathbf{h}(\mathbf{u}, \dot{\mathbf{u}}, \mathbf{r}) = \varepsilon\mathbf{h}_1(\mathbf{u}, \dot{\mathbf{u}}, \mathbf{r}) + \varepsilon^2\mathbf{h}_2(\mathbf{u}, \dot{\mathbf{u}}, \mathbf{r}) + \cdots, \quad (4.79)$$

and the post-transformed equation of motion as

$$\ddot{\mathbf{u}} + \Lambda\mathbf{u} + N_u(\mathbf{u}, \dot{\mathbf{u}}, \mathbf{r}) = P_u\mathbf{r} \quad \text{where} \quad N_u(\mathbf{u}, \dot{\mathbf{u}}, \mathbf{r}) = \varepsilon\mathbf{n}_{u1}(\mathbf{u}, \dot{\mathbf{u}}, \mathbf{r}) + \varepsilon^2\mathbf{n}_{u2}(\mathbf{u}, \dot{\mathbf{u}}, \mathbf{r}) + \cdots. \tag{4.80}$$

As before, the process of finding the near-identity transform starts with combining Eqs. (4.78), (4.79) and (4.80). Following the same procedure as for the unforced case, see Box 4.2, the balance of ε^1 terms is given by

$$\mathbf{n}_1(\mathbf{u}, \dot{\mathbf{u}}, \mathbf{r}) + \Gamma\mathbf{h}_1(\mathbf{u}, \dot{\mathbf{u}}, \mathbf{r}) + \frac{\mathbf{d}^2}{\mathbf{d}t^2}\mathbf{h}_1(\mathbf{u}, \dot{\mathbf{u}}, \mathbf{r}) = \mathbf{n}_{u1}(\mathbf{u}, \dot{\mathbf{u}}, \mathbf{r}). \tag{4.81}$$

This equation is identical to the unforced version, Eq. (4.47), except that now the vectors are functions of $\dot{\mathbf{u}}$ and $\mathbf{r}$ as well as $\mathbf{u}$. In addition there is an ε^0 equation due to the presence of forcing terms which is

$$\varepsilon^0 : P_u\mathbf{r} = P_v\mathbf{r} \tag{4.82}$$

and so $P_u = P_v$, such that the direct forcing terms are unaffected by the transform.

The trial solution for $\mathbf{u} = \mathbf{u}_p + \mathbf{u}_m$, Eq. (4.51) along with $\mathbf{r} = [\mathbf{r}_p \;\; \mathbf{r}_m]^T = [\mathrm{e}^{\mathrm{i}\Omega t} \;\; \mathrm{e}^{-\mathrm{i}\Omega t}]^T$ are now used. The vectors $\mathbf{n}_1$, $\mathbf{h}_1$ and $\mathbf{n}_{u1}$ are expressed in matrix form as

$$\begin{aligned}
\mathbf{n}_1(\mathbf{u}, \dot{\mathbf{u}}, \mathbf{r}) &= \mathbf{n}^*\mathbf{u}^*(\mathbf{u}_p, \mathbf{u}_m, \mathbf{r}),\\
\mathbf{h}_1(\mathbf{u}, \dot{\mathbf{u}}, \mathbf{r}) &= \mathbf{h}^*\mathbf{u}^*(\mathbf{u}_p, \mathbf{u}_m, \mathbf{r}),\\
\mathbf{n}_{u1}(\mathbf{u}, \dot{\mathbf{u}}, \mathbf{r}) &= \mathbf{n}_u^*\mathbf{u}^*(\mathbf{u}_p, \mathbf{u}_m, \mathbf{r}),
\end{aligned} \tag{4.83}$$

where $\mathbf{u}^*$ is a column vector (of size $L \times 1$) containing all the combinations of $\mathbf{u}_p$, $\mathbf{u}_m$ and, with the presence of forcing, $\mathbf{r}$ terms that appear within $\mathbf{n}_1(\mathbf{u})$. The $N \times L$ matrices $\mathbf{n}^*$, $\mathbf{h}^*$ and $\mathbf{n}_u^*$ contain the coefficient terms.

To proceed in finding the unknown matrices $\mathbf{h}^*$ and $\mathbf{n}_u^*$ that satisfy Eq. (4.81) the ℓth term in $\mathbf{u}^*$ is written in the general form

$$u_\ell^* = r_p^{m_{p\ell}} r_m^{m_{m\ell}} \prod_{k=1}^{N} \left(u_{pk}^{s_{pk\ell}} u_{mk}^{s_{mk\ell}} \right), \tag{4.84}$$

where $m_{p\ell}$ and $m_{m\ell}$ are power indices.

Note that this is different to the unforced case, Eq. (4.54), due to the presence of the forcing terms in $\mathbf{r}$.

Using this representation and with some manipulation, see Box 4.5, Eq. (4.81) can be rewritten as

$$\mathbf{n}_u^* = \mathbf{n}^* - \tilde{\mathbf{h}}^*, \tag{4.85}$$

where the (k, ℓ)th element in $\tilde{\mathbf{h}}^*$ is related to the corresponding element in $\mathbf{h}^*$ via

$$\tilde{h}^*_{k,\ell} = \beta^*_{k,\ell} h^*_{k,\ell}$$

$$\text{with: } \beta^*_{k,\ell} = \left(\left[m_{p\ell} - m_{ml} \right] \Omega + \sum_{n=1}^{N} \left[(s_{pn\ell} - s_{mn\ell}) \omega_{rn} \right] \right)^2 - \omega^2_{rk}, \quad (4.86)$$

and $\beta_{k,\ell}$ is the (k, ℓ)th element in matrix $\boldsymbol{\beta}^*$—the matrix that is key in determining the transform and the nonlinear terms that remain in the post-transformed equation of motion (as is discussed in the next step).

Box 4.5 *Derivation of Eq.* (4.85)
As with the unforced case, Eq. (4.81) may be simplified by considering the double differential of $\mathbf{u}^*$ with respect to time. Using the general form for the ℓth term in $\mathbf{u}^*$, Eq. (4.84) for the forced case, the first differential may be written as

$$\begin{aligned} \frac{\mathrm{d}u^*_\ell}{\mathrm{d}t} &= i\Omega \left(\frac{\partial u^*_\ell}{\partial r_p} r_p - \frac{\partial u^*_\ell}{\partial r_m} r_m \right) + \sum_{n=1}^{N} i\omega_{rn} \left(\frac{\partial u^*_\ell}{\partial u_{pn}} u_{pn} - \frac{\partial u^*_\ell}{\partial u_{mn}} u_{mn} \right) \\ &= i \left(\left[m_{p\ell} - m_{ml} \right] \Omega + \sum_{n=1}^{N} \left[(s_{pn\ell} - s_{mn\ell}) \omega_{rn} \right] \right) u^*_\ell . \end{aligned}$$

As a result

$$\frac{\mathrm{d}^2 \mathbf{u}^*}{\mathrm{d}t^2} = -\tilde{\Gamma} \mathbf{u}^*,$$

where $\tilde{\Gamma}$ is a diagonal matrix (of size $L \times L$) with the ℓth diagonal element being

$$\tilde{\Gamma}_{\ell,\ell} = \left(\left[m_{p\ell} - m_{ml} \right] \Omega + \sum_{n=1}^{N} \left[(s_{pn\ell} - s_{mn\ell}) \omega_{rn} \right] \right)^2 .$$

Equation (4.84) may now be rewritten as

$$\mathbf{n}^* - \tilde{\mathbf{h}}^* = \mathbf{n}^*_u \quad \text{where: } \tilde{\mathbf{h}}^* = \mathbf{h}^* \tilde{\Gamma} - \Gamma \mathbf{h}^* .$$

Finally $\tilde{\mathbf{h}}^*$ is linked to $\mathbf{h}^*$ on an element by element basis via matrix $\boldsymbol{\beta}^*$ using the relationship for the (k, ℓ)th element

$$\tilde{h}^*_{k,\ell} = \beta^*_{k,\ell} h^*_{k,\ell}$$

$$\text{where: } \beta^*_{k,\ell} = \left(\left[m_{p\ell} - m_{ml} \right] \Omega + \sum_{n=1}^{N} \left[(s_{pn\ell} - s_{mn\ell}) \omega_{rn} \right] \right)^2 - \omega^2_{rk} .$$

Step 3: The near-identity transform and the post-transform nonlinear terms in the equation for the dynamics may now be identified using the method outlined in the unforced case. The $\boldsymbol{\beta}^*$ matrix is used to decide, for each element in turn, whether the nonlinearity in n_1 remains in the dynamic equations and as a result is moved to n_{u1} or whether it is scaled and moved into the transform using Eqs. (4.58) and (4.59).

Step 4: With $\mathbf{n}_u$ and $\mathbf{h}$ identified, the equation of motion in $\mathbf{u}$, which may be written $\ddot{\mathbf{u}} + \Lambda\mathbf{u} + \mathbf{n}_u^*\mathbf{u}^* = P_u\mathbf{r}$ to order ε^1, can be solved using the trial solution for $\mathbf{u}$, Eq. (4.51). Finally the full response in terms of $\mathbf{x}$ can be calculated using the transforms $\mathbf{v} = \mathbf{u} + \mathbf{h}^*\mathbf{u}^*$ (to order ε^1), $\mathbf{q} = \mathbf{v} + \mathbf{e}r$ and $\mathbf{x} = \Phi\mathbf{q}$.

This method will now be applied, firstly with a near-resonant forcing to the forced Duffing equation, and then with a forcing at a frequency that is approximately half that of the resonant frequency to the escape equation (in which there is a quadratic stiffness nonlinearity).

Example 4.9 Forced one DOF oscillator with a cubic nonlinearity—the forced Duffing equation near resonance.

Problem Consider the Duffing oscillator with sinusoidal forcing close to resonance

$$\ddot{x} + 2\zeta\omega_n\dot{x} + \omega_n^2 x + \alpha x^3 = R\cos(\Omega t),$$

which can be written in the form of Eq. (4.71) using $P_x = [R/2, R/2]$ and $\mathbf{r} = \{r_p, r_m\}^T = \{e^{i\Omega t}, e^{-i\Omega t}\}^T$. Find the normal form prediction for the steady-state response of this system.

Solution As the forcing is close to resonance, the response frequency is set to be the same as the forcing frequency, $\omega_{r1} = \Omega$.

Step 1: The linear transform is trivial as the system has only one degree of freedom and so $\mathbf{q} = \mathbf{x} = x$. Using this the system can be expressed in the form of Eq. (4.43) giving

$$\ddot{\mathbf{q}} + \Lambda\mathbf{q} + N_q(\mathbf{q}, \dot{\mathbf{q}}, \mathbf{r}) = P_q\mathbf{r},$$

where: $\Lambda = \omega_n^2, \quad N_q(\mathbf{q}, \dot{\mathbf{q}}, \mathbf{r}) = 2\zeta\omega_n\dot{q}_1 + \alpha q_1^3, \quad P_q = P_x,$

where q_1 is the only element in $\mathbf{q}$. Note that as discussed around Eq. (4.72) the damping term has been included within the nonlinear vector.

Step 1f: The forcing transformation is now applied to remove non-resonant forcing terms. In this example the forcing is near-resonant for the one mode system and so the transform, $\mathbf{q} = \mathbf{v} + \mathbf{e}\mathbf{r}$, reduces to the unity transform $\mathbf{q} = \mathbf{v}$ with $\mathbf{e} = [0 \;\; 0]$—see Eqs. (4.76) and (4.77).

This results in

$$\ddot{\mathbf{v}} + \Lambda\mathbf{v} + N_v(\mathbf{v}, \dot{\mathbf{v}}, \mathbf{r}) = P_v\mathbf{r},$$

$$\text{with } N_v(\mathbf{v}, \dot{\mathbf{v}}, \mathbf{r}) = N_q(\mathbf{v}, \dot{\mathbf{v}}, \mathbf{r}) = 2\zeta\omega_n\dot{v}_1 + \alpha v_1^3, \quad P_v = P_q.$$

Step 2: Now the near-identity transform must be found. Firstly the nonlinear term is written in terms of $\mathbf{u}$ and expressed in powers of ε, as in Eq. (4.78). Assuming that both the damping and nonlinear cubic term are of order ε^1 gives

$$\mathbf{n}_1(\mathbf{u}, \dot{\mathbf{u}}, \mathbf{r}) = 2\zeta\omega_n\dot{u}_1 + \alpha u_1^3.$$

Writing $\mathbf{u} = \mathbf{u}_p + \mathbf{u}_m$, such that $u_1 = u_{p1} + u_{m1}$ and expressing $\mathbf{n}_1$ in matrix form gives

$$\mathbf{n}_1(\mathbf{u}, \dot{\mathbf{u}}, \mathbf{r}) = \mathbf{n}^*\mathbf{u}^*(\mathbf{u}_p, \mathbf{u}_m, \mathbf{r}) \;\; \text{where: } \mathbf{n}^* = \begin{bmatrix} \alpha \\ 3\alpha \\ 3\alpha \\ \alpha \\ i2\zeta\omega_n\omega_{r1} \\ -i2\zeta\omega_n\omega_{r1} \end{bmatrix}^T, \; \mathbf{u}^* = \begin{bmatrix} u_{p1}^3 \\ u_{p1}^2 u_{m1} \\ u_{p1}u_{m1}^2 \\ u_{m1}^3 \\ u_{p1} \\ u_{m1} \end{bmatrix}.$$

Using the general form for each term in $\mathbf{u}^*$ given by Eqs. (4.84), (4.86) is used to find $\boldsymbol{\beta}^*$

$$\boldsymbol{\beta}^* = \left[8\Omega^2 \; 0 \; 0 \; 8\Omega^2 \; 0 \; 0\right].$$

where $\omega_{r1} = \Omega$ has been used.

Step 3: The matrix $\boldsymbol{\beta}^*$ is used with Eqs. (4.58) and (4.59) to find the near-identity transform and the resonant nonlinear terms that remain in the dynamic equation for $\mathbf{u}$ giving

$$\boldsymbol{\beta}^* = \begin{bmatrix} 8\omega_{r1}^2 \\ 0 \\ 0 \\ 8\omega_{r1}^2 \\ 0 \\ 0 \end{bmatrix}^T \rightsquigarrow \mathbf{n}_u^* = \begin{bmatrix} 0 \\ 3\alpha \\ 3\alpha \\ 0 \\ i2\zeta\omega_n\omega_{r1} \\ -i2\zeta\omega_n\omega_{r1} \end{bmatrix}^T, \quad \mathbf{h}^* = \frac{\alpha}{\omega_{r1}^2}\begin{bmatrix} 1/8 \\ 0 \\ 0 \\ 1/8 \\ 0 \\ 0 \end{bmatrix}^T.$$

Step 4: The equation of motion in the transformed co-ordinate system and the near-identity transform may now be written as

$$\ddot{\mathbf{u}} + \Lambda\mathbf{u} + \mathbf{n}^*\mathbf{u}^* = P_u\mathbf{r}$$
$$\rightsquigarrow \ddot{\mathbf{u}} + 2\zeta\omega_n\dot{\mathbf{u}} + \omega_n^2\mathbf{u} + 3\alpha\left(u_{p1}^2 u_{m1} + u_{p1}u_{m1}^2\right) = P_u\mathbf{r}$$

and

$$\mathbf{v} = \mathbf{u} + \mathbf{n}^*\mathbf{u}^* \rightsquigarrow \mathbf{v} = \mathbf{u} + \frac{\alpha}{8\omega_{r1}^2}\left(u_{1p}^3 + u_{1m}^3\right)$$

to order ε^1, respectively. Note that for this example $\mathbf{x} = \mathbf{q} = \mathbf{v}$.

Using the trial solution, Eq. (4.51), and recalling that here $\omega_{r1} = \Omega$ and $P_u\mathbf{r} = P_x\mathbf{r} = R\cos(\Omega t)$, the equation for the dynamics may be written as

$$(\omega_n^2 - \Omega^2)\frac{U_1}{2}\left(e^{i(\Omega t-\phi_1)} + e^{-i(\Omega t-\phi_1)}\right) + 2i\zeta\omega_n\Omega\frac{U_1}{2}\left(e^{i(\Omega t-\phi_1)} - e^{-i(\Omega t-\phi_1)}\right)$$
$$+ 3\alpha\frac{U_1^3}{8}\left(e^{i(\Omega t-\phi_1)} + e^{-i(\Omega t-\phi_1)}\right) = \frac{R}{2}\left(e^{i\Omega t} + e^{-i\Omega t}\right)$$

Balancing the $e^{i(\Omega t-\phi_1)}$ and $e^{-i(\Omega t-\phi_1)}$ terms gives

$$e^{i(\Omega t-\phi_1)}\colon (\omega_n^2 - \Omega^2)U_1 + 2i\zeta\omega_n\Omega U_1 + \frac{3\alpha}{4}U_1^3 = Re^{i\phi_1}$$
$$e^{-i(\Omega t-\phi_1)}\colon (\omega_n^2 - \Omega^2)U_1 - 2i\zeta\omega_n\Omega U_1 + \frac{3\alpha}{4}U_1^3 = Re^{-i\phi_1}$$

Noting that these are a complex conjugate pair, the real and the imaginary terms can be balanced to give

$$Re\colon (\omega_n^2 - \Omega^2)U_1 + \frac{3\alpha}{4}U_1^3 = R\cos(\phi_1)$$
$$Im\colon 2\zeta\omega_n\Omega U_1 = R\sin(\phi_1)$$

Rather than balancing the complex exponential terms and then the real and imaginary ones, these two equations could have been found directly by balancing $\cos(\Omega t - \phi_1)$ and $\sin(\Omega t - \phi_1)$, however the two step approach is arguably less susceptible to algebraic error. Squaring and adding the Re and Im equations allows ϕ_1 to be eliminated to give the amplitude relationship

$$\left[(\omega_n^2 - \Omega^2)U_1 + \frac{3\alpha}{4}U_1^3\right]^2 + [2\zeta\omega_n\Omega U_1]^2 = R^2, \tag{4.87}$$

and dividing them gives the phase relationship

$$\tan(\phi_1) = \frac{2\zeta\omega_n\Omega}{\omega_n^2 - \Omega^2 + \frac{3\alpha}{4}U_1^2}.$$

Finally the trial solution and the relationship $\omega_{r1} = \Omega$ can be used to simplify the transform equation to

$$\mathbf{x} = \mathbf{q} = \mathbf{v} = U_1\cos(\Omega t - \phi_1) + \frac{\alpha}{8\Omega^2}\frac{U_1^3}{4}\cos(3[\Omega t - \phi_1]).$$

The response may be split into two components, one at Ω with amplitude $X_\Omega = U_1$ and one at 3Ω with amplitude $X_{3\Omega} = \alpha U_1^3/(32\Omega^2)$, where U_1 can be found for given forcing parameters R and Ω using Eq. (4.87). ■

To assess the accuracy of the normal form transformation technique, Fig. 4.7 shows a time-stepping simulation for the case where $\omega_n = 4\pi$, $\zeta = 0.01$, $\alpha = 200{,}000$ and with a forcing amplitude of $R = 0.03$. The simulation is started with a forcing frequency below resonance, 1.8 Hz. The frequency is gradually increased up to 2.1 Hz in a series of steps. At each frequency the steady-state response amplitude is recorded and shown as dots in the figure. The frequency is then decreased in steps back down to 1.8 Hz, and the resulting steady-state response amplitudes are shown as circles. The basin of attraction for the upper solution branch becomes smaller as frequency increases. As a result, to ensure that the numerically simulated response remains on the upper solution branch as long as possible, the step changes in frequency must be very smooth. So, the frequency changes were made after a whole number of oscillations and, to further smooth the transitions, three intermediate frequencies were used between each frequency point (each consisting of a whole number of cycles). The external forcing is held at each frequency point for approximately 500s, the first 100s of which is not included in calculating the amplitude response to ensure the system is at steady state. The intermediate frequencies are applied for around 10s.

The simulation data compares almost exactly to the normal form predictions shown as a solid line (a plot of X_Ω) in Fig. 4.7. This was calculated by observing that Eq. (4.87) is a quadratic in Ω^2 so two roots can be found for Ω^2 for any given value of response amplitude U_1, if they are positive and real then the solution is valid. It can be seen that the normal form technique predicts a longer resonance peak than is observed in the time-stepping simulations. The reason for this is that because the basin of attraction for the upper solution becomes much smaller towards the tip of the peak, and the small frequency steps made in the time-stepping approach become more likely to cause a jump to the lower solution at each increment. The amplitude of response at three times the driving frequency is comparatively small, the normal form method predicts a maximum value of approximately $X_{3\Omega} = 2.8 \times 10^{-5}$.

Example 4.10 Forced one DOF oscillator with a quadratic nonlinearity—the forced escape equation—with forcing away from resonance.

Problem Consider the escape equation, an oscillator with a quadratic stiffness nonlinearity, subjected to sinusoidal forcing at a frequency away from resonance

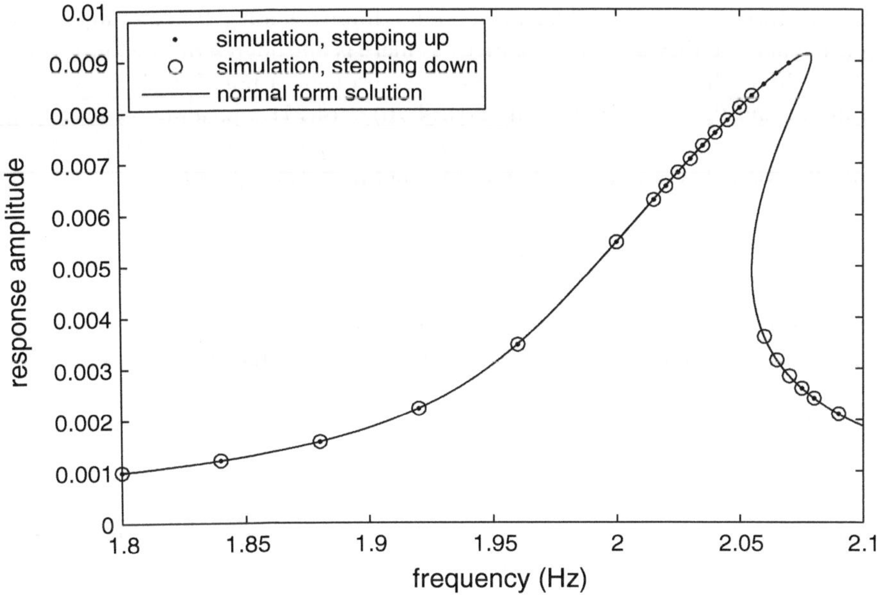

Fig. 4.7 Response of the forced Duffing equation with $\omega_n = 4\pi$, $\zeta = 0.01$, $\alpha = 200{,}000$ and with a forcing amplitude of $R = 0.03$ for a range of forcing frequencies around the natural frequency

$$\ddot{x} + 2\zeta\omega_n\dot{x} + \omega_n^2 x + \alpha x^2 = R\cos(\Omega t),$$

which can be written in the form of Eq. (4.71) using $P_x = [R/2, R/2]$ and $\mathbf{r} = \{r_p, r_m\}^T = \{\mathrm{e}^{\mathrm{i}\Omega t}, \mathrm{e}^{-\mathrm{i}\Omega t}\}^T$. Find the normal form solution for this system.

Solution As the forcing is away from resonance, the response frequency is taken to be close to the linear natural frequency, ω_n, while also being a simple multiple, or divider, of the forcing frequency. The forcing frequency can be approximately related to the linear natural frequency by $\Omega \approx a\omega_n$ where a or $1/a$ is an integer. Using this, the relationship $\Omega = a\omega_{r1}$ is used to ensure that ω_{r1}, the response frequency, is close to resonance.

Step 1: The system has only one degree of freedom and so $\mathbf{q} = \mathbf{x} = x$ and the system can be expressed in the form of Eq. (4.43) as

$$\ddot{\mathbf{q}} + \Lambda\mathbf{q} + N_q(\mathbf{q}, \dot{\mathbf{q}}, \mathbf{r}) = P_q\mathbf{r},$$

$$\text{where:}\quad \Lambda = \omega_n^2, \quad N_q(\mathbf{q}, \dot{\mathbf{q}}, \mathbf{r}) = 2\zeta\omega_n\dot{q}_1 + \alpha q_1^2, \quad P_q = P_x,$$

and where q_1 is the only element in $\mathbf{q}$.

Step 1f: The forcing transformation, $\mathbf{q} = \mathbf{v} + \mathbf{e}\mathbf{r}$, is now applied to remove the non-resonant forcing terms to give an equation in the form $\ddot{\mathbf{v}} + \Lambda\mathbf{v} + N_v(\mathbf{v}, \dot{\mathbf{v}}, \mathbf{r}) = P_v\mathbf{r}$.

The transform matrix $\mathbf{e}$ and the corresponding forcing matrix P_v are found element by element (here they are 1×2 matrices as there is one degree of freedom) by considering for each mode whether the forcing is resonant or not. The $(k, 1)$ and $(k, 2)$ terms in $\mathbf{e}$ and P_v can be found using Eqs. (4.76) or (4.77) depending on whether the forcing is resonant or not, respectively, for the kth mode. Here for the first (and only) mode the forcing is non-resonant and so using Eq. (4.77) results in

$$P_q = \begin{bmatrix} \frac{R}{2} & \frac{R}{2} \end{bmatrix} \rightsquigarrow \mathbf{e} = \frac{1}{\omega_n^2 - \Omega^2} \begin{bmatrix} \frac{R}{2} & \frac{R}{2} \end{bmatrix}, \quad P_v = \begin{bmatrix} 0\, 0 \end{bmatrix}.$$

The relationship between the nonlinear terms is given by $N_v(\mathbf{v}, \dot{\mathbf{v}}, \mathbf{r}) = N_q(\mathbf{v} + \mathbf{er}, \dot{\mathbf{v}} + \mathbf{e}W\mathbf{r}, \mathbf{r})$, Eq. (4.75), and for this example may be written as

$$N_q = 2\zeta\omega_n\dot{q}_1 + \alpha q_1^2 \rightsquigarrow N_v = 2\zeta\omega_n\left[\dot{v}_1 + \mathrm{i}\Omega e(r_p - r_m)\right] + \alpha\left[v_1 + e(r_p + r_m)\right]^2$$

where $\mathbf{e} = [e \;\; e]$ and $e = R/[2(\omega_n^2 - \Omega^2)]$.

Step 2: The near-identity transform may now be calculated by first expressing the nonlinear term N_v in terms of $\mathbf{u}$ rather than $\mathbf{v}$. Using Eq. (4.78) and taking the terms to be of order ε^1, gives

$$\mathbf{n}_1(\mathbf{u}, \dot{\mathbf{u}}, \mathbf{r}) = 2\zeta\omega_n\left[\dot{u}_1 + \mathrm{i}\Omega e(r_p - r_m)\right] + \alpha\left[u_1 + e(r_p + r_m)\right]^2$$

Expressing $\mathbf{n}_1$ in matrix form using variables u_{p1} and u_{m1} (where $u_1 = u_{p1} + u_{m1}$ from $\mathbf{u} = \mathbf{u}_p + \mathbf{u}_m$) gives $\mathbf{n}_1(\mathbf{u}, \dot{\mathbf{u}}, \mathbf{r}) = \mathbf{n}^*\mathbf{u}^*(\mathbf{u}_p, \mathbf{u}_m, \mathbf{r})$. Using the general form for each term in $\mathbf{u}^*$, Eq. (4.84), the matrix that governs whether terms are included in the near-identity transform or not, $\boldsymbol{\beta}^*$, can be found using Eq. (4.86). For this example these matrices are

$$\mathbf{u}^* = \begin{bmatrix} u_{p1}^2 \\ u_{p1}u_{m1} \\ u_{p1}r_p \\ u_{p1}r_m \\ u_{m1}^2 \\ u_{m1}r_p \\ u_{m1}r_m \\ r_p^2 \\ r_pr_m \\ r_m^2 \\ u_{p1} \\ u_{m1} \\ r_p \\ r_m \end{bmatrix}, \quad \mathbf{n}^* = \begin{bmatrix} \alpha \\ 2\alpha \\ 2e\alpha \\ 2e\alpha \\ \alpha \\ 2e\alpha \\ 2e\alpha \\ e^2\alpha \\ 2e^2\alpha \\ e^2\alpha \\ 2\zeta\omega_n\omega_{r1}\mathrm{i} \\ -2\zeta\omega_n\omega_{r1}\mathrm{i} \\ 2\zeta\omega_n\Omega e\mathrm{i} \\ -2\zeta\omega_n\Omega e\mathrm{i} \end{bmatrix}^T \rightsquigarrow \boldsymbol{\beta}^* = \omega_{r1}^2 \begin{bmatrix} 3 \\ -1 \\ a(a+2) \\ a(a-2) \\ 3 \\ a(a-2) \\ a(a+2) \\ 4a^2 - 1 \\ -1 \\ 4a^2 - 1 \\ 0 \\ 0 \\ a^2 - 1 \\ a^2 - 1 \end{bmatrix}^T,$$

where $\Omega = a\omega_{r1}$ has been used.

Step 3: Using Eqs. (4.58) and (4.59) along with matrix $\boldsymbol{\beta}^*$ the near-identity transform and the nonlinear terms in the transformed equation of motion can be identified. A term is kept in the equation of motion if the corresponding term in $\boldsymbol{\beta}$ is zero. The u_{p1} and u_{m1} terms are *unconditionally resonant* because the corresponding terms in $\boldsymbol{\beta}$ are always zero regardless of the value of a. For the cases where $a = 1/2$ and $a = 2$ there are also *conditionally resonant* terms, i.e. terms which are zero for a specific value of a. Note that $a = 1$ also results in terms in $\boldsymbol{\beta}$ equalling zero, however for this case the forcing is near-resonant and so the forcing transform would be different ($\mathbf{e} = 0$ and $P_v = P_q$).

Considering the case where $a = 1/2$, such that the forcing is approximately at half the natural frequency, results in the transform and post-transform nonlinear terms matrices

$$\boldsymbol{\beta} = \omega_{r1}^2 \begin{bmatrix} 3 \\ -1 \\ 5/4 \\ -3/4 \\ 3 \\ -3/4 \\ 5/4 \\ 0 \\ -1 \\ 0 \\ 0 \\ 0 \\ -3/4 \\ -3/4 \end{bmatrix}^T \rightsquigarrow \mathbf{n}_u^* = \begin{bmatrix} 0 \\ 0 \\ 0 \\ 0 \\ 0 \\ 0 \\ 0 \\ e^2\alpha \\ 0 \\ e^2\alpha \\ 2\zeta\omega_n\omega_{r1}\mathrm{i} \\ -2\zeta\omega_n\omega_{r1}\mathrm{i} \\ 0 \\ 0 \end{bmatrix}^T, \quad \mathbf{h}^* = \frac{1}{\omega_{r1}^2} \begin{bmatrix} \alpha/3 \\ -2\alpha \\ 8e\alpha/5 \\ -8e\alpha/3 \\ \alpha/3 \\ -8e\alpha/3 \\ 8e\alpha/5 \\ 0 \\ -2e^2\alpha \\ 0 \\ 0 \\ 0 \\ -8\zeta\omega_n\Omega e\mathrm{i}/3 \\ 8\zeta\omega_n\Omega e\mathrm{i}/3 \end{bmatrix}^T.$$

Step 4: Using these matrices, the transformed equation of motion, to order ε^1 accuracy, may be written as $\ddot{\mathbf{u}} + \Lambda\mathbf{u} + \mathbf{n}^*\mathbf{u}^* = P_u\mathbf{r}$ which gives

$$\ddot{\mathbf{u}} + 2\zeta\omega_n\dot{\mathbf{u}} + \omega_n^2\mathbf{u} + \alpha e^2\left(r_p^2 + r_m^2\right) = 0.$$

The last term on the left-hand side of this equation can be seen as an external forcing of a linear system in $\mathbf{u}$. Using the trial solution for u_1, Eq. (4.51), the definition of r_p and r_m, that they equal $\mathrm{e}^{\mathrm{i}\Omega t}$ and $\mathrm{e}^{-\mathrm{i}\Omega t}$ respectively, and recalling that $\omega_{r1} = 2\Omega$, as $a = 1/2$ gives

$$\left(\omega_n^2 - \omega_r^2\right) U_1 \cos(\omega_{r1}t - \phi_1) - 2\zeta\omega_n\omega_{r1}U_1 \sin(\omega_{r1}t - \phi_1) = -2\alpha e^2 \cos(\omega_{r1}t).$$

Balancing $\cos(\omega_{r1}t - \phi_1)$ and $\sin(\omega_{r1}t - \phi_1)$ terms, using the trigonometric identity $\cos(\omega_{r1}t) = \cos(\omega_{r1}t - \phi_1)\cos(\phi_1) - \sin(\omega_{r1}t - \phi_1)\sin(\phi)$ gives

$$\cos(\omega_{r1}t - \phi_1)\colon \left(\omega_n^2 - \omega_r^2\right) U_1 = -2\alpha e^2 \cos(\phi_1),$$
$$\sin(\omega_{r1}t - \phi_1)\colon 2\zeta\omega_n\omega_{r1}U_1 = -2\alpha e^2 \sin(\phi_1).$$

By squaring and adding these equations and by dividing them respectively gives the response amplitude and phase relationships

$$U_1 = \frac{2\alpha e^2}{\sqrt{\left(\omega_n^2 - \omega_r^2\right)^2 + (2\zeta\omega_n\omega_{r1})^2}}, \quad \tan(\phi_1) = \frac{2\zeta\omega_{n1}\omega_{r1}}{\omega_n^2 - \omega_r^2}.$$

Note that the oscillatory terms could have been kept in complex exponential form and then balanced, followed by balancing of the real and imaginary components as was done in the last example.

Based on the transforms, the solution in terms of x may be written as

$$\mathbf{x} = \Phi\mathbf{q} = \Phi(\mathbf{v} + \mathbf{e}r) = \Phi(\mathbf{u} + \mathbf{n}^*\mathbf{u}^* + \mathbf{e}r)$$

which, for this example, may be simplified to

$$\mathbf{x} = \mathbf{u} + \mathbf{n}^*\mathbf{u}^* + 2e\cos(\Omega t),$$

since $\Phi = 1$ and $\mathbf{e}r = e(\mathrm{e}^{\mathrm{i}\Omega t} + \mathrm{e}^{-\mathrm{i}\Omega t}) = 2e\cos(\Omega t)$.

Concentrating on the response at the response frequency, $\omega_{r1} = 2\Omega$, and at the forcing frequency, Ω, results firstly in

$$x_{2\Omega} = u_1 = U_1\cos(2\Omega t - \phi_1), \tag{4.88}$$

where the subscript 2Ω indicates the response at this frequency. Note that the response at the resonant frequency is fully captured in the dynamic equation for **u**. So, for a multi-mode system, for any particular mode, there are no components at the response frequency for that mode in $\mathbf{n}^*\mathbf{u}^*$. The response at the forcing frequency is made up of terms in $\mathbf{n}^*\mathbf{u}^*$ and $\mathbf{e}r$ but to the first approximation is dominated by the direct forcing term allowing us to write

$$x_\Omega \approx 2e\cos(\Omega t) \tag{4.89}$$

■

Figure 4.8 shows the results of a time-stepping simulation of the escape equation for the case where $\omega_n = 4\pi$, $\zeta = 0.001$, $\alpha = 200$ and with a forcing amplitude of $R = 1$ and a forcing frequency, Ω, that is around half the natural frequency, ω_n, such that $a = 1/2$. Adopting the same method as in Fig. 4.7, the forcing frequency is initially less than $\omega_n/2$ and is increased in steps that are sufficiently widely spaced in time for a steady-state response to be reached. The steady-state response amplitude at the forcing frequency and at the response frequency, $\omega_{r1} = 2\Omega$, are shown as dots

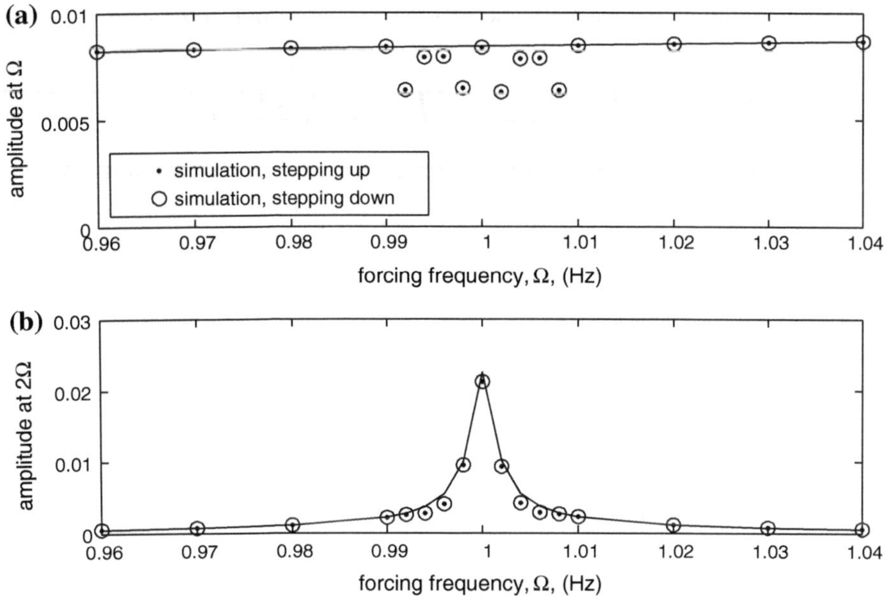

Fig. 4.8 Response at, **a** the forcing frequency and, **b** twice the forcing frequency for the forced and damped escape equation with $\omega_n = 4\pi,\ \zeta = 0.001,\ \alpha = 200$ and with a forcing amplitude of $R = 1$ for a range of forcing frequencies around half the natural frequency

in panels (a) and (b) of the figure respectively. The circles show the response at the same frequencies, however now stepping down in frequency.

The solid lines show the normal form predictions of the response amplitudes using Eqs. (4.89) and (4.88) for the Ω and 2Ω responses respectively. It can be seen that the prediction at the response frequency, in panel (b), is very good. The general trend for the response at the forcing frequency, panel (a), is also good, however the normal form equation does not capture the reduction in the response amplitude either side of $\Omega = \omega_n/2$.

4.5.4 Steady-State Stability

Finally in this section, the stability of the steady-state response solutions provided by the normal form technique is considered. To study the solution stability, the amplitude of the response must be allowed to vary slowly with time. This time dependence allows the study of the dynamics of a perturbation applied to the steady-state solution using the first-order differential equation $\mathbf{z}' = f(\mathbf{z}, T)$. Here $\mathbf{z}$ contains the slowly varying amplitudes, T represents a slow time variable over which the amplitudes vary and $\bullet'$ indicates the differential with respect to T.

The trial solution that has been used for the normal form technique is

$$u_k = u_{pk} + u_{mk} = \left(\frac{U_k}{2}\mathrm{e}^{-\mathrm{i}\phi_k}\right)\mathrm{e}^{\mathrm{i}\omega_{rk}t} + \left(\frac{U_k}{2}\mathrm{e}^{\mathrm{i}\phi_k}\right)\mathrm{e}^{-\mathrm{i}\omega_{rk}t} \tag{4.90}$$

for the kth state in $\mathbf{u}$, see Eq. (4.51). To allow the solution stability to be found a slow time dependence in U_k and ϕ_k can be introduced however the maths becomes awkward. It is better to rewrite the trial solution for the kth state as

$$u_k = u_{pk} + u_{mk} = \frac{U_{pk}}{2}\mathrm{e}^{\mathrm{i}\omega_{rk}t} + \frac{U_{mk}}{2}\mathrm{e}^{-\mathrm{i}\omega_{rk}t}$$

where

$$U_{pk} = U_k\mathrm{e}^{-\mathrm{i}\phi_k}, \quad U_{mk} = U_k\mathrm{e}^{\mathrm{i}\phi_k}. \tag{4.91}$$

Note that U_{mk} is the complex conjugate of U_{pk} and $U_{pk}U_{mk} = U_k^2$. Now adding the time dependence gives

$$u_k = u_{pk} + u_{mk} = \frac{U_{pk}(\varepsilon t)}{2}\mathrm{e}^{\mathrm{i}\omega_{rk}t} + \frac{U_{mk}(\varepsilon t)}{2}\mathrm{e}^{-\mathrm{i}\omega_{rk}t}$$

where the ε is present to indicate that the amplitude is slowly varying. As with the multiple scales derivation the slow-time variable $T = \varepsilon t$ is used.

Alterations to the normal form technique following this modification to the trial solution for $\mathbf{u}$ must now be considered. The trial solution is used twice in the derivation. Taking the forced system, it is first used to simplify Eq. (4.53) which is an equation arising from the balancing of ε^1 terms, see Box 4.2 (note that the procedure for the forced case is identical to that for the unforced one). This means that when substituting the new trial solution into Eq. (4.53) any terms that are of order ε^1 or higher should be ignored—these would effect the ε^2 balanced equation, not the ε^1 one.

This substitution is made at the beginning of Box 4.5, where $\mathrm{d}u_\ell^*/\mathrm{d}t$ is found. Using Eq. (4.84), which defines u_ℓ^*, this differential may be written as

$$\frac{\mathrm{d}u_\ell^*}{\mathrm{d}t} = \frac{\partial u_\ell^*}{\partial r_p}\frac{\mathrm{d}r_p}{\mathrm{d}t} + \frac{\partial u_\ell^*}{\partial r_m}\frac{\mathrm{d}r_m}{\mathrm{d}t} + \sum_{n=1}^{N}\left(\frac{\partial u_\ell^*}{\partial u_{pn}}\frac{\mathrm{d}u_{pn}}{\mathrm{d}t} + \frac{\partial u_\ell^*}{\partial u_{mn}}\frac{\mathrm{d}u_{mn}}{\mathrm{d}t}\right)$$

With the new trial solution $\mathrm{d}u_{pn}/\mathrm{d}t$ and $\mathrm{d}u_{mn}/\mathrm{d}t$ are now given as

$$\frac{\mathrm{d}u_{pn}}{\mathrm{d}t} = \mathrm{i}\omega_{rn}\frac{U_{pn}(T)}{2}\mathrm{e}^{\mathrm{i}\omega_{rn}t} + \frac{1}{2}\frac{\mathrm{d}U_{pn}(T)}{\mathrm{d}t}\mathrm{e}^{\mathrm{i}\omega_{rn}t},$$

$$\frac{\mathrm{d}u_{pn}}{\mathrm{d}t} = -\mathrm{i}\omega_{rn}\frac{U_{mn}(T)}{2}\mathrm{e}^{-\mathrm{i}\omega_{rn}t} + \frac{1}{2}\frac{\mathrm{d}U_{pn}(T)}{\mathrm{d}t}\mathrm{e}^{-\mathrm{i}\omega_{rn}t}.$$

The effect of having the slowly varying amplitude is to introduce the second term in both these equations, however these terms are of order ε^1, since

$$\frac{\mathrm{d}U_{pn}(T)}{\mathrm{d}t} = \varepsilon\frac{\mathrm{d}U_{pn}(T)}{\mathrm{d}T}, \quad \frac{\mathrm{d}U_{pn}(T)}{\mathrm{d}t} = \varepsilon\frac{\mathrm{d}U_{pn}(T)}{\mathrm{d}T},$$

and so may be ignored when seeking at ε^1 accurate solution. Using just the ε^0 terms the derivation of $\mathrm{d}u_\ell^*/\mathrm{d}t$, and hence $\boldsymbol{\beta}$, is identical to that presented in Box 4.5.

The second place where the trial solution is used is in solving the dynamic equation in $\mathbf{u}$, which to order ε^1 may be written as

$$\ddot{\mathbf{u}} + \Lambda\mathbf{u} + \mathbf{n}^*\mathbf{u}^* = P_u\mathbf{r},$$

where the $\mathbf{n}^*\mathbf{u}^*$ term is order ε^1 and the other terms are of order ε^0, see Step 4. It follows that, to maintain ε^1 accuracy, when substituting in the trial solution for $\mathbf{u}$, Eq. (4.92), $\dot{\mathbf{u}}$ to order ε^0 is required as velocity terms only appear in the order ε^1 nonlinear term and $\ddot{\mathbf{u}}$ to order ε^1 as the acceleration term in the equation of motion is of order ε^0. The effect of having the slowly time-varying amplitude in the trial solution is that the kth element in $\dot{\mathbf{u}}$ and $\ddot{\mathbf{u}}$ become

$$\begin{aligned}
\dot{u}_k &= \frac{1}{2}\mathrm{i}\omega_{rk}\left(U_{pk}\mathrm{e}^{\mathrm{i}\omega_{rk}t} - U_{mk}\mathrm{e}^{-i\omega_{rk}t}\right) + \mathcal{O}\{\varepsilon^1\}\\
\ddot{u}_k &= -\frac{1}{2}\omega_{rk}^2\left(U_{pk}\mathrm{e}^{\mathrm{i}\omega_{rk}t} + U_{mk}\mathrm{e}^{-i\omega_{rk}t}\right) + \varepsilon\mathrm{i}\omega_{rk}\left(\frac{\mathrm{d}U_{pk}}{\mathrm{d}T}\mathrm{e}^{\mathrm{i}\omega_{rk}t} - \frac{\mathrm{d}U_{mk}}{\mathrm{d}T}\mathrm{e}^{-\mathrm{i}\omega_{rk}t}\right) + \mathcal{O}\{\varepsilon^2\}
\end{aligned}$$

to order ε^0 and ε^1 respectively. Note that when deriving the $\ddot{u}_k$ expression to order ε^1 the ε^1 order term in the $\dot{u}_k$ expression must be retained. Dropping the ε bookkeeping aid by setting $\varepsilon = 1$ and splitting these into the u_{pk} and u_{mk} components results in

$$\begin{aligned}
u_{pk} &= \frac{1}{2}U_{pk}\mathrm{e}^{\mathrm{i}\omega_{rk}t}, \quad u_{mk} = \frac{1}{2}U_{mk}\mathrm{e}^{-\mathrm{i}\omega_{rk}t}\\
\dot{u}_{pk} &= \frac{1}{2}\mathrm{i}\omega_{rk}U_{pk}\mathrm{e}^{\mathrm{i}\omega_{rk}t}, \quad \dot{u}_{mk} = -\frac{1}{2}\mathrm{i}\omega_{rk}U_{mk}\mathrm{e}^{-\mathrm{i}\omega_{rk}t} \qquad (4.92)\\
\ddot{u}_{pk} &= \left(-\frac{\omega_{rk}^2}{2}U_{pk} + \mathrm{i}\omega_{rk}\dot{U}_{pk}\right)\mathrm{e}^{\mathrm{i}\omega_{rk}t}, \quad \ddot{u}_{mk} = \left(-\frac{\omega_{rk}^2}{2}U_{mk} - \mathrm{i}\omega_{rk}\dot{U}_{mk}\right)\mathrm{e}^{-\mathrm{i}\omega_{rk}t}.
\end{aligned}$$

To assess the stability of a steady-state response solution predicted by the normal form approach the equation $\ddot{\mathbf{u}}+\Lambda\mathbf{u}+\mathbf{n}^*\mathbf{u}^* = P_u\mathbf{r}$ can be calculated using the method described in Sect. 4.5.3. The trial solution given in Eq. (4.92) is then substituted into this equation. Previously when the constant-amplitude trial solution of Eq. (4.51) was used, the time-dependent complex exponential terms are then balanced generating two equations, one the complex conjugate of the other. Then, taking one of these equations the real and imaginary parts were balanced. When considering the stability using the slowly-varying amplitude trial solution given in Eq. (4.92), it is better to adopt a slightly different approach as the trial solution is in terms of U_{pk} and U_{mk} which are complex (and a conjugate pair) rather than U_k which is real. The positive and the negative time-dependent complex exponential terms are balanced to generate two equations. These can be written in the form $\dot{\mathbf{z}} = f(\mathbf{z}, t)$ where $\mathbf{z}$ contains all the U_{pk} and U_{mk} terms. As discussed in Sect. 4.3.2, the steady-state solutions, $\bar{\mathbf{z}}$ are then found by solving $f(\bar{\mathbf{z}}, t) = 0$. For these to be stable the real part of the eigenvalues of $Df_z(\bar{\mathbf{z}}, t)$, where $Df_z(\mathbf{z}, t)$ is the Jacobian of $f(\mathbf{z}, t)$, must be negative.

The method will now be applied to analyse the stability of steady-state solutions for the Duffing oscillator forced near resonance discussed in Example 4.10.

Example 4.11 The stability of the response of a one DOF oscillator with a cubic nonlinearity—the forced Duffing equation.

Problem Reconsidering the Duffing oscillator with sinusoidal forcing close to resonance

$$\ddot{x} + 2\zeta\omega_n\dot{x} + \omega_n^2 x + \alpha x^3 = R\cos(\Omega t),$$

that was discussed in Example 4.10, analyse the stability of the steady-state response solutions.

Solution From Example 4.10, the dynamic equation in $\mathbf{u}$ is

$$\ddot{\mathbf{u}} + 2\zeta\omega_n\dot{\mathbf{u}} + \omega_n^2\mathbf{u} + 3\alpha\left(u_{p1}^2 u_{m1} + u_{p1}u_{m1}^2\right) = \frac{R}{2}\left(\mathrm{e}^{\mathrm{i}\omega_{rk}t} + \mathrm{e}^{-\mathrm{i}\omega_{rk}t}\right). \tag{4.93}$$

Using the slowly amplitude-varying trial solution for $\mathbf{u}$, Eq. (4.92), the equation of motion, Eq. (4.93), may be written as

$$\left[\left(\omega_n^2 - \omega_{r1}^2\right)U_{p1} + 2\mathrm{i}\omega_{r1}\dot{U}_{p1}\right]\mathrm{e}^{\mathrm{i}\omega_{r1}t} + \left[\left(\omega_n^2 - \omega_{r1}^2\right)U_{m1} - 2\mathrm{i}\omega_{r1}\dot{U}_{m1}\right]\mathrm{e}^{-\mathrm{i}\omega_{r1}t}$$
$$+ 2\mathrm{i}\zeta\omega_n\omega_{r1}\left(U_{p1}\mathrm{e}^{\mathrm{i}\omega_{r1}t} - U_{m1}\mathrm{e}^{-\mathrm{i}\omega_{r1}t}\right) + \frac{3}{4}\alpha\left(U_{p1}^2 U_{m1}\mathrm{e}^{\mathrm{i}\omega_{r1}t} + U_{p1}U_{m1}^2\mathrm{e}^{-\mathrm{i}\omega_{r1}t}\right)$$
$$= R\left(\mathrm{e}^{\mathrm{i}\omega_{rk}t} + \mathrm{e}^{-\mathrm{i}\omega_{rk}t}\right).$$

Balancing the $\mathrm{e}^{\mathrm{i}\omega_{r1}t}$ and $\mathrm{e}^{-\mathrm{i}\omega_{r1}t}$ terms gives

$$\mathrm{e}^{\mathrm{i}\omega_{r1}t}\!: \quad \left(\omega_n^2 - \omega_{r1}^2\right)U_{p1} + 2\mathrm{i}\omega_{r1}\dot{U}_{p1} + 2\mathrm{i}\zeta\omega_n\omega_{r1}U_{p1} + \frac{3}{4}\alpha U_{p1}^2 U_{m1} = R,$$
$$\mathrm{e}^{-\mathrm{i}\omega_{r1}t}\!: \quad \left(\omega_n^2 - \omega_{r1}^2\right)U_{m1} - 2\mathrm{i}\omega_{r1}\dot{U}_{m1} - 2\mathrm{i}\zeta\omega_n\omega_{r1}U_{m1} + \frac{3}{4}\alpha U_{p1}U_{m1}^2 = R.$$

From this a first-order differential equation in U_{p1} and U_{m1} can be written in the form $\dot{\mathbf{z}} = f(\mathbf{z}, t)$, with $\mathbf{z} = [U_{p1} \;\; U_{m1}]^T$, as

$$\begin{pmatrix}\dot{U}_{p1}\\ \dot{U}_{m1}\end{pmatrix} = \frac{1}{2\omega_{r1}}\begin{pmatrix}\mathrm{i}(\omega_n^2 - \omega_{r1}^2)U_{p1} - 2\zeta\omega_n\omega_{r1}U_{p1} + \frac{3}{4}\mathrm{i}\alpha U_{p1}^2 U_{m1} - \mathrm{i}R\\ -\mathrm{i}(\omega_n^2 - \omega_{r1}^2)U_{m1} - 2\zeta\omega_n\omega_{r1}U_{m1} - \frac{3}{4}\mathrm{i}\alpha U_{p1}U_{m1}^2 + \mathrm{i}R\end{pmatrix} \tag{4.94}$$

To find the steady-state solutions $\dot{U}_{p1}$ and $\dot{U}_{m1}$ are set to zero resulting in

$$
\begin{aligned}
\mathrm{i}(\omega_n^2 - \omega_{r1}^2 + \frac{3}{4}\alpha U_{p1}U_{m1})U_{p1} - 2\zeta\omega_n\omega_{r1}U_{p1} &= \mathrm{i}R, \\
-\mathrm{i}(\omega_n^2 - \omega_{r1}^2 + \frac{3}{4}\alpha U_{p1}U_{m1})U_{m1} - 2\zeta\omega_n\omega_{r1}U_{m1} &= -\mathrm{i}R.
\end{aligned}
$$

These may be multiplied together to give

$$(\omega_n^2 - \omega_{r1}^2 + \frac{3}{4}\alpha U_{p1}U_{m1})^2 U_{p1}U_{m1} + (2\zeta\omega_n\omega_{r1})^2 U_{p1}U_{m1} = R^2$$

Recalling that $U_{pk} = U_k \mathrm{e}^{-\mathrm{i}\phi_k}$ and $U_{mk} = U_k \mathrm{e}^{\mathrm{i}\phi_k}$, Eq. (4.91), such that $U_{pk}U_{mk} = U_k^2$, this equation can be simplified to

$$\left[\left(\omega_n^2 - \omega_{r1}^2\right) + \frac{3}{4}\alpha U_1^2\right]^2 U_1^2 + \left[2\zeta\omega_n\omega_{r1}\right]^2 U_1^2 = R^2. \tag{4.95}$$

This is the same as that found using the constant amplitude trial solution in Example 4.10, see Eq. (4.87).

The stability is governed by the eigenvalue of the Jacobian of $f(\mathbf{z}, t)$ evaluated at the steady-state solution $\bar{\mathbf{z}}$, i.e. $Df_z(\bar{\mathbf{z}}, t)$. Using Eq. (4.94), the Jacobian may be written as

$$
\begin{aligned}
&Df_z(\mathbf{z}, t) \\
&= \frac{1}{2\omega_{r1}} \times \begin{pmatrix} -2\zeta\omega_n\omega_{r1} + \mathrm{i}(\omega_n^2 - \omega_{r1}^2 + \frac{3}{2}\alpha U_{p1}U_{m1}) & \frac{3}{4}\mathrm{i}\alpha U_{p1}^2 \\ -\frac{3}{4}\mathrm{i}\alpha U_{m1}^2 & -2\zeta\omega_n\omega_{r1} - \mathrm{i}(\omega_n^2 - \omega_{r1}^2 + \frac{3}{2}\alpha U_{p1}U_{m1}) \end{pmatrix}.
\end{aligned}
$$

From this the equation for the eigenvalues, λ, may be written as

$$(-A + \mathrm{i}B - 2\omega_{r1}\lambda)(-A - \mathrm{i}B - 2\omega_{r1}\lambda) - \frac{9}{16}\alpha^2 U_{p1}^2 U_{m1}^2 = 0$$

where the shorthand notation

$$A = 2\zeta\omega_n\omega_{r1}, \quad B = \omega_n^2 - \omega_{r1}^2 + \frac{3}{2}\alpha U_{p1}U_{m1}$$

has been used. Expanding out the brackets in the eigenvalue equation gives the quadratic in $2\omega_{r1}\lambda$

$$(2\omega_{r1}\lambda)^2 + 2A(2\omega_{r1}\lambda) + \left(A^2 + B^2 - \frac{9}{16}\alpha^2 U_{p1}^2 U_{m1}^2\right) = 0.$$

Writing this in the standard form $a\lambda^2 + b\lambda + c = 0$ and observing that both a and b are positive, it can be deduced that the system can only have an eigenvalue with a positive real component if $c \leq 0$. Using this the condition for a steady-state solution, given by Eq. (4.95), to be stable may be written as

$$(2\zeta\omega_n\omega_{r1})^2 + \left(\omega_n^2 - \omega_{r1}^2 + \frac{3}{2}\alpha U_1^2\right)^2 - \frac{9}{16}\alpha^2 U_1^4 \geq 0, \tag{4.96}$$

where $U_{p1}U_{m1} = U_1^2$ has been used. Note that the transition from a stable to an unstable solution occurs when this expression equals zero.

It is interesting to observe that the steady-state solution, Eq. (4.95), can be differentiated with respect to U_1^2 to give

$$\left(\omega_n^2 - \omega_{r1}^2 + \frac{3}{4}\alpha U_1^2\right)^2 + 2\left(\omega_n^2 - \omega_{r1}^2 + \frac{3}{4}\alpha U_1^2\right)\left(-2\omega_{r1}\frac{\mathrm{d}\omega_{r1}}{\mathrm{d}U_1^2} + \frac{3}{4}\alpha\right)U_1^2$$
$$+ (2\zeta\omega_n\omega_{r1})^2 + 2(2\zeta\omega_n)^2\omega_{r1}\frac{\mathrm{d}\omega_{r1}}{\mathrm{d}U_1^2}U_1^2 = 0. \tag{4.97}$$

Taking the stability equation, Eq. (4.96) and setting it to zero, such that it represents the points at which the stability switches, and subtract off Eq. (4.97) results in

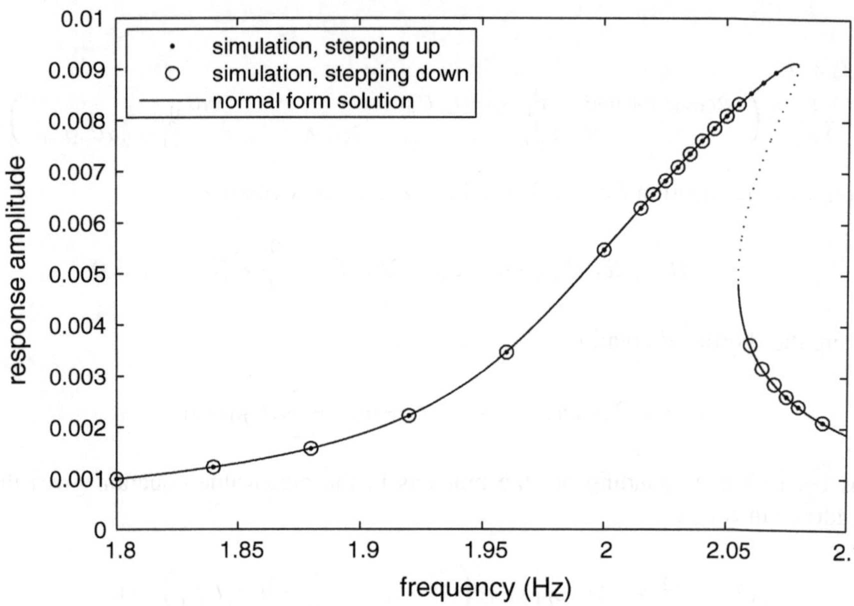

Fig. 4.9 Response of the forced Duffing equation with $\omega_n = 4\pi$, $\zeta = 0.01$, $\alpha = 200{,}000$ and with a forcing amplitude of $R = 0.03$ for a range of forcing frequencies around the natural frequency, in which the stability of the normal form solutions is indicated

$$\frac{d\omega_{r1}}{dU_1^2} = 0.$$

Hence the transition points between unstable and stable steady-state solutions occur when the gradient of the response in the U_1—ω_{r1} plane is infinite, i.e. at the fold points. ■

Figure 4.9 is a modified version of Fig. 4.7, in which the unstable steady-state response solutions predicted by the normal form approach are indicated by a dotted line. It can be seen that these correspond to the region of the curve in which there are no time-simulation data points.

This section has demonstrated how the normal form approach can be used to approximate free and forced vibration problems. The forcing case can be subdivided into forcing close to and far from resonance. Comparing the normal form results with other approximate techniques and with numerical simulations, indicates just how good an approximation these methods can give. The normal form transformation is used extensively throughout the rest of the book.

4.6 Chapter Notes

The aim of this chapter was to provide an introduction to many of the approximate nonlinear analysis techniques. A case study is provided in Chap. 7, in which the dynamics of a single mode of a cable are considered, and the various techniques introduced here are compared. Harmonic balance has long been used as a technique for approximating the response of nonlinear systems. Discussions of the harmonic balance technique and how to apply it can be found, for example, in Worden and Tomlinson (2000), Nayfeh and Mook (1995), Cartmell (1980). The averaging technique is discussed in Verhulst (1996) and Tondl et al. (2000), including details of how to use an amplitude and phase representation rather than the sine and cosine representation that has been used here. In addition, Bakri et al. (2004) compare averaging to the harmonic balance technique for a specific system. Further analysis of cable dynamics using the averaging technique may be found in Gonzalez-Buelga et al. (2008). Perturbation techniques are discussed in detail in Verhulst (1996), Strogatz (2001) and Glendinning (1994). Both Strogatz (2001) and Glendinning (1994) provide interesting discussions on extending the multiple-scales method beyond just a slow and a fast scale. The normal form approach adopted here is a second-order variant in which the equations of motion are used directly. The first-order methods in which the first transform converts the equations of motion to state-space form is discussed in Jezequel and Lamarque (1991), Nayfeh (1993), and Wagg and Neild (2009). The second-order variant is presented in Neild and Wagg (2011) and discussed in more detail in Neild (2012). The case where the forcing is away from resonance is discussed for a two degree of freedom example in Neild and Wagg (2011) and the stability of steady-state solutions in Xin et al. (2013).

Problems

4.1 Using the harmonic balance technique with a single term approximation to the response, find the nonlinear natural frequency for the following undamped, unforced oscillator with a small quadratic stiffness nonlinearity

$$\ddot{x} + \omega_n^2 x + \alpha x^2 = 0.$$

4.2 Building on Problem 4.1, now use a three-term solution along with the harmonic balance technique to find the response frequency–amplitude relationship for the oscillator with a quadratic stiffness nonlinearity

$$\ddot{x} + \omega_n^2 x + \alpha x^2 = 0.$$

In deriving this relationship it may be assumed the nonlinearity and hence the harmonics of the solution are small.

4.3 Using averaging, along with the frequency detuning and time-scaling technique, calculate the relationship between the steady-state amplitude and the frequency of oscillation for the Duffing equation

$$\ddot{x} + \omega_n^2 x + \alpha x^3 = 0.$$

4.4 Use multiple scales to calculate the steady-state response the following system

$$\ddot{x} + 2\zeta\omega_n \dot{x} + \omega_n^2 x + \alpha x^3 = f\cos(\Omega t),$$

where the forcing frequency, Ω, is close to the linear natural frequency, ω_n, and the nonlinearity can be treated as small.

4.5 Apply the normal form technique to

$$\ddot{x} + \mu(x^2 - 1)\dot{x} + \omega_n^2 x = R\cos(\Omega t),$$

in which the small damping term is nonlinear, calculate the steady-state response close to resonance. Also calculate the response at three times the forcing frequency in terms of the response at the forcing frequency.

References

Bakri, T., Nabergoj, R., Tondl, A., & Verhulst, F. (2004). Parametric excitation in non-linear dynamics. *International Journal of Non-Linear Mechanics*, *39*, 311–329.

Cartmell, M. (1990). *Introduction to linear, parametric and nonlinear vibrations*. London, Chapman and Hall.

Glendinning, P. (1994). *Stability, instability and chaos*. Cambridge: Cambridge University Press.

Gonzalez-Buelga, A., Neild, S., Wagg, D., & Macdonald, J. (2008). Modal stability of inclined cables subjected to vertical support excitation. *Journal of Sound and Vibration*, *318*, 565–579.

Jezequel, L., & Lamarque, C. H. (1991). Analysis of nonlinear dynamic systems by the normal form theory. *Journal of Sound and Vibration*, *149*(3), 429–459.

Jordan, D. W., & Smith, P. (1999). *Nonlinear ordinary differential equations: An introduction to dynamical systems* (3rd ed.). Oxford: Oxford University Press.

Nayfeh, A. H. (1993). *Method of normal forms*. New York: Wiley

Nayfeh, A. H., & Mook, D. T. (1995). *Nonlinear oscillations*. New York: Wiley.

Neild, S. A., & Wagg, D. J. (2011). Applying the method of normal forms to second-order nonlinear vibration problems. *Proceedings of the Royal Society, Part A*, *467*, 1141–1163.

Neild, S. A. (2012). Approximate methods for analyzing nonlinear structures. In D. J. Wagg & L. Virgin (Eds.), *Exploiting nonlinear behavior in structural dynamics*. New York: Springer.

Strogatz, S. H. (2001). *Nonlinear dynamics and chaos*. New York: Perseus Books Group.

Tondl, A., Ruijgrok, T., Verhulst, F., & Nabergoj, R. (2000). *Autoparametric resonance in mechanical systems*. Cambridge: Cambridge University Press

Verhulst, F. (1996). *Nonlinear differential equations and dynamical systems*. New York: Springer.

Wagg, D. J., & Neild, S. A. (2009). *Nonlinear vibration with control: For flexible and adaptive structures* (1st ed.). New York: Springer.

Worden, K., & Tomlinson, G. R. (2000). *Nonlinearity in structural dynamics*. Bristol: IOP.

Xin, Z. F., Neild, S. A., Wagg, D. J., & Zuo, Z. X. (2013). Resonant response functions for nonlinear oscillators with polynomial type nonlinearities. *Journal of Sound and Vibration*, *332*, 1777–1788.

Chapter 5
Modal Analysis for Nonlinear Vibration

Abstract Linear vibration theory uses the concept of defining a specific set of modes of vibration for the system under consideration. Physically, each mode relates to a particular geometric configuration in the system, such as two lumped masses oscillating either in- or out-of-phase with each other. For linear systems, the superposition principle means that the complete vibration response can be computed as a summation of the responses from each mode. In general terms, modal analysis has come to mean considering the response of a system by studying its vibration modes; modal decomposition is the process of transforming the system from a physical to a modal representation. This is particularly useful in linear systems, because each mode has an associated resonance, and understanding where resonances could occur in a structure is a key part of analysing vibration problems. In this chapter the use of modal analysis for nonlinear systems is also considered. First, the decomposition of discrete and continuous linear systems into modal form is reviewed and the effect of nonlinear terms on this analysis is discussed. Following this, methods for decomposing nonlinear systems are considered. Initially a brief discussion of nonlinear normal modes is given and a special case system, in which there is nonlinear but no linear coupling between two oscillators, is analysed using the harmonic balance approach. Following this, attention is turned to the main technique for carrying out nonlinear modal decomposition, which is the method of *normal forms* that was introduced in Sect. 4.5. This is a technique that transforms the system to the simplest form possible. The approach described here uses linear modal decomposition as the first step in the process. The main advantage of using normal forms is that information about nonlinear (also called internal) resonances in the system can be obtained. As a result, a normal form analysis can be used to obtain information about both linear and nonlinear resonances in a nonlinear vibration problem. The modal decomposition techniques are used to find *backbone curves* that represent the undamped vibration response of the system in the frequency domain. The reason for taking this approach is that modal analysis is most relevant for lightly damped systems, where multiple resonant peaks can occur in the response. Just like linear systems, nonlinear systems with light damping have a forced response that is determined by the underlying undamped characteristics (It's possible to define systems that don't have this property, but we will restrict our discussion to systems that do.). In the frequency domain this is captured by the backbone curves. As a result, defining the backbone curves

D. Wagg and S. Neild, *Nonlinear Vibration with Control*,
Solid Mechanics and Its Applications 218, DOI 10.1007/978-3-319-10644-1_5

for the system gives a nonlinear modal model. The nonlinear examples considered in this chapter are confined to two degrees-of-freedom, but can be extended to higher degrees-of-freedom, and a short discussion of relevant literature on this topic is given at the end of the chapter.

5.1 Modal Behaviour in Vibrating Systems

The underlying concept of modal analysis for linear vibration is that the system response can be represented as the sum of contributions from a series of *mode-shapes*. It is assumed that each of these mode-shapes is related to a specific physical configuration of the system, which is a function of spatial coordinates within the structure, but not a function of time. The contribution of the mode-shape to the overall response is represented by a *modal amplitude* which is a function of time but not of the spatial coordinates.[1] The total response of the system is then a summation of each of the modal contributions. For example, in a continuous system the displacement response $w(x, t)$ at a particular point x in the structure can be represented in modal form as

$$w(x,t) = \sum_{i=1}^{\infty} \phi_i(x) q_i(t),$$

where ϕ_i is the ith mode-shape and q_i is the corresponding modal coordinate, which represents the contribution of that mode to the overall response.

When the system is discrete (lumped mass), taking $\mathbf{x}$ to be a vector of displacements $x_1, x_2, \ldots x_n$ of an N-degrees-of-freedom system, the modal transformation in terms of the N mode-shapes $\phi_1, \ldots, \phi_N$ can be written as

$$\mathbf{x} = \begin{bmatrix} x_1 \\ x_2 \\ \vdots \\ x_n \end{bmatrix} = \Phi \mathbf{q}(t),$$

where Φ is an $N \times N$ matrix containing the N mode-shapes (that are in the form of column vectors with the elements indicating the relative movement of each of the degrees of freedom for that mode) and $\mathbf{q}(t) = [q_1(t)\, q_2(t) \, \cdots \, q_n(t)]^T$ represents a vector of modal contributions (q_i is the time-dependent contribution of the ith mode).

In linear systems, each vibration mode has an associated resonance (or natural) frequency that occurs at a clearly defined resonance peak.[2] At, or near, the point of resonance, the motion of the linear system will be dominated by the vibration

[1] These assumptions are sometimes referred to as separation of space and time variables.

[2] For each frequency value there is a single amplitude value in the resonance peak—it is a single-valued function, which increases monotonically up to the resonant frequency and then monotonically decreases after the resonant frequency.

mode which correlates to that particular resonance peak—if only a single mode is present this is called a *pure* modal response. In multi-degree-of-freedom linear systems this means that the steady-state response effectively reduces to a series of single-degree-of-freedom harmonic oscillators one corresponding to each of the modes. These oscillators are defined by the modal displacement and velocity of the resonant mode. In nonlinear systems, the shapes of the resonance peaks are typically amplitude-dependent or distorted due to nonlinear effects such as those associated with hardening or softening springs. To analyse nonlinear systems, the starting point is to consider how linear modal decomposition techniques can be used.

5.2 Modal Decomposition Using Linear Techniques

In this section, linear modal decomposition techniques are discussed for discrete and continuous dynamic systems. The effect of nonlinearity on the modal decomposition is also discussed. It is often useful to decompose weakly nonlinear systems using the modes calculated for a linearized version of the system. This decomposition decouples the linear terms, but not the nonlinear terms. For a discrete representation of the system, this transformation can be useful as the first step in de-coupling the nonlinear system. For continuous systems, applying the transformation using the linear modes allows the system to be converted from a partial differential equation into an infinite dimensional set of ordinary differential equations. Truncating this to a finite set of N equations leads to a matrix representation in terms of the linear modal coordinates. This matrix representation (which is analogous to that for the discrete system) not only allows straightforward numerical simulation of the system, but also allows the nonlinear decomposition techniques in Sect. 5.4 to be applied.

5.2.1 Discrete Linear Systems

The discussion in Chap. 1, Sect. 1.3.3 shows that the equation of motion for a discrete linear system (or a system in which the finite element approach has been applied such that the response is considered at a series of discrete locations, see for example Thompson and Dahleh (1997)) can be expressed as

$$M\ddot{\mathbf{x}} + C\dot{\mathbf{x}} + K\mathbf{x} = \mathbf{F}_E, \tag{5.1}$$

Eq. (1.23), where $\mathbf{F}_E$ is the dynamic forcing vector. The matrices M, C and K are not diagonal in general, so the equations are coupled. The aim of the modal decomposition is to apply a transformation which replaces M, C and K with equivalent diagonal matrices resulting in N uncoupled second-order differential equations. The modal decomposition can be carried out with the governing differential equations in either *second-order* or *first-order* form. These two cases will be considered in turn.

First consider the case when the governing differential equations are in second-order form (i.e. containing d^2/dt^2 terms). For lightly damped systems it is usual to consider the modal decomposition of the undamped, unforced system, $M\ddot{\mathbf{x}}+K\mathbf{x}=0$. The modes of this system can be decoupled, such that when the initial conditions are in the form of a single mode-shape the response will be at the natural frequency of that mode. In other words, assuming a solution of $\mathbf{x} = \phi_{\mathbf{k}} e^{i\omega_{nk}t}$ gives

$$-\omega_{nk}^2 M\phi_{\mathbf{k}} + K\phi_{\mathbf{k}} = 0, \tag{5.2}$$

where $\phi_{\mathbf{k}}$ and ω_{nk} are the mode-shape and natural frequency of the kth mode[3] (for the undamped system). This equation can be written in the form of an eigenvector problem

$$M^{-1}K\phi_{\mathbf{k}} = \omega_{nk}^2\phi_{\mathbf{k}}, \tag{5.3}$$

where $\phi_{\mathbf{k}}$ is an eigenvector and ω_{nk}^2 the corresponding eigenvalue of $M^{-1}K$. These eigenvectors and values can be used to generate a transformation to replace M and K with diagonal matrices, M_d and K_d. These diagonal matrices take the form

$$M_d = \Phi^T M\Phi \quad \text{and} \quad K_d = \Phi^T K\Phi.$$

An explanation of why these relationships yield diagonal matrices is given in Box 5.1.

Considering the equation of motion, Eq. (5.1), and decomposing the response into its modal components by writing $\mathbf{x} = \Phi\mathbf{q}(t)$, gives

$$M\Phi\ddot{\mathbf{q}} + C\Phi\dot{\mathbf{q}} + K\Phi\mathbf{q} = \mathbf{F}_E.$$

Multiplying this equation by Φ^T leads to

$$M_d\ddot{\mathbf{q}} + \Phi^T C\Phi\dot{\mathbf{q}} + K_d\mathbf{q} = \Phi^T\mathbf{F}_E. \tag{5.4}$$

This equation is decoupled on the left-hand side provided that the damping term $\Phi^T C\Phi$ is diagonal. It is usually assumed that damping is *linearly proportional* to the mass and stiffness matrices (proportional Rayleigh damping, see for example Clough and Penzien (1975) for a discussion on modelling damping) and, as a result, the matrix $\Phi^T C\Phi = C_d$ will be diagonal.[4] This results in a series of uncoupled second-order differential equations of the form

$$m_k\ddot{q}_k + c_k\dot{q}_k + k_k q_k = \phi_{\mathbf{k}}^T\mathbf{F}_E \tag{5.5}$$

[3] Note that a bold subscript is used to indicate that the whole term is a vector, not that the subscript is a vector.

[4] See Caughey (1963) for necessary and sufficient conditions for simultaneous diagonalisation of the M, C and K matrices.

where the subscript k indicates the kth row of a vector or the kth element on the diagonal of a matrix. It is usual to write $k_k/m_k = \omega_{nk}^2$ and $c_k/m_k = 2\zeta_k\omega_{nk}$ where ζ_k is the modal damping ratio, which means that Eq. (5.5) becomes

$$\ddot{q}_k + 2\zeta_k\omega_{nk}\dot{q}_k + \omega_{nk}^2 q_k = \phi_{\mathbf{k}}^T \frac{\mathbf{F}_E}{m_k}.$$

Note that Eq. (5.4) can be rewritten as

$$\ddot{\mathbf{q}} + M_d^{-1} C_d \dot{\mathbf{q}} + \Lambda \mathbf{q} = M_d^{-1} \Phi^T \mathbf{F}_E. \tag{5.6}$$

Here Λ is a diagonal matrix of eigenvalues, with the kth diagonal element being ω_{nk}^2. In addition, using the eigenvalue/vector equation, Eq. (5.3), the relationship $M^{-1}K\Phi = \Phi\Lambda$ can be written. Premultiplying by M and then by Φ^T results in $\Phi^T K \Phi = \Phi^T M \Phi \Lambda$. Hence $M_d^{-1} K_d = \Lambda$. It is this representation of the modal equations of motion that is used in the first step of the normal form transformation method.

When there are nonlinear terms in the equation of motion the transformation is unlikely to result in decoupling. Writing the nonlinear terms $\mathscr{N}$ as a function of the states, the equation of motion is found to be

$$M\ddot{\mathbf{x}} + C\dot{\mathbf{x}} + K\mathbf{x} + \mathscr{N}(\mathbf{x}, \dot{\mathbf{x}}) = \mathbf{F}_E, \tag{5.7}$$

which, after the transformation using the linear undamped modes, becomes

$$M_d\ddot{\mathbf{q}} + C_d\dot{\mathbf{q}} + K_d\mathbf{q} + \Phi^T \mathscr{N}(\Phi\mathbf{q}, \Phi\dot{\mathbf{q}}) = \Phi^T \mathbf{F}_E. \tag{5.8}$$

Box 5.1 *Diagonal mass and stiffness matrices*
The eigenvectors and values from Eq. (5.3) can be used to generate a transformation to replace M and K with diagonal matrices. To do this Eq. (5.2) is multiplied by the transpose of the lth mode-shape to give

$$\omega_k^2 \phi_{\mathbf{l}}^T M \phi_{\mathbf{k}} = \phi_{\mathbf{l}}^T K \phi_{\mathbf{k}}. \tag{5.9}$$

In addition, Eq. (5.2) can be written in terms of the lth mode-shape and premultiplied by the kth mode-shape to give

$$\omega_l^2 \phi_{\mathbf{k}}^T M \phi_{\mathbf{l}} = \phi_{\mathbf{k}}^T K \phi_{\mathbf{l}}.$$

Using $AB = (B^T A^T)^T$ and noting that $C^T = D^T$ is the same as $C = D$ and that M and K are symmetric (such that $M^T = M$ and $K^T = K$) allows this expression to be rewritten as

$$\omega_l^2 \phi_{\mathbf{l}}^T M \phi_{\mathbf{k}} = \phi_{\mathbf{l}}^T K \phi_{\mathbf{k}}. \tag{5.10}$$

Now subtracting Eq. (5.10) from Eq. (5.9) gives

$$(\omega_k^2 - \omega_l^2)\phi_l^T M \phi_k = \phi_l^T K \phi_k - \phi_l^T K \phi_k = 0.$$

When the natural frequencies are distinct, $\omega_k^2 \neq \omega_l^2$, it can be seen that $\phi_l^T M \phi_k = 0$ for $l \neq k$. Substituting this back into Eq. 5.9 means that $\phi_l^T K \phi_k = 0$ for $l \neq k$. Therefore diagonal mass and stiffness matrices, M_d and K_d are defined as

$$M_d = \Phi^T M \Phi \quad \text{and} \quad K_d = \Phi^T K \Phi,$$

where Φ is a matrix containing the mode-shapes in columns and the kth diagonal element in M_d is given by $\phi_k^T M \phi_k$ where ϕ_k is the kth eigenvector or mode-shape.

This equation remains coupled through the nonlinear terms. An example of such a system is discussed in Sect. 5.3.1. For full decoupling, either a new transformation must be derived, or a further transform, that can be applied after the linear modal transform, is needed. Alternatively the effect of the coupling can be assessed, for the case where the nonlinear terms are small, by using the normal form technique as is discussed later in Sect. 5.4.

5.2.2 State Space Form for Discrete Linear Systems

Now consider the case where the discrete system is written in first-order form. First-order, or state space form, has been discussed in Sect. 1.4.1. This representation is usual for control analysis but less common for vibration applications. In this case, Eq. (5.1) may be written in terms of the state vector $\boldsymbol{x} = [\mathbf{x}^T, \dot{\mathbf{x}}^T]^T$ as

$$\dot{\boldsymbol{x}} = A\boldsymbol{x} + \begin{bmatrix} 0_m \\ M^{-1}\mathbf{F}_E \end{bmatrix}, \quad A = \begin{bmatrix} 0_n & I_n \\ -M^{-1}K & -M^{-1}C \end{bmatrix}, \tag{5.11}$$

where 0_n is an $N \times N$ matrix of zeros, 0_m is a $N \times 1$ vector and I_n is an $N \times N$ identity matrix where N is the number of degrees of freedom. A transform for $\boldsymbol{x}$ is now required such that A can be replaced by a diagonal matrix such that the states become decoupled.[5] Consider the eigenvectors of A, the equation defining the kth eigenvector, ϕ_k (note the length of ϕ is $2N$ whereas in the second-order analysis it was N) and corresponding eigenvalue λ_k may be written as

$$\lambda_k \phi_k = A\phi_k.$$

Defining a matrix of the $2N$ eigenvectors Φ in which ϕ_k forms the kth column and a corresponding diagonal eigenvalue matrix Λ where the kth diagonal value is λ_k,

---

[5] There are situations where Jordan normal form is preferable to diagonal matrices, but these are not considered here.

allows this equation to be written in matrix form for all k

$$\Phi\Lambda = A\Phi \quad \rightsquigarrow \quad \Lambda = \Phi^{-1}A\Phi. \tag{5.12}$$

The equation of motion can be rewritten using the transformation $\boldsymbol{x} = \Phi\mathbf{q}$ (again noting that now $\mathbf{q}$ is of length $2N$), giving

$$\Phi\dot{\mathbf{q}} = A\Phi\mathbf{q} + \begin{bmatrix} 0_m \\ M^{-1}\mathbf{F}_E \end{bmatrix}.$$

Multiplying this equation by Φ^{-1} and using Eq. (5.12) gives

$$\dot{\mathbf{q}} = \Lambda\mathbf{q} + \Phi^{-1}\begin{bmatrix} 0_m \\ M^{-1}\mathbf{F}_E \end{bmatrix}, \tag{5.13}$$

which, since Λ is diagonal, is decoupled with respect to the states. Note that for the state space representation the mode-shapes are defined as the eigenvectors of A, whereas for the second-order representation they are the eigenvectors of $M^{-1}K$.

Notice that in this example of a system in first-order form, damping has been included in the matrix, A, which leads to complex eigenvalues and eigenvectors for underdamped vibrations.[6] In some situations, this is an advantage, for example in the application of positive position feedback, Sect. 6.3.3, Chap. 6. An example which compares the first- and second-order decompositions is considered next.

Example 5.1 State space modal decomposition of a linear system

Problem Consider the two-degree-of-freedom oscillator shown in Fig. 5.1 for the case where the spring forces are linear, such that $F_1 = kx_1$, $F_2 = k(x_2 - x_1)$, $F_3 = kx_2$ and $m_1 = m_2 = m$. Decompose both the second-order and the first-order (state space) representations of this system and compare the resulting modes. In the modal coordinates, calculate the system response to initial conditions $x_1 = 2$, $x_2 = 0$ and $\dot{x}_1 = \dot{x}_2 = 0$.

Solution The equation of motion of the system is given by

$$M\ddot{\mathbf{x}} + K\mathbf{x} = 0, \quad \text{where} \quad M = \begin{bmatrix} m & 0 \\ 0 & m \end{bmatrix}, \quad K = \begin{bmatrix} 2k & -k \\ -k & 2k \end{bmatrix}.$$

The modal decomposition for this second-order representation of the equation of motion is based on the eigenvectors of $M^{-1}K$, which are in matrix form.[7]

---

[6] Note this should not be confused with *complex modes* which arise for systems with non-proportional damping. See, Ewins (2000) for a more detailed discussion.

[7] It was noted in Sect. 1.3.3 Chap. 1 that the eigenvectors can be scaled. Here, for simplicity, they are left unscaled. For a discussion on this type of scaling, such as mass normalized modes, see Bishop et al. (2009) or Ewins (2000).

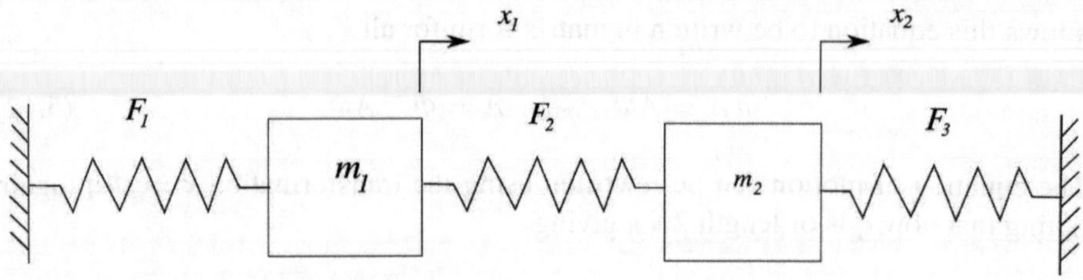

Fig. 5.1 A two-degree-of-freedom oscillator

$$\Phi = \begin{bmatrix} 1 & 1 \\ 1 & -1 \end{bmatrix}.$$

Making the transformation $\mathbf{x} = \Phi\mathbf{q}$ decouples the system into

$$M_d\ddot{\mathbf{q}} + K_d\mathbf{q} = 0 \quad \rightsquigarrow \quad \begin{aligned} \ddot{q}_1 + \omega_n^2 q_1 &= 0, \\ \ddot{q}_2 + 3\omega_n^2 q_2 &= 0, \end{aligned} \tag{5.14}$$

where $\omega_n^2 = k/m$ is the first natural frequency of the system. Using $\mathbf{q} = \Phi^{-1}\mathbf{x}$ gives the initial conditions $q_1(0) = 1$, $q_2(0) = 1$ and $\dot{q}_1(0) = \dot{q}_2(0) = 0$. Solving Eq. (5.14) using these initial conditions gives

$$\begin{aligned} q_1 &= \cos(\omega_n t) \\ q_2 &= \cos(\sqrt{3}\omega_n t) \end{aligned} \quad \rightsquigarrow \quad \begin{aligned} x_1 &= \cos(\omega_n t) + \cos(\sqrt{3}\omega_n t), \\ x_2 &= \cos(\omega_n t) - \cos(\sqrt{3}\omega_n t). \end{aligned}$$

For the state space system case, the form given by Eq. (5.11) becomes

$$\dot{\boldsymbol{x}} = A\boldsymbol{x}, \quad A = \begin{bmatrix} 0 & 0 & 1 & 0 \\ 0 & 0 & 0 & 1 \\ -2\omega_n^2 & \omega_n^2 & 0 & 0 \\ \omega_n^2 & -2\omega_n^2 & 0 & 0 \end{bmatrix},$$

where $\boldsymbol{x} = [\mathbf{x}^T, \dot{\mathbf{x}}^T]^T$. The state space formulation is decoupled using the eigenvalues and eigenvectors of A which are given by

$$\Phi\Lambda = A\Phi, \quad \Phi = \begin{bmatrix} 1 & 1 & 1 & 1 \\ 1 & 1 & -1 & -1 \\ \mathrm{i}\omega_n & -\mathrm{i}\omega_n & \mathrm{i}\sqrt{3}\omega_n & -\mathrm{i}\sqrt{3}\omega_n \\ \mathrm{i}\omega_n & -\mathrm{i}\omega_n & -\mathrm{i}\sqrt{3}\omega_n & \mathrm{i}\sqrt{3}\omega_n \end{bmatrix}, \quad \Lambda = \begin{bmatrix} \mathrm{i}\omega_n & 0 & 0 & 0 \\ 0 & -\mathrm{i}\omega_n & 0 & 0 \\ 0 & 0 & \mathrm{i}\sqrt{3}\omega_n & 0 \\ 0 & 0 & 0 & -\mathrm{i}\sqrt{3}\omega_n \end{bmatrix},$$

from Eq. (5.12). Applying the transformation $\boldsymbol{x} = \Phi\mathbf{q}$ (where $\mathbf{q}$ is a 4×1 vector) results in the decoupled equation $\dot{\mathbf{q}} = \Lambda\mathbf{q}$ using Eq. (5.13). The initial conditions can be transformed into the modal coordinates using $\mathbf{q} = \Phi^{-1}\boldsymbol{x}$ resulting in $q_1 = q_2 = q_3 = q_4 = 1/2$. Solving $\dot{\mathbf{q}} = \Lambda\mathbf{q}$ using the initial conditions gives

$$\begin{aligned} q_1 &= 0.5e^{i\omega_n t} \\ q_2 &= 0.5e^{-i\omega_n t} \\ q_3 &= 0.5e^{i\sqrt{3}\omega_n t} \\ q_4 &= 0.5e^{-i\sqrt{3}\omega_n t} \end{aligned} \quad \rightsquigarrow \quad \begin{aligned} x_1 &= \cos(\omega_n t) + \cos(\sqrt{3}\omega_n t), \\ x_2 &= \cos(\omega_n t) - \cos(\sqrt{3}\omega_n t), \end{aligned}$$

which gives the same result as the second-order form analysis. ■

If nonlinearity in the form of a function $\mathscr{N}(\mathbf{x}, \dot{\mathbf{x}})$ is present in the equation of motion, Eq. (5.13), then decoupling using the linearized transformation $\boldsymbol{x} = \Phi\mathbf{q}$ results in

$$\dot{\mathbf{q}} = \Lambda\mathbf{q} + \mathbf{f}(\mathbf{q}) + \Phi^{-1}\begin{bmatrix} 0_n \\ M^{-1}\mathbf{F}_E \end{bmatrix},$$

where

$$\mathbf{f}(\mathbf{q}) = \Phi^{-1}\begin{bmatrix} 0 \\ -M^{-1}\mathscr{N}(\Phi\mathbf{q}) \end{bmatrix}.$$

Here $\mathscr{N}(\mathbf{x}, \dot{\mathbf{x}})$ has be rewritten in terms of the state vector to give $\mathscr{N}(\boldsymbol{x})$, because $\boldsymbol{x} = [\mathbf{x}, \dot{\mathbf{x}}]^T$, and then the transformation $\boldsymbol{x} = \Phi\mathbf{q}$ has been applied to give $\mathscr{N}(\Phi\mathbf{q})$. Note that this nonlinear vector typically gives rise to cross-coupling between the linear modes.

When using the state space formulation, for lightly damped systems it is sometimes convenient to apply the linear decomposition based on the undamped and unforced system. In this case the damping terms can be included within $\mathscr{N}$ and further transforms can be used to eliminate any cross-coupling terms. An example of this using the normal forms method on a model of cable dynamics is given in the approximate methods case study in Sect. 7.4.4.

In both representations, the decoupling involves a transformation, $\mathbf{x} = \Phi\mathbf{q}$ for the second-order method and $\boldsymbol{x} = \Phi\mathbf{q}$ for the first-order method (where Φ and $\mathbf{q}$ have different dimensions in each case). The vector $\mathbf{q}$ represents the quantity of each mode that is present in the vibration response. For the second-order approach, the transformation is based on an orthogonal set of vectors, or mode-shapes, contained within matrix Φ (generated from eigenvectors of a matrix). This means that there is a unique $\mathbf{q}$ for a given $\mathbf{x}$ and the displacement response at the kth mass point will be $x_k = \phi_k^T\mathbf{q}$ for $k = 1$ to $k = N$. Orthogonality of the modes (in the second-order form) is key to the decoupling of the system, and applying this approach to a continuous system is considered next.

5.2.3 Continuous Linear Systems

For a continuous system, in which the mass is distributed continuously across the structure, the governing equations are in the form of a partial differential equation.

For example, the dynamics of a straight, uniform cross-section, rod in axial vibration can be modelled by the one-dimensional wave equation

$$\frac{\partial^2 u}{\partial t^2} = c^2 \frac{\partial^2 u}{\partial x^2}, \tag{5.15}$$

where $u(x,t)$ is the longitudinal displacement of the rod at time t and at point x and $c = \sqrt{E/\rho}$ is the wave propagation speed of the rod (see for example Thompson and Dahleh (1997) and Inman (2006)).

One method of finding the mode-shapes and natural frequencies of certain continuous systems is to use the classical separation of variables technique.[8] This technique assumes that the displacement, u, which is a function of time t, and location x, can be separated into two functions, one of x and one of t multiplied together. For the case of the axially vibrating rod, this split can be represented as an assumed response of the form

$$u(x,t) = \phi(x)q(t).$$

Note that $\phi(x)$ represents a mode-shape of the system. This equation is substituted into the equation of motion and rearranged to give two differential equations, one in location and the other in time. Under this substitution, Eq. (5.15) becomes

$$\frac{1}{c^2 q(t)} \frac{\mathrm{d}^2 q(t)}{\mathrm{d}t^2} = \frac{1}{\phi(x)} \frac{\mathrm{d}^2 \phi(x)}{\mathrm{d}x^2}.$$

In this equation, a function of time equals a function of position for all time and positions, therefore the two sides of this equation must equal a constant, which for convenience is written as $-\alpha^2$, where α could be imaginary to accommodate a negative constant. This gives the two separate ordinary differential equations

$$\frac{\mathrm{d}^2 q(t)}{\mathrm{d}t^2} + \alpha^2 c^2 q(t) = 0, \quad \frac{\mathrm{d}^2 \phi(x)}{\mathrm{d}x^2} + \alpha^2 \phi(x) = 0.$$

Solving these equations using standard techniques for linear differential equations gives the solution

$$u(x,t) = \phi(x)q(t) = \left(\bar{A}\sin(\frac{\omega_n}{c}x) + \bar{B}\cos(\frac{\omega_n}{c}x)\right)(\bar{C}\sin(\omega_n t) + \bar{D}\cos(\omega_n t)), \tag{5.16}$$

where the substitution $\omega_n = \alpha c$ has been made in which ω_n is a natural frequency. The constants $\bar{A}$, $\bar{B}$ along with an expression for the natural frequency ω_n can be found by applying the boundary conditions. An example of specific boundary conditions is now considered.

[8] This technique is often used for beams and cables, however it cannot be used for more complex structures such as plates. In these cases, approximate techniques are used, see Chap. 8.

Example 5.2 Mode-shapes for a rod with axial vibration

Problem Find the modal constants $\bar{A}$, $\bar{B}$ and ω_n in Eq. (5.16), when the rod is fixed at both ends, so that $u(0,t) = u(L,t) = 0$ where L is the length of the rod.

Solution Putting the boundary conditions into Eq. (5.16) gives $\bar{B} = 0$ and $\omega_{nk} = k\pi c/L$, where $k = 1, 2, 3, 4 \ldots$. The constant $\bar{A}$ can have any value, and $\bar{A} = 1$ is arbitrarily selected. The resulting mode-shape is $\phi_k(x) = \sin(\omega_n x/c)$ which, using $\omega_{nk} = k\pi c/L$, may be rewritten for this example as $\phi_k(x) = \sin(k\pi x/L)$ for the kth mode. By inspection it can be confirmed that these mode-shapes satisfy the boundary conditions. ■

Notice that there are in fact an infinite number of solution sets for $\bar{A}$ and ω_n, and, as a result, there are also an infinite number of mode-shapes. So the general response is actually given by

$$u(x,t) = \sum_{k=1}^{\infty} \phi_k(x) q_k(t) = \sum_{k=1}^{\infty} \sin(\frac{k\pi}{L}x) \left(\bar{C}_k \sin(\frac{k\pi c}{L}t) + \bar{D}_k \cos(\frac{k\pi c}{L}t) \right)$$

in this example. The last unknowns $\bar{C}$ and $\bar{D}$ can be found using the initial conditions.

In this derivation of the mode-shapes, the separation of variable substitution $u(x,t) = \phi(x)q(t)$ was made, however the resulting system response was in the form $u(x,t) = \sum_{n=1}^{\infty} \phi_n(x) q_n(t)$. Effectively, just one of the modes was considered by making the $u(x,t) = \phi(x)q(t)$ substitution, whereas in fact the total solution is actually the linear superposition of all possible mode-shapes that satisfy the boundary conditions.

A more general approach that uses this form of solution is the *Galerkin method*.[9] With this method the full response is assumed to be

$$u(x,t) = \sum_{k=1}^{\infty} \phi_k(x) q_k(t).$$

This expression is substituted into the equation of motion, and orthogonality conditions are used to decouple the resulting expressions. The term 'decoupling' means that a single partial differential equation is transformed into an infinite set of ordinary differential equations. Each of the ordinary differential equations corresponds to a specific mode of vibration. Consider how this would work for the axial rod example.

[9] See Finlayson (1972) for a description of the Galerkin and related methods.

Example 5.3 Galerkin method for a rod with axial vibration

Problem Use the Galerkin method to decompose the wave equation, Eq. (5.15), into a set of ordinary differential equations. Assume that the rod is fixed at both ends, so that $u(0,t) = u(L,t) = 0$ where L is the length of the rod.

Solution Substituting $u(x,t) = \sum_{k=1}^{\infty} \phi_k(x) q_k(t)$ into Eq. (5.15) gives

$$\sum_{k=1}^{\infty} \phi_k \ddot{q}_k = c^2 \sum_{k=1}^{\infty} \phi_k'' q_k \tag{5.17}$$

where $\bullet'$ and $\dot{\bullet}$ represent the derivatives with respect to position and time respectively.

In Example 5.2 the mode-shapes for a rod with axial vibration were shown to be $\phi_k(x) = \sin(k\pi x/L)$, $k = 1, 2, 3, 4 \ldots$. To decouple Eq. (5.17), first multiply by an arbitrary mode-shape ϕ_n and then integrate over the length of the rod[10] to give

$$\sum_{k=1}^{\infty} \int_0^L \phi_n \phi_k \mathrm{d}x \, \ddot{q}_k = c^2 \sum_{k=1}^{\infty} \int_0^L \phi_n \phi_k'' \mathrm{d}x \, q_k.$$

Note that q_k has been taken out of the integral as it is a function of time only. By inspection this equation may be decoupled (such that only q_n terms remain in the equation) if the integral terms for $n \neq k$ are zero, in other words

$$\int_0^L \phi_n \phi_k \mathrm{d}x = 0 \quad \text{and} \quad \int_0^L \phi_n \phi_k'' \mathrm{d}x = 0 \quad \text{for } k \neq n. \tag{5.18}$$

These are the orthogonality conditions. It can be shown that they are satisfied by the sinusoidal mode-shapes derived in Example 5.2. Using these orthogonality conditions the governing equation of motion decouples to

$$\int_0^L \phi_n^2 \mathrm{d}x \, \ddot{q}_n = c^2 \int_0^L \phi_n \phi_n'' \mathrm{d}x \, q_n, \tag{5.19}$$

for the nth mode. Substituting $\phi_n(x) = \sin(n\pi x/L)$ into the integral terms gives

$$\int_0^L \phi_n^2 \mathrm{d}x = \frac{L}{2} \quad \text{and} \quad \int_0^L \phi_n \phi_n'' \mathrm{d}x = -\frac{n^2\pi^2}{L^2} \frac{L}{2}, \tag{5.20}$$

which can be substituted into Eq. (5.19) to give an infinite number of ordinary differential equations of the form

---

[10] This is analogous to the approach used to calculate a Fourier series approximation to a function.

$$\ddot{q}_n + \omega_{nn}^2 q_n = 0. \tag{5.21}$$

Where the natural frequency of the nth mode is $\omega_{nn} = n\pi c/L$. ■

Mode-shape functions which have the properties of the sine functions in Example 5.3 are said to be *orthogonal*.[11] Note that there is no damping or forcing in Eq. (5.21), and this would need to be included to create a realistic model of axial rod vibration. Also notice that to use Eq. (5.21) as a model, the series needs to be truncated to a finite set of N equations.

The Galerkin method exploits the orthogonal nature of the assumed mode-shapes to decouple the partial differential equation into a set of ordinary differential equations. Together with a set of assumed mode-shapes, the Galerkin method can also be applied to some nonlinear equations. To consider a nonlinear example, the axial example is now extended to include a small material stiffness nonlinearity such that

$$\sigma = E\epsilon + E^*\epsilon^2,$$

where σ is stress, ϵ is strain, E is Young's modulus and the nonlinear stiffness term E^* is small. Including this relationship within the equation of motion leads to the partial differential equation

$$\frac{\partial^2 u}{\partial t^2} = c^2 \frac{\partial^2 u}{\partial x^2}\left(1 + 2\frac{E^*}{E}\frac{\partial u}{\partial x}\right). \tag{5.22}$$

Finding a modal decomposition for a nonlinear partial differential equation such as this is not always possible. However, using the linear modal basis for the underlying linear problem (i.e. when $E^* = 0$ in this case), combined with the Galerkin decomposition, can often be used to create a model. Using this modal basis ensures that the linear terms of the equations are decoupled due to orthogonality, but coupling through the nonlinear terms will typically remain.

This method involves four steps. First, the linearized equation of motion is considered and the mode-shapes are found using the separation of variables technique. Secondly, Galerkin's method is applied to the nonlinear equation of motion. Thirdly, since the mode-shapes from the linearized equation of motion are to be used as the modal set, the linearized orthogonality conditions are met and therefore can be applied to the equation of motion to decouple the linear terms. Finally, the linearized mode-shapes are substituted into the equation of motion to give a series of equations for each of the modes with nonlinear, but no linear, coupling.

Example 5.4 Galerkin method for a rod with axial vibration and nonlinear stiffness

Problem Use the Galerkin method to decompose the modified wave equation, Eq. (5.22), into a set of ordinary differential equations. Assume that the rod is fixed at both ends, so that $u(0,t) = u(L,t) = 0$ where L is the length of the rod.

---

[11] Note that if the modes are scaled such that when $k = n$ the integral equals one, the modes are said to be orthonormal.

Solution The first step of the decoupling process has already been completed in Example 5.2. From Example 5.2, the kth mode-shape for the underlying linear system (i.e. with $E^* = 0$) is given by $\phi_k(x) = \sin(k\pi x/L)$ with a corresponding natural frequency of $k\pi c/L$ and the orthogonality conditions are given by Eq. (5.18). The second step is to apply Galerkin's method by making the substitution $u(x,t) = \sum_{k=1}^{\infty} \phi_k(x) q_n(t)$ in the nonlinear equation, Eq. (5.22), then multiplying by $\phi_n(x)$ and integrating over the length to give

$$\sum_{k=1}^{\infty} \phi_k(x)\ddot{q}_k(t) = c^2 \sum_{k=1}^{\infty} \phi_k''(x) q_k(t) \left(1 + 2\frac{E^*}{E} \sum_{l=1}^{\infty} \phi_l'(x) q_l(t)\right)$$

$$\rightsquigarrow \sum_{k=1}^{\infty} \int_0^L \phi_n(x)\phi_k(x)\mathrm{d}x\, \ddot{q}_k(t) = c^2 \sum_{k=1}^{\infty} \int_0^L \phi_n(x)\phi_k''(x)\mathrm{d}x\, q_k(t)$$

$$+ 2c^2 \frac{E^*}{E} \sum_{k=1}^{\infty} \sum_{l=1}^{\infty} \int_0^L \phi_n(x)\phi_k''(x)\phi_l'(x)\mathrm{d}x\, q_k(t) q_l(t).$$

Now, in step 3, the mode-shapes of the linearized system are used to decouple the linear terms in the equation by applying the orthogonality conditions for the linearized equation mode-shapes, Eq. (5.18) to give

$$\int_0^L \phi_n^2(x)\mathrm{d}x\, \ddot{q}_n(t) = c^2 \int_0^L \phi_n(x)\phi_n''(x)\mathrm{d}x\, q_n(t)$$

$$+ 2c^2 \frac{E^*}{E} \sum_{k=1}^{\infty} \sum_{l=1}^{\infty} \int_0^L \phi_n(x)\phi_k''(x)\phi_l'(x)\mathrm{d}x\, q_k(t) q_l(t). \quad (5.23)$$

Finally, the mode-shapes of the linearized system, $\phi_n(x) = \sin(n\pi x/L)$ for this example, are substituted into Eq. (5.23) to give equations of motion that are only coupled through the nonlinear terms. Making this substitution, we find that the integrals of the linear terms reduce to

$$\int_0^L \phi_n^2(x)\mathrm{d}x = \frac{L}{2}, \quad \int_0^L \phi_n(x)\phi_n''(x)\mathrm{d}x = -\frac{n^2\pi^2}{2L}.$$

The integral in the nonlinear term becomes

$$\int_0^L \phi_n(x)\phi_k''(x)\phi_l'(x)\mathrm{d}x = \frac{k^2 l \pi^3}{2L^3} \int_0^L \left[\cos\left((n+k)\frac{\pi x}{L}\right) - \cos\left((n-k)\frac{\pi x}{L}\right)\right]$$

$$\times \cos\left(l\frac{\pi x}{L}\right)\mathrm{d}x.$$

The integral terms are non-zero only if the two cosine terms being multiplied together have equal or opposite index values, and this requires $l = n + k$ or $l = |n - k|$ for the case of the first and second terms respectively. In these cases the integral reduces to $L/2$ or $-L/2$ for the first and second terms respectively. Substituting these expressions into Eq. (5.23) gives

$$\begin{aligned}\frac{L}{2}\ddot{q}_n(t) = &- c^2\frac{n^2\pi^2}{2L}q_n(t)\\ &+ 2c^2\frac{E^*}{E}\sum_{k=1}^{\infty}\frac{k^2\pi^3}{4L^2}\left((n+k)q_k(t)q_{n+k}(t) - |n-k|q_k(t)q_{|n-k|}(t)\right).\end{aligned}$$

Rearranging this equation and using the natural frequency expression for the linearized system, $\omega_{nk} = k\pi c/L$ (where ω_{nk} is the natural frequency ω_n of the kth mode), gives the following expression for the nth modal equation

$$\ddot{q}_n(t) + \omega_{nn}^2 q_n(t) - \frac{c^2\pi^3 E^*}{L^3 E}\sum_{k=1}^{\infty}k^2\left((n+k)q_k(t)q_{n+k}(t) - |n-k|q_k(t)q_{|n-k|}(t)\right) = 0.$$

If, for example, just the first two modes of vibration were being considered the system could therefore be written in the form

$$\ddot{q}_1(t) + \omega_{n1}^2 q_1(t) + \frac{c^2\pi^3 E^*}{L^3 E}2q_1q_2 = 0, \tag{5.24}$$

$$\ddot{q}_2(t) + \omega_{n2}^2 q_2(t) + \frac{c^2\pi^3 E^*}{L^3 E}q_1^2 = 0, \tag{5.25}$$

which are a coupled set of nonlinear modal ordinary differential equations. ■

As a result of the nonlinearity and the coupled coordinates, the dynamic response of the system is complex, and it is natural to ask whether there is a coordinate transformation that would simplify the equations of motion. The ultimate aim would be to find a transform that simplified the equations of motion as much as possible, in a similar way to linear modal analysis. For nonlinear systems, looking for this type of simplest form is known as *normal form analysis*, and can lead to insights regarding the presence of nonlinear resonance such as subharmonic resonance, as was discussed in Sect. 4.5.

A different but related approach for considering the nonlinear modal behaviour is to consider the physical representation of nonlinear modes in the system. This has led to the development of *nonlinear normal mode* techniques. These ideas are discussed next as nonlinear modal decomposition is considered.

5.3 Modal Decomposition for Nonlinear Systems

Before discussing nonlinear modal decomposition using normal forms (Sect. 5.4), some of the underlying ideas are introduced.[12] The first key point is that the nonlinear modal decomposition used here will incorporate a linear modal decomposition of the system as a first step.

For example, consider the two-degree-of-freedom lumped mass system, as shown in Fig. 5.1, in which the spring forces contain linear and cubic terms with coefficients k and κ respectively for forces F_1 and F_3 and k_2 and κ_2 respectively for force F_2. Taking the masses to be equal, $m_1 = m_2 = m$, the equations of motion may be written as

$$\begin{aligned} m\ddot{x}_1 + kx_1 + k_2(x_1 - x_2) + \kappa x_1^3 - \kappa_2(x_2 - x_1)^3 &= 0, \\ m\ddot{x}_2 + kx_2 + k_2(x_2 - x_1) + \kappa x_2^3 + \kappa_2(x_2 - x_1)^3 &= 0. \end{aligned} \tag{5.26}$$

The linearized version of this system, where $\kappa = \kappa_2 = 0$, was analysed in Example 5.1 and results in the modal transform

$$\mathbf{x} = \Phi\mathbf{q} \quad \text{where} \quad \Phi = \begin{bmatrix} 1 & 1 \\ 1 & -1 \end{bmatrix}. \tag{5.27}$$

To transform the nonlinear equations of motion, they can be written in the form of Eq. (5.7) (without damping or forcing in this case)

$$M\ddot{\mathbf{x}} + K\mathbf{x} + \mathscr{N}(\mathbf{x}, \dot{\mathbf{x}}) = 0,$$

to give

$$m\begin{bmatrix} 1 & 0 \\ 0 & 1 \end{bmatrix}\ddot{\mathbf{x}} + \begin{bmatrix} k + k_2 & -k_2 \\ -k_2 & k + k_2 \end{bmatrix}\mathbf{x} + \begin{bmatrix} \kappa x_1^3 - \kappa_2(x_2 - x_1)^3 \\ \kappa x_2^3 + \kappa_2(x_2 - x_1)^3 \end{bmatrix} = 0,$$

where $\mathbf{x} = [x_1 \;\; x_2]^T$ and the last term on the left-hand side of the equation is the nonlinear term $\mathscr{N}(\mathbf{x}, \dot{\mathbf{x}})$, which in this example reduces to $\mathscr{N}(\mathbf{x})$. Applying the transformation $\mathbf{x} = \Phi\mathbf{q}$ (as detailed in Sect. 5.2) results in Eq. (5.8) which for no external forcing or damping and with no nonlinear velocity terms gives

$$M_d\ddot{\mathbf{q}} + K_d\mathbf{q} + \Phi^T\mathscr{N}(\Phi\mathbf{q}) = 0. \tag{5.28}$$

where $M_d = \Phi^T M\Phi$ and $K_d = \Phi^T K\Phi$. Using the transform given by Eq. (5.27), Eq. (5.28) may be written as

[12] Note that the related problem of nonlinear system identification is not considered here. See Kerschen et al. (2006) for a comprehensive review of these techniques.

$$m\begin{bmatrix}1 & 0\\ 0 & 1\end{bmatrix}\ddot{\mathbf{q}}+\begin{bmatrix}k & 0\\ 0 & k+2k_2\end{bmatrix}\mathbf{q}+\begin{bmatrix}\kappa(q_1^3+3q_1q_2^2)\\ \kappa(q_2^3+3q_1^2q_2)+8\kappa_2q_2^3\end{bmatrix}=0. \quad (5.29)$$

As with linear modal analysis, it is now possible to see what the linear natural frequencies for the system are. In this case they come from multiplying Eq. (5.29) by M_d^{-1} to give a dynamic matrix, $M_d^{-1}K_d$ containing the linear natural frequencies. If the system is given some forcing excitation, one would expect resonance peaks to occur at the natural frequencies, and this is one of the key pieces of information required for a vibration analysis of any mechanical or structural system.[13] However, Eq. (5.29) has additional, coupled, nonlinear terms which will mean, in practice, that the natural frequencies are not the same as the linear ones. It is also possible that the nonlinear coupling terms may cause additional *nonlinear resonances* in the system, which cannot be seen by simply looking at the form of Eq. (5.29). For example, subharmonic resonance is a form of nonlinear resonance and was discussed in Example 4.9. Other types of nonlinear resonance phenomena include parametric and auto-parametric resonances, discussed in for example Sect. 7.5. The nonlinear resonances and the nonlinear natural frequencies can be found in a systematic way by using the normal forms method, which is described in Sect. 5.4.

An alternative approach to the problem is to define *nonlinear normal modes*, as an extension of the idea of linear vibration modes into the nonlinear domain. There are several methods for finding approximate expressions for nonlinear normal modes, and a full treatment of this subject is beyond the scope of this chapter, although an example is briefly mentioned below. The interested reader can find a comprehensive overview and associated literature review in Kerschen et al. (2009) and Peeters et al. (2009).

5.3.1 Nonlinear Normal Modes

Nonlinear normal modes can be formulated in several different ways. As a starting point consider the categories of *similar* and *non-similar* nonlinear normal modes. Like linear modes, similar nonlinear normal modes can be expressed as a series of independent functions in the spatial domain. Non-similar normal modes have more complex relationships between the degrees of freedom—typically each degree of freedom is a function of the other degrees of freedom.

In this section, the system analysed will be a two-mode oscillator represented by two masses coupled by nonlinear springs. Initially the system will be general, however to proceed in the analysis the special case where the linear stiffness component of the spring joining the two masses is set to zero, $k_2 = 0$, is considered. This special case can be analysed using relatively simple approaches such as the harmonic balance, see Sect. 5.3.2. In the next section, Sect. 5.4, the Normal Form technique will

[13] The damping ratios are the other key set of quantities that is required, but in this simplified example, no damping is assumed.

be used to analyse the general case where there is a linear stiffness component in the spring coupling the two masses. This gives rise to nonlinear normal modes that are a function of amplitude of oscillation.

For the two-degree-of-freedom example shown in Fig. 5.1, governed by Eq. (5.26), similar modes give an assumed relationship of the form $x_2 = \sigma x_1$, where σ is a real constant. Substituting this relationship into the equations of motion, Eq. (5.26), gives

$$m\ddot{x}_1 + kx_1 + k_2(1-\sigma)x_1 + \kappa x_1^3 - \kappa_2(\sigma-1)^3 x_1^3 = 0,$$
$$m\sigma\ddot{x}_1 + k\sigma x_1 + k_2(\sigma-1)x_1 + \kappa\sigma^3 x_1^3 + \kappa_2(\sigma-1)^3 x_1^3 = 0.$$

Equating these two equations (an alternative energy approach is possible, see Vakakis et al. (1996)) gives the relationship for σ as

$$\kappa(\sigma^3-\sigma) + \frac{1}{x_1^2}k_2(\sigma^2-1) + \kappa_2(\sigma-1)^3(\sigma+1) = 0. \tag{5.30}$$

Noting that for similar nonlinear normal modes σ is a constant, and assuming that the linear stiffnesses are non-zero, we find two real solutions for σ, which are $\sigma = \pm 1$. For $\sigma = 1$, the masses are in-phase with each other, and for $\sigma = -1$, the masses are out-of-phase with each other.

So, assuming similar modes in this example leads to two modal solutions which are the same as the linearized version of the problem. Or, in other words, the only similar modes which exist for this system, in general, are the linear modes. However, looking at Eq. (5.30), it can be seen that if k_2 is zero then the equation governing the solutions of σ becomes independent of the state x_1. This special case has been widely studied as an example of nonlinear normal modes (see for example Vakakis et al. (1996), Rand (2005), and references therein), as it has interesting nonlinear behaviour.

In the special case where the linear stiffness between the masses is zero, $k_2 = 0$, it is possible to have additional σ value solutions to the $\sigma = \pm 1$ that have already been identified. In this case Eq. (5.30) becomes

$$\kappa(\sigma^3-\sigma) + \kappa_2(\sigma-1)^3(\sigma+1) = 0$$
$$\rightsquigarrow (1-\sigma)(1+\sigma)\left[\sigma^2 + \left(\frac{\kappa}{\kappa_2} - 2\right)\sigma + 1\right] = 0. \tag{5.31}$$

The real values of σ which satisfy this equation are

Solution $S1$: $\sigma = 1$ for all κ.

Solution $S2$: $\sigma = -1$ for all κ.

Solution $S3^+$: $\sigma = \frac{1}{2}\left[2 - \frac{\kappa}{\kappa_2} + \sqrt{\frac{\kappa}{\kappa_2}\left(\frac{\kappa}{\kappa_2} - 4\right)}\right]$ for $\kappa \geq 4\kappa_2$.

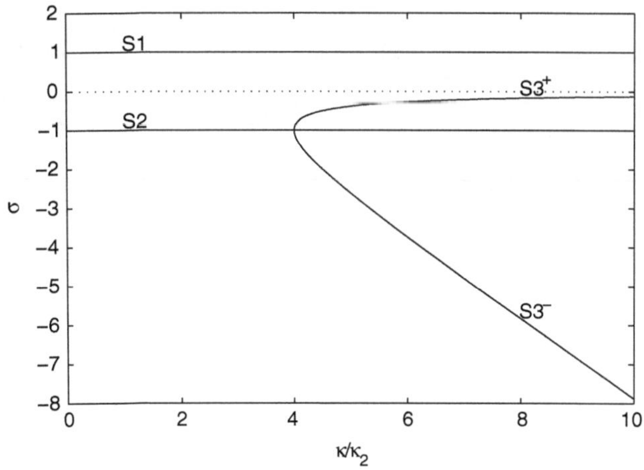

Fig. 5.2 The ratio of similar modes, σ, for the two-degree-of-freedom example. The *lines* show the four solution branches $S1$, $S2$ and $S3^{\pm}$

$$\text{Solution } S3^- : \sigma = \frac{1}{2}\left[2 - \frac{\kappa}{\kappa_2} - \sqrt{\frac{\kappa}{\kappa_2}\left(\frac{\kappa}{\kappa_2} - 4\right)}\right] \quad \text{for } \kappa \geq 4\kappa_2.$$

Note that if $\kappa < 4\kappa_2$ then the third and fourth roots are complex, and are therefore not considered to be valid solutions. The valid values of σ can be substituted back into the equations of motion to get four equations relating to the pure mode motion of each of the similar modes. However, these equations are valid only for pure mode motion in each case—cross-coupling between the modes is lost.

The σ values for the four solutions are plotted in Fig. 5.2 as a function of the stiffness ratio κ/κ_2 in the range $0 \leq \kappa/\kappa_2 \leq 10$. Physically, as κ/κ_2 increases, the spring forces F_1 and F_3 become increasingly large compared with F_2, via the nonlinear hardening term. Another way to think about this is to treat κ as fixed and then assume $\kappa_2 \to 0$ to make κ/κ_2 increase. In the range $4 < \kappa/\kappa_2 \leq 10$ solutions $S3^+$ and $S3^-$ exist in addition to solutions $S1$ and $S2$. This phenomenon has been analysed in terms of the $S2$ solution becoming unstable at the point $\kappa/\kappa_2 = 4$, as a form of local bifurcation, after which $S3^+$ and $S3^-$ become the stable solution branches—see for example Vakakis et al. (1996). No proof of the stability of the solution curves is explicitly offered here, but later in this section, the simulations shown in Figs. 5.3 and 5.4, will indicate that for this example system the $S2$ and $S3^-$ branches do not attract any numerically computed solutions, when $\kappa/\kappa_2 > 4$.

As κ/κ_2 continues to increase, the $S3^+$ solution branch tends to zero, while the $S3^-$ solution becomes increasingly large (and negative). If $\sigma \to 0$ or $\sigma \to \infty$ then the relationship between x_1 and x_2 breaks down, which means essentially that the motion of the two masses becomes decoupled: for the upper branch $x_2 = 0$ and for the lower branch $x_1 = 0$. This type of behaviour is known as *localization*, because the vibration response of each mass becomes decoupled or localized from the global

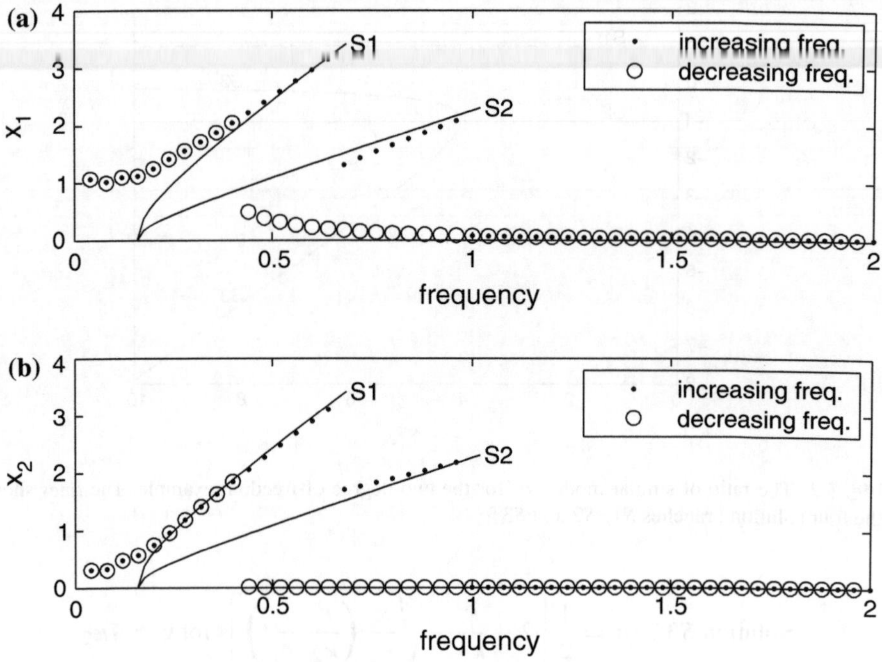

Fig. 5.3 Forced vibration with $\kappa < 4\kappa_2$. $F_1 = 3$ and $F_2 = 0$: **a** amplitude of response of mass 1 and **b** amplitude of response of mass 2

vibration response. For structures with spatially repetitive geometry (i.e. spatial periodicity), localization effects can occur, when small variations in structural period lead to a localized dynamic response. This is sometimes called *periodic structure theory*. Further details can be found, for example, in Pierre et al. (1987), Lust et al. (1995), Langley et al. (1997), Bendiksen (2000) and the references therein.

As already stated setting $k_2 = 0$ results in a special case. It is special because the resulting $S3^{\pm}$ solutions are nonlinear normal modes with a ratio between the x_1 and x_2 displacements, σ, that is a function of stiffness parameters but *not* amplitude of response. For this system if $k_2 \neq 0$, nonlinear normal modes can still exist, however they are now a function of both stiffness parameters and amplitude of response—this is discussed with the aid of the normal form technique is Sect. 5.4. However, first the special case is analysed further using the harmonic balance technique.

5.3.2 Internal Resonance

In Eq. (5.29), when $k_2 = 0$, the two linear natural frequencies in the system are the same $\omega_{n1} = \omega_{n2} = \sqrt{k/m}$. This is a form of *internal resonance*, where one part of

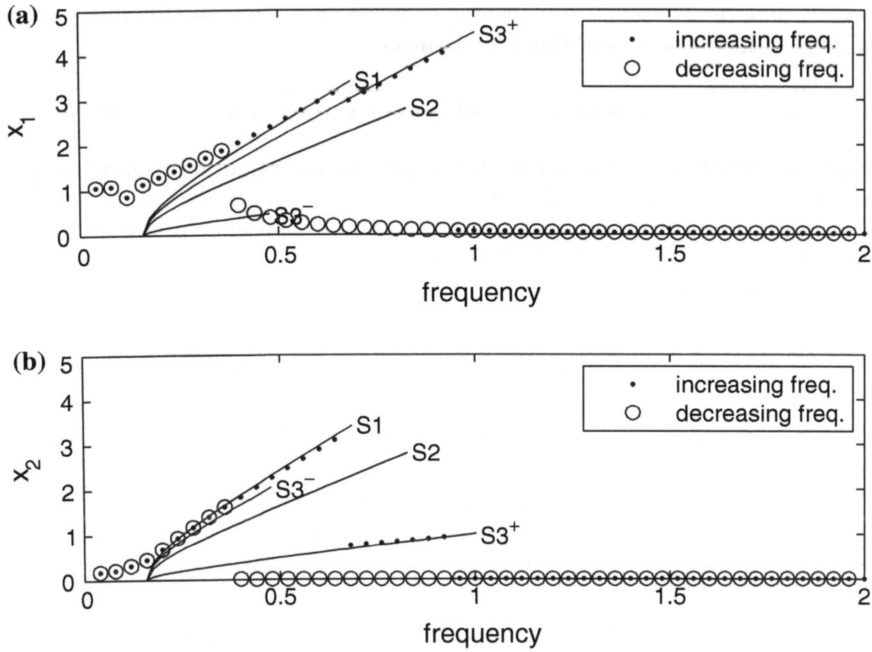

Fig. 5.4 Forced vibration with $\kappa \geq 4\kappa_2$. $F_1 = 3$ and $F_2 = 0$: **a** amplitude of response of mass 1 and **b** amplitude of response of mass 2. *Solid lines* are *backbone curves*, *dots* and *circles* are numerical simulation

the system can resonate with another.[14] For linear systems, resonance is associated with an external forcing exciting a maximum response at a certain parameter value. Internal resonance, in a nonlinear system, can be thought of as one part of the system feeding energy into another. For practical vibration problems, the system will be damped, and so external forcing will also be required to put energy into the system.

As the system is nonlinear, the actual response frequencies ω_{r1} and ω_{r2} will be amplitude dependent and equal to the underlying linear natural frequencies only when the response amplitudes are very small. So, internal resonance should more properly be defined by considering the case when $\omega_{r1} = \omega_{r2}$. In the following example, the harmonic balance method is used to find the backbone curves when Eq. (5.29) (with $k_2 = 0$) has one-to-one (1:1) internal resonance $\omega_{r1} = \omega_{r2}$, since the linear modes have close (actually identical in this case) linear natural frequencies.

Example 5.5 Backbone curves for two-degree-of-freedom nonlinear oscillator

Problem Use harmonic balance to find the backbone curves for the two-degree-of-freedom oscillator shown in Fig. 5.1, with equations of motion given by Eq. (5.29), in the special case when $k_2 = 0$.

[14] Internal resonance is a form of nonlinear resonance.

Solution For the two-degree-of-freedom example projected onto the linear modes, Eq. (5.29), the following substitutions are made

$$q_1 = Q_1 \cos(\omega_{r1} t), \quad q_2 = Q_2 \cos(\omega_{r2} t)$$

where Q_1 and Q_2 are the modal response amplitudes at nonlinear modal frequencies ω_{r1} and ω_{r2}. Due to the fact that the linear natural frequencies are identical the response frequencies are set to $\omega_{r1} = \omega_{r2}$. Using these substitutions, and the shorthands $\omega_{n1}^2 = k/m$ (the linear natural frequency), $\bar{\omega}^2 = \kappa/m$ and $\bar{\omega}_2^2 = \kappa_2/m$, the modal equations of motion become

$$Q_1\omega_{r1}^2 c_1 = Q_1\omega_{n1}^2 c_1 + \frac{3\bar{\omega}^2}{4}(3c_1 + c_3)Q_1Q_2^2 + \frac{\bar{\omega}^2}{4}(3c_1 + c_3)Q_1^3,$$

$$Q_2\omega_{r1}^2 c_1 = Q_2\omega_{n1}^2 c_1 + 2\bar{\omega}_2^2 Q_2^3(3c_1 + c_3) + \frac{\bar{\omega}^2}{4}(3c_1 + c_3)Q_2^3 + \frac{3\bar{\omega}^2}{4}(3c_1 + c_2)Q_1^2 Q_2,$$

where $c_n = \cos(n\omega_{r1}t)$. Then, using harmonic balance to equate the coefficients of the $\cos(\omega_{r1}t)$ terms in the two equations, expressions which govern the backbone curves for each mode can be derived. The harmonic balance gives

$$\begin{aligned} Q_1\left[\omega_{n1}^2 - \omega_{r1}^2 + \frac{3}{4}\bar{\omega}^2(Q_1^2 + 3Q_2^2)\right] &= 0, \\ Q_2\left[\omega_{n1}^2 - \omega_{r1}^2 + \frac{3}{4}\bar{\omega}^2(Q_2^2 + 3Q_1^2) + 6\bar{\omega}_2^2 Q_2^2\right] &= 0, \end{aligned} \tag{5.32}$$

There are several solutions to the two expressions in Eq. (5.32). Firstly there are the cases where the response is restricted to just one mode (corresponding to the cases where $\sigma = \pm 1$ in the previous discussion). For these cases Eq. (5.32) simplifies to

$$\text{S1:} \quad Q_2 = 0, \quad \omega_{r1}^2 = \omega_{n1}^2 + \frac{3}{4}\bar{\omega}^2 Q_1^2, \tag{5.33}$$

$$\text{S2:} \quad Q_1 = 0, \quad \omega_{r2}^2 = \omega_{n1}^2 + \frac{3}{4}\bar{\omega}^2 Q_2^2 + 6\bar{\omega}_2^2 Q_2^2. \tag{5.34}$$

These equations represent the backbone curves for the case where only one linear mode is present in the response.

Now consider the case where neither Q_1 or Q_2 are zero. To find a condition for this case the two expressions in Eq. (5.32) are equated to find the corresponding response amplitudes, which gives

$$Q_1^2 = Q_2^2\left(1 - 4\frac{\bar{\omega}_2^2}{\bar{\omega}^2}\right) = Q_2^2\left(1 - 4\frac{\kappa_2}{\kappa}\right). \tag{5.35}$$

or

$$Q_1 = \pm Q_2\sqrt{1 - 4\frac{\kappa_2}{\kappa}}.$$

These solutions can be expressed in terms of backbone curves by substituting the amplitude relationship, Eq. (5.35) back into Eq. (5.32) giving

$$S3^+: \quad Q_1 = Q_2\sqrt{1 - 4\frac{\kappa_2}{\kappa}}, \qquad \omega_{r1}^2 = \omega_{n1}^2 + 3\bar{\omega}^2\left(1 - \frac{\kappa_2}{\kappa}\right)Q_2^2 \tag{5.36}$$

$$S3^-: \quad Q_1 = -Q_2\sqrt{1 - 4\frac{\kappa_2}{\kappa}}, \qquad \omega_{r1}^2 = \omega_{n1}^2 + 3\bar{\omega}^2\left(1 - \frac{\kappa_2}{\kappa}\right)Q_2^2 \tag{5.37}$$

Finally using Eq. (5.27) with the assumptions that $x_2 = \sigma x_1$ and $x_1 = B\cos(\omega_{r1}t)$ it can be shown that $Q_1 = B(1+\sigma)/2$ and $Q_2 = B(1-\sigma)/2$. Making this substitution into Eq. (5.35) gives

$$\sigma^2 + (\frac{\kappa}{\kappa_2} - 2)\sigma + 1 = 0,$$

which is the same as the equation found using the similar normal mode analysis i.e. the last bracket in Eq. (5.31). This shows that the two additional solutions, $S3^+$ and $S3^-$, occur where $\omega_{r1} = \omega_{r2}$ which is when the internal resonance occurs. Note that from Eq. (5.35) it can be seen that these solutions are only valid (the amplitudes are real) if $\kappa \geq 4\kappa_2$. ■

Now consider just the internal resonance case, but compare the two cases when $\kappa < 4\kappa_2$ and $\kappa > 4\kappa_2$. To do this, a numerical simulation of the system is used alongside the backbone curve analysis. To make the simulation more physically realistic, a forced and damped version of the system in Fig. 5.1 is considered. Two viscous dampers are added such that $F_1 = kx_1 + \kappa x_1^3 + c\dot{x}_1$ and $F_3 = kx_2 + \kappa x_2^3 + c\dot{x}_2$, where c is the viscous damping coefficient. The system is forced at frequency Ω which is close to the natural frequency, such that $f_1 = P_1\cos(\Omega t)$ and $f_2 = P_2\cos(\Omega t)$. Given that the forcing is near-resonant, the response frequencies, $\omega_{r1} = \omega_{r2}$, will be the same as the forcing frequency.

Using the results from the previous example the backbone curves (the unforced, undamped response curves) for the system are governed by Eqs. (5.33) and (5.34) and, if $\kappa \geq 4\kappa_2$, Eqs. (5.36) and (5.37).

With the inclusion of damping, the extent to which the response follows the backbone curves are limited in length. The limit of the curves can be estimated by assuming that, at its peak, the modal response is 90° out-of-phase with the forcing. This assumption comes from the observation that a response that is 90° out-of-phase with the forcing is at the change-over point between in-and out-of-phase motion. For linear resonance, this occurs exactly at the resonance peak, for nonlinear systems it can usually be assumed to be at least approximately true. To locate the limit points, the forces are written as $f_1 = P_1\cos(\Omega t)$ and $f_2 = P_2\cos(\Omega t)$ while $x_1 = (Q_1 + Q_2)\sin(\Omega t)$ and $x_2 = (Q_1 - Q_2)\sin(\Omega t)$ represent the response of the masses. During steady-state vibration the energy dissipation over a cycle due to damping must equal the work done by the forcing input (as discussed in Sect. 1.2), which for this system may be written as

$$\int_{t-\pi/\Omega}^{t+\pi/\Omega} f_1\dot{x}_1 + f_2\dot{x}_2 \mathrm{d}t = \int_{t-\pi/\Omega}^{t+\pi/\Omega} (c\dot{x}_1)\dot{x}_1 + (c\dot{x}_2)\dot{x}_2 \mathrm{d}t.$$

Making the substitutions for the force and velocity terms gives the limit to the backbone curves

$$(Q_1 + Q_2)F_1 + (Q_1 - Q_2)F_2 + 2\Omega c(Q_1^2 + Q_2^2) = 0,$$

which can be solved numerically to find the limiting Ω values for each of the backbone curves.

The resulting backbone curves are shown as solid lines in Fig. 5.3 for the case when $\kappa < 4\kappa_2$ and in Fig. 5.4 for the case when $\kappa \geq 4\kappa_2$. In both figures, (a) shows the amplitude of response for mass 1 and (b) shows the amplitude of response for mass 2. These figures show the backbone curves compared with numerically computed points from the coupled equations of motion, Eq. (5.31), that are shown for both increasing and decreasing frequency to capture the hysteretic nature of the resonance peaks. The approximated backbone curves help to reveal the underlying structure of this two-degree-of-freedom vibration problem. Note that as the natural frequencies of the linear equivalent system are identical, all the backbone curves start from the same point. As amplitude increases, the backbone curves diverge from each other as the amplitude-dependent frequency relationship for each curve comes into play.

Each backbone curve can be thought of as the centre line of a nonlinear resonance peak relating to a pure modal motion of one of the four possible modes in the system. Note that solutions $S2$ and $S3^-$ in Fig. 5.4 do not attract the numerically computed solutions. A more detailed stability analysis, such as that shown in Sect. 4.5.4, would be required to find whether this indicates that these solution branches are unstable.

In both figures, as frequency is increased, the numerical points follow the first backbone curve ($S1$) and then drop onto a second backbone curve (either $S2$ or $S3^+$) at the limit of $S1$. For decreasing frequency, the numerical solution stays at a stable low amplitude response before jumping up onto the $S1$ curve at a fold bifurcation. The basins of attraction for each stable solution will determine which branch the system jumps to, when a hysteretic jump of this type occurs. Further details of basins of attraction and the fold bifurcation are given in Chap. 2.

Decomposing nonlinear systems into a modal form, and dealing with internal resonance can be done in a more systematic way by using normal form transformations. This is discussed in Sect. 5.4.

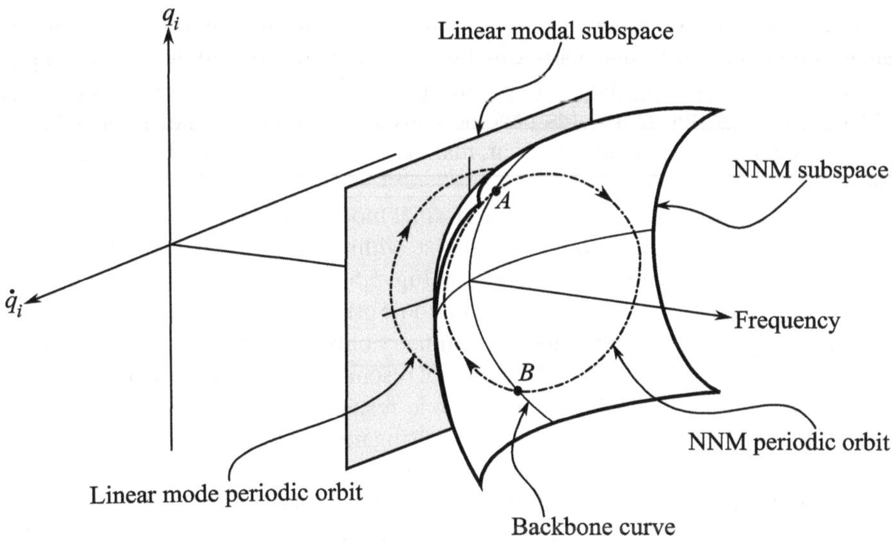

Fig. 5.5 A schematic representation of the geometry of the nonlinear response for a single mode without internal resonance. Here *NNM* denotes 'nonlinear normal mode'. Note that the phase plane is rotated 90° from its normal orientation

5.3.3 The Geometry of Nonlinear Modal Response

In order to better visualise a nonlinear modal response consider a single mode without an internal resonance that has linear modal coordinates $q_i, \dot{q}_i$. The geometry of the nonlinear response is shown schematically in Fig. 5.5. The axes in Fig. 5.5 represent the modal phase plane, $q_i, \dot{q}_i$, augmented with a third axis giving the frequency of response. There are two subspaces (or manifolds) shown in this three dimensional space relating to the undamped, unforced modal response. The first is a planar subspace denoted as the 'linear modal subspace'. This is a linear phase plane containing a family of periodic orbits corresponding to the case when the nonlinearity *is zero*. A single orbit is shown schematically to denote this in Fig. 5.5.

The second subspace in Fig. 5.5 is denoted 'NNM subspace' where NNM means 'nonlinear normal mode'. This is a non-planar manifold containing an (infinite) family of periodic orbits representing the case when nonlinearity *is not zero*. A single NNM orbit is shown schematically. The NNM orbit crosses the vertical axis at points A and B. These points coincide with a smooth backbone curve for the family of periodic NNM orbits. Note that the NNM manifold is tangent to the linear modal subspace at $q_i = \dot{q}_i = 0$. This corresponds to the observed behaviour that for very small amplitudes the nonlinear response is indistinguishable from the linear response for this type of system. Furthermore, the distortion of the NNM manifold away from the linear subspace reflects the amplitude dependence of the response, which in this case corresponds to a hardening stiffness nonlinearity, where response frequency is increasing with modal amplitude.

The concept of a nonlinear normal mode as a manifold was first developed by Shaw and Pierre (1993), and approximations to the manifold[15] allow one to compute expressions representing the nonlinear normal modes, see for example Pierre et al. (2006). However, the manifolds become extremely difficult, if not impossible, to define when internal resonances occur, making this method difficult to use for those cases.

Another way of defining a nonlinear normal mode is as an infinite set of periodic orbits. These are the same orbits that exist within the NNM manifold shown in Fig. 5.5. The important thing to notice from Fig. 5.5 is that the key information about the manifold (and therefore the family of periodic orbits as well) can be represented by the backbone curve. In fact, because of symmetry only the upper part of the backbone curve is required. Furthermore, when internal resonances occur, backbone curves can be used to represent the subsequent dynamic response in a systematic way. In the next section, backbone curves will be found using normal form transformations. This gives a more general approach for characterising multi-degree-of -freedom nonlinear systems in the displacement amplitude—frequency domain.

In Chap. 2, Sect. 2.10.1 the Fermi-Pasta-Ulam paradox was discussed and the concept of quasi-periodic motion was introduced. Quasi-periodic motion plays an important part in multi-degree-of-freedom Hamiltonian systems, and the interested reader can find further details in Arnold (1988). As the main focus of this section is using backbone curves as a model for lightly damped and forced responses, the issue of quasi-periodic response for Hamiltonian systems is not pursued further. However, it should be noted that the backbone curve analysis developed in the next section is restricted to periodic solutions. To identify any quasi-periodic solutions would require additional analysis.

5.4 Backbone Curves from Normal Form Transformations

The method of normal forms was introduced in Chap. 4, and involves finding nonlinear coordinate transforms with the objective of transforming the original governing equations into an, ideally, linear form. If a linear form cannot be achieved, the method of normal forms reduces the system to a "simpler" form, which is called a normal or standard form.[16] The resulting dynamic equations can be solved using a trial solution for the kth mode that consists just of a component at its primary frequency, ω_{rk} since the sub- and superharmonics of the response will have been removed during the transformation. Information regarding these harmonics can be found by considering the transform itself.

---

[15] In fact the manifold in Fig. 5.5 is not exactly the same as that considered by Shaw and Pierre, but the concept is similar.

[16] Note that depending on the method adopted and assumptions that are made a range of such forms can be derived, Murdock (2002).

In the analysis of linear multi-degree-of-freedom vibration problems, a modal transformation is used which achieves two things simultaneously: (i) the coordinates are transformed from physical degrees of freedom to modal ones, and (ii) the (unforced) transformed equations of motion are decoupled. Even for linear systems, certain conditions are required on the mass, stiffness and damping matrices for simultaneous diagonalisation of the system matrices to occur. In many linear examples, the damping matrix is chosen as linearly proportional to the stiffness (and/or mass) matrix specifically to allow diagonalisation and subsequent modal analysis to be carried out.[17]

The normal form method can be thought of as acting in a similar way for a nonlinear multi-degree-of-freedom vibration problem. There are two further advantages of using the normal form method for the analysis of weakly nonlinear systems. The first is that internal resonance can easily be dealt with, and this will be shown in the example below. The second is that the problem of nonlinear modal superposition can also be tackled using the normal form approach. The main drawback compared to techniques using nonlinear normal modes, is that normal forms are limited to weakly nonlinear systems.

5.4.1 Single Mode Backbone Curves

Here the discussion will concentrate on deriving the backbone curves for a system as these curves provide information regarding the structure of the response as was seen in Sect. 5.3.2. Initially the backbone curves in which only one of the linear modes is present are found. Then in Sect. 5.4.2, via a small extension to the analysis, backbone curves that contain responses from both the linear modes are identified.

Recall from Chap. 4, Sect. 4.5 that the normal form approach for an unforced system, to order ε^1 accuracy, consists of four steps. Taking the nonlinear equation of motion $M\ddot{\mathbf{x}} + K\mathbf{x} + \mathcal{N}(\mathbf{x}) = 0$, Eq. (4.42), **Step 1** is to apply the linear modal transform $\mathbf{x} = \Phi\mathbf{q}$, where Φ is a matrix of the eigenvectors of $M^{-1}K$. This results in an equation of motion for $\mathbf{q}$ given by

$$\ddot{\mathbf{q}} + \Lambda\mathbf{q} + N_q(\mathbf{q}) = 0, \quad N_q(\mathbf{q}) = (\Phi^T M \Phi)^{-1}\Phi^T \mathcal{N}(\Phi\mathbf{q}), \tag{5.38}$$

where Λ are the eigenvalues of $M^{-1}K$.

Step 2 involves finding a near-identity transform $\mathbf{q} \to \mathbf{u}$. Firstly the nonlinear term $N_q(\mathbf{q})$ is simplified to $\mathbf{n}_1(\mathbf{q})$ which consists of just the order ε^1 terms in N_q. Then $\mathbf{n}_1(\mathbf{q})$ is written in terms of $\mathbf{u}$ as $\mathbf{n}_1(\mathbf{u})$. Now $\mathbf{u}$ is split into two, $\mathbf{u} = \mathbf{u}_p + \mathbf{u}_m$ representing the positive and the negative complex exponential terms in the response respectively. This allows the nonlinear term to be written in the matrix form $\mathbf{n}_1(\mathbf{u}) = \mathbf{n}^*\mathbf{u}^*(\mathbf{u}_p, \mathbf{u}_m)$ where $\mathbf{n}^*$ is a matrix of coefficients and $\mathbf{u}^*$ is a vector containing all

[17] In the case of non-proportional damping, complex linear modes arise. See, for example, Ewins (2000) or Adhikari (2004).

the relevant nonlinear combinations of u_{pl} and u_{ml}. Finally using the general form for the lth term in $\mathbf{u}^*$, given in Eq. (4.54), the matrix $\boldsymbol{\beta}^*$, which governs the selection of the near-identity transform, can be found using Eq. (4.56).

In **Step 3** the dynamic equation for $\mathbf{q}$, $\ddot{\mathbf{q}} + \Lambda\mathbf{q} + \mathbf{n}_1(\mathbf{q}) = 0$, is transformed to

$$\ddot{\mathbf{u}} + \Lambda\mathbf{u} + \mathbf{n}_u^*\mathbf{u}^* = 0$$

using the transform

$$\mathbf{q} = \mathbf{u} + \mathbf{h}^*\mathbf{u}^*.$$

Note that here all three equations have been expressed to order ε^1 accuracy. The matrices $\mathbf{n}_u^*$ and $\mathbf{h}^*$ are found using $\boldsymbol{\beta}^*$ and $\mathbf{n}^*$ via Eqs. (4.58) and (4.59).

Finally in **Step 4** the transformed equation of motion (in $\mathbf{u}$) can be solved using the trial solution given in Eq. (4.51). This solution can then be converted back into coordinates $\mathbf{x}$ using $\mathbf{q} = \mathbf{u} + \mathbf{h}^*\mathbf{u}^*$ and $\mathbf{x} = \Phi\mathbf{q}$ if necessary. Often when backbone curves are being considered, since the system is weakly nonlinear, the sub- and superharmonic components of the response are assumed to be negligible such that the approximation $\mathbf{x} = \Phi\mathbf{q} \approx \Phi\mathbf{u}$ is made.

This technique will now be applied to an example system to demonstrate the process. Further examples of its application are given throughout the remainder of the book.

Example 5.6 Single-mode backbone curves of two-degree-of-freedom system with cubic nonlinearities

Problem Find the backbone curves for the two-degree-of-freedom system shown in Fig. 5.1, using a normal form transformation. The spring forces are $F_1 = kx_1 + \kappa x_1^3$, $F_2 = k_2(x_1 - x_2) + \kappa_2(x_1 - x_2)^3$ and $F_3 = kx_2 + \kappa x_2^3$ and $m_1 = m_2 = m$ such that the equations of motion are given by

$$\begin{aligned} m\ddot{x}_1 + (k + k_2)x_1 - k_2x_2 + \kappa x_1^3 + \kappa_2(x_1 - x_2)^3 &= 0, \\ m\ddot{x}_2 + (k + k_2)x_2 - k_2x_1 + \kappa x_2^3 - \kappa_2(x_1 - x_2)^3 &= 0. \end{aligned} \tag{5.39}$$

Assume that the mass and stiffness parameters are positive definite.

Note that if k_2 were set to zero, the natural frequencies would become identical and this system would revert to the special case considered in Sect. 5.3.1 and Example 5.5.

Solution

Step 1 Firstly the system is written in the standard form, Eq. (4.42), giving

$$M\ddot{\mathbf{x}} + K\mathbf{x} + \mathcal{N}(\mathbf{x}) = 0$$

$$\text{where: } M = \begin{bmatrix} m & 0 \\ 0 & m \end{bmatrix}, \quad K = \begin{bmatrix} k + k_2 & -k_2 \\ -k_2 & k + k_2 \end{bmatrix}, \quad \mathcal{N} = \begin{Bmatrix} \kappa x_1^3 + \kappa_2(x_1 - x_2)^3 \\ \kappa_2(x_2 - x_1)^3 + \kappa x_2^3 \end{Bmatrix}.$$

To apply the linear modal transform the eigenvalues and eigenvectors of $M^{-1}K$ are needed. These are $\lambda_1 = k/m$ with $\phi_1 = [1 \quad 1]^T$ and $\lambda_2 = (k + k_2)/m$ with $\phi_2 = [1 \quad -1]^T$. Writing these in matrix form and recognising that the eigenvalues equate to the square of the linear natural frequencies gives

$$\Phi = \begin{bmatrix} 1 & 1 \\ 1 & -1 \end{bmatrix}, \quad \Lambda = \begin{bmatrix} \omega_{n1}^2 & 0 \\ 0 & \omega_{n2}^2 \end{bmatrix}$$

where Φ is a matrix of eigenvectors in columns, Λ is a diagonal matrix of the corresponding eigenvalues, $\omega_{n1} = \sqrt{k/m}$ and $\omega_{n2} = \sqrt{(k + k_2)/m}$.

Now applying the linear modal transform $\mathbf{x} = \Phi\mathbf{q}$, see the discussion around Eq. (4.43), gives the modal equation of motion

$$\begin{aligned} &\ddot{\mathbf{q}} + \Lambda\mathbf{q} + N_q(\mathbf{q}) = 0, \\ \text{where:} \quad &N_q(\mathbf{q}) = \tfrac{\kappa}{m}\begin{Bmatrix} q_1^3 + 3q_1q_2^2 \\ 3q_1^2q_2 + \gamma q_2^3 \end{Bmatrix}, \end{aligned} \tag{5.40}$$

where $\gamma = 1 + (8\kappa_2/\kappa)$.

Step 2 Now the near-identity transform from $\mathbf{q}$ to $\mathbf{u}$ must be found through calculating matrix $\boldsymbol{\beta}$. To do this the nonlinear term N_q is firstly written as a power series of ε, $N_q(\mathbf{u}) = \varepsilon\mathbf{n}_1(\mathbf{u}) + \varepsilon^2\mathbf{n}_2(\mathbf{u}) + \cdots$. Here all the nonlinear terms will be taken to be order ε^1 so this series can be truncated to $N_q = \varepsilon\mathbf{n}_1$. Then the nonlinear terms are expressed in terms of $\mathbf{u}$ as is required in Eq. (4.47) and then the substitution $\mathbf{u} = \mathbf{u}_p + \mathbf{u}_m$ is made so that $n_1(\mathbf{u})$ can be expressed as the matrix multiplication $\mathbf{n}_1(\mathbf{u}) = \mathbf{n}^*\mathbf{u}^*(\mathbf{u}_p, \mathbf{u}_m)$ as in Eq. (4.52). The resulting matrices are

$$
\mathbf{u}^* = \begin{bmatrix} u_{p1}^3 \\ u_{p1}^2 u_{m1} \\ u_{p1} u_{m1}^2 \\ u_{m1}^3 \\ u_{p1} u_{p2}^2 \\ u_{p1} u_{p2} u_{m2} \\ u_{p1} u_{m2}^2 \\ u_{m1} u_{p2}^2 \\ u_{m1} u_{p2} u_{m2} \\ u_{m1} u_{m2}^2 \\ u_{p1}^2 u_{p2} \\ u_{p1}^2 u_{m2} \\ u_{p1} u_{m1} u_{p2} \\ u_{p1} u_{m1} u_{m2} \\ u_{m1}^2 u_{p2} \\ u_{m1}^2 u_{m2} \\ u_{p2}^3 \\ u_{p2}^2 u_{m2} \\ u_{p2} u_{m2}^2 \\ u_{m2}^3 \end{bmatrix}, \quad
\mathbf{n}^* = \frac{\kappa}{m} \begin{bmatrix} 1 & 0 \\ 3 & 0 \\ 3 & 0 \\ 1 & 0 \\ 3 & 0 \\ 6 & 0 \\ 3 & 0 \\ 3 & 0 \\ 6 & 0 \\ 3 & 0 \\ 0 & 3 \\ 0 & 3 \\ 0 & 6 \\ 0 & 6 \\ 0 & 3 \\ 0 & 3 \\ 0 & \gamma \\ 0 & 3\gamma \\ 0 & 3\gamma \\ 0 & \gamma \end{bmatrix}^T.
$$

Now, using the general form, given by Eq. (4.84), for each term in $\mathbf{u}^*$ in turn the matrix $\boldsymbol{\beta}^*$ can be calculated, using Eq. (4.86). This matrix is used to identify the resonant terms in the next step. Here it is convenient to relate the two response frequencies by writing $\omega_{r2} = r\omega_{r1}$. The resulting $\boldsymbol{\beta}$ matrix is given by

$$
\boldsymbol{\beta}^* = \Omega^2 \left[\begin{matrix} 8 & 0 & 0 & 8 & 4(r^2+r) & 0 & 4(r^2-r) & 4(r^2-r) & 0 & 4(r^2+r) & - & - & - & - & - & - & - & - & - & - \\ - & - & - & - & - & - & - & - & - & - & 4(1+r) & 4(1-r) & 0 & 0 & 4(1-r) & 4(1+r) & 8r^2 & 0 & 0 & 8r^2 \end{matrix} \right].
$$

Here a dash has been used in locations where the value of $\boldsymbol{\beta}^*$ is unimportant, i.e. where the corresponding value in $\mathbf{n}^*$ is zero.

Step 3 The matrix $\boldsymbol{\beta}^*$ is now used to identify the nonlinear transform and the post-transformed nonlinear terms in the equation of motion. Considering each term in $\boldsymbol{\beta}^*$ in turn, if the term is zero then it is resonant and Eq. (4.59), $n^*_{u,k,\ell} = n^*_{k,\ell}$ and $h^*_{k,\ell} = 0$, is used to assign values to the corresponding terms in $\mathbf{n}^*_u$ and $\mathbf{h}^*$ otherwise it is non-resonant and Eq. (4.58), $n^*_{u,k,\ell} = 0$ and $h^*_{k,\ell} = n^*_{k,\ell}/\beta^*_{k,\ell}$, is used.

Inspecting $\boldsymbol{\beta}^*$, it can be seen that some terms are always zero regardless of r and so are termed *unconditionally-resonant*. Other *conditionally-resonant* terms can be zero for specific values of r, here these exist for $r = 1$ (the case where $r = 0$, which also results in zero terms in $\boldsymbol{\beta}^*$, is not considered as this results in a response frequency of zero). These conditionally-resonant terms are reflected in the resulting matrices by using the Dirac-delta function, $\delta(r-1)$, which is unity only at $r=1$. Since here the transformed equation of motion is of most interest, the transform terms, which give information regarding sub- and super-harmonic response behaviour, won't be presented. Using $\boldsymbol{\beta}^*$ and Eqs. (4.58) and (4.59), the nonlinear terms in the equation for motion in $\mathbf{u}$ are given by

$$\boldsymbol{\beta}^* = \omega_{r1}^2 \begin{bmatrix} 8 & - \\ 0 & - \\ 0 & - \\ 8 & - \\ 4(r^2+r) & - \\ 0 & - \\ 4(r^2-r) & - \\ 4(r^2-r) & - \\ 0 & - \\ 4(r^2+r) & - \\ - & 4(1+r) \\ - & 4(1-r) \\ - & 0 \\ - & 0 \\ - & 4(1-r) \\ - & 4(1+r) \\ - & 8r^2 \\ - & 0 \\ - & 0 \\ - & 8r^2 \end{bmatrix}^T \rightsquigarrow \mathbf{n}_u^* = \frac{\kappa}{m} \begin{bmatrix} 0 & 0 \\ 3 & 0 \\ 3 & 0 \\ 0 & 0 \\ 0 & 0 \\ 6 & 0 \\ 3\delta(r-1) & 0 \\ 3\delta(r-1) & 0 \\ 6 & 0 \\ 0 & 0 \\ 0 & 0 \\ 0 & 3\delta(r-1) \\ 0 & 6 \\ 0 & 6 \\ 0 & 3\delta(r-1) \\ 0 & 0 \\ 0 & 0 \\ 0 & 3\gamma \\ 0 & 3\gamma \\ 0 & 0 \end{bmatrix}^T .$$

Step 4 The transformed equation of motion may now be written, to order ε^1 accuracy, as

$$\ddot{\mathbf{u}} + \Lambda \mathbf{u} + \mathbf{n}^* \mathbf{u}^* = 0$$

$$\rightsquigarrow \ddot{u}_1 + \omega_{n1}^2 u_1 + \frac{3\kappa}{m}\left[u_{p1}u_{m1}u_1 + 2u_{p2}u_{m2}u_1 + \delta(r-1)(u_{p1}u_{m2}^2 + u_{m1}u_{p2}^2)\right] = 0,$$
$$\ddot{u}_2 + \omega_{n2}^2 u_2 + \frac{3\kappa}{m}\left[\gamma u_{p2}u_{m2}u_2 + 2u_{p1}u_{m1}u_2 + \delta(r-1)(u_{p1}^2 u_{m2} + u_{m1}^2 u_{p2})\right] = 0, \tag{5.41}$$

where $\mathbf{u} = \mathbf{u}_p + \mathbf{u}_m$ has been used.

Backbone Curves: Only the single mode backbone curves are required in this example. From inspection of these equations it can be seen that such curves exist since all the terms in the first and second equations contain u_1 and u_2 terms respectively. Taking the first single mode backbone curve to be that when there is no second mode, i.e setting $u_2 = 0$, gives

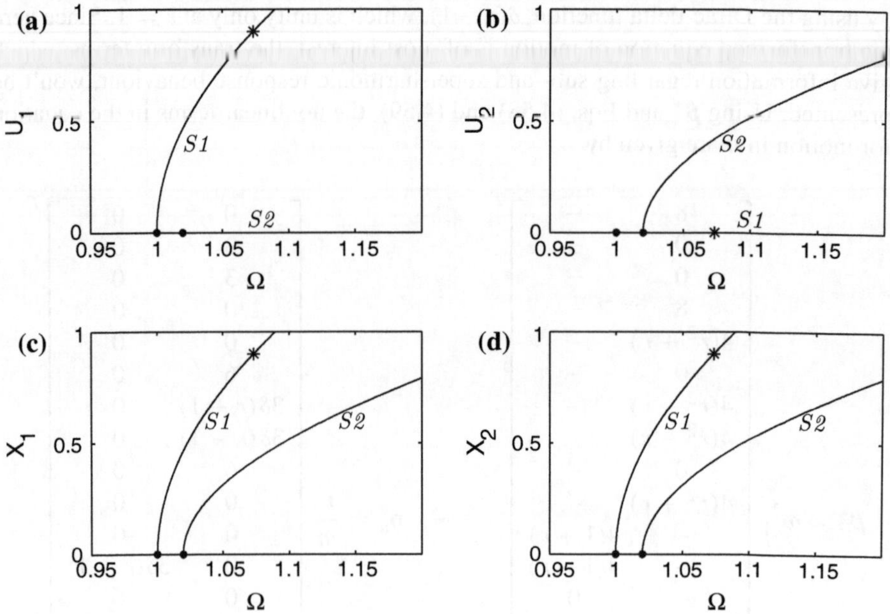

Fig. 5.6 Backbone curves of the two-degree-of-freedom considered in Example 5.6 with $\omega_{n1} = 1$, $\omega_{n2} = 1.02$, $m = 1$, $\kappa = 0.25$ and $\kappa_2 = 0.075$ such that $\kappa < 4\kappa_2$ in terms of amplitudes **a** U_1, **b** U_2, **c** X_1 and **d** X_2 (The asterisk relates to the solution shown in Fig. 5.9 and the black dots represent bifurcations of the zero solution)

$$S1: \quad \omega_{r1}^2 = \omega_{n1}^2 + \frac{3\kappa}{4m}U_1^2 = 0, \quad U_2 = 0,$$

noting that from the trial solution, Eq. (4.51), $u_{p1}u_{m1} = U_1/4$. Likewise there is a $U_1 = 0$ solution that is given by

$$S2: U_1 = 0, \quad \omega_{r2}^2 = \omega_{n2}^2 + \frac{3\kappa}{4m}\gamma U_2^2 = 0$$

These solutions are independent of r, the ratio between the response frequencies. ■

Figure 5.6 shows the backbone curves for the system considered in Example 5.6 for the case where $\kappa < 4\kappa_2$ (it will be seen in the next example that if $\kappa \geq 4\kappa_2$ additional mixed-mode backbone curve solutions may exist). Panels (a) and (b) show the response in terms of the amplitude of the linear modal coordinates, U_1 and U_2 respectively and panels (c) and (d) show the response in terms of the amplitudes of the physical coordinates, X_1 and X_2. For the physical coordinates it has been assumed that the harmonic components are small and so can be ignored (note that, if of interest, they can be calculated from the nonlinear near-identity transform).

5.4.2 Multi-mode Backbone Curves and Bifurcations

Having considered the single mode backbone curves, in this section the case where backbone curves that consist of a response containing multiple linear modes (i.e. for the example system when $u_1 \neq 0$ and $u_2 \neq 0$) is considered.[18] This leads to addition backbone curve solution branches that bifurcate from the $S1$ and $S2$ branches. Continuing Example 5.6 it will be seen that such curves exist if $\kappa \geq 4\kappa_2$ and $r = 1$, indicating that the response frequencies are exact which requires that the linear natural frequencies are close but not necessarily equal.

Example 5.7 Multi-mode backbone curves of two-degree-of-freedom system with cubic nonlinearities

Problem Find the multi-mode backbone curves for the two-degree-of-freedom system in Example 5.6, using a normal form transformation.

Solution The solution follows that for Example 5.6 until **Step 4**.

Step 5 In Example 5.6, the transformed equations of motion

$$\ddot{u}_1 + \omega_{n1}^2 u_1 + \frac{3\kappa}{m}\left[u_{p1}u_{m1}u_1 + 2u_{p2}u_{m2}u_1 + \delta(r-1)(u_{p1}u_{m2}^2 + u_{m1}u_{p2}^2)\right] = 0, \tag{5.42}$$

$$\ddot{u}_2 + \omega_{n2}^2 u_2 + \frac{3\kappa}{m}\left[\gamma u_{p2}u_{m2}u_2 + 2u_{p1}u_{m1}u_2 + \delta(r-1)(u_{p1}^2 u_{m2} + u_{m1}^2 u_{p2})\right] = 0, \tag{5.43}$$

where derived.

Starting at this point a substitution for u_k may be made using the solution

$$u_k = u_{pk} + u_{mk} = \left(\frac{U_k}{2}\mathrm{e}^{-\mathrm{i}\phi_k}\right)\mathrm{e}^{\mathrm{i}\omega_{rk}t} + \left(\frac{U_k}{2}\mathrm{e}^{\mathrm{i}\phi_k}\right)\mathrm{e}^{-\mathrm{i}\omega_{rk}t},$$

from Eq. (4.51). Using this the first equation of motion, Eq. (5.42), may be written as

$$\begin{aligned}
&\left(\omega_{n1}^2 - \omega_{r1}^2\right)U_1\left(\mathrm{e}^{\mathrm{i}(\omega_{r1}t-\phi_1)} + \mathrm{e}^{-\mathrm{i}(\omega_{r1}t-\phi_1)}\right) \\
&+\frac{3\kappa}{4m}U_1\left[(U_1^2 + 2U_2^2)\left(\mathrm{e}^{\mathrm{i}(\omega_{r1}t-\phi_1)} + \mathrm{e}^{-\mathrm{i}(\omega_{r1}t-\phi_1)}\right)\right. \\
&\qquad\left. + \delta(r-1)U_2^2\left(\mathrm{e}^{-\mathrm{i}(\omega_{r1}t-\phi_1)}\mathrm{e}^{-\mathrm{i}2(\phi_1-\phi_2)} + \mathrm{e}^{\mathrm{i}(\omega_{r1}t-\phi_1)}\mathrm{e}^{\mathrm{i}2(\phi_1-\phi_2)}\right)\right] = 0,
\end{aligned}$$

[18] The term mode refers to a mode of the linearised system, whereas a nonlinear normal mode, or NNM, is for the full nonlinear system. A mixed-mode backbone curve defines a response made up of multiple linear modes.

Here the $\delta(r-1)$ term, which is only non-zero when $r = 1$, has been simplified using $\omega_{r2} = \omega_{r1}$ which is valid for the case where $r = 1$ (r was defined using $\omega_{r2} = r\omega_{r1}$).

Balancing the $e^{i(\omega_{r1}t-\phi_1)}$ term gives

$$\left\{\omega_{n1}^2 - \omega_{r1}^2 + \frac{3\kappa}{4m}\left[U_1^2 + U_2^2\left(2 + \delta(r-1)e^{i2(\phi_1-\phi_2)}\right)\right]\right\} U_1 = 0. \quad (5.44)$$

Note that balancing the $e^{-i(\omega_{r1}t-\phi_1)}$ term gives the complex conjugate of this expression and so yields no extra information. Applying the same process to the second equation of motion, Eq. (5.43), albeit now balancing the $e^{i(\omega_{r2}t-\phi_2)}$ term, gives

$$\left\{\omega_{n2}^2 - r^2\omega_{r1}^2 + \frac{3\kappa}{4m}\left[U_1^2\left(2 + \delta(r-1)e^{-i2(\phi_1-\phi_2)}\right) + \gamma U_2^2\right]\right\} U_2 = 0, \quad (5.45)$$

where $\omega_{r2} = r\omega_{r1}$ has been used.

Backbone Curves: These normal form solutions, Eqs. (5.44) and (5.45), can now be used to derive the backbone curves. The backbone curves relating to responses in purely the first or the second mode of the linearised system have already been calculated in Example 5.6 and are summarised as

$$S1:\ \omega_{r1}^2 = \omega_{n1}^2 + \frac{3\kappa}{4m}U_1^2 = 0, \quad U_2 = 0,$$

$$S2:\ U_1 = 0, \quad \omega_{r2}^2 = \omega_{n2}^2 + \frac{3\kappa}{4m}\gamma U_2^2 = 0.$$

Note that for both these solutions the backbone curves are unaffected by the value of r.

The mixed mode backbone curves relate to when both $U_1 \neq 0$ and $U_2 \neq 0$. These solutions require that terms in the braces are zero in Eqs. (5.44) and (5.45) such that

$$\omega_{r1}^2 = \omega_{n1}^2 + \frac{3\kappa}{4m}\left[U_1^2 + U_2^2\left(2 + \delta(r-1)e^{i2(\phi_1-\phi_2)}\right)\right],$$

$$r^2\omega_{r1}^2 = \omega_{n2}^2 + \frac{3\kappa}{4m}\left[U_1^2\left(2 + \delta(r-1)e^{-i2(\phi_1-\phi_2)}\right) + \gamma U_2^2\right].$$

Note that the terms with the exponentials in them impose a phase relationship between the the responses of the two linear modal components u_1 and u_2 (ignoring the higher harmonics). Both these terms are present only when $r = 1$. So, for the case where $r \neq 1$ the linear modal components that make up the response are not linked in phase—this indicates that their interaction is non-resonant.

Therefore, the $r = 1$ case will be considered for the remainder of the analysis. This corresponds to having $\omega_{r1} = \omega_{r2}$ and so requires $\omega_{n1} \approx \omega_{n2}$. Setting $r = 1$, to satisfy Eqs. (5.44) and (5.45) when $U_1 \neq 0$ and $U_2 \neq 0$ requires

$$\omega_{r1}^2 = \omega_{r2}^2 = \omega_{n1}^2 + \frac{3\kappa}{4m}\left[U_1^2 + U_2^2(2+p)\right] = \omega_{n2}^2 + \frac{3\kappa}{4m}\left[U_1^2(2+p_c) + \gamma U_2^2\right]. \tag{5.46}$$

noting that $\omega_{r2} = \omega_{r1}$ when $r = 1$. Here $p = \mathrm{e}^{\mathrm{i}2(\phi_1-\phi_2)}$ has been used and subscript c indicates the complex conjugate. Since the response frequency ω_{r1} is real for physically meaningful solutions p must be real, which requires that $\sin(2(\phi_1-\phi_2)) = 0$. This results in the phase requirement $\phi_1 - \phi_2 = n\pi/2$ where n is an integer and so p may be simplified to

$$p = \mathrm{e}^{\mathrm{i}2(\phi_1-\phi_2)} = \begin{cases} +1 & \text{if: } \phi_1 - \phi_2 = \cdots - \pi, 0, \pi, \ldots \\ -1 & \text{if: } \phi_1 - \phi_2 = \cdots - \pi/2, \pi/2, 3\pi/2, \ldots \end{cases}$$

For the case where $p = 1$ the two linear modal contributions are in-phase (or in anti-phase) whereas when $p = -1$ the contributions are 90° out-of-phase.

Firstly considering the case where $p = -1$, Eq. (5.46) may rewritten as

$$\omega_{r1}^2 = \omega_{r2}^2 = \omega_{n1}^2 + \frac{3\kappa}{4m}\left[U_1^2 + U_2^2\right] = \omega_{n2}^2 + \frac{3\kappa}{4m}\left[U_1^2 + \gamma U_2^2\right].$$

where $\gamma = 1 + 8\kappa_2/\kappa$ has been used. A valid solution to these expressions requires that

$$U_2^2 = -\frac{m}{6\kappa_2}(\omega_{n2}^2 - \omega_{n1}^2).$$

As κ_2 is positive in this example, it can be seen that there is no physically meaningful solution since the second linear natural frequency is, by definition, larger (or at least equal to) the first natural frequency.

Now considering the case where $p = 1$, Eq. (5.46) may be simplified to

$$\omega_{r1}^2 = \omega_{r2}^2 = \omega_{n1}^2 + \frac{3\kappa}{4m}\left[U_1^2 + 3U_2^2\right] = \omega_{n2}^2 + \frac{3\kappa}{4m}\left[3U_1^2 + \gamma U_2^2\right]. \tag{5.47}$$

which, using $\gamma = 1 + 8\kappa_2/\kappa$, can be rearranged to give

$$U_1^2 = \left(1 - 4\frac{\kappa_2}{\kappa}\right)U_2^2 - \frac{2m}{3\kappa}\left(\omega_{n2}^2 - \omega_{n1}^2\right). \tag{5.48}$$

As both κ and κ_2 are positive in this example the second term on the right-hand side of this equation is negative. Since U_1 must be real for a valid solution it follows that the first term on the right-hand side must both be positive and have a magnitude exceeding that of the second term. The first of these conditions may be expressed as

$$\kappa \geq 4\kappa_2.$$

This is consistent with the nonlinear normal modes solution in Example 5.5, which considers the same system but with the additional constraint that $\omega_{n2} = \omega_{n1}$ from the fact that $k_2 = 0$. The second condition requires that

$$U_2^2 \geq U_{2,\text{min}}^2, \quad \text{where} \quad U_{2,\text{min}}^2 = \frac{2m}{3(\kappa - 4\kappa_2)}(\omega_{n2}^2 - \omega_{n1}^2).$$

This condition shows that $U_2 = 0$ is only a valid solution if $\omega_{n2} = \omega_{n1}$, whereas $U_1 = 0$ is always a possible solution regardless of the linear natural frequencies ($U_1 = 0$ occurs when the two sides of the inequality are equal so $U_2 = U_{2,\text{min}}$). So for the $p = 1$ case, possible solutions only exist over the range $U_1 \geq 0$, which correspond to the range $U_2 \geq U_{2,\text{min}}$.

Taking the $p = 1$ equation, Eq. (5.47), and, this time, rearranging it to eliminate U_1 gives

$$\omega_{r1}^2 = \omega_{r2}^2 = \frac{3\omega_{n1}^2 - \omega_{n2}^2}{2} + \frac{3(\kappa - \kappa_2)}{m} U_2^2$$

Provided $\kappa \geq 4\kappa_2$, this, along with Eq. (5.48), results in two further backbone curves $S3^+$ and $S3^-$. These may be expressed as

$$S3^\pm : \ \omega_{r1}^2 = \frac{3\omega_{n1}^2 - \omega_{n2}^2}{2} + \frac{3(\kappa - \kappa_2)}{m} U_2^2$$
$$U_1 = \sqrt{\left(1 - 4\frac{\kappa_2}{\kappa}\right) U_2^2 - \frac{2m}{3\kappa}\left(\omega_{n2}^2 - \omega_{n1}^2\right)}. \qquad (5.49)$$

The difference between the two solutions is the phase between the linear modal components with

$$S3^+ : \quad \phi_1 - \phi_2 = \cdots - 2\pi, 0, 2\pi, \ldots,$$
$$S3^- : \quad \phi_1 - \phi_2 = \cdots - \pi, \pi, 3\pi, \ldots,$$

corresponding to the linear modal components being in-phase and in anti-phase respectively.

These solutions correspond to the nonlinear normal modes of the system, with the $S3^\pm$ solutions being nonlinear modes which are made up of both modes of the linear system. ■

Considering the starting point of the $S3^\pm$ backbone branches, setting $U_1 = 0$ (and so $U_2 = U_{2,\text{min}}$), the response frequency relationship for the $S3^\pm$ solutions may be rewritten. Using the $U_{2,\text{min}}$ expression to eliminate ω_{n1} in the response frequency equation, Eq. (5.49), results in $\omega_{r2} = \omega_{n2}^2 + 3\kappa\gamma U_2^2/(4m)$. Here it has been recalled that $\omega_{r1} = \omega_{r2}$ as $r = 1$ and that $\gamma = 1 + 8\kappa_2/\kappa$. It can be seen that this response frequency relationship is the same as that for $S2$ and hence the start of the $S3^\pm$ solution curves branch from (or bifurcate off) the $S2$ curve.

Figure 5.7 shows the backbone curves for the system where κ_2 has been chosen such that $\kappa \geq 4\kappa_2$. Here $S3^\pm$ solution curves exist and, as shown in the analysis in Example 5.7, they branch off the $S2$ solution curve. This bifurcation results in the $S2$ branch becoming unstable (indicated as a dashed line) for response amplitudes above the bifurcation point. Note that the $S3^+$ and $S3^-$ lines are overlaid in the U_1 and U_2 plots and, due to the symmetry of the system, are reversed when comparing

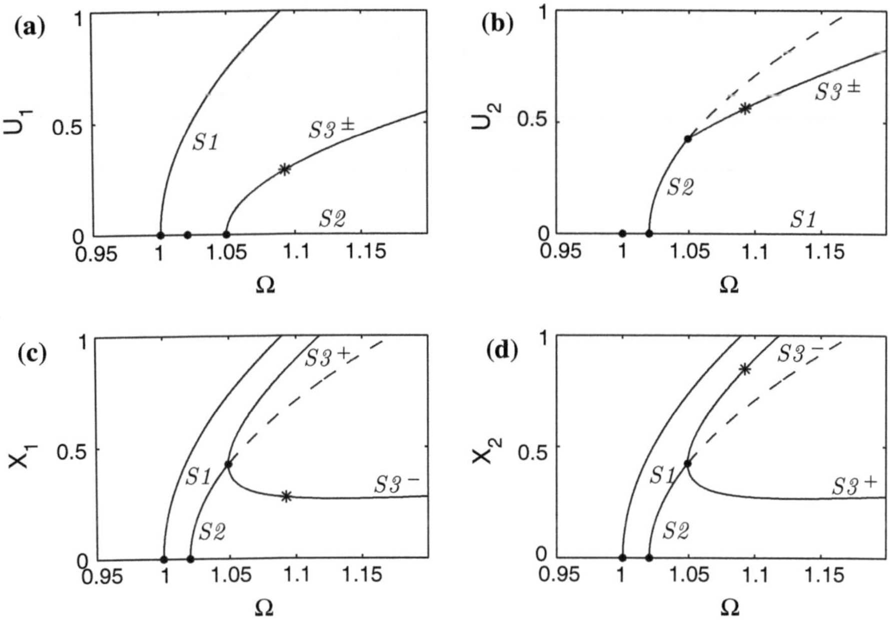

Fig. 5.7 Backbone curves of the two-degree-of-freedom considered in Example 5.7 with $\omega_{n1} = 1$, $\omega_{n2} = 1.02$, $m = 1$, $\kappa = 0.25$ and $\kappa_2 = 0.025$ such that $\kappa \geq 4\kappa_2$ in terms of amplitudes **a** U_1, **b** U_2, **c** X_1 and **d** X_2 (The asterisk relates to the solution shown in Fig. 5.10 and the black dots represent bifurcations)

the X_1 plot to the X_2 one. In the discussion above it was shown that for the special case where $\omega_{n2} = \omega_{n1}$ the $S3^{\pm}$ solutions start at $U_2 = 0$, this can be seen in Fig. 5.4.

When the system is forcing in the second mode shape, the $S3^{\pm}$ solutions can be thought of as an internal resonance. Initially the forced response follows the $S2$ branch, as would be expected from linear theory, and then follows the $S3^{\pm}$ branch indicating that components of both the first and the second linear modes are present in the response. The point at which this transition occurs is the point at which the zero response of the first mode becomes unstable—an example of this behaviour is discussed in Sect. 7.5.3, in which a cable is excited vertically and through nonlinear interactions between the modes of the linear system and a sway motion is triggered.

5.4.3 Nonlinear Mode Shape Analysis

Backbone curves indicate where resonant behaviour can occur in the frequency domain, but they don't give information about their exact spatial configurations. This can be found by analysing the mode shapes, or the *nonlinear normal modes*, of the nonlinear system that are associated with the various backbone curves. Before considering in detail the NNMs observed in Examples 5.6 and 5.7, NNMs are considered in general. This is done by considering a projection of the motion onto the u_1

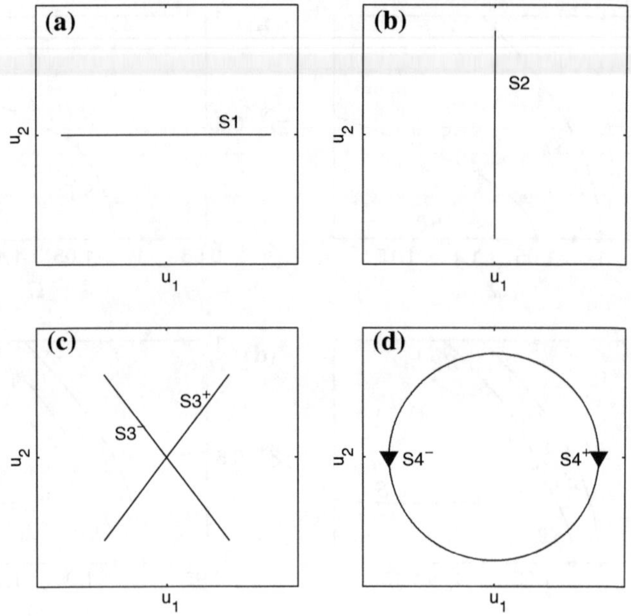

Fig. 5.8 Spatial configuration of possible modes for the two degree-of-freedom system considered in Examples 5.6 and 5.7. Each plot shows a schematic representation the relative modal motion for **a** $S1$ backbone curves, **b** $S2$ backbone curves **c** $S3^{\pm}$ backbone curves with the phase between u_1 and u_2 being 0 or π for $S3^+$ and $S3^-$ respectively and **d** $S4^{\pm}$ backbone curves in which the phase between q_1 and q_2 is $\pi/2$ or $-\pi/2$ for $S4^+$ and $S4^-$ respectively

vs u_2 plane. In this chapter backbone curves consisting of a single linear mode have been observed, these appear in the plane as vertical or horizontal lines in the plane, see Fig. 5.8a, b for the mode shapes of the $S1$ ($U_2 = 0$) and $S2$ ($U_1 = 0$) solutions respectively. In this projection the frequency of oscillation is not seen, but it should be noted that while the mode shape remains the same all along the backbone curve, the nonlinear natural frequency does not.

In Example 5.7 the p and p_c terms in Eq. 5.46 lead to the possibility of 4 further types of backbone curve which contain components from both linear modes. The two that occur for $p = 1$, relating to backbone curve branches $S3^{\pm}$, are shown in Fig. 5.8c. Note here that the angle of the line is dependent on the ratio between U_1 and U_2 which is typically a function of U_1, for the system considered in the examples—see Eq. (5.49). In addition, note that if the projection was shown in q_1 versus q_2 the mode shapes would not be exactly straight lines as harmonics are present in the full solution, however the effect of these harmonics are small for the examples considered in Figs. 5.6 and 5.7. Finally there are two solutions relating to $p = -1$, which for the examples were non-physical but do occur in other systems such as the cable dynamics considered in Sect. 7.5.3 and for the system considered here under certain parameter conditions if both κ_2 and κ are allowed to be negative. The response shape for this type of backbone curve is shown in Fig. 5.8d.

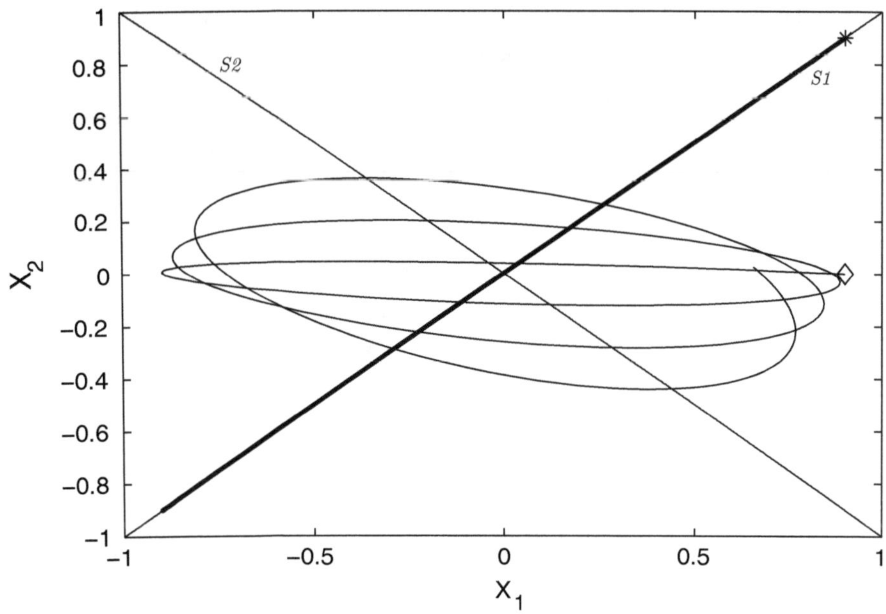

Fig. 5.9 Plot of initial displacement conditions (with the initial velocities set to zero) for the case where $\kappa < 4\kappa_2$ (the full list of parameters is given in Fig. 5.6). The two thick lines show responses of the system when released over about three cycles. When released from the asterisk on the S1 solution line (see also the asterisks in Fig. 5.6) the response frequency is ω_{r1} which is governed by the $S1$ equation. When released from the diamond, the response does not sit on a nonlinear normal mode solution

The specific mode shapes for Examples 5.6 and 5.7 may be examined in more detail, by considering a plane of initial conditions for x_1 and x_2 (with the initial velocities set to zero). Assuming that harmonics of the response are small, the $S1$ and $S2$ backbone solutions may be plotted in this projection and appear as straight lines through the origin at gradients of $+1$ and -1 respectively. These solution lines are shown in Fig. 5.9 for the case where $\kappa < 4\kappa_2$. If the system is set off from a point on one of these lines then, ignoring harmonics, the response will oscillate along the line the point lies on at the response frequency—an example of this is shown as a thick line in the plot with the starting point (or initial conditions) marked with an asterisk. This is the mode shape for this set of initial conditions (a 45° rotation from Fig. 5.8 due to the linear modal transform $\mathbf{x} = \boldsymbol{\Phi}\mathbf{q}$). Releasing the system from a point away from one of the solution lines results in a response that contain significant components at two frequencies—an example of this is also shown in the plot (a thin line with the starting point marked with a diamond). This response may be periodic over a long enough period depending of the ratio of the response frequencies ω_{r1} and ω_{r2} however this does not need to be the case.

The same plane is plotted for the case where $\kappa \geq 4\kappa_2$ in Fig. 5.10. In addition to $S1$ and $S2$ solutions (which appear the same in this plane to those for the $\kappa < 4\kappa_2$ case, although for $S2$ their response frequencies differ) are the $S3^{\pm}$ solution, which

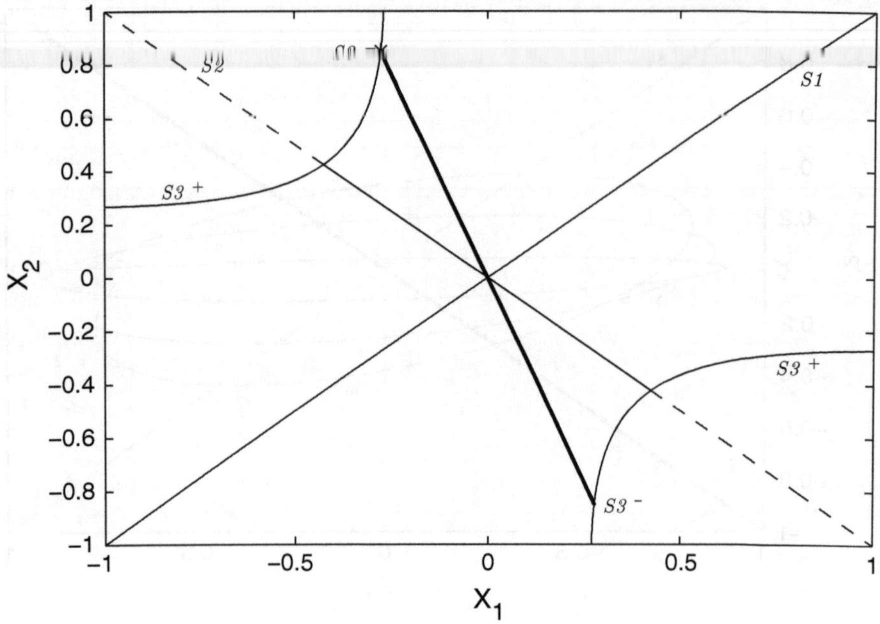

Fig. 5.10 Plot of initial displacement conditions (with the initial velocities set to zero) for the case where $\kappa < 4\kappa_2$ (the full list of parameters is given in Fig. 5.7). The thick line shows a response of the system when released from the asterisk on the $S3^-$ solution (see also the asterisks in Fig. 5.6) in which the response frequency is ω_{r1} and is governed by the $S3$ equation

branch off the $S2$ solution. For these $S3$ solutions the nonlinear normal mode contains contributions from both linear modes. As the ratio of the contributions from the two linear modes also differs the shape of the nonlinear normal mode differs with amplitude. An example response, starting on the $S3^-$ nonlinear normal mode at the asterisk, is shown as a thick line in the figure. It can be seen that the response is periodic, indicated by the fact that the repeated cycles lie on top of each other (in contrast to the response of the system in Fig. 5.9 when released from a point that does not sit on a nonlinear normal mode solution—the thick line starting at the diamond). In addition it can be seen that moving along the $S3^{\pm}$ solution lines will result in a change to the mode shape.

5.4.4 Backbone Curves in the Symmetry Breaking Case

An important case for engineering systems, is when the symmetry of the system is broken. In all the previous examples in this chapter the systems are symmetric in terms of both their linear and their nonlinear characteristics. In the next example the system considered in Example 5.6 is now revisited and the symmetry of the nonlinear terms is broken.

Example 5.8 Symmetry breaking two-degree-of-freedom system with cubic nonlinearities and one-to-one resonance

Problem Consider the system analysed in Example 5.6 but with an increased cubic stiffness term attached between ground and the first mass, so $F_1 = kx_1 + \kappa(1+\delta)x_1^3$ where δ is a small parameter, and a reduced cubic stiffness between attached between ground and the second mass, so $F_3 = kx_3 + \kappa(1-\delta)x_2^3$, such that the equations of motion become

$$m\ddot{x}_1 + (k + k_2)x_1 - k_2x_2 + \kappa(1 + \delta)x_1^3 + \kappa_2(x_1 - x_2)^3 = 0,$$
$$m\ddot{x}_2 + (k + k_2)x_2 - k_2x_1 + \kappa(1 - \delta)x_2^3 - \kappa_2(x_1 - x_2)^3 = 0.$$

Using normal form analysis, calculate the new terms in the equations of motion in the transformed coordinates and find the backbone curves when one-to-one resonance occurs.

Solution Firstly the equation of motion may be written in the standard form $M\ddot{\mathbf{x}} + K\mathbf{x} + \mathcal{N}(\mathbf{x}) = 0$. The linear terms are the same as in the previous example and due to the additional nonlinear stiffness term the nonlinear vector becomes

$$\mathcal{N} = \begin{Bmatrix} \kappa x_1^3 + \kappa_2(x_1 - x_2)^3 \\ \kappa_2(x_2 - x_1)^3 + \kappa x_2^3 \end{Bmatrix} + \begin{Bmatrix} \kappa\delta x_1^3 \\ -\kappa\delta x_2^3 \end{Bmatrix},$$

where the second term on the right-hand side may be thought of as the perturbation away from the symmetric system due to the symmetry breaking stiffness term.

Since the linear dynamics remain the same as those for the previous example, the linear mode shapes are unaltered by the introduction of the $\kappa\delta x_1$ force term. Therefore the linear modal transform, $\mathbf{x} = \Phi\mathbf{q}$, is identical and the resulting transformed equation of motion only differs in its nonlinear vector. This vector becomes

$$N_q(\mathbf{q}) = \frac{\kappa}{m}\begin{Bmatrix} q_1^3 + 3q_1q_2^2 \\ 3q_1^2q_2 + \gamma q_2^3 \end{Bmatrix} + \frac{\kappa\delta}{m}\begin{Bmatrix} 3q_1^2q_2 + q_2^3 \\ q_1^3 + 3q_1q_2^2 \end{Bmatrix},$$

where again the last term represents the perturbation from the symmetric system.

Now considering the nonlinear near-identity transform and proceeding in a similar way as that laid out in **Step 2** in the previous two examples gives the $\mathbf{n}^*$ matrix

$$\mathbf{u}^* = \begin{bmatrix} u_{p1}^3 \\ u_{p1}^2 u_{m1} \\ u_{p1} u_{m1}^2 \\ u_{m1}^3 \\ u_{p1} u_{p2}^2 \\ u_{p1} u_{p2} u_{m2} \\ u_{p1} u_{m2}^2 \\ u_{m1} u_{p2}^2 \\ u_{m1} u_{p2} u_{m2} \\ u_{m1} u_{m2}^2 \\ u_{p1}^2 u_{p2} \\ u_{p1}^2 u_{m2} \\ u_{p1} u_{m1} u_{p2} \\ u_{p1} u_{m1} u_{m2} \\ u_{m1}^2 u_{p2} \\ u_{m1}^2 u_{m2} \\ u_{p2}^3 \\ u_{p2}^2 u_{m2} \\ u_{p2} u_{m2}^2 \\ u_{m2}^3 \end{bmatrix}, \quad \mathbf{n}^* = \frac{\kappa}{m} \begin{bmatrix} 1 & 0 \\ 3 & 0 \\ 3 & 0 \\ 1 & 0 \\ 3 & 0 \\ 6 & 0 \\ 3 & 0 \\ 3 & 0 \\ 6 & 0 \\ 3 & 0 \\ 0 & 3 \\ 0 & 3 \\ 0 & 6 \\ 0 & 6 \\ 0 & 3 \\ 0 & 3 \\ 0 & \gamma \\ 0 & 3\gamma \\ 0 & 3\gamma \\ 0 & \gamma \end{bmatrix}^T + \frac{\kappa\delta}{m} \begin{bmatrix} 0 & 1 \\ 0 & 3 \\ 0 & 3 \\ 0 & 1 \\ 0 & 3 \\ 0 & 6 \\ 0 & 3 \\ 0 & 3 \\ 0 & 6 \\ 0 & 3 \\ 3 & 0 \\ 3 & 0 \\ 6 & 0 \\ 6 & 0 \\ 3 & 0 \\ 3 & 0 \\ 1 & 0 \\ 3 & 0 \\ 3 & 0 \\ 1 & 0 \end{bmatrix}^T.$$

Note that with symmetry broken the nonlinear coefficient matrix, $\mathbf{n}^*$, is now fully populated.

While the values of $\boldsymbol{\beta}^*$ calculated in the previous example remain unaltered (they are not a function of the nonlinear coefficient matrix) as $\mathbf{n}^*$ is now fully populated $\boldsymbol{\beta}^*$ must also be fully populated. The Problem asks for only the case where there the response frequencies of the two linear modes match, $r = 1$, to be considered. The matrix $\boldsymbol{\beta}^*$ and the resulting matrix of nonlinear terms in the transformed equation (following the method discussed in **Steps 2** and **3**) are

$$\boldsymbol{\beta}^* = \omega_{r1}^2 \begin{bmatrix} 8\,8 \\ 0\,0 \\ 0\,0 \\ 8\,8 \\ 8\,8 \\ 0\,0 \\ 0\,0 \\ 0\,0 \\ 0\,0 \\ 8\,8 \\ 8\,8 \\ 8\,8 \\ 0\,0 \\ 0\,0 \\ 0\,0 \\ 8\,8 \\ 8\,8 \\ 0\,0 \\ 0\,0 \\ 8\,8 \end{bmatrix}^T \rightsquigarrow \quad \mathbf{n}_u^* = \frac{\kappa}{m}\begin{bmatrix} 0 & 0 \\ 3 & 0 \\ 3 & 0 \\ 0 & 0 \\ 0 & 0 \\ 6 & 0 \\ 3 & 0 \\ 3 & 0 \\ 6 & 0 \\ 0 & 0 \\ 0 & 0 \\ 0 & 3 \\ 0 & 6 \\ 0 & 6 \\ 0 & 3 \\ 0 & 0 \\ 0 & 0 \\ 0 & 3\gamma \\ 0 & 3\gamma \\ 0 & 0 \end{bmatrix}^T + \frac{\kappa\delta}{m}\begin{bmatrix} 0\,0 \\ 0\,3 \\ 0\,3 \\ 0\,0 \\ 0\,0 \\ 0\,6 \\ 0\,3 \\ 0\,3 \\ 0\,6 \\ 0\,0 \\ 0\,0 \\ 3\,0 \\ 6\,0 \\ 6\,0 \\ 3\,0 \\ 0\,0 \\ 0\,0 \\ 3\,0 \\ 3\,0 \\ 0\,0 \end{bmatrix}^T .$$

Note that the two columns in $\boldsymbol{\beta}^*$ are identical since $r = 1$.

From this, **Step 4** is to write the transformed equations of motion, which, to order ε^1 accuracy, are

$$\begin{aligned}
&\ddot{\mathbf{u}} + \Lambda\mathbf{u} + \mathbf{n}^*\mathbf{u}^* = 0 \\
&\rightsquigarrow \ddot{u}_1 + \omega_{n1}^2 u_1 + \frac{3\kappa}{m}\left[u_{p1}u_{m1}u_1 + 2u_{p2}u_{m2}u_1 + u_{p1}u_{m2}^2 + u_{m1}u_{p2}^2\right] \\
&\quad + \frac{3\kappa\delta}{m}\left[u_{p1}^2u_{m2} + 2u_{p1}u_{m1}u_2 + u_{m1}^2u_{p2} + u_{p2}u_{m2}u_2\right] = 0, \\
&\ddot{u}_2 + \omega_{n2}^2 u_2 + \frac{3\kappa}{m}\left[\gamma u_{p2}u_{m2}u_2 + 2u_{p1}u_{m1}u_2 + u_{p1}^2u_{m2} + u_{m1}^2u_{p2}\right] \\
&\quad + \frac{3\kappa\delta}{m}\left[u_{p1}u_{m1}u_1 + 2u_{p2}u_{m2}u_1 + u_{p1}u_{m2}^2 + u_{m1}u_{p2}^2\right] = 0,
\end{aligned} \tag{5.50}$$

where $\mathbf{u} = \mathbf{u}_p + \mathbf{u}_m$ has been used.

Now, as discussed in the second half of **Step 4** in the previous example, substitutions of the form $u_{p1} = (U_1/2)\mathrm{e}^{\mathrm{i}(\omega_{rk}t - \phi_k)}$ are made into these equations and then terms are balanced (exactly, as opposed to harmonic balance, where some terms are typically ignored). Balancing the $\mathrm{e}^{\mathrm{i}(\omega_{r1}t-\phi_1)}$ and the $\mathrm{e}^{\mathrm{i}(\omega_{r2}t-\phi_2)}$ terms for the first and second equations respectively gives

$$\omega_{r1}^2 = \omega_{n1}^2 + \frac{3\kappa}{4m}\left[U_1^2 + U_2^2\left(2 + \mathrm{e}^{\mathrm{i}2(\phi_1-\phi_2)}\right)\right]$$
$$+\frac{3\kappa\delta}{4m}\left[\left(2\mathrm{e}^{\mathrm{i}(\phi_1-\phi_2)} + \mathrm{e}^{-\mathrm{i}(\phi_1-\phi_2)}\right)U_1U_2 + \mathrm{e}^{\mathrm{i}(\phi_1-\phi_2)}\frac{U_2^3}{U_1}\right],$$
$$\omega_{r1}^2 = \omega_{n2}^2 + \frac{3\kappa}{4m}\left[U_1^2\left(2 + \mathrm{e}^{-\mathrm{i}2(\phi_1-\phi_2)}\right) + \gamma U_2^2\right]$$
$$+\frac{3\kappa\delta}{4m}\left[\mathrm{e}^{-\mathrm{i}(\phi_1-\phi_2)}\frac{U_1^3}{U_2} + \left(2\mathrm{e}^{-\mathrm{i}(\phi_1-\phi_2)} + \mathrm{e}^{\mathrm{i}(\phi_1-\phi_2)}\right)U_1U_2\right],$$

where as $r = 1$, $\omega_{r2} = \omega_{r1}$ has been used. Note that balancing the $\mathrm{e}^{-\mathrm{i}(\omega_{r1}t-\phi_1)}$ and $\mathrm{e}^{-\mathrm{i}(\omega_{r2}t-\phi_2)}$ terms respectively result in the complex conjugate of these equations and so yields no extra information.

As the right-hand side of these equations must be real the phase requirements $\sin(2(\phi_1 - \phi_2)) = 0$ and $\sin(\phi_1 - \phi_2) = 0$ must be satisfied and so $\phi_1 - \phi_2 = n\pi$, where n is an integer. The result is two sets of equations relating to the backbone curves corresponding to different phase relationships, just as with the symmetric case. This time they are

$$\phi_1 - \phi_2 = \cdots, -2\pi, 0, 2\pi, \ldots \rightarrow \quad q = \mathrm{e}^{-\mathrm{i}(\phi_1-\phi_2)} = 1, \quad \mathrm{e}^{-\mathrm{i}2(\phi_1-\phi_2)} = 1$$
$$\phi_1 - \phi_2 = \cdots, \ -\pi, \pi, 3\pi, \ldots \rightarrow \quad q = \mathrm{e}^{-\mathrm{i}(\phi_1-\phi_2)} = -1, \quad \mathrm{e}^{-\mathrm{i}2(\phi_1-\phi_2)} = 1$$

where in this case $q = 1$ corresponds to the two linear modal contributions being in-phase and $q = -1$ corresponds to them being in anti-phase. In this example there is no possibility for a 90° out of phase response (the $p = -1$ response in the previous example).

Using this the equations become

$$\omega_{r1}^2 = \omega_{n1}^2 + \frac{3\kappa}{4m}\left[U_1^2 + 3U_2^2\right] + q\frac{3\kappa\delta}{4m}\left[3U_1U_2 + \frac{U_2^3}{U_1}\right],$$
$$\omega_{r1}^2 = \omega_{n2}^2 + \frac{3\kappa}{4m}\left[\gamma U_2^2 + 3U_1^2\right] + q\frac{3\kappa\delta}{4m}\left[3U_1U_2 + \frac{U_1^3}{U_2}\right].$$

It can be seen that for the asymmetric case the pure response in one or other of the modes of the linear system (solutions $S1$ and $S2$ in the pervious example) do not exist. Instead, when δ is small, there are solutions that are dominated by one or other of the modes of the linear system but both modes are always present. To complicate things further these solutions could potentially exist for $q = +1$ and $q = -1$. ■

Figure 5.11 shows the backbone curves for the asymmetric system considered in Example 5.8. In this case the parameters are the same as the symmetric system shown in Fig. 5.7 and in addition the asymmetric parameter $\delta = 0.04$. Comparing Fig. 5.11 with Fig. 5.10, it can be seen that the bifurcation in Fig. 5.10 no longer exists, in fact

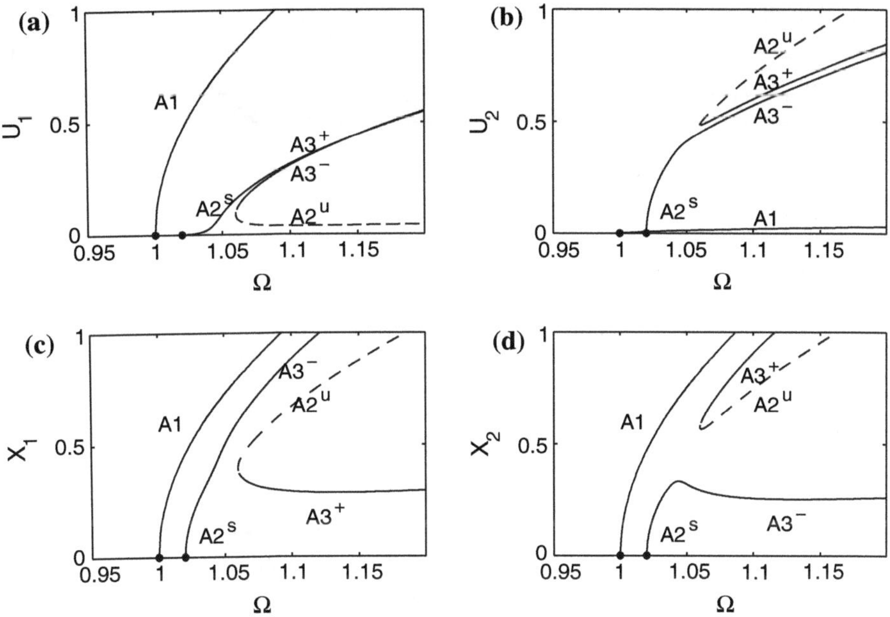

Fig. 5.11 Backbone curves of the two-degree-of-freedom considered in Example 5.8, with $\omega_{n1} = 1$, $\omega_{n2} = 1.02$, $m = 1$, $\kappa = 0.25$ and $\kappa_2 = 0.025$ and asymmetric parameter $\delta = 0.04$, in terms of amplitudes **a** U_1, **b** U_2, **c** X_1 and **d** X_2

there is now an imperfect bifurcation. To aid comparison between the figures, the curves in Fig. 5.11 are labelled to indicate on which solution branch the response would return to if the asymmetry were reduced to zero—note A is used to indicate a solution of the asymmetric solution and superscript u and s indicate the stable and unstable parts of a solution. In breaking the symmetry the symmetric single linear mode response solution $S1$ is slightly modified in terms of U_1 but now contains a non-zero U_2 component indicated by the line close to the $U_2 = 0$ axis in panel (b) which joins the axis at $\Omega = 1$, $A1$ in Fig. 5.11b. The $S2$ solution curve in the symmetric solution is largely similar to that for the asymmetric system in terms of U_2, however, as already stated, the bifurcation to the $S3^{\pm}$ solution is now an imperfect bifurcation. Note that again the $A2$ solution in the asymmetric system has a small U_1 component—see the curve close to the $U_1 = 0$ axis in the range $1.02 < \Omega < 1.05$ and the unstable (dashed) curve beyond, $A2^s$ and $A2^u$ respectively in Fig. 5.11a. The $S3^{\pm}$ solutions are identical in the U_1 and U_2 plots for the symmetric system, however for the asymmetric $A3^{\pm}$ ones they are slightly separated—these are the two curves that look similar to (but are not) a forced response in panel (a).

Considering the physical co-ordinates, panels (c) and (d), it can be seen that, due to the imperfect bifurcation, there is a backbone curve in the asymmetric system that is separated from the others. Part of this curve, labelled $A2^u$, is unstable, corresponding to the unstable section of $S2$ following the bifurcation in the symmetric system,

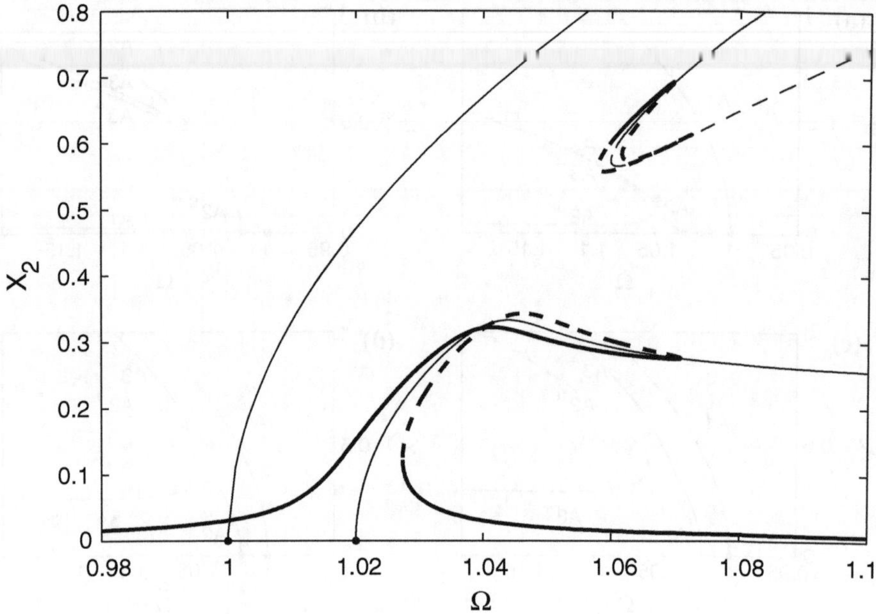

Fig. 5.12 Backbone curves of the two-degree-of-freedom considered in Example 5.8, with $\omega_{n1} = 1$, $\omega_{n2} = 1.02$, $m = 1$, $\kappa = 0.25$ and $\kappa_2 = 0.025$ and asymmetric parameter $\delta = 0.04$, in terms of amplitudes **a** U_1, **b** U_2, **c** X_1 and **d** X_2

however there is also a stable part, $A2^s$, indicating the possibility of an isola associated with the stable part of the disconnected branch.

To demonstrate how the isola may exist, we consider the same system with forcing and damping. The forcing is applied in the spatial configuration of the second mode at a sufficiently large level for an isola to appear, so we will consider the response of X_2 as shown in Fig. 5.11d. The resulting forced damped response is shown in Fig. 5.12. The forced response was calculated using numerical continuation and so in addition to stable solutions (thick solid lines), unstable solutions have been identified (thick dashed lines). It can be seen that the response is not effected by the left-hand backbone curve, this is because the forcing is in the second linear mode whereas this backbone curve corresponds to a mode shape that is close to that of the first linear mode. In contrast to this, the response closely follows the right-hand backbone curve, in which there is a significant component of the second linear mode. In addition there is a response isola around the upper backbone curve that contains a small region of stable response. This example demonstrates how backbone curves determine the underlying dynamic behaviour of the forced system when the damping is small.

5.5 Application to Larger Scale Systems

Until recently, use of techniques such as nonlinear normal modes had been limited to academic study. However, recent work has begun to do two important things. Firstly an increasing number of research studies have started to use experimental results. Secondly, several researchers have considered how these techniques could be applied to much larger models, particularly finite element models. One of the earliest works in this area is that carried out by McEwan et al. (2001), where nonlinear stiffness forces were estimated using static load cases in the finite element model. More recently there have been several studies on creating reduced order nonlinear modal models from finite element models Hollkamp and Gordon (2008) and Kuether and Allen (2014) and a review of these techniques is given by Mignolet et al. (2013).

5.6 Chapter Notes

Linear modal decomposition is covered in most texts on linear vibration theory. For example Clough and Penzien (1975), Thompson and Dahleh (1997), Ewins (2000) and Inman (2006) all give a detailed treatment of the subject matter. Nonlinear normal modes have a wide literature, see for example Shaw and Pierre (1993), Vakakis et al. (1996), Nayfeh et al. (1999), Lacarbonara and Camillacci (2004), Bellizzi and Bouc (2005), Jiang et al. (2005a, b), Pierre et al. (2006) and Burton (2007). A comprehensive overview of nonlinear normal modes and associated literature can be found in Kerschen et al. (2009) and Peeters et al. (2009). The method of normal forms, applied to nonlinear equations of motion in the form of second order differential equations is based on the work of Neild and Wagg (2011) which has been extended to study modal interactions by Xin et al. (2012, 2013), Neild and Wagg (2013), Hill et al. (2014), and Cammarano et al. (2014). In the first edition of this book, the normal forms method was applied to first order state-space versions of the equations. This type of approach is also discussed in the literature, see for example Jezequel and Lamarque (1991), Murdock (2002), Touze et al. (2004) and Touze and Amabili (2006). Application to finite element models is considered by McEwan et al. (2001), Hollkamp and Gordon (2008), Kuether and Allen (2014), and Mignolet et al. (2013).

References

Arnold, V. I. (1988). *Geometrical methods in the theory of ordinary differential equations*. New York: Springer.

Adhikari, S. (2004). Optimal complex modes and an index of damping non-proportionality. *Mechanical Systems and Signal Processing*, *18*(1), 1–27.

Bellizzi, S., & Bouc, R. (2005). A new formulation for the existence and calculation of nonlinear normal modes. *Journal of Sound and Vibration, 287*(3), 545–569.

Bendiksen, O. O. (2000). Localization phenomena in structural dynamics. *Chaos Solitons & Fractals, 11*(10), 1621–1660.

Bishop, R. E. D., Gladwell, G. M. L., & Michaelson, S. (2009). *The matrix analysis of vibrations*. UK: Cambridge University Press.

Burton, T. D. (2007). Numerical calculation of nonlinear normal modes in structural systems. *Nonlinear Dynamics, 49*(3), 425–441.

Cammarano, A., Hill, T. L., Neild, S. A., & Wagg, D. J. (2014). Bifurcations in backbone curves for systems of coupled nonlinear oscillators. *Nonlinear Dynamics, 77*, 311–320.

Clough, R. W., & Penzien, J. (1975). *Dynamics of structures*. New York: McGraw-Hill.

Caughey, T. K. (1963). *General theory of vibration of damped linear dynamic systems*. California Institute of Technology.

Ewins, D. J. (2000). *Modal testing*. Taunton: Research Studies Press.

Finlayson, B. A. (1972). *The method of weighted residuals*. New York: Academic Press.

Hill, T., Cammarano, A., Neild, S., & Wagg, D. (2014). Towards a technique for nonlinear modal reduction. In *Special topics in structural dynamics*, Vol. 6. (pp. 121–128). Springer International Publishing.

Hollkamp, J. J., & Gordon, R. W. (2008). Reduced-order models for nonlinear response prediction: Implicit condensation and expansion. *Journal of Sound and Vibration, 318*(4), 1139–1153.

Inman, D. J. (2006). *Vibration with control*. New York: Wiley.

Jezequel, L., & Lamarque, C. H. (1991). Analysis of nonlinear dynamic systems by the normal form theory. *Journal of Sound and Vibration, 149*(3), 429–459.

Jiang, D., Pierre, C., & Shaw, S. W. (2005a). The construction of non-linear normal modes for systems with internal resonance. *International Journal of Non-Linear Mechanics, 40*(5), 729–746.

Jiang, D., Pierre, C., & Shaw, S. W. (2005b). Nonlinear normal modes for vibratory systems under harmonic excitation. *Journal of Sound and Vibration, 288*(4–5), 791–812.

Kerschen, G., Peeters, M., Golinval, J.-C., & Vakakis, A. F. (2009). Nonlinear normal modes, part i: A useful framework for the structural dynamicist. *Mechanical Systems and Signal Processing, 23*(1), 170–194.

Kerschen, G., Worden, K., Vakakis, A. F., & Golinval, J. C. (2006). Past, present and future of nonlinear system identification in structural dynamics. *Mechanical Systems and Signal Processing, 20*(3), 505–592.

Kuether, R. J. & Allen, M. S. (2014). A numerical approach to directly compute nonlinear normal modes of geometrically nonlinear finite element models. *Mechanical Systems and Signal Processing*.

Lacarbonara, W., & Camillacci, R. (2004). Nonlinear normal modes of structural systems via asymptotic approach. *International Journal of Solids and Structures, 41*(20), 5565–5594.

Langley, R. S., Bardell, N. S., & Loasby, P. M. (1997). The optimal design of near-periodic structures to minimize vibration transmission and stress levels. *Journal of Sound and Vibration, 207*(5), 627–646.

Lust, S. D., Friedmann, P. P., & Bendiksen, O. O. (1995). Free and forced response of multi-span beams and multi-bay trusses with localized modes. *Journal of Sound & Vibration, 180*(2), 313–332.

McEwan, M., Wright, J., Cooper, J., & Leung, A. (2001). A combined modal/finite element analysis technique for the dynamic response of a non-linear beam to harmonic excitation. *Journal of Sound and Vibration, 243*(4), 601–624.

Mignolet, M. P., Przekop, A., Rizzi, S. A., & Spottswood, S. M. (2013). A review of indirect/non-intrusive reduced order modeling of nonlinear geometric structures. *Journal of Sound and Vibration, 332*(10), 2437–2460.

Murdock, J. (2002). *Normal forms and unfoldings for local dynamical systems*. New York: Springer.

Nayfeh, A. H., Lacarbonara, W., & Chin, C.-M. (1999). Nonlinear normal modes of buckled beams: Three-to-one and one-to-one internal resonances. *Nonlinear Dynamics*, *18*, 253–273.

Neild, S. A., & Wagg, D. J. (2011). Applying the method of normal forms to second-order nonlinear vibration problems. *Proceedings of the Royal Society of London A*, *467*(2128), 1141–1163.

Neild, S. A., & Wagg, D. J. (2013). A generalized frequency detuning method for multidegree-of-freedom oscillators with nonlinear stiffness. *Nonlinear Dynamics*, *73*, 649–663.

Peeters, M., Viguié, R., Sérandour, G., Kerschen, G., & Golinval, J.-C. (2009). Nonlinear normal modes, part ii: Toward a practical computation using numerical continuation techniques. *Mechanical Systems and Signal Processing*, *23*(1), 195–216.

Pierre, C., Tang, D. M., & Dowell, E. H. (1987). Localized vibrations of disordered multispan beams: theory and experiment. *AIAA Journal*, *25*(9), 1249–1257.

Pierre, C., Jiang, D. Y., & Shaw, S. (2006). Nonlinear normal modes and their application in structural dynamics. *Mathematical Problems in Engineering*, 10847.

Rand, R. H. (2005). *Lecture notes on nonlinear vibrations*. www.math.cornelledu/~rand/randdocs/nlvibe52.pdf.

Shaw, S. W., & Pierre, C. (1993). Normal modes for non-linear vibratory systems. *Journal of Sound and Vibration*, *164*(1), 85–124.

Thompson, W. T. & Dahleh, M. D. (1997). *Theory of vibration with applications* (4th ed.). Prentice Hall: London.

Touze, C., & Amabili, M. (2006). Nonlinear normal modes for damped geometrically nonlinear systems: Application to reduced-order modelling of harmonically forced structures. *Journal of Sound and Vibration*, *298*(4–5), 958–981.

Touze, C., Thomas, O., & Chaigne, A. (2004). Hardening/softening behaviour in non-linear oscillations of structural systems using non-linear normal modes. *Journal of Sound and Vibration*, *273*(1–2), 77–101.

Vakakis, A. F., Manevitch, L. I., Mikhlin, Y. V., Pilipchuk, V. N., & Zevin, A. A. (1996). *Normal modes and localization in nonlinear systems*. New York: Wiley.

Xin, Z., Zuo, Z., Feng, H., Wagg, D. J., & Neild, S. A. (2012). Higher order accuracy analysis of the second-order normal form method. *Nonlinear Dynamics*, *70*(3), 2175–2185.

Xin, Z. F., Neild, S. A., Wagg, D. J., & Zuo, Z. X. (2013). Resonant response functions for nonlinear oscillators with polynomial type nonlinearities. *Journal of Sound and Vibration*, *332*(7), 1777–1788.

Chapter 6
Beams

Abstract This chapter discusses methods for modelling the nonlinear vibration of beams. The starting point is to consider the physics of beams, for both small and large deflections. The resulting partial differential equations are then decomposed, using the techniques discussed in Chap. 5, to give a set of ordinary differential equations which can be analysed. Large deflections lead to nonlinear governing equations for the beam vibrations. Another important case in practice is when the beam is axially loaded, which also leads to nonlinearities in the governing expressions. In the final part of the chapter, control of beam vibrations using modal control is discussed.

6.1 Small-Deflection Beam Theory

As a starting point, consider the classical small-deflection (i.e. linear) approach to analysing a cantilever beam. A full account of this type of analysis can be found in many classic engineering texts, for example Bishop and Johnson (1960), Timoshenko et al. (1974) and Meirovitch (1976). The beam is assumed to be homogeneous and isotropic, such that for a constant cross-sectional area, the distributed mass and elasticity of the beam are constant along its length. It is also assumed to be slender, so that bending deformation dominates shear deformation, which is neglected. Any such continuum requires an infinite number of coordinates in order to specify the position of every particle in the beam as it vibrates. So, in this sense the beam possesses an infinite number of degrees of freedom and its governing equation of motion is a partial differential equation: the Euler-Bernoulli equation.[1]

For continua, partial differential equations account for the fact that the vibration depends not just on time, but also on which point of the structure is being considered. As a result, when dealing with continua, the spatial aspects of the problem need to be considered, as discussed in Chap. 5. For vibration problems, this usually means answering the following questions. (i) At what points on the structure is knowledge

[1] Note that more complex formulations exist, such as the *Timoshenko* beam equation, but these are not considered here.

D. Wagg and S. Neild, *Nonlinear Vibration with Control*,
Solid Mechanics and Its Applications 218, DOI 10.1007/978-3-319-10644-1_6

about the vibration response required? (ii) What type of *boundary conditions* does the structure have? (iii) What type of vibration shapes (i.e. modes) will the structure naturally tend to vibrate in? The boundary conditions describe the way the structure is attached to its surrounding environment, i.e. clamped, hinged, or free.

There are a wide variety of methods for solving partial differential equations. For an introduction see, for example, Kreyszig (1993) and for a more in depth discussion see King et al. (2003). In this book, the discussion will be restricted to structural components with relatively simple geometry, where a governing partial differential equation can be approximated by a set of ordinary differential equations. This will allow an analytical treatment to be developed, to demonstrate the underlying format of the problems being considered. For larger structures and particularly those with more complex geometry, purely numerical techniques must be used, see for example Vichnevetsky (1981) and Press et al. (1994). Of the techniques available, the finite element method is by far the most powerful for structural vibration problems, but a detailed discussion is beyond the scope of this book, see instead Zienkiewicz and Taylor (1991) and Crisfield (1997) for an introduction to the method.

The two solution methods discussed in this chapter have been chosen for their relevance to vibration problems. First is the *Galerkin* method, which is an approach based on *separation of variables* (introduced in Sect. 5.2 of Chap. 5). For the (two-dimensional) beam example there is just one spatial variable (length along the beam), so the Galerkin method assumes that the solution can be approximated by some combination of spatial functions and time (or temporal) functions. The second method uses direct spatial discretisation of the beam, and is called the *collocation method.* In this approach, a set of discrete points along the beam is selected and the spatial and temporal functions are made to fit at these points.[2]

6.1.1 The Euler-Bernoulli Equation

When a load acts on a beam or column it will deflect into a deformed shape with a certain curvature. To derive a *mathematical model* of the beam behaviour, a series of assumptions must be made about the basic physical behaviour of the beam. By considering statics, the basic modelling assumption for beam bending is that the bending moment, M, at any point in the beam, is proportional to the curvature. The usual way of representing this relationship is in the form

$$M = EI\frac{1}{R} = EI\frac{d\psi}{ds}, \tag{6.1}$$

[2] In fact both these methods are part of a wider class called *weighted residual methods*, which includes the Rayleigh-Ritz method and finite element methods. See Hagedorn and DasGupta (2007) for an introduction to how these methods can be applied to beams and other continuous structures.

Fig. 6.1 Curvature of an element of the beam

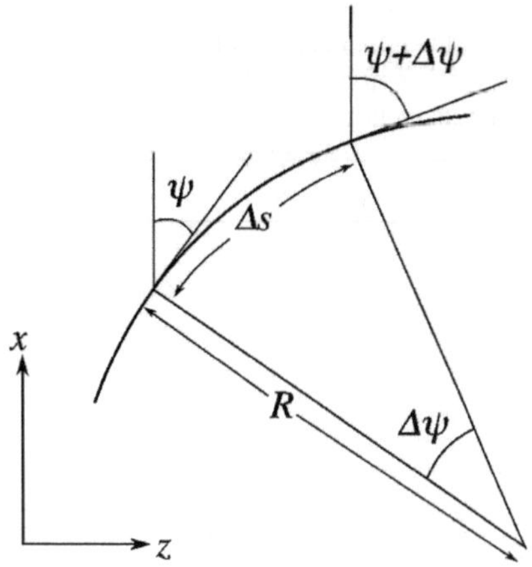

where $E(\mathrm{N/m^2})$ is the Young's modulus, $I(\mathrm{m^4})$ is the second moment of area and R (m) is the radius of curvature. The slope of the beam is defined as ψ and hence the curvature (which equals $1/R$) is given by $\mathrm{d}\psi/\mathrm{d}s$ where s (m) is the length along the beam, as shown in Fig. 6.1. The term *EI* represents the *flexural rigidity* of the beam. Further assumptions are normally made when using Eq. (6.1) to derive the vibration response of a beam. These are that E and I are constant along the length of the beam, which physically corresponds to a beam of uniform cross-sectional area $\hat{A}$, uniform material Young's modulus E and uniform mass distribution. In addition it is assumed that the beam is slender, such that shear deformation can be neglected.[3]

Now the loading on a small element of beam will be considered. This analysis is the one most commonly described in vibration textbooks.[4] It is assumed that the beam deflection in the z direction, $w(x, t)$, is small and in the x direction (the axial direction) is negligible, $u(x, t) = 0$. It is assumed that the rotary inertia of the element can be ignored. Consider the beam element in Fig. 6.2 which has an external forcing function[5] per unit length, $F(x, t)$, acting along its length (but no axial loading). If the force per unit length at position x is F and the force per unit length at position $x + \Delta x$ is $F + \Delta F$, the element experiences a force of $(F + \Delta F/2)\Delta x$ acting through its centre of mass. The shear force on the element at a distance x along the beam is V, and the bending moment is M. Over a small axial distance, Δx, the shear force and bending moment change to $V + \Delta V$ and $M + \Delta M$ respectively.

Force equilibrium in the z direction including the inertia of the element gives

$$\rightarrow_z \qquad -(V + \Delta V) + V + \left(F + \frac{\Delta F}{2}\right)\Delta x = \rho\hat{A}\Delta s\frac{\partial^2 w}{\partial t^2}, \qquad (6.2)$$

---

[3] For beams which cannot be considered as slender a Timonshenko beam analysis could be used (Timoshenko et al. 1974).

[4] Note that we derive the equations of motion for a beam aligned in the vertical position. An equivalent derivation can also be carried out for a horizontal beam, see for example Clough and Penzien (1993).

[5] Moving loads are not considered here, but a good overview is given by Ouyang (2011).

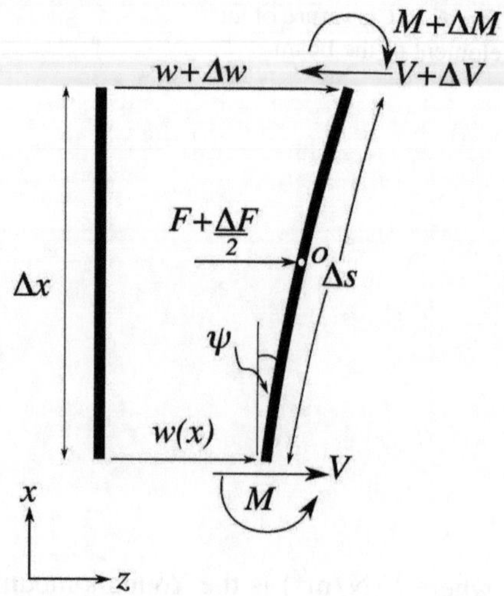

Fig. 6.2 Element of a vertical bending beam with a forcing function applied along its length

where ρ is the density and $\hat{A}$ is the cross-sectional area. Note that here the external force is written in terms of the span of the beam element along the x-axis, Δx, but it can also be expressed as a function of the beam length Δs. For example, if the external force, $F(x, t)$, is the self-weight due to gravity (where gravity is acting in the z-direction) then $F\Delta x = \rho\hat{A}g\Delta s$ where $\rho\hat{A}g$ is the weight per unit length of beam. Here the deflections are taken to be small, so that $\Delta s \approx \Delta x$ is assumed. Using this approximation and taking the limit as $\Delta x \to 0$, we can write Eq. (6.2) as

$$-\frac{\partial V}{\partial x} + F(x,t) = \rho\hat{A}\frac{\partial^2 w}{\partial t^2}, \tag{6.3}$$

where the $\Delta F \Delta x$ term has become negligibly small in the limit.

Moment equilibrium about point O in Fig. 6.2, neglecting the rotary inertia, gives

$$\circlearrowleft_O \quad (V+\Delta V)\frac{\Delta x}{2} + V\frac{\Delta x}{2} + M - (M+\Delta M) = 0.$$

Dividing by Δx and taking the limit as $\Delta x \to 0$ gives

$$V = \frac{\partial M}{\partial x}, \tag{6.4}$$

where the $\Delta V \Delta x$ term has become negligibly small in the limit.

Now substituting Eq. (6.4) into Eq. (6.3) gives

$$\frac{\partial^2 M}{\partial x^2} + \rho\hat{A}\frac{\partial^2 w}{\partial t^2} = F(x,t). \tag{6.5}$$

Finally the moment relationship, Eq. (6.1), may be approximated based on the small-deflection assumptions by writing $\psi \approx \Delta w/\Delta x$ and noting that $\Delta s \approx \Delta x$ to give

$$M = EI\frac{\mathrm{d}\psi}{\mathrm{d}s} \quad \leadsto \quad M = EI\frac{\mathrm{d}^2 w}{\mathrm{d}x^2}.$$

Using this relationship to eliminate M from Eq. (6.5) gives

$$EI\frac{\partial^4 w}{\partial x^4} + \rho\hat{A}\frac{\partial^2 w}{\partial t^2} = F(x, t), \tag{6.6}$$

which is the linear (undamped) Euler-Bernoulli equation for beam vibration with a time-varying distributed load $F(x, t)$ applied along its length. This equation can be used to model the transverse (z direction) vibration of the beam for small amplitude displacements.

It should be noted that damping in continuous systems such as beams, plates and cables is not well understood. The exact physics of the damping process has yet to be fully explained, and as a result it is quite usual to derive equations of motion and ignore damping effects completely. However, for the model to be physically realistic, damping will normally need to be added to the model at some stage, and throughout this book it is done after the partial differential equation has been reduced to a set of ordinary differential equations. Physical damping terms can be added into the partial differential equation, but, in order to decouple the system into a series of ordinary differential equations, the terms should be proportional to either stiffness or mass, as discussed for lumped mass systems in Sect. 1.3.3—see Lazan (1968), Clough and Penzien (1993), and Jones (2001), for more detailed discussions. In fact, it can be shown that a series summation of proportional terms can be used to give any linear modal damping coefficients that may be required—Clough and Penzien (1993). This, combined with the uncertainty of the underlying physics, makes it somewhat irrelevant to derive damping terms in the partial differential equation; this will rarely happen in this text.

There are various methods for carrying out the reduction from partial to ordinary differential equations, which are often collectively known as *modal decomposition* methods. Here we will use the Galerkin method to undertake the modal decomposition, which is discussed in detail next.

6.1.2 The Galerkin Method

Many partial differential equations can be reduced to an infinite series of ordinary differential equations by making a substitution involving separation of space and time variables. Here we consider a general form of this separation known as the Galerkin method, see Chap. 5, Sect. 5.2, Finlayson (1972) or Fletcher (1984). For a beam, the transverse displacement is approximated by

$$w(x,t) = \sum_{j=1}^{\infty} \phi_j(x) q_j(t),$$

where $\phi_j(x)$ represent the spatial functions and $q_j(t)$ represent the temporal part of the solution. For linear vibration problems, the space functions can be found by treating the Euler-Bernoulli equation as an eigenvalue problem, see Timoshenko et al. (1974). The eigenvalues relate to the natural frequencies of the beam and the eigenvectors relate to the *normal mode-shapes*.[6] The q_j coordinates are known as the *modal displacements*, which can be determined in a variety of ways.

Substituting for $w(x,t)$ into the Euler-Bernoulli equation, Eq. (6.6), gives

$$EI \sum_{j=1}^{\infty} \frac{d^4\phi_j(x)}{dx^4} q_j(t) + \rho\hat{A} \sum_{j=1}^{\infty} \phi_j(x)\ddot{q}_j(t) = F(x,t), \tag{6.7}$$

where, because of the separation of variables, the partial derivatives have become ordinary derivatives.

The normal mode-shapes have the special property of being an orthogonal basis of functions, in a similar way to the discrete multi-degree-of-freedom system discussed in Sect. 1.3.3 and Chap. 5. This property of orthogonality can be used to decouple Eq. (6.7). First, the equation is multiplied by an arbitrary mode shape $\phi_i(x),\ i \neq j$ and then integrated over the length of the beam, L, to obtain

$$EI \sum_{j=1}^{\infty} \int_0^L \frac{d^4\phi_j}{dx^4} \phi_i dx q_j(t) + \rho\hat{A} \sum_{j=1}^{\infty} \int_0^L \phi_j\phi_i dx \ddot{q}_j(t) = \int_0^L F(x,t)\phi_i dx. \tag{6.8}$$

Orthogonality conditions, such that Eq. (6.8) is decoupled, may now be written as

$$\int_0^L \frac{d^4\phi_j}{dx^4} \phi_i dx = 0 \quad \text{and} \quad \int_0^L \phi_j\phi_i dx = 0 \tag{6.9}$$

for the case where $i \neq j$. These orthogonality conditions depend on the *boundary conditions*. For example, for a cantilever beam the conditions are that one end of the beam is clamped and the other is free. These conditions can be used to formulate an eigenvalue problem, as discussed in Chap. 3—see also Timoshenko et al. (1974) and Blevins (1979). Solving the eigenvalue problem gives a set of mode-shapes, ϕ_j, and dimensionless frequency parameters ξ_j. An example of this is discussed in Sect. 5.2.

By using the orthogonality conditions, we can express Eq. (6.8) as an infinite number of independent equations of which the ith mode is given by

---

[6] Sometimes also called the eigenfunctions, modal basis or linear modes.

$$EI\int_0^L \frac{\mathrm{d}^4\phi_i}{\mathrm{d}x^4}\phi_i \mathrm{d}x q_i(t) + \rho\hat{A}\int_0^L \phi_i\phi_i \mathrm{d}x\ddot{q}_i(t) = \int_0^L F(x,t)\phi_i \mathrm{d}x. \tag{6.10}$$

The definitions of mode-shapes set out in Blevins (1979), when $i = j$ for a cantilever beam results in

$$\int_0^L \phi_i\phi_i \mathrm{d}x = L \quad \text{and} \quad \int_0^L \phi_i''''\phi_i \mathrm{d}x = \xi_i^4/L^3 \tag{6.11}$$

for all i, where $\bullet'$ represents differentiation with respect to x and $\xi_i^4 = \omega_{ni}^2\rho\hat{A}L^4/EI$. Equation (6.11) may be used to simplify the equation for the ith mode, Eq. (6.10), to give

$$\ddot{q}_i(t) + \omega_{ni}^2 q_i(t) = \frac{1}{\rho\hat{A}L}Q_i(t),$$

where ω_{ni} is the natural frequency of the ith mode and $Q_i(t)$ is the generalized force given by

$$Q_i(t) = \int_0^L F(x,t)\phi_i \mathrm{d}x. \tag{6.12}$$

For beams, solving the eigenvalue problem takes account of the boundary conditions, for example different combinations of beam end conditions being clamped, pinned or free. This means that ξ_i and ϕ_j will vary depending on the boundary conditions. The possible cases are derived in most standard texts on linear vibration, see for example Timoshenko et al. (1974), and are tabulated in Blevins (1979) and Bishop and Johnson (1960).[7]

From a vibration perspective, the model must include some damping and this can be done in a number of ways. The simplest way is to add a damping term to each mode of vibration *after* the Galerkin method has been applied. This approach is appropriate because of the limitations of our current understanding of damping in continua. As the exact mechanisms are not known, a simple viscous damping term is added to each mode. This approach is usually very effective as a modelling technique, especially where structures are lightly damped, but will clearly have limitations in applications where the damping is non-viscous.[8] The result is that for the ith mode

$$\ddot{q}_i(t) + 2\zeta_i\omega_{ni}\dot{q}_i(t) + \omega_{ni}^2 q_i(t) = \frac{1}{\rho\hat{A}L}Q_i(t). \tag{6.13}$$

---

[7] For a discussion on alternative modal scaling, such as mass normalized modes, see Clough and Penzien (1993) or Ewins (2000).

[8] Forms of non-viscous damping often appear in vibration problems, see for example Jones (2001).

This equation is essentially the same as Eq. (1.16) in Chap. 1 (and as Eq. (1.24), the modal equation for a lumped mass multi degree of freedom system). Equation (6.13) has an exact solution, which is derived in Sect. 1.3.1.

Having reduced the partial differential equation to an infinite set of ordinary differential equations, we need to select a finite set of N equations as the model. This process is often referred to as *modal truncation*. A detailed discussion of modal truncation can be found in Inman (2006). If detailed experimental measurements are available, then N can be estimated by selecting only the significant resonance peaks in the frequency spectrum, see Ewins (2000) and Worden and Tomlinson (2000). Another approach is to start with a small number of modes, then add additional higher modes until no change is detected in the response. Whatever approach is used, a degree of judgement needs to be exercised in choosing N. In the following discussion it will be assumed that N has been selected using appropriate judgement for the problem at hand.

6.1.3 Initial Conditions and Forcing

The initial conditions for the motion of the beam can be determined from

$$w(x,0) = \sum_{j=1}^{N} \phi_j(x) q_j(0),$$

and

$$\dot{w}(x,0) = \sum_{j=1}^{N} \phi_j(x) \dot{q}_j(0).$$

Applying the orthogonality principle gives

$$q_i(0) = \frac{1}{L} \int_0^L w(x,0)\phi_i \mathrm{d}x, \tag{6.14}$$

and

$$\dot{q}_i(0) = \frac{1}{L} \int_0^L \dot{w}(x,0)\phi_i \mathrm{d}x. \tag{6.15}$$

If the initial conditions are known, or can be calculated by Eqs. (6.14) and (6.15), the only unknown is the forcing function $Q_j(t)$ given by Eq. (6.12). The forcing function (per unit length) $F(x,t)$ is assumed to be separable such that $F(x,t) =$

Table 6.1 Modal participation factors for a cantilever beam to 4 decimal places

j	1	2	3	4	5
α_j	0.7830	0.4339	0.2544	0.1819	0.1415

$g(x)h(t)/L$ where $h(t)$ is a function of time only and has units of force, and $g(x)$ is a function of position only and has no units.[9] With this expression, Eq. (6.12) can be expressed as

$$Q_i(t) = \frac{h(t)}{L}\int_0^L g(x)\phi_i \mathrm{d}x = \alpha_i h(t). \tag{6.16}$$

It can be seen from this expression that the time-dependent forcing $h(t)$ is applied to the whole beam, such that the integral term, or α_i term, acts as a *modal distribution* or *participation factor* for the ith mode (and has no units), and hence distributes the effect of the forcing $h(t)$ across the N modes considered in the analysis.

When the distribution across the beam is uniform, such that $g(x) = 1$, the modal distributions α_i become

$$\alpha_i = \frac{1}{L}\int_0^L \phi_i \mathrm{d}x.$$

For common boundary conditions, these participation factors are tabulated, see for example Blevins (1979). The first five values of modal participation factor are given in Table 6.1 for a cantilever.[10] The relative size of the participation factors is important when considering which modes must be targeted in an active control scheme.

If however, the forcing $h(t)$ is applied at a single point, $x = c$, on the beam, $g(x)$ becomes the Dirac-Delta function $g(x) = \delta(x - c)$ and evaluating Eq. (6.16) gives

$$Q_i(t) = h(t)\frac{\phi_i(c)}{L}. \tag{6.17}$$

This equation shows that if c is a node point of mode i, then this mode will not be directly excited, since $\phi_i(c) = 0$ for a node point. However, if c is not a node point, $\phi_i(c)$ represents the modal distribution for the point forcing at $x = c$.

If $g(x)$ is a cantilever mode-shape per unit length ϕ_j/L then $Q_i(t) = h(t)$ for $i = j$, and $Q_i(t) = 0$ for $i \neq j$. With this condition the beam would theoretically vibrate in the jth mode only. Now consider an example where a beam has a moment applied between two points along its length by a piezoelectric actuator.

Example 6.1 Modal decomposition of a cantilever beam with applied moment

[9] Note that it is equally valid to give the functions $h(t)$ and $g(x)$ other units provided the resulting $F(x, t)$ has units of force per length.

[10] See Blevins (1979), p. 455, for further details.

Problem Find the modal decomposition of the linear (undamped) Euler-Bernoulli equation for a cantilever beam vibration with an applied external moment per unit length, Q, from a piezoelectric actuator bonded onto the beam. This applied moment is a constant applied between two points on the beam, x_1 to x_2. The equation of motion neglecting damping (which will be derived in Sect. 6.3.2) is given by

$$EI\frac{\partial^4 w}{\partial x^4} + \frac{\partial Q}{\partial x} + \rho\hat{A}\frac{\partial^2 w}{\partial t^2} = 0.$$

The moment function is given by $Q = CV_a[H(x - x_1) - H(x - x_2)]$, where $H(\bullet)$ is the Heaviside function, V_a is the actuator voltage and C is a constant.[11]

Solution Substituting $w(x, t) = \sum_{j=1}^{N} \phi_j(x) q_j(t)$ to separate the space and time dependence of the transverse displacement $w(x, t)$ gives

$$EI\sum_{j=1}^{N}\frac{\mathrm{d}^4\phi_j(x)}{\mathrm{d}x^4}q_j(t) + \rho\hat{A}\sum_{j=1}^{N}\phi_j(x)\ddot{q}_j(t) = -CV_a\frac{\partial}{\partial x}[H(x - x_1) - H(x - x_2)],$$

where $\phi_j(x)$ represent the normal mode-shapes and $q_j(t)$ represent the modal coordinates.

The principle of orthogonality can now be used to decouple the equation. First the equation is multiplied by an arbitrary mode shape, $\phi_i(x)$, and then integrated over the length of the beam to give

$$EI\sum_{j=1}^{N}\int_0^L\frac{\mathrm{d}^4\phi_j}{\mathrm{d}x^4}\phi_i\mathrm{d}x q_j(t) + \rho\hat{A}\sum_{j=1}^{N}\int_0^L\phi_j\phi_i\mathrm{d}x\ddot{q}_j(t)$$
$$= -\int_0^L CV_a\phi_i\frac{\partial}{\partial x}[H(x - x_1) - H(x - x_2)]\mathrm{d}x.$$

Applying the orthogonality conditions for a cantilever, Eq. (6.9), we uncouple this equation to give

$$EI\int_0^L\phi_i''''\phi_i\mathrm{d}x q_i(t) + \rho\hat{A}\int_0^L\phi_i\phi_i\mathrm{d}x\ddot{q}_i(t) = -\int_0^L CV_a\phi_i\frac{\partial}{\partial x}[H(x - x_1) - H(x - x_2)]\mathrm{d}x.$$

Using the integrals in Eq. (6.11), $\int_0^L \phi_i\phi_i\mathrm{d}x = L$ and $\int_0^L \phi_i''''\phi_i\mathrm{d}x = \xi_i^4/L^3$, and dividing by $\rho\hat{A}L$ gives

[11] See Fanson and Caughey (1990) for more details on modelling moments generated by piezoelectric actuators.

$$\ddot{q}_i(t) + \omega_{ni}^2 q_i(t) = -\frac{CV_a}{\rho\hat{A}L}\int_0^L \phi_i \frac{\partial}{\partial x}[H(x - x_1) - H(x - x_2)]\mathrm{d}x$$

for the ith mode, where $\omega_{ni}^2 = EI\xi_i^4/\rho\hat{A}L^4$.

Now using the fact that $\frac{\partial H(\bullet)}{\partial x} = \delta(\bullet)$, where δ is the Dirac-Delta function, gives

$$\ddot{q}_i(t) + \omega_{ni}^2 q_i(t) = -\frac{CV_a}{\rho\hat{A}L}\int_0^L \phi_i[\delta(x - x_1) - \delta(x - x_2)]\mathrm{d}x. \tag{6.18}$$

Since $\int_0^L f(x)\delta(x - \bullet)\mathrm{d}x = f(\bullet)$, Eq. (6.18) becomes

$$\ddot{q}_i(t) + \omega_{ni}^2 q_i(t) = -\frac{CV_a}{\rho\hat{A}L}[\phi_i(x_2) - \phi_i(x_1)]. \tag{6.19}$$

The term in brackets on the right-hand side of Eq. (6.19) is now just a constant, dependent on the slope of the mode-shape

$$D_i = \phi_i(x_2) - \phi_i(x_1),$$

so finally we have for the ith mode

$$\ddot{q}_i(t) + \omega_{nj}^2 q_i(t) = -aD_iV_a, \tag{6.20}$$

where $a = C/\rho\hat{A}L$. ■

This decomposition will be used in Sect. 6.3.2 for moment-based actuation to control vibrations in a beam.

6.1.4 Collocation Method

An alternative approach to the Galerkin method is the collocation method. In the collocation method we consider discrete spatial locations along the beam and find spatial and temporal functions that describe the motion at these locations. This transforms the partial differential equation of motion into a matrix equation representing the motion of discrete locations along the beam.

In the collocation method, the structural component is divided into a series of segments. For a beam this means selecting N collocation points along the length of the beam. Consider the Euler-Bernoulli equation with a damping term and external forcing

$$EI\frac{\partial^4 w}{\partial x^4} + \eta\frac{\partial w}{\partial t} + \rho\hat{A}\frac{\partial^2 w}{\partial t^2} = F(x,t), \tag{6.21}$$

where E is the Young's modulus, ρ density, $\hat{A}$ cross-sectional area, η the damping constant and I the second moment of area for the beam of length L.

Now, as with the Galerkin method, assume that there is a series solution to the Euler-Bernoulli equation given by

$$w(x,t) = \sum_{j=1}^{N} \phi_j(x) q_j(t), \tag{6.22}$$

where $\phi_j(x)$ are the normal mode-shapes of the beam, and $q_j(t)$ are the modal coordinates. Then substituting Eq. (6.22) into the Euler-Bernoulli equation, Eq (6.21), gives

$$\sum_{j=1}^{N} (\phi_j \ddot{q}_j(t) + \beta\phi_j \dot{q}_j(t) + \alpha\phi_j'''' q_j(t)) = \gamma F(x,t), \tag{6.23}$$

where $\bullet'$ and $\dot{\bullet}$ represents differentiation with respect to x and t respectively, $\alpha = EI/\rho\hat{A}$, $\beta = \eta/\rho\hat{A}$ and $\gamma = 1/\rho\hat{A}$.

If the normal linear beam modes based on zero damping are being used, Eq. (6.23) can be rewritten in terms of natural frequencies of each mode to give

$$\sum_{j=1}^{N} (\phi_j \ddot{q}_j(t) + \beta\phi_j \dot{q}_j(t) + \omega_{nj}^2 \phi_j q_j(t)) = \gamma F(x,t), \tag{6.24}$$

where ω_{nj} is the jth natural frequency. In making this modification the standard relationship $\phi_j'''' = (\xi_j^4/L^4)\phi_j$ is used, where as before

$$\xi_j^4 = \omega_{nj}^2 \frac{p\hat{A}L^4}{EI},$$

see Blevins (1979).[12]

Now N collocation points $x_1, x_2, \ldots, x_N$ are chosen along the length of the beam. Collocation points are usually chosen at evenly spaced intervals. At the N discrete collocation points, Eq. (6.24) can be represented in a matrix form

$$\boldsymbol{\Phi}\ddot{\mathbf{q}} + \beta\boldsymbol{\Phi}\dot{\mathbf{q}} + \boldsymbol{\Phi}\hat{\omega}_{nj}^2\mathbf{q} = \gamma\mathbf{F}, \tag{6.25}$$

where

[12] If normal linear beam modes are not used, collocation can still be applied providing the fourth derivative of the shape function ϕ_j can be computed for each collocation point.

$$\Phi = \begin{bmatrix} \phi_1(x_1) & \phi_2(x_1) & \dots & \phi_N(x_1) \\ \phi_1(x_2) & \phi_2(x_2) & \dots & \phi_N(x_2) \\ \vdots & \vdots & \dots & \vdots \\ \phi_1(x_N) & \phi_2(x_N) & \dots & \phi_N(x_N) \end{bmatrix},$$

$\mathbf{q} = [q_1, q_2 \dots q_N]^T$, $\hat{\omega}^2_{nj} = diag\{\omega^2_{n1}, \omega^2_{n2} \dots \omega^2_{nN}\}$ and $\mathbf{F} = [F(x_1, t), F(x_2, t) \dots, F(x_N, t)]^T$. Multiplying Eq. (6.25) by Φ^{-1} and putting it into first-order form gives

$$\dot{\mathbf{z}} = A\mathbf{z} + \hat{\mathbf{F}}, \tag{6.26}$$

where $\mathbf{z} = [\mathbf{q}, \dot{\mathbf{q}}]^T$, $\hat{\mathbf{F}} = [0_N, \ \gamma\Phi^{-1}\mathbf{F}]^T$ and

$$A = \begin{bmatrix} 0_N & I_N \\ -\hat{\omega}^2_{nj} & -\beta I_N \end{bmatrix},$$

where 0_N is an $N \times N$ zero matrix and I_N is an $N \times N$ identity matrix. Equation (6.26) can now be integrated forward in time from a set of initial conditions using a suitable time-stepping method—for example a fourth-order Runge-Kutta method, see for example Press et al. (1994).[13]

The collocation method can be used to model problems where a nonlinear event occurs at a specific point on an otherwise linearly vibrating beam. For example, Fig. 6.3 shows a sinusoidally forced cantilever beam subjected to a motion limiting constraint (an impact stop) near the tip. As the beam vibrates it has an impact each time $w(b, t) = a$, where $w(x, t)$ is the transverse displacement of the beam. Although the beam has linear dynamics, the nonlinear effect of the impact makes the problem nonlinear. A common way to model this type of impact system is to assume that the velocity of the beam at impact is reversed and reduced by a coefficient of restitution, e, such that $\dot{w}(b, t)_{after} = -e\dot{w}(b, t)_{before}$. To make this type of model work, it is usually assumed that only the first mode of the beam is significant and that the impact is effectively instantaneous. An alternative model is to use the Hertz impact law of the form $f_I = C\delta^{\frac{3}{2}}$, where f_I is the impact force, δ is the axial displacement of the impact stop and C is a constant depending on the material and geometric properties of the impact stop. A description of the experiments and modelling originally carried out for this problem can be found in Moon and Shaw (1983).

In the next example, we consider the collocation method for a vibrating beam with an impact constraint.

Example 6.2 Collocation method for a beam with an impact constraint

Problem Apply the collocation method to the beam with an impact constraint shown in Fig. 6.3. Assume that the impact occurs at $w(b, t) = a$ and a coefficient of restitution rule of the form

[13] This assumes that the system is not numerically stiff, see Press et al. (1994) for more details.

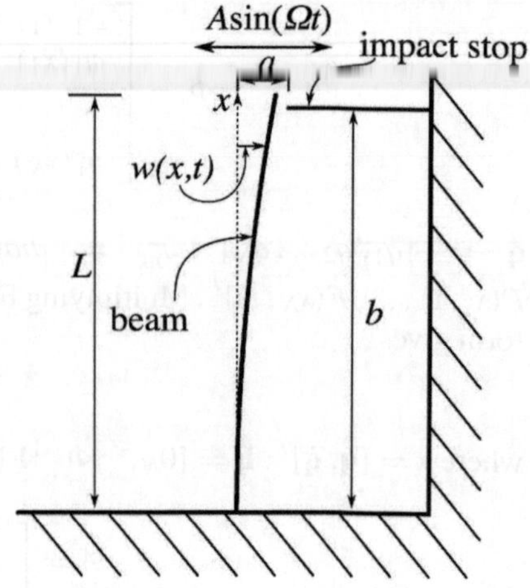

Fig. 6.3 Vertical beam with an impact constraint

$$\left.\begin{aligned}\dot{w}(b,t_+) &= -e\dot{w}(b,t_-)\\ w(b,t_+) &= w(b,t_-)\end{aligned}\right\} \quad \text{when} \quad w(b,t_-) = a \tag{6.27}$$

is applied, where t_- is the time just before impact, t_+ is the time just after impact and $e \in [0, 1]$ is the coefficient of restitution.[14] It is assumed that the velocities are normal to the beam centre line, and that the tangential velocity component at impact is negligible. Equation (6.27) is applied instantaneously such that $t_- = t_+$, and a nonsmooth discontinuity in velocity occurs at impact.

Away from the impact constraint, the beam is assumed to be governed by the Euler-Bernoulli equation, Eq. (6.21).

Solution A key requirement for this problem is that the point of contact, $x = b$, is at a collocation point. At the point the impact occurs, however, for a continuous structural component such as a beam, the velocity is a continuous function of distance along the beam. Therefore, in order to apply the nonsmooth impact condition, Eq. (6.27), at $w = a$, the velocity components for the non-impacting part of the beam $x \neq b$ remain unaffected such that

$$\dot{w}(x \neq b, t_+) = \dot{w}(x \neq b, t_-) \quad \text{when} \quad w(b,t_-) = a, \tag{6.28}$$

applies. The combination of Eqs. (6.27) and (6.28) are essentially a nonsmooth representation of the physical impact process for the beam. In the physical beam system, the contact time will be finite (though small for materials with high stiffness) and the velocity reversal will propagate outwards from the point of impact, a process which is captured with this type of model.

To apply the nonsmooth impact condition, a coefficient of restitution matrix, R, is defined using Eqs. (6.27) and (6.28). Equation (6.27) applies to the collocation

[14] See Stronge (2000) for a detailed description of impact problems and definitions of the coefficient of restitution.

point, $x = b$, where impact occurs, and Eq. (6.28) applies to all other collocation points. For example, for a choice of N collocation points with the impact at point N (i.e. at the beam tip, $b = L$, in Fig. 6.3) the coefficient of restitution matrix is

$$\hat{e} = \begin{bmatrix} 1 & 0 & \dots & 0 \\ 0 & 1 & \dots & 0 \\ \vdots & \vdots & \dots & \vdots \\ 0 & 0 & \dots & -e \end{bmatrix}. \tag{6.29}$$

Now the beam vibration can be simulated using Eq. (6.26) combined with Eq. (6.29) at impact. To do this a time-stepping method is used, and at each time step the condition for the beam having an impact, $w(b, t) > a$, is checked. Once an impact is detected, a root finding method is used to find the exact time at which $w(b, t) = a$. Then the modal velocities are updated according to the modal matrix coefficient of restitution rule, using Eq. (6.27), given by

$$\dot{\mathbf{q}}(t_+) = [\Phi]^{-1}[\hat{e}][\Phi]\dot{\mathbf{q}}(t_-),$$

and time-stepping begins again (see Wagg and Bishop 2002 and Wagg 2003). Figure 6.4 compares results for the Galerkin and collocation methods using the first four beam modes (taken from Wagg 2003). ■

Notice that in Fig. 6.4 the Galerkin and collocation methods give broadly, but not exactly, the same time series signals. At the start of the simulations (time zero) the two signals are identical, but as time evolves, small differences in the simulations grow—see Wagg (2003) for more detail. To apply the Galerkin method to a system with impacts, the same approach is used for root finding and updating the modal velocity matrix. A more general analysis of nonlinear beam vibration is considered next.

6.2 Nonlinear Beam Vibration

First consider an entirely static problem. The example shown in Fig. 6.5a consists of a vertically clamped light cantilever which is subjected to a large deflection due to a vertical load, P. The deflected shape is defined by considering a point a distance s along the beam from the root, which corresponds to a position $x = s$ when undeformed, moving to the point $(x, z) = (x(s), z(s))$ once loaded. Taking moments about this point while considering the section of beam to the tip, shown in Fig. 6.5b, and using Eq. (6.1) gives

$$M = EI\frac{\mathrm{d}\psi}{\mathrm{d}s} = P(z(L) - z(s)),$$

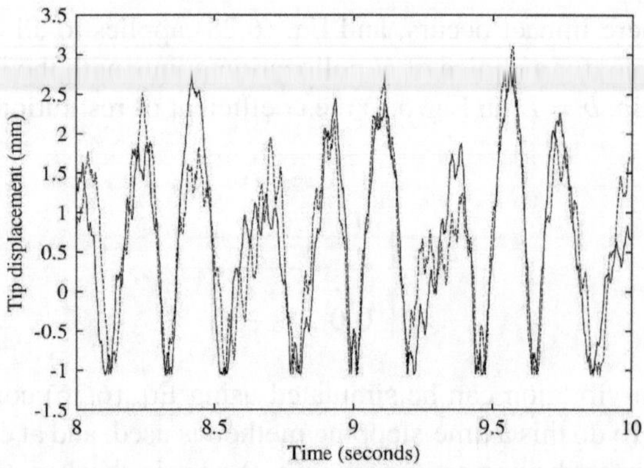

Fig. 6.4 Impacting beam simulation, showing: *solid line*, collocation; *dashed line*, Galerkin. Reproduced with kind permission of Elsevier from Wagg (2003)

where L is the length of the beam, which for large deflections is different from the span of the beam along the x-axis, defined as ℓ. Differentiating with respect to s leads to an expression

$$\frac{\mathrm{d}^2\psi}{\mathrm{d}s^2} + C^2\frac{\mathrm{d}z(s)}{\mathrm{d}s} = 0, \tag{6.30}$$

where $C^2 = P/EI$. For a small element of beam, with length Δs as shown in Fig. 6.5c, a triangle can be assumed which, once the $\Delta s \to 0$ limit is taken, gives

$$\frac{\mathrm{d}z}{\mathrm{d}s} = \sin\psi.$$

So as a result $\mathrm{d}z/\mathrm{d}s$ can be eliminated from Eq. (6.30) to give

$$\frac{\mathrm{d}^2\psi}{\mathrm{d}s^2} + C^2\sin\psi = 0, \tag{6.31}$$

(see for example the derivations in Frish-Fay 1962).

Equation (6.31) is a nonlinear differential equation which defines the curvature of the column with vertical tip load P. There are no vibrations in this model, purely the static deformation which is nonlinear because of large deformations (a major source of nonlinear structural vibration). In fact, Eq. (6.31) is analogous to the equation of motion of the pendulum, $\ddot{\theta} + \omega^2\sin\theta = 0$, discussed in Sect. 1.2.2, Chap. 1 and Example 2.2 in Chap. 2. This dynamical analogy[15] equates length along the beam to time and curvature to pendulum angle. The dynamic analogy can be used to apply techniques from dynamical systems theory to problems of nonlinear elastic structural components, sometimes called *elastica*.

[15] Often known as Kirchhoff's dynamical analogy after G.R. Kirchhoff.

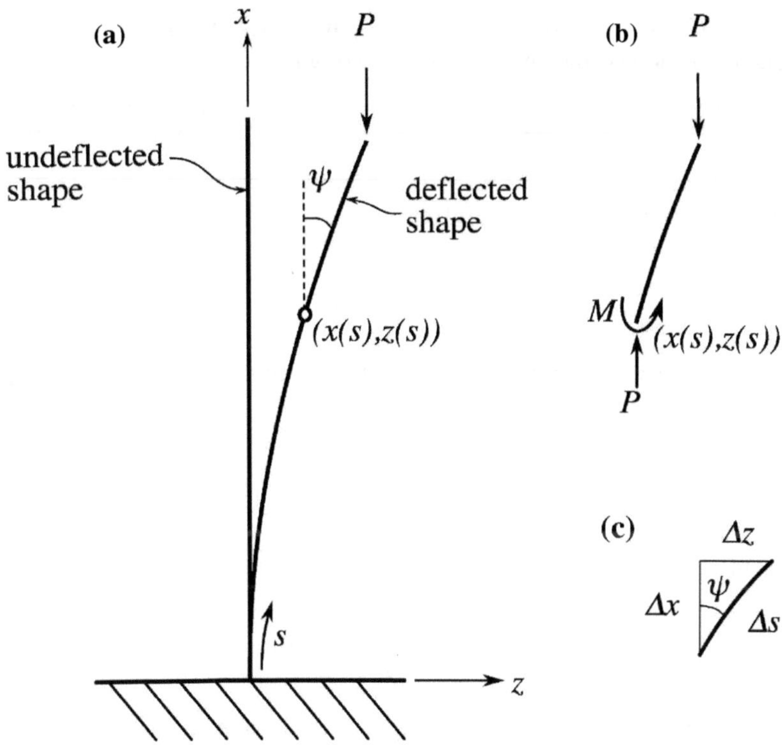

Fig. 6.5 **a** Large deflection of a vertical cantilever, **b** free-body diagram for tip of cantilever to the point $(x, z) = (x(s), z(s))$ and **c** small segment of the beam

In terms of nonlinear vibrations, there are several important physical scenarios which regularly occur. These are discussed in the following subsections.

6.2.1 Large Deflections of Thin Beams

One of the key assumptions used in deriving the Euler-Bernoulli equation for beam vibration (discussed in detail in Sect. 6.1.1) is that the curvature may be approximated as

$$\frac{d\psi}{ds} \approx \frac{d\psi}{dx} \approx \frac{d^2z}{dx^2}.$$

For anything other than small deflections, these simplifying assumptions quickly lead to significant errors in the predicted behaviour of the beam.

To derive a general curvature relationship, consider again the element of beam shown in Fig. 6.5c. The length of the element, Δs, may be approximated as

$$\Delta s \approx (\sqrt{\Delta x^2 + \Delta z^2}) \quad \rightsquigarrow \quad \frac{ds}{dx} = \left[1 + \left(\frac{dz}{dx}\right)^2\right]^{\frac{1}{2}}, \tag{6.32}$$

where, to derive the relationship on the right-hand side, the limit as $\Delta s \to 0$ has been taken. In addition the angle ψ may be written as

$$\psi = \arctan(\frac{\Delta z}{\Delta x}) \quad \rightsquigarrow \quad \psi = \arctan(\frac{\mathrm{d}z}{\mathrm{d}x}).$$

Differentiating ψ with respect to x gives

$$\frac{\mathrm{d}\psi}{\mathrm{d}x} = \frac{1}{1 + (\mathrm{d}z/\mathrm{d}x)^2}(\frac{\mathrm{d}^2 z}{\mathrm{d}x^2}),$$

using the relationship $\mathrm{d}\{\arctan(u)\}/\mathrm{d}u = 1/(1 + u^2)$. Now the curvature can be written as

$$\frac{\mathrm{d}\psi}{\mathrm{d}s} = \frac{\mathrm{d}\psi/\mathrm{d}x}{\mathrm{d}s/\mathrm{d}x} = \frac{\mathrm{d}^2 z/\mathrm{d}x^2}{[1 + (\mathrm{d}z/\mathrm{d}x)^2]^{\frac{3}{2}}}. \tag{6.33}$$

Note that Eq. (6.32) can be written as

$$\frac{\mathrm{d}x}{\mathrm{d}s} = \cos\psi = \frac{1}{[1 + (\mathrm{d}z/\mathrm{d}x)^2]^{\frac{1}{2}}}. \tag{6.34}$$

The analysis so far has taken no account of the fact that, as a beam bends, distortions will occur, especially if the deflections are large. Another set of assumptions is required. The first point to note is that the analysis presented is for a beam vibrating in the x, z plane. We assume that, as the beam bends, any cross-section of the beam normal to the s axis remains plane, i.e. plane sections remain plane. For thicker beams this assumption breaks down, where distorting effects, particularly due to shear, occur during bending. To proceed it is usual to assume that the beam section is sufficiently thin (or slender) such that issues of distortion can be neglected. Models for thicker beams are captured by *Timoshenko beam theory*[16] which includes corrections for both shear deformation and rotary inertia (but not necessarily large deflections). The reason for neglecting these features here is that flexible structural elements, by their nature, tend to have low stiffness and therefore tend to be slender.

6.2.2 Nonlinear Beam Equations with Axial Loading

Now consider the dynamics of a beam with axial loading. A small element of bent beam with large deflections and axial loading is shown in Fig. 6.6, where vertical load N, horizontal load V (which if the deflection were small would correspond to axial load and shear forces respectively) and moments M are shown, acting at the bottom

[16] After S.P. Timoshenko (1878–1972).

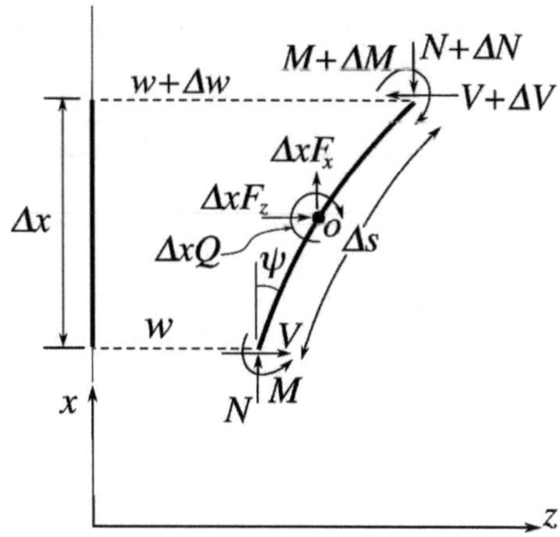

Fig. 6.6 Element of a bending beam which spans Δx along the x-axis with applied horizontal and vertical forcing functions and an applied moment function

of the element. The changes in the vertical force, horizontal force and moment over the length of the element, which spans Δx along the x-axis, are represented by the small increments ΔN, ΔV and ΔM respectively. In addition, the beam element is subject to force functions F_x and F_z in the x and z directions respectively and a moment function Q. These force and moment functions can include applied forces and inertial terms. They are defined as forces (and moment) per unit length along the x-axis, so the forces on the element are $F_x \Delta x$ and $F_z \Delta x$ in the x direction and z direction respectively and the moment $Q\Delta x$, as shown in Fig. 6.6. Note that the forces and moment should be written in the form $(F_x + \Delta F_x/2)\Delta x$ where F_x is the force function at the lower end of the element and ΔF_x represents the change in force function over the length of the element in the x-axis, however when the limit as $\Delta x \to 0$ are taken this additional $\Delta F_x/2$ term becomes negligible.

In the analysis it is assumed that, while the deflections are not small, they are not very large. As a result the dynamics in the x-direction will be ignored in this analysis. The deflection of the beam is therefore considered in terms of just the z deflection, w, and the angle ψ. Taking force equilibrium vertically and horizontally gives

$$\uparrow_x \qquad F_x \Delta x = \Delta N \quad \rightsquigarrow \quad F_x = \frac{\partial N}{\partial x}, \tag{6.35}$$

$$\rightarrow_z \qquad F_z \Delta x = \Delta V \quad \rightsquigarrow \quad F_z = \frac{\partial V}{\partial x} \tag{6.36}$$

where, for the right-hand equations, the limit as $\Delta x \to 0$ has been taken, which results in partial derivatives as N and V could be functions of both position and time.[17] Now taking moments about point O in Fig. 6.6 gives

[17] Note that deflection w now replaces z, which was used earlier when statics were being considered.

$$\circlearrowleft_O \quad (M + \Delta M - M) + (N + \Delta N + N)\frac{\Delta w}{2} - (V + \Delta V + V)\frac{\Delta x}{2} + Q\Delta x = 0$$
$$\rightsquigarrow \frac{\partial M}{\partial x} = V - N\frac{\partial w}{\partial x} - Q, \tag{6.37}$$

where the second-order Δ^2 terms become negligibly small as $\Delta x \to 0$.

Now consider the case where there is no external loading applied along the length of the beam except a compressive tip and root load, P, in the x-direction. The inertial forces in the axial beam direction are ignored in this analysis, and as a result the x-axis force function becomes

$$F_x = P\delta(x - 0) - P\delta(x - L). \tag{6.38}$$

This expression utilizes the Dirac-Delta function to indicate that the forcing function is zero except at the root $x = 0$ and the tip $x = L$.[18] Using Eq. (6.35) this gives

$$N = \int_0^x \{P\delta(x - 0) - P\delta(x - L)\}\mathrm{d}x = P, \tag{6.39}$$

where the integral has been evaluated for $0 < x < L$, i.e. within the beam. This expression indicates that, in this case, N is a constant. Using this fact and combining Eqs. (6.36) and (6.37) we find an equation of motion for the transverse beam deflection where N has been eliminated, such that

$$\frac{\partial^2 M}{\partial x^2} = F_z - P\frac{\partial^2 w}{\partial x^2} - \frac{\partial Q}{\partial x}. \tag{6.40}$$

As there are no external moments or horizontal forces applied to the beam, F_z and Q consist of just the horizontal and rotational inertia of the element (recall that in general F_z and Q include both inertial and external loading). Since the length of the beam element is Δs these inertial loading terms may be written as

$$F_z\Delta x = -\rho\hat{A}\Delta s\frac{\partial^2 w}{\partial t^2}, \qquad Q\Delta x = -\rho I\Delta s\frac{\partial^2 \psi}{\partial t^2}, \tag{6.41}$$

where $\rho(\mathrm{kg/m^3})$ is the density, $\hat{A}(\mathrm{m^2})$ is the cross-sectional area and I is the second moment of area. Substituting these relationships (after taking the limit as $\Delta x \to 0$) into Eq. (6.40) gives

[18] Note that the Dirac-Delta function has arbitrary units, in this case they are length^{-1}.

$$\frac{\partial^2 M}{\partial x^2} + \rho \hat{A} \frac{\partial s}{\partial x} \frac{\partial^2 w}{\partial t^2} - \rho I \frac{\partial}{\partial x} \left(\frac{\partial s}{\partial x} \frac{\partial^2 \psi}{\partial t^2} \right) + P \frac{\partial^2 w}{\partial x^2} = 0.$$

Finally the bending moment to curvature relationship, Eq. (6.1), may be substituted to give

$$EI \frac{\partial^2}{\partial x^2} \left(\frac{\partial \psi}{\partial s} \right) + \rho \hat{A} \frac{\partial s}{\partial x} \frac{\partial^2 w}{\partial t^2} - \rho I \frac{\partial}{\partial x} \left(\frac{\partial s}{\partial x} \frac{\partial^2 \psi}{\partial t^2} \right) + P \frac{\partial^2 w}{\partial x^2} = 0. \tag{6.42}$$

Note that the full derivative in Eq. (6.1) has been converted into a partial derivative, $M = EI\partial\psi/\partial s$, as ψ is now a function of location and time. Equation (6.42) is the governing equation of motion for an axially-loaded beam with large transverse deflections subject to a constant axial loading force P. Note that Eq. (6.42) contains no damping or additional dynamic forcing terms.

In addition to Eq. (6.42), relationships for the deflection and angle with respect to the element length have already been derived in Eqs. (6.33) and (6.34). For this analysis, the expressions are modified so that the deflection in the z direction is now w, and derivatives become partial as the deflection is a function of time as well as position along the beam. This gives

$$\frac{\partial \psi}{\partial s} = \frac{\partial^2 w/\partial x^2}{\left[1 + (\partial w/\partial x)^2\right]^{\frac{3}{2}}}, \tag{6.43}$$

$$\frac{\partial x}{\partial s} = \cos\psi = \frac{1}{\left[1 + (\partial w/\partial x)^2\right]^{\frac{1}{2}}}.$$

Depending on the system being considered, it is sometimes convenient to write the equation of motion as a partial differential equation in t and vertical position x and at other times in terms of t and length along the beam s. In the second case it is helpful to derive two further geometric equations. Referring to Fig. 6.5 while recalling that the deflection is now w rather than y, we have

$$\frac{\Delta x}{\Delta s} = \cos\psi = \frac{\sqrt{\Delta s^2 - \Delta w^2}}{\Delta s} \quad \rightsquigarrow \quad \frac{\partial x}{\partial s} = \cos\psi = \sqrt{1 - \left(\frac{\partial w}{\partial s}\right)^2}, \tag{6.44}$$

$$\sin\psi = \frac{\Delta w}{\Delta s} \quad \rightsquigarrow \quad \frac{\partial \psi}{\partial s} \cos\psi = \frac{\partial^2 w}{\partial s^2}, \tag{6.45}$$

where in deriving the right-hand expression in Eq. (6.45) the derivative with respect to s has been taken in addition to the limit as $\Delta s \to 0$.

From these expressions, two cases commonly occur in applications: (i) small amplitude vibrations with axial load, and (ii) large amplitude vibration without axial load. Examples of both cases are considered now.

Example 6.3 An axially-loaded beam with small deflections

Problem Assuming small deflections, derive the governing equations of motion for a slender axially-loaded beam from Eq. (6.42) for the case where the beam is subject to external forcing per unit length $F(x, t)$ acting in the z direction.

Solution For small amplitude vibrations it is assumed that $\Delta s \approx \Delta x$, such that $\partial x/\partial s \approx 1$. In addition ψ is small, so $\sin\psi \approx \psi$. This leads to the approximation that $\psi = \sin\psi = \Delta w/\Delta s = \Delta w/\Delta x$ and therefore in the limit $\psi = \partial w/\partial x$. Substituting these expressions into Eq. (6.42), and setting the rotational inertia term to zero (as the beam is slender and the deflections are small), i.e. $\partial(\partial^2\psi/\partial t^2)/\partial x \approx 0$, gives

$$EI\frac{\partial^4 w}{\partial x^4} + \rho\hat{A}\frac{\partial^2 w}{\partial t^2} + P\frac{\partial^2 w}{\partial x^2} = 0 \qquad (6.46)$$

when no external loading is present. To include the external forcing, Eq. (6.41) is modified to

$$F_z\Delta x = F(x, t)\Delta x - \rho\hat{A}\Delta s\frac{\partial^2 w}{\partial t^2},$$

which results in the modification to Eq. (6.46) to give

$$EI\frac{\partial^4 w}{\partial x^4} + \rho\hat{A}\frac{\partial^2 w}{\partial t^2} + P\frac{\partial^2 w}{\partial x^2} = F(x, t).$$

This equation is the linear (undamped) Euler-Bernoulli equation for beam vibration with an axial compressive load P and transverse forcing $F(x, t)$.[19] ■

Note that, even assuming small deflections, the axial force, P, can lead to nonlinear behaviour, an example of which will be discussed in Sect. 6.2.3.

In the next example a beam subject to large deflections but no axial load is considered. The resulting equation of motion will be nonlinear and the vertical motion of the tip cannot be assumed to be small. For these reasons it is convenient to derive the equation of motion in terms of position, s, along the beam (and in terms of t) rather than the vertical height x. This is because to transform the partial differential equation of motion into a series of modal equations, a technique such as the Galerkin method is used in which, after making the substitution $w = \sum_i \phi_i q_i$, the equation is multiplied by a mode-shape ϕ_n and then integrated over the length of the beam. If the partial differential equation is in terms of s, this integral is with respect to s and the limits are from 0 to L (the length of the beam). However, if x is used, the limits are from 0 to ℓ where ℓ is the span of the beam in the x direction and, if the deflection is not small, is both unknown and dependent on the amplitude of deflection. Note that when the deflection is small $\ell \approx L$.

[19] This type of equation can be derived in a variety of ways, see for example Virgin (2007), Timoshenko et al. (1974) and Clough and Penzien (1993).

Example 6.4 A cantilever beam with large deflections and no axial load

Problem Assuming large deflections, derive the approximate governing equations of motion for the case where there is no tip loading (i.e. $P = 0$) and where the rotational inertia term can be ignored.

Solution Using Eq. (6.42) and setting the vertical tip load and rotational inertial terms to zero gives

$$EI\frac{\partial^2}{\partial x^2}\left(\frac{\partial \psi}{\partial s}\right) + \rho\hat{A}\frac{\partial s}{\partial x}\frac{\partial^2 w}{\partial t^2} = 0. \tag{6.47}$$

By using Eq. (6.44), and noting that

$$\frac{\partial}{\partial x} = \frac{\partial s}{\partial x}\frac{\partial}{\partial s},$$

we can eliminate x from Eq. (6.47) to give

$$EI\frac{1}{\cos\psi}\frac{\partial}{\partial s}\left(\frac{1}{\cos\psi}\frac{\partial^2 \psi}{\partial s^2}\right) + \rho\hat{A}\frac{1}{\cos\psi}\frac{\partial^2 w}{\partial t^2} = 0.$$

Now ψ can be eliminated using Eq. (6.45) to give

$$EI\frac{\partial}{\partial s}\left(\frac{1}{\cos\psi}\frac{\partial}{\partial s}\left(\frac{1}{\cos\psi}\frac{\partial^2 w}{\partial s^2}\right)\right) + \rho\hat{A}\frac{\partial^2 w}{\partial t^2} = 0, \tag{6.48}$$

where $\cos\psi = \sqrt{1-(\partial w/\partial s)^2}$ from Eq. (6.44). This equation is an approximate (undamped, unforced) governing equation for the large transverse deflection of a beam. ■

Equation (6.48) is the two-dimensional nonlinear equivalent to the Euler-Bernoulli beam equation for large transverse deflections with no axial load (or damping or forcing). This equation cannot directly be reduced to a set of ordinary differential equations by using separation of variables (i.e. Galerkin or similar). However, for the case where the deflections are large, but not very large, a two term binomial series expansion $(1+x)^n \approx 1 + nx + \cdots$ can be used to simplify the $1/\cos\psi$ terms and the resulting equation can then be reduced to coupled ordinary differential equations using the Galerkin method. The binomial expansion is applied to $1/\cos\psi = 1/\sqrt{1-(\partial w/\partial s)^2}$ to give the two term approximation $1/\cos\psi \approx 1 + (1/2)(\partial w/\partial s)^2$.

Physically this can be seen as making a less restrictive approximation for the angle of deflection. For small amplitude vibrations it is assumed that $\cos\psi \approx 1$ and hence $1/\cos\psi = 1$, which requires that $(1/2)(\partial w/\partial s)^2 \ll 1$. Whereas, in using $1/\cos\psi \approx 1 + (1/2)(\partial w/\partial s)^2$ the assumption that $(3/8)(\partial w/\partial s)^4 \ll 1$ (the third term in the binomial series is much less than unity) is made, which is a less restrictive assumption. Approximately, $\partial w/\partial s$ may be thought of as the angle of deflection (strictly $\partial w/\partial s = \sin\psi$) and so including the second binomial term in the binomial expansion of $1/\cos\psi$ relaxes the restriction on the angle of deflection. An example using the binomial expansion is considered next.

Example 6.5 Galerkin decomposition for a cantilever beam with large deflections

Problem Find the modal decomposition of Eq. (6.48), using a binomial series expansion and the Galerkin method, to give a series of approximate coupled modal equations for modelling large deflections in beams.

Solution First, the two-term binomial expansion $1/\cos\psi \approx 1 + (1/2)(\partial w/\partial s)^2$ is used to simplify Eq. (6.48) to give

$$EI\frac{\partial}{\partial s}\left(\left[1+\frac{1}{2}\left(\frac{\partial w}{\partial s}\right)^2\right]\frac{\partial}{\partial s}\left(\left[1+\frac{1}{2}\left(\frac{\partial w}{\partial s}\right)^2\right]\frac{\partial^2 w}{\partial s^2}\right)\right)+\rho\hat{A}\frac{\partial^2 w}{\partial t^2}=0.$$

It is expected that $\left(\frac{1}{2}\partial w/\partial s\right)^2$ is small because, when making the binomial expansion approximation, it has been assumed that $(3/8)(\partial w/\partial s)^4 \ll 1$. Therefore, an ε symbol is introduced to the $\frac{1}{2}(\partial w/\partial s)^2$ terms as a book-keeping aid to indicate that these terms are small. This allows simplifications later when ε^2 terms are ignored as these have been generated by the multiplication of two small terms. Adding in the ε terms and expanding the inner partial derivative with respect to s gives

$$EI\frac{\partial}{\partial s}\left(\left[1+\frac{\varepsilon}{2}\left(\frac{\partial w}{\partial s}\right)^2\right]\left[\frac{\partial^3 w}{\partial s^3}+\frac{\varepsilon}{2}\frac{\partial^3 w}{\partial s^3}\left(\frac{\partial w}{\partial s}\right)^2+\varepsilon\left(\frac{\partial^2 w}{\partial s^2}\right)^2\frac{\partial w}{\partial s}\right]\right)+\rho\hat{A}\frac{\partial^2 w}{\partial t^2}=0.$$

Now the [•][•] term can be expanded out, while ignoring ε^2 terms as small, giving

$$EI\frac{\partial}{\partial s}\left(\frac{\partial^3 w}{\partial s^3}+\varepsilon\frac{\partial^3 w}{\partial s^3}\left(\frac{\partial w}{\partial s}\right)^2+\varepsilon\left(\frac{\partial^2 w}{\partial s^2}\right)^2\frac{\partial w}{\partial s}\right)+\rho\hat{A}\frac{\partial^2 w}{\partial t^2}=0.$$

Finally, the partial derivative with respect to s can be evaluated to give the approximate partial differential equation

$$EI\frac{\partial^4 w}{\partial s^4}+\varepsilon EI\left[\frac{\partial^4 w}{\partial s^4}\left(\frac{\partial w}{\partial s}\right)^2+4\frac{\partial^3 w}{\partial s^3}\frac{\partial^2 w}{\partial s^2}\frac{\partial w}{\partial s}+\left(\frac{\partial^2 w}{\partial s^2}\right)^3\right]+\rho\hat{A}\frac{\partial^2 w}{\partial t^2}=0. \tag{6.49}$$

It can be seen from Eq. (6.49) that if the ε term is ignored the equation of motion reduces to the unforced Euler-Bernoulli equation, Eq. (6.21), with forcing $F = 0$ and no damping. As the Euler-Bernoulli equation is a linear approximation to Eq. (6.49), it will be used to define the mode-shapes used in the Galerkin decomposition. The mode-shapes for the Euler-Bernoulli equation can be found using the separation of variables technique, discussed in Sect. 5.2, along with the boundary conditions. The orthogonality conditions for these mode-shapes have already been derived and are given in Eq. (6.9) and repeated here

$$\int_0^L \frac{\mathrm{d}^4\phi_j}{\mathrm{d}x^4}\phi_i \mathrm{d}x = 0 \quad \text{and} \quad \int_0^L \phi_j\phi_i \mathrm{d}x = 0 \quad \text{for} \quad i \neq j. \tag{6.50}$$

Making the Galerkin substitution $w(s,t) = \sum_{j=1}^{N} \phi_j(s) q_j(t)$, to separate the space and time dependence of the transverse displacement $w(s,t)$, into Eq. (6.49) and dropping the ε book-keeping aids, gives

$$\begin{aligned}
&\rho\hat{A}\sum_{j=1}^{N}\phi_j(s)\ddot{q}_j(t) + EI\sum_{j=1}^{N}\phi_j''''(s)q_j(t) \\
&+ EI\Bigg[\sum_{j=1}^{N}\phi_j''''(s)q_j(t)\sum_{k=1}^{N}\phi_k'(s)q_k(t)\sum_{l=1}^{N}\phi_l'(s)q_l(t) \\
&\qquad + 4\sum_{j=1}^{N}\phi_j'''(s)q_j(t)\sum_{k=1}^{N}\phi_k''(s)q_k(t)\sum_{l=1}^{N}\phi_l'(s)q_l(t) \\
&\qquad + \sum_{j=1}^{N}\phi_j''(s)q_j(t)\sum_{k=1}^{N}\phi_k''(s)q_k(t)\sum_{l=1}^{N}\phi_l''(s)q_l(t)\Bigg] = 0,
\end{aligned} \tag{6.51}$$

where $\phi_j(x)$ and $q_j(t)$ represent the jth normal mode-shape and modal coordinate respectively and $\bullet'$ and $\dot{\bullet}$ represent the derivative with respect to position along the beam, s, and time, t, respectively. It is important to note that when applying the Galerkin method to nonlinear terms, each approximation of w needs to have a separate set of indices. This is to ensure that all possible *modal cross-coupling terms* are captured in the model.

The next step is to apply the orthogonality conditions. To do this, Eq. (6.51) is multiplied by an arbitrary mode-shape, $\phi_n(s)$, and integrated over the length of the beam: $s = 0$ to $s = L$. The resulting governing equation can be written as

$$\rho\hat{A}\sum_{j=1}^{N}\int_0^L \phi_j\phi_n \mathrm{d}s\,\ddot{q}_j + EI\sum_{j=1}^{N}\int_0^L \phi_j''''\phi_n \mathrm{d}s\, q_j$$

$$+EI\Bigg[\sum_{j=1}^{N}\sum_{k=1}^{N}\sum_{l=1}^{N}\int_0^L \phi_j''''\phi_k'\phi_l'\phi_n \mathrm{d}s\, q_j q_k q_l$$

$$+4\sum_{j=1}^{N}\sum_{k=1}^{N}\sum_{l=1}^{N}\int_0^L \phi_j'''\phi_k''\phi_l'\phi_n \mathrm{d}s\, q_j q_k q_l$$

$$+\sum_{j=1}^{N}\sum_{k=1}^{N}\sum_{l=1}^{N}\int_0^L \phi_j''\phi_k''\phi_l''\phi_n \mathrm{d}s\, q_j q_k q_l\Bigg] = 0.$$

Now the orthogonality conditions, Eq. (6.50), can be applied to give the equation for the nth mode

$$\rho\hat{A}\int_0^L \phi_n\phi_n \mathrm{d}s\,\ddot{q}_n + EI\int_0^L \phi_n''''\phi_n \mathrm{d}s\, q_n$$

$$+EI\Bigg[\sum_{j=1}^{N}\sum_{k=1}^{N}\sum_{l=1}^{N}\int_0^L \phi_j''''\phi_k'\phi_l'\phi_n \mathrm{d}s\, q_j q_k q_l$$

$$+4\sum_{j=1}^{N}\sum_{k=1}^{N}\sum_{l=1}^{N}\int_0^L \phi_j'''\phi_k''\phi_l'\phi_n \mathrm{d}s\, q_j q_k q_l$$

$$+\sum_{j=1}^{N}\sum_{k=1}^{N}\sum_{l=1}^{N}\int_0^L \phi_j''\phi_k''\phi_l''\phi_n \mathrm{d}s\, q_j q_k q_l\Bigg] = 0.$$

This equation can be simplified slightly by using the mode-shape integrals for a cantilever $\int_0^L \phi_i\phi_i \mathrm{d}s = L$ and $\int_0^L \phi_i''''\phi_i \mathrm{d}s = \xi_i^4/L^3$, as given in Eq. (6.11). Using this, and dividing through by $\rho\hat{A}L$, we find the following equation for the nth mode

$$\ddot{q}_n(t) + \omega_{nn}^2 q_n(t) + \frac{EI}{\rho\hat{A}L}\sum_{j=1}^{N}\sum_{k=1}^{N}\sum_{l=1}^{N}\mu_{jkln} q_j(t) q_k(t) q_l(t) = 0, \qquad (6.52)$$

where ω_{nn} is the natural frequency of the nth mode for the linearized system, $\omega_{ni}^2 = EI\xi_i^4/(L^4\rho A)$ and

$$\mu_{jkln} = \int_0^L (\phi_j''''(s)\phi_k'(s)\phi_l'(s) + 4\phi_j'''(s)\phi_k''(s)\phi_l'(s) + \phi_j''(s)\phi_k''(s)\phi_l''(s))\phi_n(s)\mathrm{d}s, \tag{6.53}$$

which can be evaluated for the cantilever mode-shapes, to give a series of coefficients for the nonlinear cubic terms. ■

Typically, Eq. (6.52) leads to cubic terms of the form $q_j q_k q_l$, etc., depending on the nonlinear modal coupling coefficients of the beam. So, it can be seen that these ordinary differential equations are similar to the Duffing-type oscillators with modal coupling.

Example 6.6 Backbone curves for (moderately) large deflections of cantilever beam

Problem Using the normal form analysis developed in Chap. 5 find approximate expressions for the backbone curves $S1$ and $S2$ of a two mode model of a 0.5m long cantilever beam when the beam modal deflections are in the region of 0.01m (which we denote as moderately large). Assume the system is unforced and undamped. The modal equations are given by Eq. 6.52 and the beam parameters are $E = 70\,\text{Gpa}$, $\rho = 2,700\,\text{kg/m}^3$, $I = 2.44 \times 10^{-13}\,\text{m}^4$, $\hat{A} = 1.24 \times 10^{-5}\,\text{m}^2$ and $L = 0.5\,\text{m}$. It can be assumed that the modal displacements for both modes are of a similar amplitude.

Solution First to find a two mode model via Eq. 6.52 the nonlinear coefficients need to be determined. To do this, first compute the natural frequencies of the system using $\omega_{ni}^2 = EI\xi_i^4/(L^4\rho A)$ to give $\omega_{n1} = 10\,\text{rad/s}$ and $\omega_{n2} = 63\,\text{rad/s}$. Then, using Eq. 6.53, the values for the nonlinear coefficients can be computed. The results are shown in Tabie 6.2.

In Table 6.2, the constant $\hat{M} = EI/\rho\hat{A}L$ is defined from the material properties of the beam. Recall from Chap. 4 that the size of the nonlinear term is relative to the size of the displacements. So, in the third column of Table 6.2, $\hat{M}\mu_{jkln} \times (0.01)^3$, the size of the term for an arbitrary cubic modal displacement is shown. These values can be compared to the linear terms $\omega_{n1}^2 q_1$ and $\omega_{n2}^2 q_2$ which for $q_n = 0.01$ are 1 and 40 respectively.

In addition, it can be seen from Table 6.2 that there are multiple coefficients for similar modal terms. Therefore a two mode (undamped and unforced) nonlinear model is given by

$$\begin{aligned} \ddot{q}_1 + \omega_{n1}^2 q_1 + \alpha_{11} q_1^3 + \alpha_{12} q_1^2 q_2 + \alpha_{13} q_2^2 q_1 + \alpha_{14} q_2^3 = 0, \\ \ddot{q}_2 + \omega_{n2}^2 q_2 + \alpha_{21} q_1^3 + \alpha_{22} q_1^2 q_2 + \alpha_{23} q_2^2 q_1 + \alpha_{24} q_2^3 = 0, \end{aligned} \tag{6.54}$$

with the nonlinear coefficient values for the first and second modes given in Tables (6.3) and (6.4) respectively.

Table 6.2 Nonlinear coefficients for Example 6.6 computed using Eq. 6.53

$jkln$	$q_j q_k q_l$	μ_{jkln}	$\hat{\mu}_{jkln} \times (0.01)^3$
1111	q_1^3	1294.161207	0.001320
1121	$q_1^2 q_2$	−9196.125423	−0.009383
1211	$q_1^2 q_2$	14490.630939	0.014785
2111	$q_1^2 q_2$	−15119.961495	−0.015427
1221	$q_2^2 q_1$	9576.273869	0.009771
2121	$q_2^2 q_1$	93850.527350	0.095757
2211	$q_2^2 q_1$	−23902.459786	−0.024388
2221	q_2^3	−69697.017530	−0.071113
1112	q_1^3	−3274.288015	−0.003341
1122	$q_1^2 q_2$	−3513.856126	−0.003585
1212	$q_1^2 q_2$	32652.828913	0.033316
2112	$q_1^2 q_2$	50371.056296	0.051394
1222	$q_2^2 q_1$	10359.835360	0.010570
2122	$q_2^2 q_1$	−189720.533701	−0.193574
2212	$q_2^2 q_1$	−29662.202516	−0.030265
2222	q_2^3	429407.589715	0.438129

Table 6.3 Coefficients for $n = 1$ in Example 6.6

Term	Coefficient	Assigned label	Value
q_1^3	μ_{1111}	α_{11}	1294
$q_1^2 q_2$	$\mu_{1121} + \mu_{1211} + \mu_{2111}$	α_{12}	−19825
$q_2^2 q_1$	$\mu_{1221} + \mu_{2121} + \mu_{2211}$	α_{13}	79524
q_2^3	μ_{2221}	α_{14}	−69697

Table 6.4 Coefficients for $n = 2$ in Example 6.6

Term	Coefficient	Assigned label	Value
q_1^3	μ_{1111}	α_{21}	−3274
$q_1^2 q_2$	$\mu_{1122} + \mu_{1212} + \mu_{2112}$	α_{22}	79510
$q_2^2 q_1$	$\mu_{1222} + \mu_{2122} + \mu_{2212}$	α_{23}	−209020
q_2^3	μ_{2222}	α_{24}	429408

Now using the analysis from Chap. 4, Sect. 4.5

$$\ddot{\mathbf{q}} + \boldsymbol{\Lambda}\mathbf{q} + \mathbf{N}_q = 0 \quad \text{where:} \tag{6.55}$$

$$\boldsymbol{\Lambda} = \begin{bmatrix} \omega_{n1}^2 & 0 \\ 0 & \omega_{n2}^2 \end{bmatrix}, \quad \text{and} \tag{6.56}$$

$$\mathbf{N}_q = \mu \begin{pmatrix} \alpha_{11} q_1^3 + \alpha_{12} q_1^2 q_2 + \alpha_{13} q_2^2 q_1 + \alpha_{14} q_2^3 \\ \alpha_{21} q_1^3 + \alpha_{22} q_1^2 q_2 + \alpha_{23} q_2^2 q_1 + \alpha_{24} q_2^3 \end{pmatrix}. \tag{6.57}$$

Here $\mathbf{\Lambda}$ is a diagonal matrix of the squares of the linearised natural frequencies ω_{n1} and ω_{n2} defined by Eq. (6.68). Substituting $q_n \rightarrow u_n = u_{pn} + u_{mn}$ gives the functional form of the $\mathbf{u}^*$ vector and coefficient vector $\mathbf{n}_u$ as

$$\mathbf{u}^* = \begin{bmatrix} u_{1p}^3 \\ u_{1p}^2 u_{1m} \\ u_{1p} u_{1m}^2 \\ u_{1m}^3 \\ u_{1p} u_{2p}^2 \\ u_{1p} u_{2p} u_{2m} \\ u_{1p} u_{2m}^2 \\ u_{1m} u_{2p}^2 \\ u_{1m} u_{2p} u_{2m} \\ u_{1m} u_{2m}^2 \\ u_{1p}^2 u_{2p} \\ u_{1p} u_{1m} u_{2p} \\ u_{1m}^2 u_{2p} \\ u_{1p}^2 u_{2m} \\ u_{1p} u_{1m} u_{2m} \\ u_{1m}^2 u_{2m} \\ u_{2p}^3 \\ u_{2p}^2 u_{2m} \\ u_{2p} u_{2m}^2 \\ u_{2m}^3 \end{bmatrix}, \quad \mathbf{n}_u^T = \mu \begin{bmatrix} \alpha_{11} & \alpha_{21} \\ 3\alpha_{11} & 3\alpha_{21} \\ 3\alpha_{11} & 3\alpha_{21} \\ \alpha_{11} & \alpha_{21} \\ \alpha_{13} & \alpha_{23} \\ 2\alpha_{13} & 2\alpha_{23} \\ \alpha_{13} & \alpha_{23} \\ \alpha_{13} & \alpha_{23} \\ 2\alpha_{13} & 2\alpha_{23} \\ \alpha_{13} & \alpha_{23} \\ \alpha_{12} & \alpha_{22} \\ 2\alpha_{12} & 2\alpha_{22} \\ \alpha_{12} & \alpha_{22} \\ \alpha_{12} & \alpha_{22} \\ 2\alpha_{12} & 2\alpha_{22} \\ \alpha_{12} & \alpha_{22} \\ \alpha_{14} & \alpha_{24} \\ 3\alpha_{14} & 3\alpha_{24} \\ 3\alpha_{14} & 3\alpha_{24} \\ \alpha_{14} & \alpha_{24} \end{bmatrix}. \tag{6.58}$$

Now, values for $\beta_{k,l}$ can be computed using Eq. 4.56 from Chap. 4, as shown in Table 6.5. Using these expressions, and assuming for $\omega_{r1} \approx \omega_{n1}$ and $\omega_{r2} \approx \omega_{n2}$, the $\boldsymbol{\beta}$ matrix can be calculated to be

$$\beta^T = \begin{bmatrix} 800 & -3069 \\ 0 & -3869 \\ 0 & -3869 \\ 800 & -3069 \\ 18369 & 14527 \\ 0 & -3869 \\ 13356 & 9487 \\ 13356 & 9487 \\ 0 & -3869 \\ 18369 & 14527 \\ 6789 & 2920 \\ 3869 & 0 \\ 1749 & -2120 \\ 1749 & -2120 \\ 3869 & 0 \\ 6789 & 2920 \\ 35621 & 31752 \\ 3869 & 0 \\ 3869 & 0 \\ 35621 & 31752 \end{bmatrix}.$$

From $\boldsymbol{\beta}$ it can be seen that there are eight unconditionally-resonant terms (i.e. zeros), namely elements [1, 2], [1, 3], [1, 6], [1, 9], [2, 12], [2, 15], [2, 18] and [2, 19]. For all resonant terms we set the corresponding terms in $\mathbf{n}_u$ as equal to those in $\mathbf{n}$, the resulting dynamic equations are

$$\begin{aligned} &\ddot{u}_1 + \omega_{n1}^2 u_1 + 3\alpha_{11}[u_{1p}^2 u_{1m} + u_{1p} u_{1m}^2] + 2\alpha_{13}[u_{1p} u_{2p} u_{2m} + u_{1m} u_{2p} u_{2m}] = 0, \\ &\ddot{u}_2 + \omega_{n2}^2 u_2 + 3\alpha_{24}\left[u_{2p} u_{2m}^2 + u_{2p}^2 u_{2m}\right] + 2\alpha_{22}\left[u_{1p} u_{1m} u_{2m} + u_{1p} u_{1m} u_{2p}\right] = 0. \end{aligned} \tag{6.59}$$

Making the substitutions $u_{ip} = (U_i/2)\mathrm{e}^{j(\omega_{ri}t - \phi_i)}$ and $u_{im} = (U_i/2)\mathrm{e}^{-j(\omega_{ri}t - \phi_i)}$, we can write Eq. (6.59) in the form

$$\chi_i \mathrm{e}^{j\omega_{ri}t} + \tilde{\chi}_i \mathrm{e}^{-j\omega_{ri}t} = 0, \tag{6.60}$$

where χ_i and $\tilde{\chi}_i$ are complex conjugates. Inspecting the χ_i components of Eq. (6.59) results in the time-independent equations

$$\left[-\omega_{r1}^2 + \omega_{n1}^2 + 3\alpha_{11} U_1^2 + 2\alpha_{13} U_2^2\right] U_1 = 0, \tag{6.61a}$$

$$\left[-\omega_{r2}^2 + \omega_{n2}^2 + 3\alpha_{24} U_2^2 + 2\alpha_{22} U_1^2\right] U_2 = 0. \tag{6.61b}$$

Two solutions to Eq. (6.61a, 6.61b) can be found be setting U_1 and then U_2 to zero in turn. These single mode responses are

$$S1: \quad U_2 = 0, U_1 \neq 0, \qquad \omega_{r1}^2 = \omega_{n1}^2 + 3\alpha_{11}U_1^2, \tag{6.62}$$

$$S2: \quad U_1 = 0, U_2 \neq 0, \qquad \omega_{r2}^2 = \omega_{n2}^2 + 3\alpha_{24}U_2^2. \tag{6.63}$$

In addition it is possible to have solutions in which both modes are present by setting both the expressions in the square brackets in Eq. (6.61a, 6.61b) to zero. However, these are not resonant relationships in this case. See Chap. 5 for more details. ■

Figure 6.7 shows the backbone curves defined by Eqs. (6.62) and (6.63), from Example 6.8. Modal displacement amplitudes are shown up to 30 mm which is three times larger than the assumed 10 mm considered in Example 6.8. Despite this is can be seen that the $S1$ backbone curve is hardly distorted from straight, meaning that there is almost no change in the first response frequency due to moderately large deflections. This is not true of the $S2$ curve, where significant distortions occur, even for relatively small modal displacements. This is explained because the α_{11} is very small. In fact the α_{24} coefficient is approximately 330 times larger than the α_{11} coefficient in this example which explains why the backbone curve distortion is greater for the higher frequency resonance. Remember also that to obtain the physical displacement for the cantilever beam, the relationship $w(x, t) = \sum_{i=1}^{N} \phi_i(x)q_i(t)$ will need to be used.

Table 6.5 Values for $\beta_{k,l}$ computed using Eq. 4.56 from Chap. 4

u_l^*	s_{l1p}	s_{l1m}	s_{l2p}	s_{l2m}	$\beta_{k,l}$
u_{1p}^3	3	0	0	0	$(3\omega_{r1})^2 - \omega_{rk}^2$
$u_{1p}^2u_{1m}$	2	1	0	0	$(\omega_{r1})^2 - \omega_{rk}^2$
$u_{1p}u_{1m}^2$	1	2	0	0	$(-\omega_{r1})^2 - \omega_{rk}^2$
u_{1m}^3	0	3	0	0	$(-3\omega_{r1})^2 - \omega_{rk}^2$
$u_{1p}u_{2p}^2$	1	0	2	0	$(\omega_{r1} + 2\omega_{r2})^2 - \omega_{rk}^2$
$u_{1p}u_{2p}u_{2m}$	1	0	1	1	$(\omega_{r1})^2 - \omega_{rk}^2$
$u_{1p}u_{2m}^2$	1	0	0	2	$(\omega_{r1} - 2\omega_{r2})^2 - \omega_{rk}^2$
$u_{1m}u_{2p}^2$	0	1	2	0	$(-\omega_{r1} + 2\omega_{r2})^2 - \omega_{rk}^2$
$u_{1m}u_{2p}u_{2m}$	0	1	1	1	$(-\omega_{r1})^2 - \omega_{rk}^2$
$u_{1m}u_{2m}^2$	0	1	0	2	$(-\omega_{r1} - 2\omega_{r2})^2 - \omega_{rk}^2$
$u_{1p}^2u_{2p}$	2	0	1	0	$(2\omega_{r1} + \omega_{r2})^2 - \omega_{rk}^2$
$u_{1p}u_{1m}u_{2p}$	1	1	1	0	$(\omega_{r2})^2 - \omega_{rk}^2$
$u_{1m}^2u_{2p}$	0	2	1	0	$(-2\omega_{r1} + \omega_{r2})^2 - \omega_{rk}^2$
$u_{1p}^2u_{2m}$	2	0	0	1	$(2\omega_{r1} - \omega_{r2})^2 - \omega_{rk}^2$
$u_{1p}u_{1m}u_{2m}$	1	1	0	1	$(-\omega_{r2})^2 - \omega_{rk}^2$
$u_{1m}^2u_{2m}$	0	2	0	1	$(-2\omega_{r1} - \omega_{r2})^2 - \omega_{rk}^2$
u_{2p}^3	0	0	3	0	$(3\omega_{r2})^2 - \omega_{rk}^2$
$u_{2p}^2u_{2m}$	0	0	2	1	$(\omega_{r2})^2 - \omega_{rk}^2$
$u_{2p}u_{2m}^2$	0	0	1	2	$(-\omega_{r2})^2 - \omega_{rk}^2$
u_{2m}^3	0	0	0	3	$(3\omega_{r2})^2 - \omega_{rk}^2$

Axial loading (or stretching) of a beam also leads to governing equations with cubic nonlinearities, and this is considered next.

6.2.3 Stretching of a Constrained Beam

When a beam is constrained by two immovable supports, axial stretching can occur during vibrations. The example of a pinned-pinned flexible beam is shown in Fig. 6.8. At equilibrium the beam has the same length as the distance between supports, ℓ, and so is not in tension. As it deflects, the beam extends to length L and, as a result, a tension force is generated. As before, for a small length of beam

$$\Delta s \approx \left(\sqrt{\Delta x^2 + \Delta w^2}\right) \rightsquigarrow \frac{\partial s}{\partial x} = \sqrt{1 + \left(\frac{\partial w}{\partial x}\right)^2}.$$

Therefore the total beam length of the deflected beam, L, is

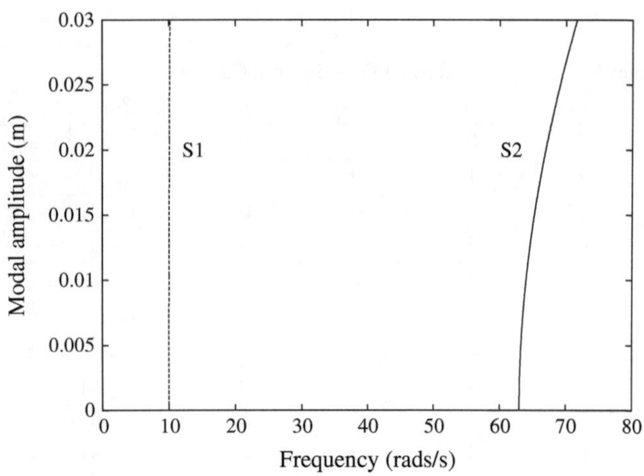

Fig. 6.7 Backbone curves for a cantilever beam with moderate displacements. The curves are defined by Eqs. (6.62) and (6.63), see Example 6.8

Fig. 6.8 Axial stretching in a clamped-clamped flexible beam

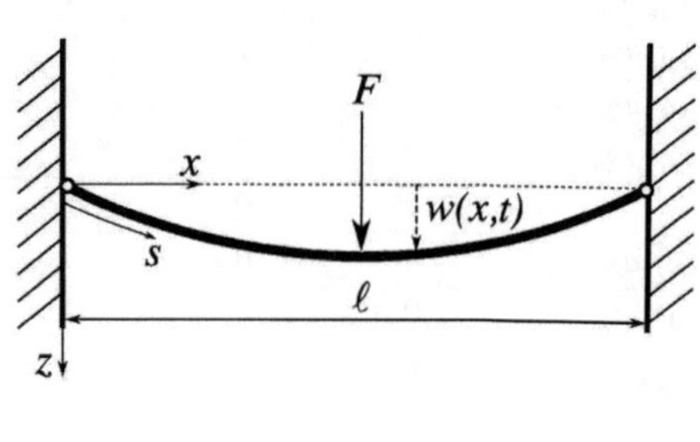

$$L = \int_0^{\ell} \sqrt{1 + \left(\frac{\partial w}{\partial x}\right)^2} \, dx.$$

Assuming the transverse displacement is small in comparison to the length of the beam, the integral can be approximated by

$$L \approx \int_0^{\ell} 1 + \frac{1}{2}\left(\frac{\partial w}{\partial x}\right)^2 dx = \ell + \frac{1}{2}\int_0^{\ell} \left(\frac{\partial w}{\partial x}\right)^2 dx,$$

using a two-term binomial series expansion for $(1+x)^n \approx 1+nx+\cdots$. The change in length can then be written as

$$L - \ell = \frac{1}{2}\int_0^{\ell} \left(\frac{\partial w}{\partial x}\right)^2 dx.$$

The tension force, T, acting at the root of the beam is equal to $E\hat{A}(L-\ell)/\ell$, where $\hat{A}$ is the cross-sectional area of the beam,[20] E is Young's modulus and $(L-\ell)/\ell$ is the axial strain. So, the tension is given by

$$T = \frac{E\hat{A}}{2\ell}\int_0^{\ell} \left(\frac{\partial w}{\partial x}\right)^2 dx. \tag{6.64}$$

The compressive loading of the beam at the root and tip, defined as P in Eq. (6.38), is given by $P = -T\cos(\psi_0)$ where $\psi_0 = \psi(x=0)$ is the rotation at the root. For small deflections this may be approximated by $P = -T$. Now it is assumed that there is no external loading or significant dynamics in the x direction, such that $F_x = 0$ except at the ends where F_x is equal to P. Note that in general, Eq. (6.35) gives the force balance in the x direction. When F_x is only non-zero at the ends of the beam Eq. (6.35) simplifies to Eq. (6.39), which may be written as $N = P = -T$. This type of scenario is discussed further when cables are considered, see Sect. 7.1.2 where the derivation of the cable equation of motion is presented.

Equation (6.40) is the result of a force and moment balance for a beam subjected to a load in the x direction at the end of the beam. Ignoring rotational inertia by setting $Q = 0$ and making the substitution that $P = -T$, where T is given by Eq. (6.64), we find

$$\frac{\partial^2 M}{\partial x^2} = F_z + T\frac{\partial^2 w}{\partial x^2}, \tag{6.65}$$

[20] Note that as the beam stretches, the cross-sectional area of the beam will decrease due to the Poisson's ratio effect. This effect is not considered in the current analysis.

where F_z represents the external loading and inertial effects in the z-axis (per unit length Δx). In this example, F_z must include external loading $F(t)$ acting at the mid-span of the beam, as shown in Fig. 6.8. Combining external and inertial forces gives

$$F_z = -\rho\hat{A}\frac{\partial s}{\partial x}\frac{\partial^2 w}{\partial t^2} + F(t)\delta(x - \frac{\ell}{2}),$$

where the applied force $F(t)\delta(x - \ell/2)$ consists of the time-dependent forcing $F(t)$ acting at the mid-span, and δ is the Dirac-Delta function. Substituting this expression into Eq. (6.65) and making the small-deflection approximation $\partial s/\partial x = 1$ gives

$$\rho\hat{A}\frac{\partial^2 w}{\partial t^2} + \frac{\partial^2 M}{\partial x^2} - T\frac{\partial^2 w}{\partial x^2} = F(t)\delta(x - \frac{\ell}{2}).$$

Finally, M and T may be eliminated from this equation, using the small-deflection bending moment relationship, $M = \partial^2 w/\partial x^2$, and Eq. (6.64) respectively, to give

$$EI\frac{\partial^4 w}{\partial x^4} - \frac{E\hat{A}}{2\ell}\int_0^{\ell}\left(\frac{\partial w}{\partial x}\right)^2 \mathrm{d}x\left(\frac{\partial^2 w}{\partial x^2}\right) + \rho\hat{A}\frac{\partial^2 w}{\partial t^2} = F\delta(x - \ell/2). \tag{6.66}$$

In the following example this governing equation for a stretched beam is decomposed into a set of ordinary differential equations using a Galerkin decomposition.

Example 6.7 Modal decomposition of beam with axial stretching

Problem Find the modal decomposition of the (undamped) Euler-Bernoulli equation for the vibration of a pinned-pinned beam with an external load F applied at the mid-span as shown in Fig. 6.8; include the effects of axial tension. The governing equation of motion for this beam is given by Eq. (6.66). In the Galerkin decomposition, assume the following modes

$$\phi_j(x) = \sin\left(j\pi\frac{x}{\ell}\right), \tag{6.67}$$

for $j = 1, 2, 3, 4, \ldots, N$. These modes correspond to the mode-shapes for a simply-supported beam in which axial tension is ignored; the corresponding natural frequencies are given by

$$\omega_{nj} = \left(\frac{j\pi}{\ell}\right)^2\sqrt{\frac{EI}{\rho\hat{A}}}. \tag{6.68}$$

Solution The governing equation can be decoupled using the Galerkin method. First, substitute $w(x,t) = \sum_{j=1}^{N}\phi_j(x)q_j(t)$ into Eq. (6.66), to separate the space and time dependence of the transverse displacement $w(x,t)$:

$$\sum_{j=1}^{N}\left\{EI\frac{\mathrm{d}^4\phi_j}{\mathrm{d}x^4}q_j-\frac{E\hat{A}}{2\ell}\sum_{k=1}^{N}\sum_{l=1}^{N}\int_0^{\ell}\left(\frac{\mathrm{d}\phi_k}{\mathrm{d}x}q_k\right)\left(\frac{\mathrm{d}\phi_l}{\mathrm{d}x}q_l\right)\mathrm{d}x\left(\frac{\mathrm{d}^2\phi_j}{\mathrm{d}x^2}\right)q_j\right.$$
$$\left.+\ \rho\hat{A}\phi_j\ddot{q}_j\right\}=F\delta(x-\ell/2), \tag{6.69}$$

where $\phi_{j,k,l}(x)$ represent the normal mode-shapes and $q_{j,k,l}(t)$ the modal coordinates. As mentioned before, when applying the Galerkin to nonlinear terms, each approximation of w needs to have a separate set of indices. This is to ensure that all possible modal cross-coupling terms are captured in the model.

Using Eq. (6.67) it can be shown that

$$\int_0^{\ell}\left(\frac{\mathrm{d}\phi_k}{\mathrm{d}x}\right)\left(\frac{\mathrm{d}\phi_l}{\mathrm{d}x}\right)\mathrm{d}x=\int_0^{\ell}\left(\frac{\mathrm{d}\phi_k}{\mathrm{d}x}\right)^2\mathrm{d}x=\frac{k^2\pi^2}{2\ell}\quad\text{for } l=k,$$

and zero when $l \neq k$, so the summation over l can be eliminated in this case as there are no cross-coupling terms between l and k. Note also that $\frac{\mathrm{d}^2\phi_j}{\mathrm{d}x^2}=-\left(\frac{j\pi}{\ell}\right)^2\phi_j(x)$, so that Eq. (6.69) becomes

$$\sum_{j=1}^{N}\left\{EI\frac{\mathrm{d}^4\phi_j}{\mathrm{d}x^4}q_j+\sum_{k=1}^{N}\frac{E\hat{A}}{2\ell}\left(\frac{k^2j^2\pi^4}{2\ell^3}\right)\phi_j q_k^2 q_j+\rho\hat{A}\phi_j\ddot{q}_j\right\}=F\delta(x-\ell/2). \tag{6.70}$$

Next, to decouple the linear terms on the left-hand side of Eq. (6.70) and produce an equation for the nth mode of vibration, Eq. (6.70) is multiplied by an arbitrary mode, ϕ_n, and integrated over the length of the span to give

$$EI\sum_{j=1}^{N}\int_0^{\ell}\phi_n\frac{\mathrm{d}^4\phi_j}{\mathrm{d}x^4}\mathrm{d}x\,q_j+\frac{E\hat{A}\pi^4}{4\ell^4}\sum_{j=1}^{N}\sum_{k=1}^{N}k^2j^2\int_0^{\ell}\phi_n\phi_j\mathrm{d}x\,q_k^2q_j$$
$$+\rho\hat{A}\sum_{j=1}^{N}\int_0^{\ell}\phi_n\phi_j\mathrm{d}x\,\ddot{q}_j=\int_0^{\ell}F\delta(x-\ell/2)\phi_n\mathrm{d}x. \tag{6.71}$$

The mode-shapes used are for the case where axial tension is ignored, i.e. the second term in Eq. (6.66) is removed. The orthogonality conditions for this case are $\int_0^{\ell}\phi_j\phi_n\mathrm{d}x=\ell/2$ and $\int_0^{\ell}\phi_j''''\phi_n\mathrm{d}x=(\ell/2)(n\pi/\ell)^4$ if $j=n$ and zero otherwise. This allows Eq. (6.71) to be written as

$$\ddot{q}_j(t) + \omega_{nj}^2 q_j(t) + \sum_{k=1}^{N} \frac{Ek^2 j^2 \pi^4}{4\rho\ell^4} q_k^2 q_j = \frac{2F}{\rho A\ell} \phi_j \left(\frac{L}{2}\right) \quad j, k = 1, 2, 3 \ldots N, \tag{6.72}$$

where the forcing term has been evaluated using Eq. (6.17). ■

Note that the form of the modal equation given by Eq. (6.72) is that of an oscillator with cross-coupled cubic stiffness. When $k = j$ this is similar to the Duffing oscillator.

Example 6.8 Backbone curves for two-mode model of a pinned-pinned beam with axial stretching

Problem Using the normal form analysis developed in Chap. 5 find approximate expressions for the backbone curves $S1$ and $S2$ of a two mode model of a pinned-pinned beam with axial stretching. Assume the system is unforced and undamped. The equations of motion were defined in Example 6.7, and the modal equations are given by Eq. 6.72.

Solution From Eq. 6.72, with $N = 2$ we can define the undamped, unforced model as

$$\begin{aligned} \ddot{q}_1 + \omega_{n1}^2 q_1 + \mu(q_1^3 + 4q_2^2 q_1) = 0, \\ \ddot{q}_2 + \omega_{n2}^2 q_2 + \mu(16q_2^3 + 4q_1^2 q_2) = 0, \end{aligned} \tag{6.73}$$

where $\mu = \dfrac{E\pi^4}{4\rho\ell^4}$. Now using the analysis from Chap. 4, Sect. 4.5

$$\ddot{\mathbf{q}} + \mathbf{\Lambda}\mathbf{q} + \mathbf{N}_q = 0 \quad \text{where:} \tag{6.74}$$

$$\mathbf{\Lambda} = \begin{bmatrix} \omega_{n1}^2 & 0 \\ 0 & \omega_{n2}^2 \end{bmatrix}, \quad \text{and} \tag{6.75}$$

$$\mathbf{N}_q = \mu \begin{pmatrix} q_1^3 + 4q_2^2 q_1 \\ 16q_2^3 + 4q_1^2 q_2 \end{pmatrix}. \tag{6.76}$$

Here $\mathbf{\Lambda}$ is a diagonal matrix of the squares of the linearised natural frequencies ω_{n1} and ω_{n2} defined by Eq. (6.68). The ratio between the backbone curve response frequencies, is defined such that $\omega_{r2} = r\omega_{r1}$. Note that from Eq. (6.68) $\omega_{n2} = 4\omega_{n1}$, and as $\omega_{rj} \approx \omega_{nj}$ we take $r = 4$. Using this, the $\boldsymbol{\beta}$ matrix can be calculated using Eq. 4.86 from Chap. 4 giving

$$\mathbf{u}^* = \begin{bmatrix} u_{1p}^3 \\ u_{1p}^2 u_{1m} \\ u_{1p} u_{1m}^2 \\ u_{1m}^3 \\ u_{1p} u_{2p}^2 \\ u_{1p} u_{2p} u_{2m} \\ u_{1p} u_{2m}^2 \\ u_{1m} u_{2p}^2 \\ u_{1m} u_{2p} u_{2m} \\ u_{1m} u_{2m}^2 \\ u_{1p}^2 u_{2p} \\ u_{1p} u_{1m} u_{2p} \\ u_{1m}^2 u_{2p} \\ u_{1p}^2 u_{2m} \\ u_{1p} u_{1m} u_{2m} \\ u_{1m}^2 u_{2m} \\ u_{2p}^3 \\ u_{2p}^2 u_{2m} \\ u_{2p} u_{2m}^2 \\ u_{2m}^3 \end{bmatrix}, \quad \mathbf{n}^T = \mu \begin{bmatrix} 1 & 0 \\ 3 & 0 \\ 3 & 0 \\ 1 & 0 \\ 4 & 0 \\ 8 & 0 \\ 4 & 0 \\ 4 & 0 \\ 8 & 0 \\ 4 & 0 \\ 0 & 4 \\ 0 & 8 \\ 0 & 4 \\ 0 & 4 \\ 0 & 8 \\ 0 & 4 \\ 0 & 16 \\ 0 & 48 \\ 0 & 48 \\ 0 & 16 \end{bmatrix}, \quad \boldsymbol{\beta}^T = \omega_{r1}^2 \begin{bmatrix} 8 & - \\ 0 & - \\ 0 & - \\ 8 & - \\ 80 & - \\ 0 & - \\ 48 & - \\ 48 & - \\ 0 & - \\ 80 & - \\ - & 20 \\ - & 0 \\ - & -12 \\ - & -12 \\ - & 0 \\ - & 20 \\ - & 128 \\ - & 0 \\ - & 0 \\ - & 128 \end{bmatrix}$$

where in $\boldsymbol{\beta}$ a dash has been used where the corresponding value in $\mathbf{n}$ is zero and hence the value in $\boldsymbol{\beta}$ is of no importance. From $\boldsymbol{\beta}$ it can be seen that there are eight unconditionally-resonant terms, namely elements [1, 2], [1, 3], [1, 6], [1, 9], [2, 11], [2, 14], [2, 18] and [2, 19]. For all resonant terms we set the corresponding terms in $\mathbf{n}_u$ as equal to those in $\mathbf{n}$, the resulting dynamic equations are

$$\begin{aligned} \ddot{u}_1 + \omega_{n1}^2 u_1 + 3\mu \left[u_{1p}^2 u_{1m} + u_{1p} u_{1m}^2 + \tfrac{8}{3} u_{1p} u_{2p} u_{2m} + \tfrac{8}{3} u_{1m} u_{2p} u_{2m} \right] &= 0, \\ \ddot{u}_2 + \omega_{n2}^2 u_2 + 48\mu \left[u_{2p} u_{2m}^2 + u_{2p}^2 u_{2m} + \tfrac{1}{6} u_{1p} u_{1m} u_{2m} + \tfrac{1}{6} u_{1p} u_{1m} u_{2p} \right] &= 0. \end{aligned} \tag{6.77}$$

Making the substitutions $u_{ip} = (U_i/2)\mathrm{e}^{j(\omega_{ri}t - \phi_i)}$ and $u_{im} = (U_i/2)\mathrm{e}^{-j(\omega_{ri}t - \phi_i)}$, we can write Eq. (6.77) in the form

$$\chi_i \mathrm{e}^{j\omega_{ri}t} + \tilde{\chi}_i \mathrm{e}^{-j\omega_{ri}t} = 0, \tag{6.78}$$

where χ_i and $\tilde{\chi}_i$ are complex conjugates. Inspecting the χ_i components of Eq. (6.77) results in the time-independent equations

$$\left[-\omega_{r1}^2 + \omega_{n1}^2 + 3\mu \left\{ \frac{U_1^2}{4} + \frac{2}{3} U_2^2 \right\} \right] U_1 = 0, \tag{6.79a}$$

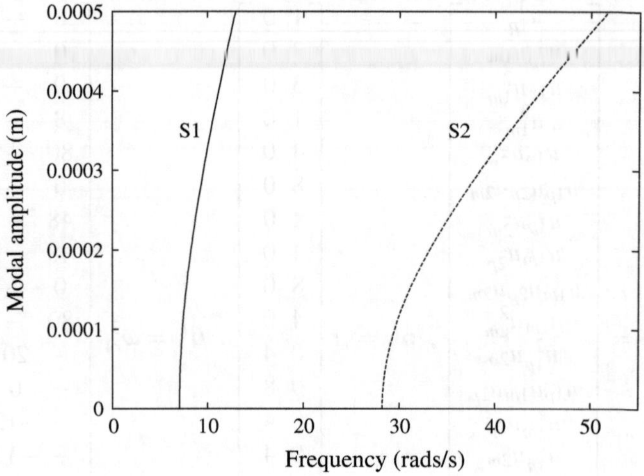

Fig. 6.9 Backbone curves for a pinned-pinned beam with axial stretching. The curves are defined by Eqs. (6.80) and (6.81), see also Example 6.8

$$\left[-\omega_{r2}^2 + \omega_{n2}^2 + 48\mu\left\{\frac{U_2^2}{4} + \frac{1}{24}U_1^2\right\}\right]U_2 = 0. \tag{6.79b}$$

There are two solutions to Eq. (6.79a, 6.79b) which can be found be setting U_1 and then U_2 to zero in turn. These single mode responses are

$$S1: \quad U_2 = 0, U_1 \neq 0, \qquad \omega_{r1}^2 = \omega_{n1}^2 + \frac{3\mu}{4}U_1^2, \tag{6.80}$$

$$S2: \quad U_1 = 0, U_2 \neq 0, \qquad \omega_{r2}^2 = \omega_{n2}^2 + 12\mu U_2^2. \tag{6.81}$$

In addition it is possible to have solutions in which both modes are present by setting both the expressions in the square brackets in Eq. 6.79a, 6.79b) to zero. However, these are not requested in the question. See Chap. 5 for more details. ■

Figure 6.9 shows the backbone curves for Example 6.8 that are defined by Eqs. (6.80) and (6.81) with the following parameters $E = 70\,\text{Gpa}$, $\rho = 2,700\,\text{kg/m}^3$, $I = 2.44 \times 10^{-13}\,\text{m}^4$, $A = 1.24 \times 10^{-5}\,\text{m}^2$ and $L = 1.0\,\text{m}$. With these parameters it can be found that $\omega_{n1} = 7\,\text{rads/s}$, $\omega_{n2} = 28.2\,\text{rads/s}$ and $\mu = 6.3136 \times 10^8$. Then computing the backbone curves from Eqs. (6.80) and (6.81), it can be seen that for modal amplitudes up to 0.5 mm, there are significant distortions in the natural frequency. This is a much stronger form of backbone curve distortion than occurred in Example 6.6 for large displacements of a cantilever beam, and is what would be expected when trying to stretch a beam that is much stiffer in the axial direction than it is in the transverse direction.

6.3 Case Study of Modal Control Applied to a Cantilever Beam

Modal control is a control strategy which aims to modify the dynamics of single or multiple modes of vibration. Often a desirable property of a modal controller is for the control strategy to affect the dynamics of the modes being targeted while leaving the dynamics of the other modes unchanged. When the controller affects untargeted modes, *control spillover* is said to have occurred, see Balas (1978). In this section, first, the effect of adding a controller designed to target a specific resonance is discussed by considering the modal response of the overall system. This is examined for a general case in which an actuator forcing can be applied to a beam. Then a piezoelectric actuation system is considered, in which the actuator applies a moment to the beam. Finally the use of positive position feedback (PPF) control is studied in conjunction with a piezoelectric actuator.

6.3.1 Modal Control of a Beam

Consider a beam subject to small deflections due to an unknown applied forcing, consisting of a *disturbance*,[21] and a control forcing. The modal equation of motion was derived in Sect. 6.1.2 and the modal equation of motion for the nth mode is given by

$$\ddot{q}_n(t) + 2\zeta_n\omega_{nn}\dot{q}_n(t) + \omega_{nn}^2 q_n(t) = \frac{1}{p\hat{A}L}\int_0^L F(x,t)\phi_n \mathrm{d}x, \tag{6.82}$$

from Eqs. (6.13) and (6.12) and where $F(x,t)$ consists of the disturbance and control forcing. In the derivation of Eq. (6.13), the orthogonality conditions for the mode-shapes of a cantilever beam given by Eq. (6.9) were used.

The system is actively controlled through the application of an actuator force (per unit length) F_a which acts in opposition to the disturbance forcing. The resulting forcing function $F(x,t)$ may be written as $F(x,t) = F_d(x,t) - F_a(x,t)$. The steady-state response of Eq. (6.82) can be considered in the Laplace domain. Substituting the forcing relation $F = F_d - F_a$, taking the Laplace transform and assuming zero initial conditions means that Eq. (6.82) can be written as

$$(s^2 + 2\zeta_n\omega_{nn}s + \omega_{nn}^2)Q_n(s) = \frac{1}{\rho\hat{A}L}\int_0^L \bar{F}_d(x,s)\phi_n \mathrm{d}x - \frac{1}{\rho\hat{A}L}\int_0^L \bar{F}_a(x,s)\phi_n(x)\mathrm{d}x, \tag{6.83}$$

where $\bar{F}_d(x,s)$, $\bar{F}_a(x,s)$ and $Q_n(s)$ are the Laplace transforms of $F_d(x,t)$, $F_a(x,t)$ and $q_n(t)$ respectively.

[21] Disturbance just means any unwanted, and usually unknown, signal.

Consider the case where just the kth mode is being targeted by the controller. A method of targeting this one mode without modifying other modes is to use the control law

$$\bar{F}_a(x, s) = C(s)\phi_k(x)Q_k(s),$$

where $C(s)$ represents the dynamics of the controller in the Laplace domain. Substituting this control law into Eq. (6.83) gives

$$(s^2+2\zeta_n\omega_{nn}s+\omega_{nn}^2)Q_n(s) = \frac{1}{\rho\hat{A}L}\int_0^L \bar{F}_d(x, s)\phi_n\mathrm{d}x - \frac{C(s)}{\rho\hat{A}L}\int_0^L \phi_k\phi_n\mathrm{d}xQ_k(s). \quad (6.84)$$

By considering the second orthogonality condition given in Eq. (6.9), $\int_0^L \phi_j\phi_j\mathrm{d}x = 0$ for $i \neq j$, it can be seen that, for all $k \neq n$, the controller term (the second term on the right-hand side) in Eq. (6.84) disappears and so the modal response is identical to the case where no control force is applied. However, for the kth mode, using Eq. (6.11) which states that $\int_0^L \phi_k^2\mathrm{d}x = L$ for a cantilever, the response may be written as

$$\left(s^2 + 2\zeta_k\omega_{nk}s + \omega_{nk}^2 + \frac{C(s)}{\rho\hat{A}}\right)Q_k(s) = \frac{1}{\rho\hat{A}L}\int_0^L \bar{F}_d(x, s)\phi_n\mathrm{d}x.$$

In this case the disturbance forcing remains unaltered. However the response is modified by the dynamics of the controller $C(s)$. For example, if a negative velocity feedback controller is used such that $C(s) = \kappa s$ (where $\kappa > 0$) it can be seen that the controlled mode is transformed from the uncontrolled dynamics

$$Q_k(s) = \frac{1/(\rho\hat{A}L)}{s^2 + 2\zeta_k\omega_{nk}s + \omega_{nk}^2}\int_0^L \bar{F}_d(x, s)\phi_n\mathrm{d}x,$$

to the controlled dynamics

$$Q_k(s) = \frac{1/(\rho\hat{A}L)}{s^2 + (2\zeta_k\omega_{nk} + \kappa/p\hat{A})s + \omega_{nk}^2}\int_0^L \bar{F}_d(x, s)\phi_n\mathrm{d}x.$$

Using this implementation, the controller has the effect of increasing the damping of the controlled mode without affecting the dynamics of the other modes.

Unfortunately, this control law is not easily implemented. First, the actuator force must be applied along the whole length of the beam (in the shape of the kth mode) and, secondly, the on-line measurement of q_k is not straightforward as transducers measure the displacement, velocity or acceleration at specific locations rather than the response of one particular mode. A modal measurement would require many transducers distributed along the beam.

These limitations will now be addressed by considering more realistic control implementations. First, a controller is considered that is limited to applying a force at a single location, but has modal measurement information. In this case the control law can be written as

$$\bar{F}_a(x, s) = C(s)\delta(x - x_a)Q_k(s),$$

where x_a is the location of the actuator. With this control law and Eq. (6.83), for $k \neq n$, the closed-loop modal dynamics may be written

$$Q_n(s) = \frac{1/(\rho\hat{A}L)}{s^2 + 2\zeta_n\omega_{nn}s + \omega_{nn}^2}\left(\int_0^L \phi_n(x)\bar{F}_d(x, s)\mathrm{d}x - \phi_n(x_a)C(s)Q_k(s)\right).$$

It can be seen that, while the modal dynamics remain unaltered by the controller, an additional forcing is now present in the modal response. This feature is called *control spillover* (*control* as it is due to non-modal application of the control forcing). For the $k = n$ mode, the closed-loop dynamics are given by

$$Q_k(s) = \frac{1/(\rho\hat{A}L)}{s^2 + 2\zeta_k\omega_{nk}s + \omega_{nk}^2 + C(s)\phi_k(x_a)}\int_0^L \phi_n(x)\bar{F}_d(x, s)\mathrm{d}x,$$

where the controller affects the closed-loop dynamics, provided that x_a is not at a node of ϕ_k.

Finally, we consider the control strategy that is the most convenient to implement. In this strategy the control law is based on only a single transducer measurement $w(x_t)$, where x_t is the transducer location, rather than on a modal measurement. Recalling that the displacement may be written as $w(x, t) = \sum_{j=1}^{\infty}\phi_j(x)q_j(t)$, we may write the transducer displacement as a Laplace transform to give

$$w(x_t, t) = \sum_{j=1}^{\infty}\phi_j(x_t)q_j(t) \quad \rightsquigarrow \quad W(x_t, s) = \sum_{j=1}^{\infty}\phi_j(x_t)Q_j(s),$$

where W is the Laplace transform of w. The control law may then be written as

$$\bar{F}_a(x, s) = C(s)\delta(x - x_a)W(x_t, s) = C(s)\delta(x - x_a)\sum_{j=1}^{\infty}\phi_j(x_t)Q_j(s).$$

Using Eq. (6.83) the modal dynamics for the nth mode are given by

$$(s^2 + 2\zeta_n\omega_{nn}s + \omega_{nn}^2)Q_n(s) = \frac{1}{\rho\hat{A}L}\int_0^L \bar{F}_d(x,s)\phi_n \mathrm{d}x - \frac{C(s)}{\rho\hat{A}L}\phi_n(x_a)\sum_{j=1}^{\infty}\phi_j(x_t)Q_j(s)$$

Rearranging this equation to separate the Q_n term from the forcing terms gives

$$Q_n(s) = \frac{1/(\rho\hat{A}L)}{s^2 + 2\zeta_n\omega_{nn}s + \omega_{nn}^2 + \phi_n(x_a)\phi_n(x_t)C(s)/(\rho\hat{A}L)} \times \left(\int_0^L \bar{F}_d(x,s)\phi_n \mathrm{d}x - \frac{C(s)}{\rho\hat{A}L}\phi_n(x_a)\sum_{j=1, j\neq n}^{\infty}\phi_j(x_t)Q_j(s)\right).$$

This equation shows that, in addition to additional forcing due to the controller, the closed-loop dynamics of all modes are affected by the controller. This is due to the presence of the $\phi_n(x_a)\phi_n(x_t)C(s)/(\rho\hat{A}L)$ term in the denominator of the right-hand side. Both control and *observer spillover* is said to have occurred (*observer* due to the non-modal measurement used to generate the control force)—see the discussion in Chap. 3.

The potential effect of spillover on the modal dynamics can be seen by again considering a negative velocity feedback controller, in which $C(s) = \kappa s$ and $\kappa > 0$. This results in the modal dynamics

$$Q_n(s) = \frac{1/(\rho\hat{A}L)}{s^2 + (2\zeta_n\omega_{nn} + \phi_n(x_a)\phi_n(x_t)\kappa/(\rho\hat{A}L))s + \omega_{nn}^2} \times \left(\int_0^L \bar{F}_d(x,s)\phi_n \mathrm{d}x - \frac{\kappa s}{\rho\hat{A}L}\phi_n(x_a)\sum_{j=1, j\neq n}^{\infty}\phi_j(x_t)Q_j(s)\right).$$

If the product $\phi_n(x_a)\phi_n(x_t)$ is positive then the controller increases the modal damping for the nth mode. Often the actuator and sensor are collocated, such that $x_a = x_t$, to ensure that this is the case. However if $\phi_n(x_a)\phi_n(x_t)$ is negative then the controller will have the undesirable effect of reducing the modal damping.

In this case, where the modal equations are coupled, a modal analysis ceases to be an effective way of analysing the system. With modal cross-coupling, the system can be written in the displacement domain, as opposed to the modal domain, and then the modal properties can be found by using the state space approach. This state space approach and the effect of spillover for a more complex control strategy, the positive position feedback control, are considered in Sect. 6.3.3.

6.3.2 Vibration Suppression Using Piezoelectric Actuation

Active vibration suppression can be carried out for a range of flexible continuous structural elements using piezoelectric actuation devices. These actuators apply a moment to the surface to which they are attached, usually in just one direction.

First consider the case where a moment is applied to a beam subjected to small displacements. Equation (6.40) is the partial differential equation governing the vibration of a beam with a vertical tip force P. The term F_z in Eq. (6.40) represents the transverse applied force and inertia terms while Q represents the applied moment and rotational inertia terms. With the small-deflection assumptions, the rotational inertia can be ignored, such that Q represents just the applied moment. In addition, as $\Delta x \approx \Delta s$, the moment may be written $M = EI\mathrm{d}^2 w/\mathrm{d}x^2$. For the application of a moment from the piezo-actuator with no additional loading, $P = 0$, and F_z represents just the transverse inertia which is given by the first expression in Eq. (6.41). Incorporating all these assumptions into Eq. (6.40) leads to the equation of motion

$$EI\frac{\partial^4 w}{\partial x^4} + \frac{\partial Q}{\partial x} + \rho\hat{A}\frac{\partial^2 w}{\partial t^2} = 0. \tag{6.85}$$

This represents the equation of motion of a beam (without damping) in which an external moment per unit length $Q(x, t)$ is applied to the beam and where the deflection is assumed to be small.

For modal control of the beam with an applied moment, the modal decomposition of Eq. (6.85) needs to be found. The case when the applied moment acts between two points on the beam, x_1 to x_2, corresponding to the position of the actuator providing the control moment, is of particular interest. Following Fanson and Caughey (1990), we take the moment function $Q = CV_a[H(x - x_1) - H(x - x_2)]$, where $H(\bullet)$ is the Heaviside function, V_a is the actuator voltage and C is a constant.

Substituting $w(x, t) = \sum_{i=1}^{\infty} \phi_i(x)q_i(t)$ to separate the space and time dependence of the transverse displacement $w(x, t)$ in Eq. (6.85) gives

$$EI\sum_{i=1}^{\infty}\frac{d^4\phi_i(x)}{dx^4}q_i(t) + \rho\hat{A}\sum_{i=1}^{\infty}\phi_i(x)\ddot{q}_i(t) = -CV_a\frac{\partial}{\partial x}[H(x - x_1) - H(x - x_2)],$$

where $\phi_i(x)$ represent the normal mode-shapes and $q_i(t)$ represent the modal coordinates.

The modal decomposition of this equation has already been discussed in Example 6.1. Adding a modal damping term to the resulting modal equation, Eq. (6.20), gives

$$\ddot{q}_i(t) + 2\zeta_i\omega_{ni}\dot{q}_i(t) + \omega_{ni}^2 q_i(t) = -aD_iV_a, \tag{6.86}$$

where ζ_i is the modal damping ratio, $a = C/(\rho\hat{A}L)$ and

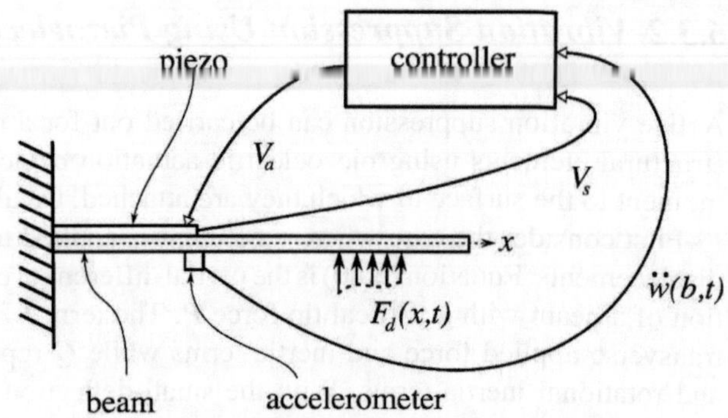

Fig. 6.10 Vibration control of a cantilever beam with piezoelectric actuator at the root and an additional accelerometer

$$D_i = \partial_i(x_2) - \partial_i(x_1).$$

The piezoelectric actuator can also be used as a sensor. The sensor voltage is a function of the beam curvature

$$V_s = b\frac{\partial^2 w}{\partial x^2},$$

where b is a constant. By making the modal substitution $w = \sum_{j=1}^{N} \phi_j q_j$ this becomes

$$V_s = b\sum_{j=1}^{N} q_j \frac{\partial^2 \phi_j}{\partial x^2} = b\sum_{j=1}^{N} q_j B_j.$$

These expressions will now be used to model the modal control of a beam using positive position feedback (PPF).

6.3.3 Positive Position Feedback

Figure 6.10 shows an example of a cantilever beam with an active vibration-control system. In this scenario, a piezoelectric sensor-actuator is positioned at the root of the cantilever with the ends corresponding to $x_1 = 0$ and $x_2 = b$ along the beam. An accelerometer is positioned at point $x = b$, so that the acceleration signal $\ddot{w}(b, t)$ can also be recorded. It is assumed that the controller can be implemented so that the piezoelectric patch can in effect be simultaneously used as both a sensor and actuator.[22] The voltage signal read by the controller as a sensor input is V_s and the actuator voltage sent to the piezoelectric is V_a. The beam is subject to an external force, $F_d(x, t)$, acting along all (or some part) of its length.

The control objective is to reduce vibration of the modes using an input to the piezoelectric actuator. As the beam is flexible, the response will typically consist of

[22] See Leo (2007) for a physical explanation of how this can be done using piezoelectric materials.

several modes of vibration. Ideally, the vibration mode with the largest response will be *targeted* by the controller to reduce the vibration.

In this section a technique which has been developed as an alternative to direct velocity feedback is discussed. The technique is called *positive position feedback* (PPF), and has been developed from linear modal analysis, which is the starting point used here. For the beam with an applied moment, each mode of vibration is governed by Eq. (6.86). In the active vibration control problem the beam will be subject to the external force $F_d(x,t) = g(x)h(t)$ per unit length (see Sect. 6.1.3), which means that Eq. (6.86) for each mode becomes

$$\ddot{q}_j(t) + 2\zeta_j\omega_{nj}\dot{q}_j(t) + \omega_{nj}^2 q_j(t) = \hat{\alpha}_j h(t) - aD_jV_a,$$

where $\hat{\alpha}_j = \alpha_j/(\rho\hat{A}L)$ is the modal participation factor for the forcing divided by mass. Note that the control input is such that the controller is in the standard negative feedback form, i.e. the sign of the control input is negative when the input is included on the right-hand side of the equation.

To apply PPF, first re-write the voltage input term $aD_jV_a = -\sigma_j\omega_{nj}^2 u$, where ω_{nj} is the modal frequency, σ_j is the *modal gain* and u is the control input. The modal gain is similar to the modal participation factor for the control actuator. In practice, the modal gains must be identified by giving an input to the control actuator, and observing the steady-state system response for each mode of vibration.

So how does feeding a position-like signal, u, back to the structure add damping? The idea is to treat u as a position coordinate of a more highly damped mode in the system, so that the governing expression for u is given by

$$\ddot{u}(t) + 2\zeta_c\omega_c\dot{u}(t) + \omega_c^2 u(t) = \omega_c^2 g q_j,$$

where ζ_c is the controller damping parameter, ω_c the controller frequency and g is the control gain. Then by coupling this controller mode to the structural modes in the system, extra damping can be added to the structural modes. The main advantage is that the controller mode acts like a filter, with fast roll-off at frequencies above the resonance, which means that spillover effects are reduced compared to direct velocity feedback.

Consider the case when the system response is approximated with just one vibration mode so that $N = 1$ and $q_1 = q$. Then a new composite system can be written as

$$\begin{bmatrix} \ddot{q} \\ \ddot{u} \end{bmatrix} + \begin{bmatrix} 2\zeta\omega_n & 0 \\ 0 & 2\zeta_c\omega_c \end{bmatrix}\begin{bmatrix} \dot{q} \\ \dot{u} \end{bmatrix} + \begin{bmatrix} \omega_n^2 & -\omega_n^2\sigma \\ -\omega_c^2 g & \omega_c^2 \end{bmatrix}\begin{bmatrix} q \\ u \end{bmatrix} = \begin{bmatrix} \hat{\alpha}h(t) \\ 0 \end{bmatrix}, \qquad (6.87)$$

where the j subscripts have been dropped as there is only a single mode. From a control perspective, this combined system can be thought of as a closed-loop system. The closed-loop system will be stable if the frequency matrix is positive definite. This condition is satisfied if $\omega_n^2 > 0$, which is always true, and

$$\omega_n^2\omega_c^2 - \omega_c^2\omega_n^2\sigma g > 0,$$

which reduces to the condition $g < \frac{1}{\sigma}$.

From a vibration perspective the combined system is now two coupled modes, and this coupled system will have different frequencies and damping ratios from the two uncoupled systems. To find the new values, set $h(t) = 0$ and write the coupled system as

$$\ddot{\mathbf{p}} + D\dot{\mathbf{p}} + K\mathbf{p} = 0,$$

where $\mathbf{p} = [q,\ u]^T$. In first-order form this becomes $\dot{\boldsymbol{x}} = A\boldsymbol{x}$ with

$$A = \begin{bmatrix} 0 & I \\ -K & -D \end{bmatrix},$$

and $\boldsymbol{x} = [\mathbf{p}, \dot{\mathbf{p}}]^T$. The eigenvalues of A can be used to compute the new combined system damping ratios and natural frequencies.

Note that the maximum damping effect occurs if the controller frequency is set to be the same as the frequency of the mode of interest. An example is considered next.

Example 6.9 Single mode PPF

Problem Design a PPF controller for the single-mode system given by Eq. (6.87). In this case the structural mode frequency and damping ratio have been estimated as $\omega_n = 25.1$, $\zeta = 0.0242$. In addition $\sigma = 0.1148$ has been estimated from the experimental system identification tests.[23] The control parameters $\omega_c = 25$, $g = 1$ and $\zeta_c = 0.5$ have been selected. Check that the PPF system is stable, and find the new closed-loop damping ratios and frequencies.

Solution The stability condition requires that $g < \frac{1}{\sigma}$, and for $\sigma = 0.1148$ the maximum g allowable is $g = 8.71$. So when $g = 1$ is selected the system will be stable. To check the new closed-loop damping ratios and frequencies, the matrix A is formed to give

$$A = \begin{bmatrix} 0 & 0 & 1 & 0 \\ 0 & 0 & 0 & 1 \\ -630.0 & 72.33 & -1.22 & 0 \\ 625.0 & -625.0 & 0 & -25.0 \end{bmatrix}.$$

From Matlab, the eigenvalues of A are found to be $\lambda_{1,2}\ -2.2940 \pm i25.3827$ and $\lambda_{3,4}\ -10.8134 \pm i20.4862$, which are complex and stable. The first pair of eigenvalues corresponds to the oscillatory solution of the new structural vibration

[23] These data are taken from the modal identification of an aluminium beam with a piezo-actuator attached, see Chap. 3 of Malik (2009).

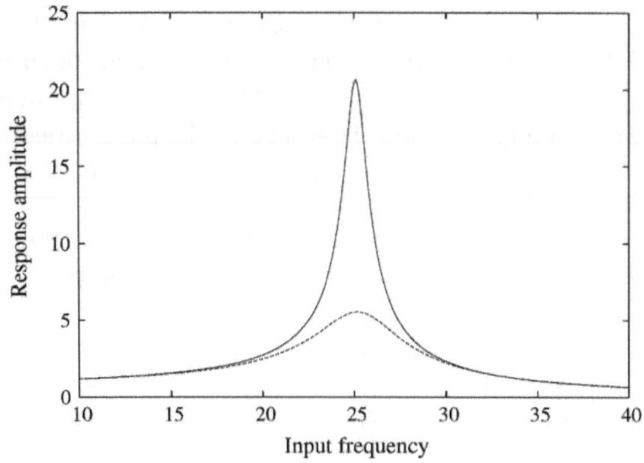

Fig. 6.11 Positive position feedback applied to a single-mode model of a beam. *Solid line* $\zeta = 0.0242$, *dashed line* $\zeta = 0.0904$

mode, for which the new frequency is $\omega_n = 25.3827$ and the damping ratio is $\zeta = 2.2940/25.3827 = 0.0904$. This damping ratio is approximately 3.7 times larger than the original one. The new controller frequency is $\omega_f = 20.4862$ and controller damping ratio $\zeta_c = 10.8134/20.4862 = 0.5278$. The reduction in the modal resonance peak is shown in Fig. 6.11. ■

The same approach can be applied to a system with more than a single mode, in which case the composite system, given by Eq. (6.87) for the one-mode case, can be written as

$$\begin{bmatrix}\ddot{\mathbf{q}}\\ \ddot{u}\end{bmatrix} + \begin{bmatrix}\begin{bmatrix}2\zeta_j\omega_{n1} & \cdots & 0\\ 0 & \ddots & 0\\ 0 & \cdots & 2\zeta_j\omega_{nN}\end{bmatrix} & 0\\ 0 & 2\zeta_c\omega_c\end{bmatrix}\begin{bmatrix}\dot{\mathbf{q}}\\ \dot{u}\end{bmatrix} + \begin{bmatrix}\begin{bmatrix}\omega_{n1}^2 & \cdots & 0\\ 0 & \ddots & 0\\ 0 & \cdots & \omega_{nN}^2\end{bmatrix} & \begin{bmatrix}-\omega_{n1}^2\sigma_1\\ \vdots\\ -\omega_{nN}\sigma_N\end{bmatrix}\\ -\omega_c^2[g_1,\ldots,g_N] & \omega_c^2\end{bmatrix}\begin{bmatrix}\mathbf{q}\\ u\end{bmatrix} = \begin{bmatrix}\begin{bmatrix}\hat{\alpha}_1\\ \cdots\\ \hat{\alpha}_N\end{bmatrix} h(t)\\ 0\end{bmatrix}. \qquad (6.88)$$

To target a single mode, all the g_i values are set to zero, except the mode of interest. Note that this is only possible if the modal responses q_i can be measured, the implications of using a point deflection measurement have already been discussed in Sect. 6.3.1. In the next example, a three-mode model of a beam is used, and the second mode is targeted using PPF.

Example 6.10 Single-mode PPF for a three-mode system

Problem Design a PPF controller which targets the second mode of a three-mode system taking the form of Eq. (6.88). In this case the structural mode frequencies and damping ratios have been estimated as $\omega_{n1} = 25.1$, $\omega_{n2} = 120.1$, $\omega_{n3} = 284.5$,

$\zeta_1 = 0.0242$, $\zeta_2 = 0.0211$, $\zeta_3 = 0.0237$. In addition, the modal gain factors $\sigma_1 = 0.1148$, $\sigma_2 = 0.0346$, $\sigma_3 = 0.0247$ have been estimated from the experimental system identification tests.[24] The control parameters $\omega_c = 120$, $g_1 = 0$, $g_2 = 6$, $g_3 = 0$ and $\zeta_c = 0.8$ have been selected to specifically target the second mode of vibration. Find the new closed-loop damping ratios and frequencies.

Solution To calculate the new closed-loop damping ratios and frequencies, the matrix A, Eq. (6.88) is formed to give for this example

$$A = \begin{bmatrix} 0 & 0 & 0 & 0 & 1 & 0 & 0 & 0 \\ 0 & 0 & 0 & 0 & 0 & 1 & 0 & 0 \\ 0 & 0 & 0 & 0 & 0 & 0 & 1 & 0 \\ 0 & 0 & 0 & 0 & 0 & 0 & 0 & 1 \\ -630.01 & 0 & 0 & 72.33 & -1.21 & 0 & 0 & 0 \\ 0 & -1442.4 & 0 & 499.07 & 0 & -5.07 & 0 & 0 \\ 0 & 0 & -80940.25 & 1999.22 & 0 & 0 & -13.49 & 0 \\ 0 & 86400.0 & 0 & -14400.0 & 0 & 0 & 0 & -192.0 \end{bmatrix}$$

Using Matlab, the eigenvalues of A are found to be $\lambda_{1,2} = -0.61 \pm i25.09$, $\lambda_{3,4} = -11.04 \pm i121.34$, $\lambda_{5,6} = -6.74 \pm i284.42$ and $\lambda_{7,8} = -87.49 \pm i58.59$, which correspond to modes 1, 2, 3 and controller respectively. So the closed-loop structural frequencies are $\omega_{n1} = 25.09$, $\omega_{n2} = 121.34$, $\omega_{n3} = 284.42$, which are very close to the original frequencies. The closed-loop damping ratios are $\zeta_1 = 0.61/25.09 = 0.0243$, $\zeta_2 = 11.04/121.34 = 0.0910$, $\zeta_3 = 6.74/284.42 = 0.0237$. The targeted mode (mode 2) now has a damping ratio which is approximately 4.3 times greater than the uncontrolled damping ratio. The modes which were not targeted have damping ratios which are almost exactly the same as the original system indicating that there is no significant spillover. The reduction in the second modal resonance peak is shown in Fig. 6.12. ■

These types of PPF techniques can be extended for use with multiple sensors and to target multiple modes, see for example Moheimani and Fleming (2006). Dealing with nonlinear modal vibration presents a different set of challenges, and this is considered next.

6.3.4 PPF for Nonlinear Vibration

When the vibrations are nonlinear there are two approaches which can be used to apply the PPF control. First, it is possible to modify the control signal to linearize the

[24] These data are taken from the modal identification of an aluminium beam with a piezo-actuator attached, see Chap. 3 of Malik (2009).

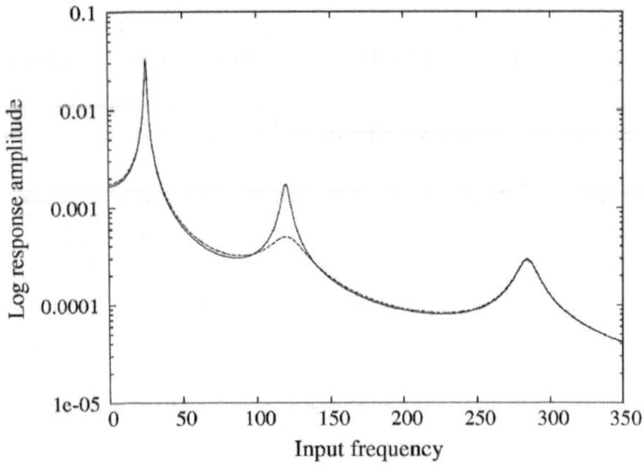

Fig. 6.12 Positive position feedback applied to a three-mode model of a beam: *solid line*, original frequency response; *dashed line*, with PPF applied to the second mode

system as well as apply a PPF control.[25] The second approach is to use an adaptive PPF strategy.

Details of feedback linearisation control are given in Sect. 3.5, in Chap. 3. For example, consider a modal equation of the form

$$\ddot{q}_j(t) + 2\zeta_j\omega_{nj}\dot{q}_j(t) + \omega_{nj}^2 q_j(t) + \mathcal{N}(q_j, \dot{q}_j) = \hat{\alpha}_j h(t) - aD_j V_a,$$

where $\hat{\alpha}_j = \alpha_j/(\rho\hat{A}L)$ is the modal participation factor divided by mass and $\mathcal{N}(\bullet)$ is the nonlinear term. Then if it is possible to set $aD_j V_a = -\sigma_j\omega_{nj}^2 u - \mathcal{N}(\bullet)$, the mode can be linearized and PPF control applied in the same way as the linear case. Consider the following example.

Example 6.11 Single-mode PPF for a nonlinear modal oscillator

Problem Design a PPF controller for the single-mode nonlinear system given by

$$\ddot{q}(t) + 2\zeta_j\omega_{nj}\dot{q}(t) + \omega_{nj}^2 q(t) + \mu q^3 = \hat{\alpha}_j h(t) - aD_j V_a$$

The structural mode frequency and damping ratio have been estimated as $\omega_n = 25.1$, $\zeta = 0.0242$, also $\sigma = 0.118$ and $\mu = 0.012$. The control parameters $\omega_c = 25$, $g = 1$ and $\zeta_c = 0.5$ have been selected to target the primary resonance. Check the PPF system is stable and find the new closed-loop damping ratios and frequencies.

Solution Setting $aD_j V_a = -\sigma_j\omega_{nj}^2 u - \mu q^3$ will linearize the system, so that the closed-loop PPF system takes exactly the same form as the system considered in Example 6.9. The linear steady-state stability has been considered in Example 6.9,

[25] Theoretically linearisation is possible.

but the transient stability could be an issue, depending on the form of the forcing. A time simulation when $h(t) = \sin(25.1t)$ is shown in Fig. 6.13 with $q(0) = 0.1$. As the underlying linear system is stable, it would be expected that the forced system would be bounded stable, as shown in Fig. 6.13. ■

This idea of feedback linearisation PPF can be extended to include more modes in the structural model. Providing the modal displacements and velocities are obtainable in real time, then they can be used to design a feedback linearisation PPF controller.

The second approach is to use adaptive control to deal with parameter uncertainties, which can be caused by nonlinearities or other effects. This is a particular problem when changes occur in the modal frequency over time. The effectiveness of the PPF controller will reduce significantly as the difference between ω_c and ω_{nj} increases. This can be avoided by implementing a real-time fast Fourier transform to identify changes in the dominant frequency response of the system. To do this, ω_{nj} is identified at regular intervals, so in effect it becomes a function of time, $\omega_{nj}(t)$. Then $\omega_c = \omega_{nj}(t)$ is set to ensure the best performance of the PPF control even if ω_{nj} varies. See Creasy et al. (2008) and Malik (2009) for more details of this adaptive process.

6.4 Chapter Notes

This chapter focused on the nonlinear vibration and control of beams. General introductions to linear vibrations of beams are covered in classic texts such as Bishop

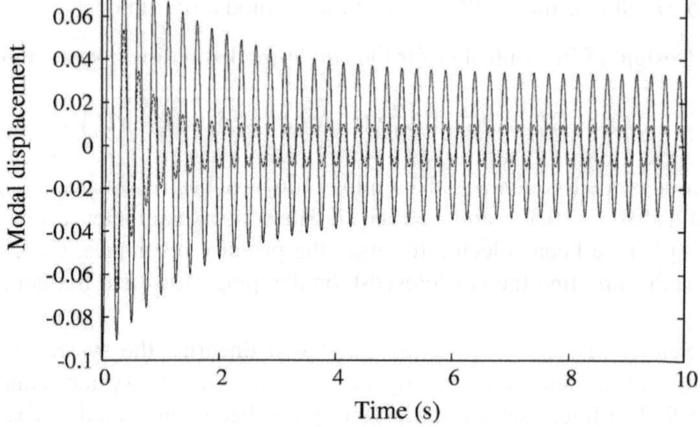

Fig. 6.13 Positive position feedback applied to a single-mode model of a beam with cubic nonlinearity: *solid line*, original time series; *dashed line*, with PPF applied and feedback linearized

and Johnson (1960), Timoshenko et al. (1974) and Meirovitch (1976). Discussion of nonlinear vibration due to large deflections and axial stretching follows the approach taken by Frish-Fay (1962). A more up-to-date treatment of these topics is given in Virgin (2007) and Hagedorn and DasGupta (2007). An insightful discussion of the back ground to variational methods, particularly the Rayleigh-Ritz method is given by Leissa (2005). Galerkin methods are described in Fletcher (1984). An example of backbone curves for nonlinear beam vibration is detailed by Lewandowski (1994). A good introduction to the topic of modal control and PPF for linear systems is given by Leo (2007). These topics are also detailed in Inman (2006) and Moheimani et al. (2003). A general overview of spatial control is given by Moheimani and Fleming (2006).

References

Ayres, F. Jr. (1964). *Calculus*. New York: McGraw Hill.

Balas, M. J. (1978). Feedback control of flexible systems. *IEEE Transactions on Automatic Control, 23*(4), 673–679.

Bishop, R. E. D. & Johnson, D. C. (1960). *The mechanics of vibration*. Cambridge: Cambridge University Press.

Blevins, R. D. (1979). *Formulas for natural frequency and mode shape*. New York: Van Nostrand Reinhold.

Clough, R. W., & Penzien, J. (1993). *Dynamics of structures* (2nd ed.). New York: McGraw-Hill.

Creasy, M. A., Leo, D. J., & Farinholt, K. M. (2008). Adaptive positive position feedback for actively absorbing energy in acoustic cavities. *Journal of Sound and Vibration, 311*, 461–472.

Crisfield, M. A. (1997). *Non-linear finite element analysis of solids and structures*. Vol. 2: *Advanced topics*. New York: Wiley.

Ewins, D. J. (2000). *Modal testing*. Baldock: Research Studies Press.

Fanson, J., & Caughey, T. (1990). Positive position feedback control for large space structures. *AIAA, 28*, 717–724.

Finlayson, B. A. (1972). *The method of weighted residuals*. New York: Academic Press.

Fletcher, C. A. J. (1984). *Computational Galerkin methods*. New York: Springer.

Frish-Fay, R. (1962). *Flexible bars*. London: Butterworths.

Hagedorn, P., & DasGupta, A. (2007). *Vibrations and waves in continuous mechanical systems*. New York: Wiley.

Inman, D. J. (2006). *Vibration with control*. New York: Wiley.

Jones, D. I. G. (2001). *Handbook of viscoelastic vibration damping*. Sussex: Wiley-Black well.

King, A. C., Billingham, J., & Otto, S. R. (2003). *Differential equations*. Cambridge: Cambridge University Press.

Kreyszig, E. (1993). *Advanced engineering mathematics* (7th ed.). New York: Wiley.

Lazan, B. J. (1968). *Damping of materials and members in structural mechanisms*. Oxford: Pergamon.

Leissa, A. W. (2005). The historical bases of the rayleigh and ritz methods. *Journal of Sound and Vibration, 287*(4), 961–978.

Leo, D. J. (2007). *Smart material systems*. New York: Wiley.

Lewandowski, R. (1994). Solutions with bifurcation points for free vibration of beams: An analytical approach. *Journal of Sound and Vibration, 177*, 239–249.

Malik, N. S. (2009). *Adaptive vibration control of flexible structures using piezoelectric actuators*. Ph.D. thesis, University of Bristol.

Meirovitch, L. (1976). *Elements of vibration analysis*. New York: McGraw-Hill.

Moheimani, S. O. R., & Fleming, A. J. (2006). *Piezoelectric transducers for vibration control and damping*. Berlin: Springer.

Moheimani, S. O. R., Halim, D., & Fleming, A. J. (2003). *Spatial control of vibration*. New York: World Scientific.

Moon, F. C., & Shaw, S. W. (1983). Chaotic vibrations of a beam with non-linear boundary conditions. *International Journal of Non-Linear Mechanics*, *18*(6), 465–477.

Ouyang, H. (2011). Moving-load dynamic problems: A tutorial (with a brief overview). *Mechanical Systems and Signal Processing*, *25*, 2039–2060.

Press, W. H., Teukolsky, S. A., Vettering, W. T., & Flannery, B. P. (1994). *Numerical recipes in C* (2nd ed.). Cambridge: Cambridge University Press.

Stronge, W. J. (2000). *Impact mechanics*. Cambridge: Cambridge University Press.

Timoshenko, S. P., Young, D., & Weaver, W., Jr. (1974). *Vibration problems in engineering*. New York, USA: Van Nostrand.

Vichnevetsky, R. (1981). *Computer methods for partial differential equations*. Englewood Cliffs: Prentice Hall.

Virgin, L. N. (2007). *Vibration of axially-loaded structures*. Cambridge: Cambridge University Press.

Wagg, D. J. (2003). A note on using the collocation method for modelling the dynamics of a flexible continuous beam subject to impacts. *Journal of Sound and Vibration*, *276*(3–5), 1128–1134.

Wagg, D. J., & Bishop, S. R. (2002). Application of nonsmooth modelling techniques to the dynamics of a flexible impacting beam. *Journal of Sound and Vibration*, *256*(5), 803–820.

Worden, K., & Tomlinson, G. R. (2000). *Nonlinearity in structural dynamics*. Bristol: IOP.

Zienkiewicz, O. C., & Taylor, R. L. (1991). *The finite element method*. Vol. 2: *Solid and fluid mechanics, dynamics and non-linearity* (4th ed.). New York: McGraw-Hill.

Chapter 7
Cables

Abstract In this chapter, the vibration behaviour of cables is considered. The starting point is to consider horizontal cables, which are initially assumed to be inextensible. Of particular importance is cable *sag*, the static displacement of a cable due to gravity. Sag results in cables having complex dynamic behaviour. This is seen when the nonlinear equations of motion for an inclined cable are developed. Inclined cables are important for applications such as cable-stayed bridges. Galerkin's method is used to convert the nonlinear partial differential equations into a set of modal equations in which the linear terms are decoupled. However, modal coupling remains in the nonlinear terms. These nonlinear coupled terms lead to internal resonance, such as autoparametric resonance. This type of resonance can be observed for cable-stay bridges when certain combinations of external excitation frequency, deck frequency and cable mode frequency occur. In the final part of the chapter, two case studies of cable vibration are considered. Firstly the techniques of averaging, multiple scales and normal forms are compared when applied to the analysis of a single mode of vibration of an inclined cable. Then the modal interaction between the second in- and out-of-plane modes is considered when the cable is subjected to a vertical support motion close to the second natural frequency. This second case study uses the normal forms technique to find the backbone curves for the system and then to identify a region in which there is response from both the directly excited second in-plane mode and the auto-parametrically excited out-of-plane mode.

7.1 Horizontal Cable Vibration

In this section, the free vibration of a horizontal taut cable is examined. A cable is a structural element which is under axial tension but is usually assumed to have no bending stiffness. A normal starting point is to derive equations of motion ignoring gravitational effects on the distributed mass of the cable. When gravitational effects are neglected in this way, the cable is usually referred to as a *string*. The analysis

D. Wagg and S. Neild, *Nonlinear Vibration with Control*,
Solid Mechanics and Its Applications 218, DOI 10.1007/978-3-319-10644-1_7

of strings leads to a relationship where the frequency at which the string vibrates depends on the tension applied. This scenario occurs in musical instruments such as the violin or guitar, where the vibration frequency gives the musical note. As there is no gravity or bending stiffness, the restoring force is entirely dependent on the tension, and increased tension increases the rate at which the string is pulled back toward its resting position, leading to a higher frequency oscillation.

Including the effects of gravity results in a static displaced shape for a horizontal cable. This is analysed in two stages. First, the static deflected shape due to gravity is derived, and then the equation of motion including the effects of this static deflection is developed from this.

7.1.1 Cable Sag

For a straight string without any sag, the equation of motion can be developed by using the equation of motion for a beam without bending stiffness as developed in Sect. 6.2.3. For example, neglecting the bending stiffness term and setting $P = -T$, where T is the tension in the string in Eq. (6.46), gives

$$-T\frac{\partial^2 w}{\partial x^2} + \rho\hat{A}\frac{\partial^2 w}{\partial t^2} = 0. \tag{7.1}$$

This is often written as

$$\frac{\partial^2 w}{\partial t^2} = c^2\frac{\partial^2 w}{\partial x^2}, \tag{7.2}$$

where $c = \sqrt{T/(\rho\hat{A})}$ is the propagation speed of a wave in the string. Equation (7.2) is usually referred to as the *wave equation*, which can be applied to *taut* strings, meaning those without sag (or any bending stiffness). The wave equation has a range of applications in classical mechanics—see for example the classic texts (Rayleigh 1894a, b).

Vibrations in the string are dependent on the balance between the inertia force and the restoring force, and these effectively correspond to the two terms in the wave equation (multiplied by T to give force per unit length). Notice also that the restoring force term, the first term in Eq. (7.1), is proportional to $\frac{\partial^2 w}{\partial x^2}$, the curvature in the z direction.

As the mass of the string increases, sag becomes a significant factor and needs to be included in the model. Tension elements with sag, but without significant bending stiffness, are normally referred to as cables. These are an important class of structural element used in a wide variety of applications.

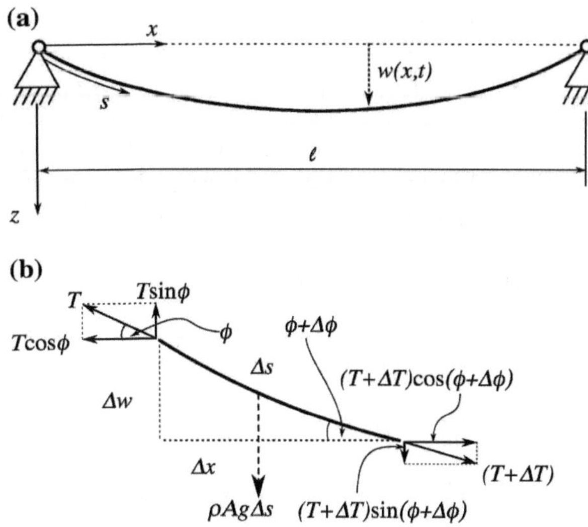

Fig. 7.1 **a** Planar stretched cable with sag due to gravity forces, **b** forces on a small element

An example of a stretched cable with sag is shown in Fig. 7.1. In Fig. 7.1a a horizontal cable is shown (inclined cables will be considered in Sect. 7.2). A small element of cable is shown in Fig. 7.1b. T is the tension in the cable, w is the cable deflection from the chord line between the two supports a distance ℓ apart. In the following analysis both the tension and the deflection will be split into a static and a dynamic component, $T = T_s + T_d$ and $w = w_s + w_d$. First, a static cable will be considered, then a dynamic analysis will be carried out. Both these analyses will assume that the deflection is in the vertical plane. Finally, the analysis will be extended to an inclined cable and include horizontal motion and support motion, which has important applications to the analysis of cable-stayed bridge dynamics.

7.1.2 Static Deflection Due to Sag

Now the *static* deflection of an inextensible cable under its own self weight is considered. To fully capture the dynamics of cables, this static deflection must be considered when deriving the equation of motion.

Assume that a small element of cable takes the form shown in Fig. 7.1b, where for the static case the cable tension is $T = T_s$ and the cable deflection is $w = w_s$. The horizontal force equilibrium is given by

$$\rightarrow_x \quad -T_s \cos\phi + (T_s + \Delta T_s)\cos(\phi + \Delta\phi) = 0, \tag{7.3}$$

and vertical force equilibrium by

$$\downarrow_z \quad -T_s \sin\phi + (T_s + \Delta T_s)\sin(\phi + \Delta\phi) + \rho A g \Delta s = 0, \tag{7.4}$$

where ρ is the density, g acceleration due to gravity and A the cross sectional area. Expanding the $\cos(\phi + \Delta\phi)$ term in Eq. (7.3) using the standard trigonometric relationship and using the approximations $\cos(\Delta\phi) \simeq 1$ and $\sin(\Delta\phi) \simeq \Delta\phi$, gives

$$\Delta T_s \cos\phi - T_s \Delta\phi \sin\phi - \Delta T_s \Delta\phi \sin\phi = 0. \tag{7.5}$$

Dividing by Δx and taking the limit as $\Delta x \to 0$ gives

$$\frac{\mathrm{d}T_s}{\mathrm{d}x}\cos\phi - T_s \sin\phi \frac{\mathrm{d}\phi}{\mathrm{d}x} = 0,$$

(noting that the third term in Eq. (7.5) disappears when the limit is taken), which can be written as

$$\frac{\mathrm{d}}{\mathrm{d}x}\{T_s \cos\phi\} = 0. \tag{7.6}$$

Using the same method, Eq. (7.4) can be expressed as

$$\frac{\mathrm{d}}{\mathrm{d}x}\{T_s \sin\phi\} + \rho A g \frac{\mathrm{d}s}{\mathrm{d}x} = 0. \tag{7.7}$$

To solve these equations to give the deflected shape, Eq. (7.6) is first integrated with respect to x

$$T_s \cos\phi = \text{constant} = T_{sx}, \tag{7.8}$$

where T_{sx} is the horizontal force in the cable and is constant over the length of the cable. Substituting Eq. (7.8) into the vertical force balance equation, Eq. (7.7), to eliminate T_s gives

$$T_{sx}\frac{\mathrm{d}}{\mathrm{d}x}\{\tan\phi\} + \rho A g \frac{\mathrm{d}s}{\mathrm{d}x} = 0. \tag{7.9}$$

Here it has been noted that T_{sx} is constant so can be taken outside the differential operator. For static deflection $\tan\phi = \frac{\mathrm{d}w_s}{\mathrm{d}x}$, where w_s is the static deflection in the z-axis. Substituting for the static deflection in Eq. (7.9) gives

$$T_{sx}\frac{\mathrm{d}^2 w_s}{\mathrm{d}x^2} + \rho A g \frac{\mathrm{d}s}{\mathrm{d}x} = 0. \tag{7.10}$$

To proceed, the $\frac{\mathrm{d}s}{\mathrm{d}x}$ term must be replaced to eliminate s. This is done by considering the length of the cable element, Δs, see Fig. 7.1b. This length may be written as

$$\Delta s = \sqrt{\Delta x^2 + \Delta w^2},$$

which, taking the limit as $\Delta x \to 0$ and then applying a Taylor series expansion, gives

$$\frac{\mathrm{d}s}{\mathrm{d}x} = \sqrt{1 + \left(\frac{\mathrm{d}w_s}{\mathrm{d}x}\right)^2} \simeq 1 + \frac{1}{2}\left(\frac{\mathrm{d}w_s}{\mathrm{d}x}\right)^2, \qquad (7.11)$$

assuming the sag is small. Using Eq. (7.11) to eliminate s gives

$$T_{sx}\frac{\mathrm{d}^2 w_s}{\mathrm{d}x^2} + \rho A g\sqrt{1 + \left(\frac{\mathrm{d}w_s}{\mathrm{d}x}\right)^2} = 0, \qquad (7.12)$$

which is a second-order nonlinear differential equation governing the sag of the cable. Solving Eq. (7.12) is a classical problem in applied mechanics and results in a function known as the *catenary*, see for example von Kármán and Biot (1940).

Typically, cables used in structures are put under high tensile loads with the result that the sag can often be considered small, $w_s \ll \ell$. This can be used to considerably simplify the analysis, as the static curve can then be reasonably approximated as parabolic. In this case, the approximation $\frac{\mathrm{d}s}{\mathrm{d}x} \approx 1$ is made in Eq. (7.10) giving

$$\frac{\mathrm{d}^2 w_s}{\mathrm{d}x^2} = -\frac{\rho A g}{T_{sx}}. \qquad (7.13)$$

Double integrating this equation and applying the boundary conditions that the static deflection w_s is zero at $x = 0$ and $x = \ell$, to find the integration constants gives

$$w_s = \frac{\rho A g}{T_{sx}}\left(\frac{\ell x}{2} - \frac{x^2}{2}\right). \qquad (7.14)$$

Equation (7.14) defines a parabolic curve of static deflection for the cable with small sag.

The slope at the centre of the horizontally suspended cable is zero at its mid-span due to symmetry which is consistent with Eq. (7.14). This point is where maximum sag occurs. Using Eq. (7.14) the maximum sag is given by

$$w_s^{\max} = \frac{\rho A g \ell^2}{8 T_{sx}}.$$

One further piece of information is required to calculate the sag and corresponding horizontal tension for an inextensible cable, and that is the length of the cable.

The cable length L may be calculated by integrating the expression for the length of a cable element, Eq. (7.11), over the range $x = 0$ to $x = \ell$, giving

$$L = \int_0^\ell 1 + \frac{1}{2}\left(\frac{\mathrm{d}w_s}{\mathrm{d}x}\right)^2 \mathrm{d}x.$$

Eliminating w_s using Eq. (7.14) gives

$$L = \ell\left[1 + \frac{1}{24}\left(\frac{\rho g A \ell}{T_{sx}}\right)^2\right]. \tag{7.15}$$

This equation in conjunction with Eq. (7.14) allows for the calculation of the deflected shape for a given length of cable L and span ℓ. Note that Eq. (7.15) is a form of compatibility equation, which for real materials is usually in the from $\sigma = E\epsilon$ where ϵ is the strain, σ is stress and E is Young's modulus. As it is assumed that the cable is inextensible, this reduces to $\epsilon = 0$ and hence the cable length must equal the distance along the cable from $x = 0$ to $x = \ell$.

In reality a cable will be extensible. However if the sag is small and the Young's modulus is high, such as in steel, then there is good agreement between the deflected shape and the deflected shape of an inextensible cable, see Irvine and Caughey (1974).

7.1.3 Dynamic Deflection

A common method for modelling cable vibration is to consider the total deflection from the chord position, shown as a dotted line in Fig. 7.1a. The total deflection consists of a dynamic deflection added to the static deflection as defined above, $w(x, t) = w_s(x) + w_d(x, t)$, where w_s is the (time-independent) static deflection and w_d is the vibration (or dynamic) displacement. The tension is represented in the same way, $T(x, t) = T_s(x) + T_d(x, t)$. In the following analysis, it is assumed that the cable tension is high and so the deflection is low. This means that the approximations $\frac{ds}{dx} \simeq 1$, $\cos\phi \simeq 1$ and $\sin\phi \simeq \tan\phi \simeq \phi$ can be made.

First, consider the vertical direction. The force equilibrium equation, Eq. (7.7), may be modified to include the additional dynamic tension and deflection:

$$\frac{\partial}{\partial x}\{(T_s + T_d)\sin\phi\} + \rho A g = \rho A \frac{\partial^2}{\partial t^2}\{w_s + w_d\} = \rho A \frac{\partial^2 w_d}{\partial t^2}. \tag{7.16}$$

Notice that the derivatives are now partial as T_d and w_d are functions of time and x.

Now consider the horizontal direction. Because the sag is assumed to be small, the stiffness in the horizontal direction is very high in comparison to the vertical stiffness. Therefore the frequencies of vibration of the horizontal modes will be much higher. It is therefore assumed that over the frequency range where the transverse modes are excited, the longitudinal modes of the cable are in the quasi-static range and hence the longitudinal inertia can be ignored. Modifying Eq. (7.6) to include both the static and dynamic tension gives

$$\frac{\partial}{\partial x}\{(T_s + T_d)\cos\phi\} = 0. \tag{7.17}$$

The final modification that must be made is to express the relationship between rotation and displacement as

$$\tan\phi = \frac{\partial}{\partial x}\{w_s + w_d\}. \tag{7.18}$$

Note that in the dynamic analysis ϕ is a function of time as well as position x, since w_d is time-varying.

The method of simplifying these equations to give an equation that governs the transverse vibration of the cable is very similar to the method of deriving the static shape. The assumption that the sag is small also implies that ϕ is small, and this assumption is introduced at this point in the derivation (in the static analysis it is made after the nonlinear sag equation, Eq. (7.12), is derived). Equation (7.17) is integrated to give

$$(T_s + T_d)\cos\phi = T_{sx} + T_{dx}(t), \tag{7.19}$$

where the right-hand side is now a function of time but not position x. This can be simplified by noting that $\cos\phi \simeq 1$ as the sag is small. This is similar to the static analysis in which $T_s\cos\phi = T_{sx}$ was simplified to $T_s = T_{sx}$ using the small sag assumption, where T_{sx} is the static tension in the cable in the horizontal direction.

Using Eq. (7.19), the term $(T_s + T_d)$ in Eq. (7.16) may be replaced to give

$$(T_{sx} + T_{dx})\frac{\partial\phi}{\partial x} + \rho A g = \rho A\frac{\partial^2 w_d}{\partial t^2},$$

where $\sin\phi \simeq \phi$ as the sag is small. Equation (7.18) with $\tan\phi \simeq \phi = \frac{\partial}{\partial x}\{w_s + w_d\}$, is now used to eliminate ϕ to give

$$(T_{sx} + T_{dx})\frac{\partial^2}{\partial x^2}\{w_s + w_d\} + \rho A g = \rho A\frac{\partial^2 w_d}{\partial t^2}. \tag{7.20}$$

From the static analysis, once it was assumed the sag is small, it was found that $T_{sx}\frac{\mathrm{d}^2 w_s}{\mathrm{d}x^2} = -\rho A g$, Eq. (7.13). Eliminating this term from Eq. (7.20) gives

$$T_{dx}\frac{\partial^2 w_s}{\partial x^2} + (T_{sx} + T_{dx})\frac{\partial^2 w_d}{\partial x^2} = \rho A\frac{\partial^2 w_d}{\partial t^2}. \tag{7.21}$$

Equation (7.21) governs the vertical vibration of the cable shown in Fig. 7.1. Note, however, that the equation contains both horizontal tension and vertical deflection terms. A further piece of information is required to relate these terms. This is the compatibility equation which relates stress and strain (and hence tension). This will be discussed in the next section.

Up to now the vertical vibration of a cable has been considered. However, vibrating the cable in the vertical plane also affects the out-of-plane vibration (or sway motion) of the cable. In fact, much of the nonlinear modelling of cables is focused on the coupling between vertical in-plane vibration and out-of-plane vibration. To derive a model for this, an additional degree of freedom in the out-of-plane direction is required. This is discussed next. In addition, two other features are added to the model that are relevant for practical applications. One is to consider the cable inclined to the horizontal, the second is to include a support motion, which provides a form of dynamic excitation to the cable.

In the next section, the derivation is extended to enable an inclined cable forced via vertical motion of the lower support to be analysed. In practice, this model can be used to understand the dynamics of cables used in cable-stayed bridges when excited by deck motion.

7.2 Inclined Cable Vibration

Now the model of a horizontal cable developed in the previous section is extended to allow modelling of inclined cables subject to motion of the lower support. This is done by considering the forces on a cable element, which leads to partial differential equations including a dynamic tension term, and then the cable strain which provides an equation for the dynamic tension term. Together these two equations capture the free vibration cable dynamics. Following this, the excitation due to the lower support motion is considered by splitting the dynamic response into a *quasi-static* component, which satisfies the changing boundary conditions due to the support motion but does not include dynamic terms, and the *modal motion*, which captures the dynamic response of the cable. During this derivation, restrictive conditions are made to ensure the dynamics are linear. In the next section, Sect. 7.3, these restrictive conditions are relaxed, resulting in modal expressions for the nonlinear dynamics of an inclined cable.

An example of an inclined cable is shown in Fig. 7.2. Notice that, in the inclined case, there are now x, y and z coordinates and that this frame of reference is rotated by the angle of inclination, θ. The y direction now corresponds to the out-of-plane motion, $v(x, t)$. As before, the cable can also vibrate along the x direction, but these vibrations are usually excluded from models, because the frequency of oscillations is much faster (and of smaller amplitude) than that in the w and v directions. In addition

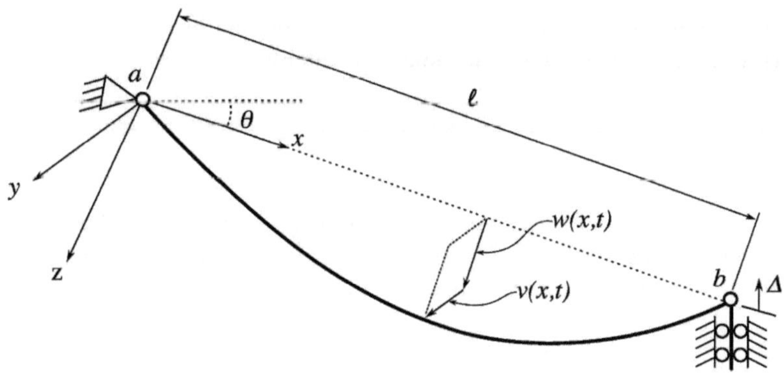

Fig. 7.2 An inclined cable with in-plane and out-of-plane motion and support input amplitude Δ

it is assumed that the cable sag is small in comparison to the length of the cable and hence the cable is taut. Now consider the motion of the cable. The quantities u, v and w are the axial, out-of-plane transverse and in-plane transverse displacements of the cable, in the x, y and z directions respectively, as shown in Fig. 7.2.

7.2.1 Force Balance

First, the static force balance and dynamic equations of motion must be modified to account for the inclination to the horizontal. Considering the small element shown in Fig. 7.1b, along the z-axis the gravitational term $\rho A g \Delta s$ is now replaced by $\rho A g \Delta s \cos\theta$. In addition, there is now a gravitational term $\rho A g \Delta s \sin\theta$ acting in the positive x direction.

In the x direction, the static force balance, Eq. (7.6), now becomes

$$\rightarrow_x \quad \frac{\mathrm{d}}{\mathrm{d}x}\{T_s \cos\phi\} = -\rho g A \sin\theta.$$

Integrating and redefining T_{sx} as the static stress acting in the x-axis direction at $x = 0$ gives

$$T_s \cos\phi = T_{sx} - \rho g A x \sin\theta.$$

It is assumed that the cable sag is small, therefore the static tension must be large. As a result, the gravitational term, which at its limit is equal to the cable weight, may be assumed to be small in comparison with T_{sx} and can therefore be neglected. This results in the same x-axis relationship as derived for the horizontal cable. By the same argument the gravitational component may be ignored for the dynamic analysis such that Eq. (7.19), $T_s + T_d = T_{sx} + T_{dx}(t)$, remains unaltered. Note that as before T_{sx} is a constant and $T_{dx}(t)$ is a function of time but not of x.

Modifying the gravitational term in the z direction, the resulting static displaced shape (Eq. (7.11) for the horizontal cable) is given by

$$w_s = \frac{\gamma A}{T_{sx}}\left(\frac{\ell x}{2} - \frac{x^2}{2}\right), \tag{7.22}$$

where $\gamma = \rho g \cos\theta$. This gives the sag displacement relationship. The z-axis equation of motion for the dynamic analysis (Eq. (7.21) for the horizontal cable) remains unchanged except that the static deflection is altered

$$\downarrow_z \quad T_{dx}\frac{\partial^2 w_s}{\partial x^2} + (T_{sx} + T_{dx})\frac{\partial^2 w_d}{\partial x^2} = \rho A \frac{\partial^2 w_d}{\partial t^2}. \tag{7.23}$$

Now that the cable is being considered in three-dimensions, a further equation of motion must be derived for the y-direction. This equation is derived in the same way as the z-axis equation of motion, except in this case there is no static deflection. Hence using Eq. (7.21), the equation of motion may be written as

$$\swarrow_y \quad (T_{sx} + T_{dx})\frac{\partial^2 v_d}{\partial x^2} = \rho A \frac{\partial^2 v_d}{\partial t^2}, \tag{7.24}$$

where v_d is the dynamic deflection in the y-axis direction (the corresponding static deflection v_s is zero).

7.2.2 Strain

If the cable remains within its linear elastic range, the stress can be expressed as a linear function of strain $\epsilon(x, t)$ and the Young's modulus of the cable E. As the strain is small in materials of high Young's modulus such as steel, the dynamic strain can be written in terms of the dynamic and static lengths giving

$$T_{dx} = AE\epsilon_d,$$

where

$$\epsilon_d = \frac{\Delta s_d - \Delta s_s}{\Delta s_s},$$

and Δs_s and Δs_d are defined in Fig. 7.3. These lengths may be written as

$$\Delta s_s = \sqrt{\Delta x^2 + \Delta w_s^2},$$
$$\Delta s_d = \sqrt{(\Delta x + \Delta u_d)^2 + (\Delta v_d)^2 + (\Delta w_s + \Delta w_d)^2}.$$

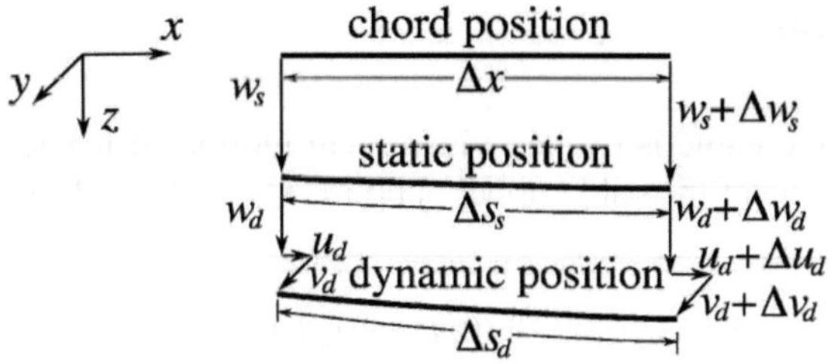

Fig. 7.3 The deflections and extensions of a small section of cable from the chord position to the static sag position and then to the dynamic position

Taking the limit as $\Delta x \to 0$ gives

$$T_{dx} = AE\left[\frac{\sqrt{\left(1+\frac{\partial u_d}{\partial x}\right)^2+\left(\frac{\partial v_d}{\partial x}\right)^2+\left(\frac{\mathrm{d}w_s}{\mathrm{d}x}+\frac{\partial w_d}{\partial x}\right)^2}}{\sqrt{1+\left(\frac{\mathrm{d}w_s}{\mathrm{d}x}\right)^2}}-1\right].$$

Expanding out the squared terms and then using the Taylor series expansions $\sqrt{1+\delta} = 1+\delta/2-\delta^2/8+\cdots$ and $1/(\sqrt{1+\delta}) = 1-\delta/2+\cdots$ gives

$$\begin{aligned}T_{dx} = AE\Bigg\{&\Bigg[1+\frac{\partial u_d}{\partial x}+\frac{1}{2}\left(\frac{\partial v_d}{\partial x}\right)^2+\frac{1}{2}\left(\frac{\mathrm{d}w_s}{\mathrm{d}x}\right)^2+\frac{1}{2}\left(\frac{\partial w_d}{\partial x}\right)^2\\&+\frac{\mathrm{d}w_s}{\mathrm{d}x}\frac{\partial w_d}{\partial x}+\mathscr{O}\left(\frac{\partial \bullet^3}{\partial x}\right)\Bigg]\\&\times\left[1+\frac{1}{2}\left(\frac{\mathrm{d}w_s}{\mathrm{d}x}\right)^2+\mathscr{O}\left(\frac{\partial \bullet^4}{\partial x}\right)\right]-1\Bigg\},\end{aligned}$$

where $\partial\bullet/\partial x$ indicates the partial differential of u_d, v_d, w_s or w_d with respect to x. Note that, in the Taylor series expansion for the $\sqrt{1+\delta}$ type term, the $\delta^2/8$ term is required as it produces one partial differential term with order less that 3. Finally, expanding the terms gives

$$T_{dx} = AE\left[\frac{\partial u_d}{\partial x}+\frac{1}{2}\left(\frac{\partial v_d}{\partial x}\right)^2+\frac{1}{2}\left(\frac{\partial w_d}{\partial x}\right)^2+\frac{\mathrm{d}w_s}{\mathrm{d}x}\frac{\partial w_d}{\partial x}+\mathscr{O}\left(\frac{\partial \bullet^3}{\partial x}\right)\right], \quad (7.25)$$

where it is worth noting again that, due to the assumptions that there is no significant dynamic response along the x-axis and that the sag is small, T_{dx} is a function of time but not position x.

7.2.3 Excitation

In this analysis, the cable is excited via vertical motion of the lower support. The boundary conditions at the cable upper support a, $x = 0$, and lower cable support b, $x = \ell$, are:

$$\begin{aligned} u(0,t) &= 0, \qquad v(0,t) = 0, \; w(0,t) = 0, \\ u(\ell,t) &= u_b(t), \; v(\ell,t) = 0, \; w(\ell,t) = w_b(t). \end{aligned}$$

To satisfy these time-varying boundary conditions, it is assumed that the cable response can be separated into two components, a quasi-static component, which satisfies the boundary conditions but has no significant dynamic response (although they are time-varying as the boundary conditions are time-varying) and a modal component which represents the dynamic response of a cable that has static supports at both ends. This split is represented as

$$\begin{aligned} u_d(x,t) &= u_q(x,t) + u_m(x,t), \\ v_d(x,t) &= v_q(x,t) + v_m(x,t), \qquad (7.26) \\ w_d(x,t) &= w_q(x,t) + w_m(x,t), \\ T_{dx}(t) &= T_{qx}(t) + T_{mx}(t), \qquad (7.27) \end{aligned}$$

where subscripts q and m indicate the quasi-static and the modal-dynamic components. These expressions can now be substituted into the equations of motion and the compatibility expression. For complete decoupling of the quasi-static and modal terms these expressions must be linearized. For the equations of motion, this requires the assumption that the tension due to the dynamics is small compared to the static tension. As a result, using these expressions the equations of motion for the cable, Eqs. (7.23) and (7.24), may be split into quasi-static and modal terms

$$T_{qx}\frac{\partial^2 w_s}{\partial x^2} + T_{sx}\frac{\partial^2 w_q}{\partial x^2} = 0, \quad T_{mx}\frac{\partial^2 w_s}{\partial x^2} + T_{sx}\frac{\partial^2 w_m}{\partial x^2} = \rho A\frac{\partial^2 w_m}{\partial t^2}, \qquad (7.28)$$

where $\frac{\partial^2 w_q}{\partial t^2}$ is ignored as w_q is quasi-static, and

$$T_{sx}\frac{\partial^2 v_q}{\partial x^2} = 0, \quad T_{sx}\frac{\partial^2 v_m}{\partial x^2} = \rho A\frac{\partial^2 v_m}{\partial t^2}, \qquad (7.29)$$

where again the dynamic term $\frac{\partial^2 v_q}{\partial t^2}$ is ignored.

For the strain relationship, Eq. (7.25), the $\frac{\partial \bullet}{\partial x}^2$-type terms result in cross-coupled terms $\frac{\partial w_q}{\partial x}\frac{\partial w_m}{\partial x}$ and $\frac{\partial v_q}{\partial x}\frac{\partial v_m}{\partial x}$. The strain relationship is linearized by assuming that the $\frac{\partial \bullet}{\partial x}^2$-type terms are small to give the linearized relationship

$$T_{dx} = AE\left[\frac{\partial u_d}{\partial x} + \frac{1}{2}\left(\frac{\partial v_d}{\partial x}\right)^2 + \frac{1}{2}\left(\frac{\partial w_d}{\partial x}\right)^2 + \frac{\mathrm{d}w_s}{\mathrm{d}x}\frac{\partial w_d}{\partial x}\right]$$

$$\rightsquigarrow T_{dx} = AE\left[\frac{\partial u_d}{\partial x} + \frac{\mathrm{d}w_s}{\mathrm{d}x}\frac{\partial w_d}{\partial x}\right]. \qquad (7.30)$$

This requires that w_d and v_d must be much smaller than the sag displacement w_s and therefore that the vibrations must be very small. Using the Eq. (7.27) relationships gives

$$T_{qx} = AE\left[\frac{\partial u_q}{\partial x} + \frac{\mathrm{d}w_s}{\mathrm{d}x}\frac{\partial w_q}{\partial x}\right], \quad T_{mx} = AE\left[\frac{\partial u_m}{\partial x} + \frac{\mathrm{d}w_s}{\mathrm{d}x}\frac{\partial w_m}{\partial x}\right]. \qquad (7.31)$$

This is the strain compatibility condition. Later in the analysis, in Sect. 7.3, the full equations of motion and strain will be considered.

7.2.4 *Quasi-Static Motion*

This motion is due to the displacement of the cable supports; the resulting quasi-static motion may be viewed as the movement of a mass-less elastic tendon stretched between the supports.

Applying a double integration to the y-axis relationship, Eq. (7.29), and noting that T_{sx} is not a function of x, results in

$$T_{sx}\frac{\partial^2 v_q}{\partial x^2} = 0 \quad \rightsquigarrow \quad v_q = c_1 x + c_2,$$

where c_1 and c_2 are integration constants. Applying the boundary conditions, Eq. (7.26), gives $v_q = 0$. For the z-axis relationship, Eq. (7.28), noting that T_{sx} and T_{qx} are not functions of x, eliminating the sag displacement using Eq. (7.22), and integrating twice with respect to x gives

$$T_{qx}\frac{\partial^2 w_s}{\partial x^2} + T_{sx}\frac{\partial^2 w_q}{\partial x^2} = 0 \quad \rightsquigarrow \quad w_q = \frac{\gamma A T_{qx}}{2T_{sx}^2}x^2 + c_3 x + c_4,$$

where c_3 and c_4 are integration constants. Applying the boundary conditions gives

$$w_q = \frac{\gamma A T_{qx}\ell^2}{2T_{sx}^2}\left[\left(\frac{x}{\ell}\right)^2 - \frac{x}{\ell}\right] + w_b\frac{x}{\ell}.$$

This leaves T_{qx} and u_q, which are unknown. These can be found by considering the strain equation. Substituting expressions for w_s and w_q into Eq. (7.31) results in

$$T_{qx} = AE\left[\frac{\partial u_q}{\partial x} + \frac{\gamma A\ell}{T_{sx}}\left(\frac{1}{2} - \frac{x}{\ell}\right)\left(\frac{w_b}{\ell} + \frac{T_{qx}\gamma A\ell}{T_{sx}^2}\left(\frac{x}{\ell} - \frac{1}{2}\right)\right)\right].$$

Integrating with respect to x and rearranging gives

$$u_q = \frac{T_{qx}}{EA}x - \frac{\gamma A\ell^2}{T_{sx}}\left[\frac{w_b}{2}\left(\frac{x}{\ell} - \left(\frac{x}{\ell}\right)^2\right) + \frac{T_{qx}\gamma A\ell}{4T_{sx}^2}\left(2\left(\frac{x}{\ell}\right)^2 - \frac{4}{3}\left(\frac{x}{\ell}\right)^3 - \frac{x}{\ell}\right)\right] + e,$$

where e is an integration constant. Applying the two boundary conditions gives $e = 0$ and

$$u_b = \frac{\ell}{EA}T_{qx} + \frac{\gamma^2 A^2 \ell^3}{12T_{sx}^3}T_{qx}.$$

This can be rearranged to give an expression for T_{qx} in terms of the support motion u_b

$$T_{qx} = AE_q\frac{u_b}{\ell}, \quad E_q = \frac{E}{1+\lambda^2/12}, \quad \lambda^2 = \frac{E\gamma^2\ell^2A^3}{T_{sx}^3},$$

where λ^2 is Irvine's parameter. Finally, this expression for the tension can be substituted into the equations for w_q and u_q to give

$$\begin{aligned} u_q &= \frac{E_q u_b}{E}\left(\frac{x}{\ell}\right) - \frac{\gamma A\ell w_b}{2T_{sx}}\left[\left(\frac{x}{\ell}\right) - \left(\frac{x}{\ell}\right)^2\right] \\ &\quad + \frac{\lambda^2 E_q u_b}{4E}\left[\left(\frac{x}{\ell}\right) - 2\left(\frac{x}{\ell}\right)^2 + \frac{4}{3}\left(\frac{x}{\ell}\right)^3\right], \end{aligned} \tag{7.32}$$

$$v_q = 0, \tag{7.33}$$

$$w_q = w_b\left(\frac{x}{\ell}\right) - \frac{\gamma E_q \ell A^2 u_b}{2T_{sx}^2}\left[\left(\frac{x}{\ell}\right) - \left(\frac{x}{\ell}\right)^2\right]. \tag{7.34}$$

Next the dynamic, modal, motion of the cable is considered.

7.2.5 Modal Motion

The dynamic behaviour of the cables is represented by the modal displacements u_m, v_m and w_m which must satisfy simply-supported boundary conditions at both ends of the cable (at $x = 0$ and $x = \ell$, $u_m = v_m = w_m = 0$ for all time). As with the horizontal cable, the axial dynamic displacement is assumed to be small. In line with the Galerkin method, the transverse displacements are separated into functions of position x and time to give

$$u_m(x,t) = 0, \quad v_m(x,t) = \sum_{n=1}^{\infty} \phi_n(x) y_n(t), \quad w_m(x,t) = \sum_{n=1}^{\infty} \psi_n(x) z_n(t), \tag{7.35}$$

where the spatial functions $\phi(x)$ and $\psi(x)$ are the in-plane and out-of-plane linear modes of a cable with fixed ends, and $y_n(t)$ and $z_n(t)$ their corresponding time-dependent generalized coordinates.

In this discussion, the linearized equation for strain, Eq. (7.30), is used to investigate the mode-shapes $\phi(x)$ and $\psi(x)$ and their corresponding frequencies. The more realistic nonlinear strain equation is used later, in the development of the nonlinear equations of motion, see Sect. 7.3.

First, the dynamics in the y-direction are considered. Substituting the modal expression for v_m into Eq. (7.29) gives

$$T_{sx} \sum_{n=1}^{\infty} \frac{\mathrm{d}^2 \phi_n}{\mathrm{d}x^2} y_n = \rho A \sum_{n=1}^{\infty} \phi_n \frac{\mathrm{d}^2 y_n}{\mathrm{d}t^2}. \tag{7.36}$$

Letting $\ddot{y}_n = -\omega_{yn}^2 y_n$, then considering mode n gives

$$T_{sx} \frac{\mathrm{d}^2 \phi_n}{\mathrm{d}x^2} + \rho A \omega_{yn}^2 \phi_n = 0 \tag{7.37}$$

for all n. Solving this linear differential equation and applying the boundary conditions $v_m(0,t) = v_m(\ell,t) = 0$ leads to

$$\phi_n = \sin\left(n\pi \frac{x}{\ell}\right), \quad \omega_{yn} = \frac{n\pi}{\ell}\sqrt{\frac{T_{sx}}{\rho A}}. \tag{7.38}$$

Note that, since the dynamics in this case are linear, the modal decomposition has been achieved using separation of variables. As a result, the mode-shapes are orthogonal. The same result can be achieved using the Galerkin method. The Galerkin method applied to Eq. (7.36) gives the orthogonality conditions

$$\int_0^{\ell} \phi_n \phi_k dx = \int_0^{\ell} \frac{\mathrm{d}^2 \phi_n}{\mathrm{d}x^2} \phi_k dx = 0, \qquad n \neq k, \tag{7.39}$$

which is the same relationship found for linear beams in Chap. 6. In effect, the assumption is that a series linear mode-shapes is being added to the static deflection to give the overall response of the cable. (Note also that these modes do not affect the compatibility condition, Eq. (7.31).)

Now consider the z-direction modal response. Here both the dynamics and compatibility must be considered. Substituting the modal expression into Eq. (7.28) gives

$$\frac{\mathrm{d}^2 w_s}{\mathrm{d}x^2}\sum_{n=1}^{\infty} T_{nx} + T_{sx}\sum_{n=1}^{\infty}\frac{\mathrm{d}^2\psi_n}{\mathrm{d}x^2}z_n = \rho A\sum_{n=1}^{\infty}\psi_n\frac{\mathrm{d}^2 z_n}{\mathrm{d}t^2},$$

where the dynamic tension has been split into a summation of modal components $T_{mx}(t) = \sum_{n=1}^{\infty} T_{nx}(t)$. Considering the nth mode, writing $\ddot{z}_n = -\omega_{zn}^2 z_n$ and using the sag displacement relationship, Eq. (7.22), results in

$$T_{sx}\frac{\mathrm{d}^2\psi_n}{\mathrm{d}x^2}z_n + \rho A\omega_{zn}^2\psi_n z_n = \frac{\gamma A}{T_{sx}}T_{nx}, \tag{7.40}$$

for all n. The compatibility equation, Eq. (7.31), can also be split into individual modal components. The nth may be expressed as

$$T_{nx}(t) = AE\left[\frac{\partial\theta_n}{\partial x}u_n(t) + \frac{\mathrm{d}w_s}{\mathrm{d}x}\frac{\mathrm{d}\psi_n}{\mathrm{d}x}z_n(t)\right], \tag{7.41}$$

where the axial dynamic displacement has been split into modal components $u_m = \sum_{n=1}^{\infty}\theta_n(x)u_n(t)$. It has already been stated that this displacement can be assumed to be zero, but it is useful to keep it in the expression for a moment. To apply boundary conditions to this expression, it is integrated with respect to x from $x = 0$ to $x = \ell$ to give

$$T_{nx}(t)\frac{\ell}{AE} = [\theta_n]_0^{\ell}u_n(t) + \left[\frac{\mathrm{d}w_s}{\mathrm{d}x}\psi_n\right]_0^{\ell} z_n(t) - \int_0^{\ell}\frac{\mathrm{d}^2 w_s}{\mathrm{d}x^2}\psi_n\mathrm{d}x\, z_n(t), \tag{7.42}$$

where integration by parts has been applied to the last term in Eq. (7.41). Applying the boundary conditions and eliminating the sag displacement relationship using Eq. (7.22), gives

$$T_{nx}(t) = \frac{\gamma A^2 E}{T_{sx}\ell}\int_0^{\ell}\psi_n\mathrm{d}x\, z_n(t). \tag{7.43}$$

Note that finding ψ_n expressions that satisfy this condition is not sufficient to satisfy compatibility. Rather, Eq. (7.41) must then be used to find a corresponding expression for θ_n and u_n in terms of ψ_n and z_n to ensure $T_{nx}(t)$ is not a function of x. However, this step is not taken in this analysis since, for cables with small sag, these expressions for u_n are small.

Equation (7.43) shows that anti-symmetric modes will have $T_{nx}(t) = 0$. This simplifies Eq. (7.40) such that it has the same form as the modal dynamics in the y-direction, Eq. (7.37). Therefore the modes are of the same form, namely

$$\psi_n = \sin\left(n\pi\frac{x}{\ell}\right), \quad \omega_{zn} = \frac{n\pi}{\ell}\sqrt{\frac{T_{sx}}{\rho A}}, \tag{7.44}$$

except that in the z-direction these expressions are valid only for $n = 2, 4, 6 \ldots$. This is because the odd values of n correspond to symmetric modes and hence $T_{nx} \neq 0$. Since the equation of motion for the even, or anti-symmetric, in-plane modes are the same as for the out-of-plane modes, the orthogonality conditions are also the same:

$$\int_0^\ell \psi_n \psi_k dx = \int_0^\ell \frac{d^2\psi_n}{dx^2}\psi_k dx = 0, \qquad n \neq k. \tag{7.45}$$

For the symmetric z-direction modes (which are labelled modes $n = 1, 3, 5 \ldots$ to match the y-direction symmetric modes) the linear differential equation, Eq. (7.40), can be solved to give

$$\psi_n(x)z_n(t) = \frac{\gamma}{T_{sx}\omega_{zn}^2\rho}T_{nx}(t) + a_{cf}(t)\sin\left(B_n\frac{x}{\ell}\right) + b_{cf}(t)\cos\left(B_n\frac{x}{\ell}\right),$$

where the first term on the right-hand side is the particular integral, the other two terms make up the complementary function, $a_{cf}(t)$ and $b_{cf}(t)$ are unknowns and

$$B_n^2 = \frac{\rho A \ell^2}{T_{sx}}\omega_{zn}^2. \tag{7.46}$$

The boundary condition $\psi_n(0, t) = 0$ fixes $b_{cf}(t)$ to give

$$\psi_n(x)z_n(t) = \frac{\gamma}{T_{sx}\omega_{zn}^2\rho}T_{nx}(t)\left[1 - \cos\left(B_n\frac{x}{\ell}\right)\right] + a_{cf}(t)\sin\left(B_n\frac{x}{\ell}\right).$$

The remaining boundary condition $\psi_n(\ell, t) = 0$ gives

$$a_{cf}(t) = \frac{\gamma}{T_{sx}\omega_{zn}^2\rho}T_{nx}(t)\frac{\cos B_n - 1}{\sin B_n} = -\frac{\gamma}{T_{sx}\omega_{zn}^2\rho}T_{nx}(t)\tan\left(\frac{B_n}{2}\right),$$

where the trigonometric identities $\tan(B_n/2) = (\sin B_n)/(1+\cos B_n)$ and $\cos^2 B_n + \sin^2 B_n = 1$ have been used. This gives

$$w_n(x, t) = \psi_n(x)z_n(t) = \frac{\gamma}{T_{sx}\omega_{zn}^2\rho}T_{nx}(t)\left[1 - \cos\left(B_n\frac{x}{\ell}\right) - \tan\left(\frac{B_n}{2}\right)\sin\left(B_n\frac{x}{\ell}\right)\right], \tag{7.47}$$

where the resulting mode-shape may be written in the form

$$1 - \cos\left(B_n \frac{x}{\ell}\right) - \tan(\frac{B_n}{2}) \sin\left(B_n \frac{x}{\ell}\right).$$

For convenience, this can be written in a normalized form, such that the peak mode-shape deflection is unity, as

$$\psi_n(x) = \left[1 - \sec\left(\frac{B_n}{2}\right)\right]^{-1} \left[1 - \cos\left(B_n \frac{x}{\ell}\right) - \tan\left(\frac{B_n}{2}\right) \sin\left(B_n \frac{x}{\ell}\right)\right]. \quad (7.48)$$

Finally, the corresponding natural frequency, ω_{zn}, can be found by deriving an expression for B_n, which is related to ω_{zn} via Eq. (7.46). This is achieved by substituting the expression for w_n, Eq. (7.47), into the compatibility relationship, Eq. (7.43), to give

$$\tan\left(\frac{B_n}{2}\right) = \frac{B_n}{2} - \frac{B_n^3}{2\lambda^2}. \quad (7.49)$$

This is the transcendental frequency equation. In deriving this expression, the relationships $\tan(B_n/2) = (\sin B_n)/(1 + \cos B_n)$ and $\lambda^2 = E\gamma^2\ell^2 A^3/T_{sx}^3$ are used.

The orthogonality of these symmetric modes can be checked by using the Galerkin method. Multiplying Eq. (7.40) by ψ_k and integrating over the length gives

$$T_{sx} \int_0^\ell \psi_k(x) \frac{\mathrm{d}^2\psi_n(x)}{\mathrm{d}x^2} \mathrm{d}x \; z_n(t) - \frac{\gamma A}{T_{sx}} T_{nx}(t) \int_0^\ell \psi_k(x) \mathrm{d}x = \rho A \int_0^\ell \psi_k(x)\psi_n(x) \mathrm{d}x \; \ddot{z}_n(t),$$

where it is noted that T_{nx} is not a function of x. Substituting the equation for T_{nx}, Eq. (7.43), gives

$$\left[T_{sx} \int_0^\ell \psi_k(x) \frac{\mathrm{d}^2\psi_n(x)}{\mathrm{d}x^2} \mathrm{d}x - \frac{\gamma^2 A^3 E}{T_{sx}^2 \ell} \int_0^\ell \psi_n(x) \mathrm{d}x \int_0^\ell \psi_k(x) \mathrm{d}x\right] z_n(t)$$

$$= \left[\rho A \int_0^\ell \psi_k(x)\psi_n(x) \mathrm{d}x\right] \ddot{z}_n(t). \quad (7.50)$$

For orthogonality of the modes, the expressions in the square brackets must be zero when $k \neq n$, giving the conditions

$$\int_0^\ell \psi_n \psi_k dx = 0, \quad n \neq k, \quad (7.51)$$

$$\int_0^{\ell} \psi_k(x)\frac{\mathrm{d}^2\psi_n(x)}{\mathrm{d}x^2}\mathrm{d}x - \frac{\lambda^2}{\ell^3}\int_0^{\ell} \psi_n(x)\mathrm{d}x \int_0^{\ell} \psi_k(x)\mathrm{d}x = 0, \quad n \neq k. \tag{7.52}$$

These conditions are met by the mode-shape given in Eq. (7.48) along with the transcendental frequency equation, Eq. (7.49). It is convenient to note that these orthogonality conditions hold true for the even modes as well. This is due to the form of ψ_n for the even modes, given by Eq. (7.44), which results in $\int_0^{\ell} \psi_n(x)\mathrm{d}x = 0$ for even values of n. As a result, these orthogonality conditions reduce to those given in Eq. (7.45).

The orthogonality conditions and Eq. (7.50), allow the decoupled equation of motion for the nth in-plane mode, for both even and odd n, to be written in terms of its mode-shape

$$T_{sx}\left[\int_0^{\ell} \psi_n(x)\frac{\mathrm{d}^2\psi_n(x)}{\mathrm{d}x^2}\mathrm{d}x - \frac{\lambda^2}{\ell^3}\left(\int_0^{\ell} \psi_n(x)\mathrm{d}x\right)^2\right] z_n(t) = \rho A \int_0^{\ell} \psi_n^2(x)\mathrm{d}x \, \ddot{z}_n(t), \tag{7.53}$$

where ψ_n is given by Eq. (7.48) for odd n and Eq. (7.44) for even n. Alternatively, the equation of motion for the nth mode may be written in terms of the natural frequency ω_{zn}

$$\ddot{z}_n(t) + \omega_{zn}^2 z_n(t) = 0, \tag{7.54}$$

where ω_{zn} can be calculated from Eqs. (7.46) and (7.49) for odd n and from Eq. (7.44) for even n. These two equations given an expression for the natural frequency in terms of the mode-shapes:

$$\omega_{zn}^2 = \frac{T_{sx}}{\rho A}\frac{\frac{\lambda^2}{\ell^3}\left(\int_0^{\ell} \psi_n(x)\mathrm{d}x\right)^2 - \int_0^{\ell} \psi_n(x)\psi_n''(x)\mathrm{d}x}{\int_0^{\ell} \psi_n^2(x)\mathrm{d}x}, \tag{7.55}$$

where $\bullet'$ is the derivative with respect to x.[1]

To calculate the natural frequency for symmetric, $n = 1, 3, 5 \ldots$, in-plane modes, the transcendental equation, Eq. (7.49), must be solved. Figure 7.4 shows a graphical solution to this equation for a range of values of λ^2. For each value of λ^2 the points

[1] To validate this equation for odd n, note that $\phi_n'' = (B_n/\ell)^2[(1 - \sec(B_n/2))^{-1} - \phi_n]$ and that $\int_0^{\ell} \phi_n \mathrm{d}x = [1 - \sec(B_n/2)]^{-1}(\ell B_n^2/\lambda^2)$. Substituting these into Eq. (7.55) gives $\omega_n^2 = [T_{sx} B_n^2/(\rho A \ell^2)](\int_0^{\ell} \phi_n^2 \mathrm{d}x)/(\int_0^{\ell} \phi_n^2 \mathrm{d}x)$. So $\omega_n^2 = T_{sx} B_n^2/(\rho A \ell^2)$ which gives the same relationship for B_n as its definition given in Eq. (7.46).

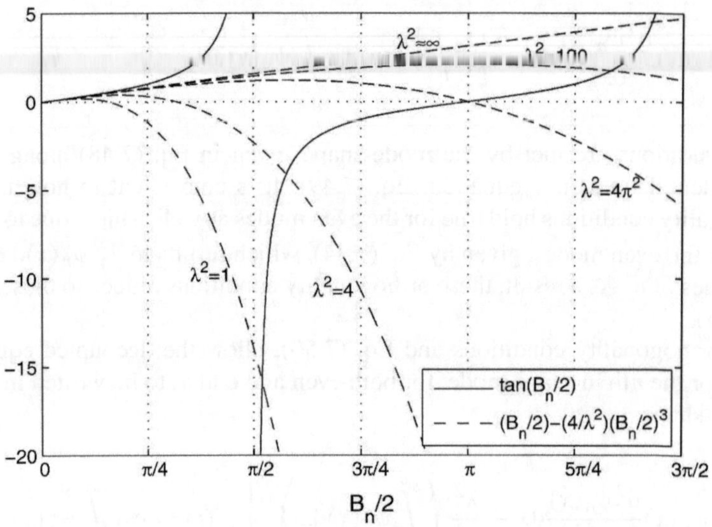

Fig. 7.4 Graphical solution to the transcendental equation for B_n, in which $\tan(B_n/2)$ and $(B_n/2)-(4\lambda^2)(B_n/2)^3$ are plotted and the crossing points mark valid solutions to Eq. (7.49), for a range of values of λ^2

where the $B_n/2 - B_n^3/(2\lambda^2)$ curve crosses the $\tan(B_n/2)$ curve marks a solution to the equation. The natural frequency can then be found using $\omega_n = B_n/\ell\sqrt{T_{sx}/(\rho A)}$, Eq. (7.46). It can be seen that for low values of λ^2, the $B_n/2 - (4/\lambda^2)(B_n/2)^3$ curve has an increasingly steep negative gradient with increasing $B_n/2$. As a result, the crossing points for the higher n values will be very close to $B_n/2 = n\pi/2$, which corresponds to the out-of-plane natural frequencies (and the frequencies if sag is ignored). Note that the second crossing point corresponds to $n = 3$ and the third to $n = 5$ etc. In fact, for λ^2 of order unity or less, only $n = 1$ will have a natural frequency significantly different from that of the out-of-plane modes.

Once the transcendental equation is solved to give the natural frequencies, the mode-shapes can be found using Eq. (7.48). Figure 7.5 shows the first in-plane mode-shape for a range of λ^2 values. For comparison, the first out-of-plane modeshape, $\sin(n\pi x/\ell)$ is also included. It can be seen that for low values of λ^2, say $\lambda^2 \leq 4$, the first in-plane mode-shape is almost identical to the first out-of-plane mode-shape. For the higher modes, low values of λ^2 imply that the mode-shapes will be almost exactly represented by $\sin(n\pi x/\ell)$. This can be seen by considering the normalized mode-shape equation, Eq. (7.48), and writing $B_n/2 = n\pi/2$ which, as discussed in the previous paragraph, closely matches the correct value of $B_n/2$ for $n = 3, 5, \ldots$.

By approximating the mode-shape as a sinusoid for odd values of n, an estimate of the natural frequency can be found using Eq. (7.55). Substituting in $\phi = \sin(n\pi x/\ell)$ for all n gives

$$\omega_{zn}^2 = \frac{T_{sx}}{\rho A \ell^2} n^2 \pi^2 (1 + k_n), \quad k_n = \frac{2\lambda^2}{n^4 \pi^4} (1 - (-1)^{n+1})^2. \tag{7.56}$$

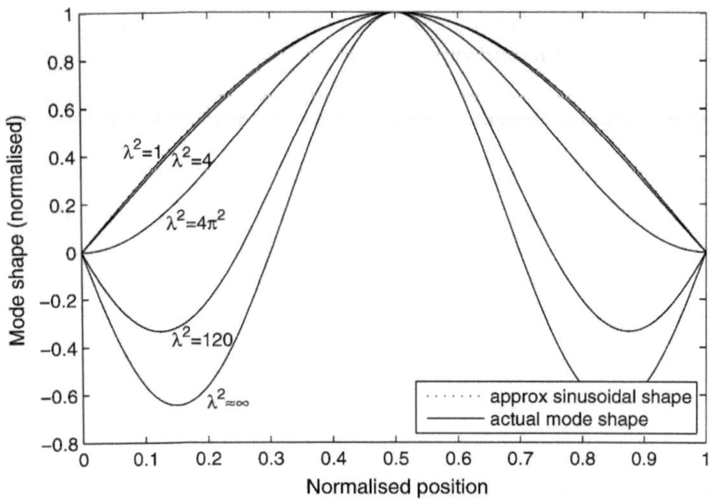

Fig. 7.5 Mode-shape for $n = 1$ for a range of values of λ^2, using Eq. (7.48)

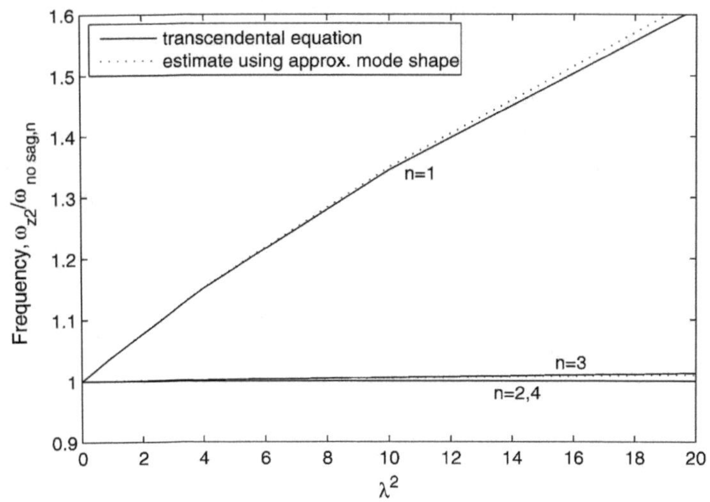

Fig. 7.6 Natural frequencies of the out-of-plane modes

The accuracy of this approximation for odd values of n can be assessed by comparing it to the natural frequency using the true mode-shapes, which can be found numerically by solving Eq. (7.49) in conjunction with Eq. (7.46). For the even modes the mode-shape is sinusoidal so Eq. (7.56) is not an approximation. This comparison is shown in Fig. 7.6 for the first four natural frequencies for a range of λ^2 values. The natural frequency is plotted in the form $\omega_{zn}/\omega_{nosag,n}$ where the natural frequency if there were no sag is given by $\omega_{nosag,n} = (n\pi/\ell)\sqrt{T_{sx}/(\rho A)}$ and matches the natural frequencies for the out-of-plane modes.

Note that to calculate the full cable response the quasi-static motion derived in Sect. 7.2.1 must also be included

$$u_d(x,t) = u_q(x,t), \quad v_d(x,t) = v_m(x,t), \quad w_d(x,t) = w_q(x,t) + w_m(x,t),$$

(noting that $u_m = v_q = 0$).

In this section the response of an inclined cable excited via vertical motion of the lower support has been derived for the case where the amplitude of response is small compared to the static sag due to gravity. This was achieved by firstly deriving force balance and strain equations for an element of cable. Then the cable response to the support excitation was considered by splitting the response into the sum of two parts, a quasi-static response which satisfies the time-dependent boundary conditions and a modal response which captures the dynamic response of the cable.

7.3 Nonlinear Cable Dynamics

In the previous section, the linearized equations of motion for the nth mode of vibration of an inclined cable, subject to vertical motion of the lower support were derived. These equations assumed that the modal and quasi-static motions were small compared with the static sag.

Now nonlinear equations of motion will be derived which include nonlinear dynamic tension terms, and the cross-coupling between quasi-static and modal terms. In the following derivation it is assumed that the quasi-static motion is small in comparison with the static sag. The justification for this is that the important dynamics occur near resonance, and at resonance the modal response is large for small support motion (and hence quasi-static motions). Note that, in contrast to the linear analysis, the modal motion is now *not* assumed to be small compared with the static sag.

Once the equation of motion is calculated, as with the linear analysis, it is desirable to find an orthogonal modal set. As a first approximation, the mode-shapes derived for the linear system will be used such that the modal components of the response are given by Eq. (7.35)

$$u_m(x,t) = 0, \quad v_m(x,t) = \sum_{n=1}^{\infty} \phi_n(x) y_n(t), \quad w_m(x,t) = \sum_{n=1}^{\infty} \psi_n(x) z_n(t),$$

where from Eqs. (7.38), (7.44) and (7.48) the mode-shapes are given by

$$\phi_n = \sin(n\pi \tfrac{x}{\ell}) \quad \text{for } n = 1, 2, 3, \ldots,$$

$$\psi_n = \begin{cases} \left[1 - \sec\left(\frac{B_n}{2}\right)\right]\left[1 - \cos\left(B_n \frac{x}{\ell}\right) - \tan\left(\frac{B_n}{2}\right)\sin\left(B_n \frac{x}{\ell}\right)\right] & \text{for } n = 1, 3, 5, \ldots, \\ \sin\left(n\pi \frac{x}{\ell}\right) & \text{for } n = 2, 4, 6, \ldots. \end{cases}$$

Since these mode-shapes are not an orthogonal set for the nonlinear equations, full decoupling of the modes will not be achieved, instead modal cross-coupling will be present (discussed in the final part of this section). Note that, for the full dynamic motion, the quasi-static motions given by Eqs. (7.32–7.34) must be added to the modal motions.

7.3.1 Compatibility

Consider the tension first. The nonlinear expression is given by Eq. (7.25) and is modified by replacing the dynamic deflections with the quasi-static and modal components, $u_d = u_q$, $v_d = v_m$ and $w_d = w_q + w_m$ to give

$$T_{dx} = AE\left[\frac{\partial u_q}{\partial x} + \frac{\mathrm{d}w_s}{\mathrm{d}x}\left(\frac{\partial w_q}{\partial x} + \frac{\partial w_m}{\partial x}\right) + \frac{1}{2}\left(\frac{\partial v_m}{\partial x}\right)^2 + \frac{1}{2}\left(\left(\frac{\partial w_q}{\partial x}\right)^2 + 2\frac{\partial w_q}{\partial x}\frac{\partial w_m}{\partial x} + \left(\frac{\partial w_m}{\partial x}\right)^2\right)\right],$$

where the $(\partial w_q/\partial x)^2$ term can be ignored as small, since the quasi-static deflections are taken to be small compared to the modal motion near resonance. Since T_{dx} is a function of time only, this equation can be integrated over x from $x = 0$ to $x = \ell$ and the boundary conditions can be applied, as in Eq. (7.42):

$$T_{dx} = EA\left[\frac{E_q u_b}{E\ell} + \frac{1}{2\ell}\int_0^\ell \sum_{k=1}^{\infty}\sum_{n=1}^{\infty}(\phi_k'\phi_n' y_k y_n + \psi_k'\psi_n' z_k z_n)\mathrm{d}x + \left(1 + 2\frac{E_q u_b}{\ell T_{sx}}\right)\frac{\gamma A}{\ell T_{sx}}\int_0^\ell \sum_{k=1}^{\infty}\psi_k z_k \mathrm{d}x\right],$$

where $\bullet'$ represents the derivative with respect to x.

In the equation for the dynamics in both the in- and out-of-plane directions, the dynamic tension appears only in summation with the static tension. In the linear analysis, the dynamic tension was ignored as small compared to the static tension. However, here it is not removed, but remains as a small nonlinear effect. Since the dynamic tension is small, small terms within the dynamic tension will have little effect on the overall dynamics. One such term exists for a taut cable where the sag is small: it is assumed that the tension due to the quasi-static motion is small in comparison with the static tension, $T_q = E_q A u_b/\ell \ll T_s$. Therefore the $E_q u_b/\ell T_{sx}$ component of the last term can be removed, to give

$$T_{dx} = EA\left[\frac{E_q u_b}{E\ell} + \frac{1}{2\ell}\int_0^\ell \sum_{k=1}^{\infty}\sum_{n=1}^{\infty}(\phi_k'\phi_n' y_k y_n + \psi_k'\psi_n' z_k z_n)\,\mathrm{d}x + \frac{\gamma A}{\ell T_{sx}}\int_0^\ell \sum_{k=1}^{\infty}\psi_k z_k \mathrm{d}x\right]. \tag{7.57}$$

7.3.2 Out-of-Plane Motion

Now the out-of-plane motion can be considered. The equation of motion was derived in Eq. (7.24) as:

$$(T_{sx} + T_{dx})\frac{\partial^2 v_m}{\partial x^2} = \rho A \frac{\partial^2 v_m}{\partial t^2},$$

where the substitution $v_d = v_m$ has been made ($v_q = 0$). In the linear analysis, the dynamic tension was assumed to be small compared to the static tension. However, now this is relaxed; it is assumed that the quasi-static tension contribution to the dynamic tension is small in comparison to the static tension. Applying Galerkin's method, multiplying the equation by ϕ_n, then integrating over $0 \leq x < \ell$ (noting that the tension is not a function of x) gives

$$(T_{sx} + T_{dx})\int_0^\ell \sum_{k=1}^{\infty} \phi_n \phi_k'' y_k \mathrm{d}x = \rho A \int_0^\ell \sum_{k=1}^{\infty} \phi_n \phi_k \ddot{y}_k \mathrm{d}x, \tag{7.58}$$

where v_m has been expressed as a summation of the modal contributions with their corresponding mode-shapes $v_m = \sum_{k=1}^{\infty} \phi_k y_k$, where ϕ_k is the mode-shape for the kth mode of the linearized system.

In the linear analysis, it has already been shown that the mode-shapes of the linear system have two orthogonality conditions, see Eq. (7.39). When $n = k$, substitute $\phi_n = \sin(n\pi x/\ell)$, and the expression for the mode-shape (corresponding to pinned-pinned boundary conditions) becomes:

$$\int_0^\ell (\phi_n)^2 \mathrm{d}x = \frac{\ell}{2}, \quad \int_0^\ell \phi_n'' \phi_n \mathrm{d}x = -\frac{n^2\pi^2}{2\ell}.$$

These equations simplify Eq. (7.58), so that

$$\frac{\rho A \ell}{2}\ddot{y}_n + (T_{sx} + T_{dx})\frac{n^2\pi^2}{2\ell} y_n = 0.$$

The last step is to substitute Eq. (7.57) in the expression for the dynamic tension,

$$\frac{\rho A \ell}{2}\ddot{y}_n + \left(T_{sx} + \frac{AE_q u_b}{\ell} + \frac{EA}{2\ell}\int_0^\ell \sum_{k=1}^{\infty}\sum_{n=1}^{\infty} (\phi_k' \phi_n' y_k y_n + \psi_k' \psi_n' z_k z_n) \mathrm{d}x \right.$$
$$\left. + \frac{\gamma EA^2}{\ell T_{sx}} \int_0^\ell \sum_{k=1}^{\infty} \psi_k z_k \mathrm{d}x \right) \frac{n^2\pi^2}{2\ell} y_n = 0,$$

then insert the mode-shapes and evaluate the integrals that the expression for the dynamic tension contains.

In evaluating the integrals within the dynamic tension, for ease of computation, the odd in-plane mode-shapes are assumed to be sinusoidal; $\psi_n \approx \sin(n\pi x/\ell)$ for $n = 1, 3, 5, \ldots$. The justification for this is that, first, for a taut cable, it is reasonable to assume that the dynamic tension is small compared to the static tension (although in this nonlinear analysis it is still taken to be significant) and, secondly, it has been shown that these mode-shapes are approximately sinusoidal for taut cables (see the discussion regarding Fig. 7.5). Therefore, this assumption introduces small errors into terms that are already small. The even in-plane and all the out-of-plane mode-shapes are sinusoidal regardless of this assumption. Substituting in sinusoidal representations for the mode-shapes gives

$$\frac{\rho A\ell}{2}\ddot{y}_n + \left[T_{sx} + \frac{AE_q u_b}{\ell} + \frac{EA\pi^2}{4\ell^2}\sum_{k=1}^{\infty} k^2(y_k^2 + z_k^2) \right.$$
$$\left. + \frac{\gamma EA^2}{T_{sx}\pi}\sum_{k=1}^{\infty}\frac{(1+(-1)^{k+1})}{k}z_k\right]\frac{n^2\pi^2}{2\ell}y_n = 0.$$

Rearranging this equation gives

$$m_{yn}(\ddot{y}_n + 2\zeta_{yn}\omega_{yn}\dot{y}_n + \omega_{yn}^2 y_n) + \sum_{k=1}^{\infty}\nu_{nk}y_n(y_k^2+z_k^2) + \sum_{k=1}^{\infty}2\beta_{nk}y_n z_k + 2\eta_n u_b y_n = 0, \tag{7.59}$$

where

$$m_{yn} = \frac{\rho A\ell}{2}, \qquad \omega_{yn} = \frac{n\pi}{\ell}\sqrt{\frac{T_{sx}}{\rho A}}, \qquad \nu_{nk} = \frac{EA\pi^4 n^2 k^2}{8\ell^3},$$
$$\beta_{nk} = \frac{\gamma EA^2\pi n^2}{4\ell T_{sx}}\left(\frac{1+(-1)^{k+1}}{k}\right), \qquad \eta_n = \frac{E_q A\pi^2 n^2}{4\ell^2},$$

and modal damping, with a damping ratio ζ_{yn}, has been introduced. Equation (7.59) is now in the "standard form" as defined by Warnitchai et al. (1995). This form will be used throughout as the key modal equation for out-of-plane cable vibrations.

7.3.3 In-Plane Motion

The equation of motion for the in-plane motion was derived in Eq. (7.23):

$$T_{dx}\frac{\partial^2 w_s}{\partial x^2} + (T_{sx} + T_{dx})\frac{\partial^2 (w_q + w_m)}{\partial x^2} = \rho A\frac{\partial^2 (w_q + w_m)}{\partial t^2},$$

where the substitution $w_d = w_q + w_m$ has been made. Expressing w_m as a summation of the modal contributions with their corresponding mode shapes $w_m = \sum_{k=1}^{\infty} \psi_k z_k$, where ψ_k is the mode-shape for the kth mode of the linearized system, gives

$$T_{dx}\left(w_s'' + w_q'' + \sum_{k=1}^{\infty} \psi_k'' z_k\right) + T_{sx}(w_q'' + \sum_{k=1}^{\infty} \psi_k'' z_k) = \rho A(\ddot{w}_q + \sum_{k=1}^{\infty} \psi_k \ddot{z}_k), \quad (7.60)$$

where $\dot{\bullet}$ and $\bullet'$ represent the derivatives with respect to time and x respectively. This expression can be simplified slightly by applying the assumption that the quasistatic deflections are small compared to the static sag. This means that the second term in the first bracket can be removed. Applying Galerkin's method, multiplying Eq. (7.60) by ψ_n and integrating over $0 \leq x < \ell$ gives

$$T_{dx}\left(w_s'' \int_0^\ell \psi_n \mathrm{d}x + \sum_{k=1}^{\infty} z_k \int_0^\ell \psi_n \psi_k'' \mathrm{d}x\right) + T_{sx}\left(w_q'' \int_0^\ell \psi_n \mathrm{d}x + \sum_{k=1}^{\infty} z_k \int_0^\ell \psi_n \psi_k'' \mathrm{d}x\right)$$
$$= \rho A\left(\int_0^\ell \psi_n \ddot{w}_q \mathrm{d}x + \sum_{k=1}^{\infty} \ddot{z}_k \int_0^\ell \psi_n \psi_k \mathrm{d}x\right).$$

In deriving this equation it has been noted that the tensions, w_s'' and w_q'' are not function of x, $w_s'' = -\gamma A / T_{sx}$ from Eq. (7.22) and $w_q'' = \gamma E_q A^2 u_b / (T_{sx}^2 \ell)$ from Eq. (7.34). Using the nonlinear compatibility expression, Eq. (7.57), to eliminate the dynamic tension and substituting in the expressions for w_s'' and w_q'' gives

$$\left(\frac{E_q A u_b}{\ell} + \frac{EA}{2\ell} \int_0^\ell S \mathrm{d}x + \frac{\gamma E A^2}{T_{sx}\ell} \sum_{k=1}^{\infty} z_k \int_0^\ell \psi_k \mathrm{d}x\right)\left(-\frac{\gamma A}{T_{sx}} \int_0^\ell \psi_n \mathrm{d}x + \sum_{k=1}^{\infty} z_k \int_0^\ell \psi_n \psi_k'' \mathrm{d}x\right)$$
$$+ T_{sx}\left(\frac{\gamma E_q A^2 u_b}{T_{sx}^2 \ell} \int_0^\ell \psi_n \mathrm{d}x + \sum_{k=1}^{\infty} z_k \int_0^\ell \psi_n \psi_k'' \mathrm{d}x\right)$$
$$= \rho A\left(\int_0^\ell \psi_n \ddot{w}_q \mathrm{d}x + \sum_{k=1}^{\infty} \ddot{z}_k \int_0^\ell \psi_n \psi_k \mathrm{d}x\right), \quad (7.61)$$

where

$$S = \sum_{k=1}^{\infty} \sum_{n=1}^{\infty} (\phi_k' \phi_n' y_k y_n + \psi_k' \psi_n' z_k z_n). \quad (7.62)$$

Expanding out the brackets in the first term of Eq. (7.61), rearranging and using the definition of Irvine's parameter $\lambda^2 = \gamma^2 E\ell^2 A^3/T_{sx}^3$ gives

$$T_{sx}\left(1+\frac{E_q A u_b}{T_{sx}\ell}\right)\sum_{k=1}^{\infty} z_k \int_0^\ell \psi_n \psi_k'' \mathrm{d}x - T_{sx}\frac{\lambda^2}{\ell^3}\sum_{k=1}^{\infty} z_k \int_0^\ell \psi_k \mathrm{d}x \int_0^\ell \psi_n \mathrm{d}x$$
$$-\frac{\gamma E A^2}{2T_{sx}\ell}\int_0^\ell \psi_n \mathrm{d}x \int_0^\ell S\mathrm{d}x + \frac{\gamma E A^2}{T_{sx}\ell}\sum_{k=1}^{\infty} z_k \int_0^\ell \psi_k \mathrm{d}x \sum_{k=1}^{\infty} z_k \int_0^\ell \psi_n \psi_k'' \mathrm{d}x$$
$$+\frac{EA}{2\ell}\int_0^\ell S\mathrm{d}x \sum_{k=1}^{\infty} z_k \int_0^\ell \psi_n \psi_k'' \mathrm{d}x = \rho A\left(\int_0^\ell \psi_n \ddot{w}_q \mathrm{d}x + \sum_{k=1}^{\infty} \ddot{z}_k \int_0^\ell \psi_n \psi_k \mathrm{d}x\right). \tag{7.63}$$

Note that the second term in the first bracket, $E_q A u_b/(T_{sx}\ell)$ is small compared to the first term in the first bracket, since the static tension is much larger than the quasi-static tension $E_q A u_b/\ell$. When deriving the compatibility equation for the dynamic tension, such a term was removed because the dynamic tension was already small in comparison to the static tension. So the small contribution of the $E_q A u_b/(T_{sx}\ell)$ term to the dynamic tension will have a small effect on a term that is already small and so can be ignored. However, here the term is not deleted because the unity term is a linear term and so the second term in the bracket is potentially of the same order of significance as the other nonlinear terms.

The orthogonality conditions for the linearized mode-shapes, Eqs. (7.51) and (7.52), can now be applied. The last term in Eq. (7.63) can be simplified using Eq. (7.51). In addition, the first two terms in Eq. (7.63) can be simplified using Eq. (7.52). The combined result is

$$\rho A \int_0^\ell (\psi_n)^2 \mathrm{d}x \ddot{z}_n + T_{sx}\left[\frac{\lambda^2}{\ell^3}\left(\int_0^\ell \psi_n \mathrm{d}x\right)^2 - \int_0^\ell \psi_n \psi_n'' \mathrm{d}x\right] z_n$$
$$-\frac{E_q A u_b}{\ell}\sum_{k=1}^{\infty} z_k \int_0^\ell \psi_n \psi_k'' \mathrm{d}x - \frac{\gamma E A^2}{T_{sx}\ell}\sum_{k=1}^{\infty} z_k \int_0^\ell \psi_k \mathrm{d}x \sum_{k=1}^{\infty} z_k \int_0^\ell \psi_n \psi_k'' \mathrm{d}x$$
$$+\frac{\gamma E A^2}{2T_{sx}\ell}\int_0^\ell \psi_n \mathrm{d}x \int_0^\ell S\mathrm{d}x - \frac{EA}{2\ell}\int_0^l S\mathrm{d}x \sum_{k=1}^{\infty} z_k \int_0^\ell \psi_n \psi_k'' \mathrm{d}x + \rho A \int_0^\ell \psi_n \ddot{w}_q \mathrm{d}x = 0.$$

The first line of this equation represents the linearized dynamics of the nth mode of vibration, as derived in Eq. (7.53). By dividing through by $\rho A \int_0^\ell (\psi_n)^2 \mathrm{d}x$, these linear terms can be expressed in the form of Eq. (7.54), where the natural frequency of the linearized system, ω_{zn} is given by Eq. (7.55):

$$\ddot{z}_n + \omega_{zn}^2 z_n + \frac{1}{\rho A \int_0^\ell (\psi_n)^2 \mathrm{d}x} \left(\frac{E_q A u_b}{\ell} \sum_{k=1}^{\infty} z_k \int_0^\ell \psi_n \psi_k'' \mathrm{d}x \right.$$

$$- \frac{\gamma E A^2}{T_{sx}\ell} \sum_{k=1}^{\infty} z_k \int_0^\ell \psi_k \mathrm{d}x \sum_{k=1}^{\infty} z_k \int_0^\ell \psi_n \psi_k'' \mathrm{d}x + \frac{\gamma E A^2}{2T_{sx}\ell} \int_0^\ell \psi_n \mathrm{d}x \int_0^\ell S \mathrm{d}x$$

$$\left. - \frac{EA}{2\ell} \int_0^\ell S\mathrm{d}x \sum_{k=1}^{\infty} z_k \int_0^\ell \psi_n \psi_k'' \mathrm{d}x + \rho A \int_0^\ell \psi_n \ddot{w}_q \mathrm{d}x \right) = 0. \tag{7.64}$$

Since all the terms in the large round brackets are nonlinear, they are assumed to be small compared to the linear terms. So, as with the out-of-plane analysis, the in-plane odd mode-shapes are approximated as sinusoidal, $\psi_n \approx \sin(n\pi x/\ell)$ for $n = 1, 3, 5, \ldots$. Substituting in sinusoidal mode-shapes for all the modes allows the evaluation of each of the nonlinear terms. To simplify the resulting expressions the following relationships are used

$$\int_0^\ell \psi_n \mathrm{d}x = \tfrac{\ell}{n\pi}(1+(-1)^{n+1}), \quad \int_0^\ell (\psi_n)^2 \mathrm{d}x = \tfrac{\ell}{2},$$

$$\int_0^\ell \psi_n' \psi_k' \mathrm{d}x = 0, \quad \int_0^\ell (\psi_n')^2 \mathrm{d}x = \tfrac{\pi^2 n^2}{2\ell},$$

$$\int_0^\ell \psi_n \psi_k'' \mathrm{d}x = 0, \quad \int_0^\ell \psi_n \psi_n'' \mathrm{d}x = -\tfrac{n^2\pi^2}{2\ell},$$

$$\int_0^\ell \psi_n \tfrac{x}{\ell} \mathrm{d}x = \tfrac{\ell}{n\pi}(-1)^{n+1}, \quad \int_0^\ell \psi_n \left[\tfrac{x}{\ell} - \left(\tfrac{x}{\ell}\right)^2\right] \mathrm{d}x = \tfrac{2\ell}{n^3\pi^3}(1+(-1)^{n+1}),$$

where $n \neq k$ and it is noted that $\phi_n = \psi_n$. Using these expressions, S, defined in Eq. (7.62), simplifies to

$$S = \sum_{k=1}^{\infty} \sum_{n=1}^{\infty} (\phi_k' \phi_n' y_k y_n + \psi_k' \psi_n' z_k z_n) = \sum_{k=1}^{\infty} \frac{\pi^2 k^2}{2\ell} (y_k^2 + z_k^2).$$

Using the equation for quasi-static deflection w_q, Eq. (7.34), and these relationships, Eq. (7.64) can be rewritten as

$$\ddot{z}_n + \omega_{zn}^2 z_n + \frac{2}{\rho A \ell} \left[\frac{E_q A}{\ell} \frac{n^2\pi^2}{2\ell} u_b z_n + \frac{\gamma E A^2}{T_{sx}\ell} \sum_{k=1}^{\infty} z_k \frac{\ell}{k\pi} \left(1+(-1)^{k+1}\right) \frac{n^2\pi^2}{2\ell} z_n \right.$$

$$+ \frac{\gamma E A^2}{2T_{sx}\ell} \frac{\ell}{n\pi} (1+(-1)^{n+1}) \sum_{k=1}^{\infty} \frac{\pi^2 k^2}{2\ell} (y_k^2 + z_k^2) + \frac{EA}{2\ell} \sum_{k=1}^{\infty} \frac{\pi^2 k^2}{2\ell} (y_k^2 + z_k^2) \frac{n^2\pi^2}{2\ell} z_n$$

$$\left. + \rho A \left(\frac{\ell}{n\pi} (-1)^{n+1} \ddot{w}_b - \frac{\gamma E_q \ell A^2}{2T_{sx}^2} \frac{2\ell}{n^3\pi^3} (1+(-1)^{n+1}) \ddot{u}_b \right) \right] = 0.$$

Finally, this equation can be rearranged into a standard form for in-plane modes (Warnitchai et al. 1995), to give

$$m_{zn}(\ddot{z}_n + 2\zeta_{zn}\omega_{zn}\dot{z}_n + \omega_{zn}^2 z_n) + 2\eta_n u_b z_n + 2\beta_{nk}\sum_{k=1}^{\infty} z_k z_n$$

$$+ \beta_{kn}\sum_{k=1}^{\infty}(y_k^2 + z_k^2) + \nu_{nk}\sum_{k=1}^{\infty}(y_k^2 + z_k^2)z_n + \chi_n(-1)^{n+1}\ddot{w}_b - \alpha_n\ddot{u}_b = 0, \quad (7.65)$$

where

$$m_{zn} = \frac{\rho A\ell}{2}, \qquad \chi_n = \frac{\rho A\ell}{n\pi}, \qquad \alpha_n = \frac{\gamma E_q \ell^2 A^3}{T_{sx}^2 n^3 \pi^3}(1 + (-1)^{n+1}),$$

and modal damping, with a damping ratio ζ_{zn}, has been added. Next the effect of interaction between in-plane and out-of-plane modes is considered.

7.3.4 Modal Interaction

The in- and out-of-plane nonlinear equations of motion, Eqs. (7.65) and (7.59) respectively, have modal coupling via the nonlinear terms. These nonlinear coupling terms can cause energy to move from one mode to another. One result of this is that even though the support excitation is purely in-plane, an out-of-plane response can be excited.

Consider the case where the frequency of the vertical support excitation is very close to the second linear in-plane natural frequency, $\omega_{z2} = (2\pi/\ell)\sqrt{T_{sx}/\rho A}$. If the system were linear, the response would consist almost entirely of the second in-plane mode (if the excitation was exactly at the natural frequency, then the response would consist purely of second mode). However, under certain excitation levels, the nonlinear terms can lead to first or second mode, out-of-plane response. Considering just the second in-plane mode and the first and second out-of-plane modes, using Eqs. (7.65) and (7.59), the equations of motion, based on the linear modes, can be written as

$$\ddot{y}_1 + 2\zeta_{y1}\omega_1\dot{y}_1 + \omega_1^2 y_1 + W_{11}y_1^3 + W_{12}y_1(y_2^2 + z_2^2) + N_1\delta y_1 = 0 \quad (7.66)$$

$$\ddot{y}_2 + 2\zeta_{y2}\omega_2\dot{y}_2 + \omega_2^2 y_2 + W_{21}y_2 y_1^2 + W_{22}y_2(y_2^2 + z_2^2) + N_2\delta y_2 = 0 \quad (7.67)$$

$$\ddot{z}_2 + 2\zeta_{z2}\omega_2\dot{z}_2 + \omega_2^2 z_2 + W_{21}z_2 y_1^2 + W_{22}z_2(y_2^2 + z_2^2) + N_2\delta z_2 = B\ddot{\delta}, \quad (7.68)$$

where $W_{nk} = \nu_{nk}/m$, $N_n = 2\eta_n \sin(\theta)/m$, $B = \chi_2\cos(\theta)/m$ and $m = m_{y1} = m_{y2} = m_{z2}$. Note that the β_{nk} terms have disappeared as the only in-plane mode being considered is mode 2 hence the only β_{nk} terms present are β_{n2} terms and $\beta_{n2} = 0$ for all n.

These equations show that the second in-plane mode, z_2, is directly excited by the support motion via the $B\ddot{\delta}$ term, as would be the case for the linearized version of the equations. For this mode there are also nonlinear terms which cause interactions with the forcing, $N_2\delta z_2$ and the out of plane modes, $W_{21}z_2y_1^2 + W_{22}z_2y_2^2$. In addition, there is a nonlinear cubic stiffness term $W_{22}z_2^3$. There is no direct excitation of the out-of-plane modes as the excitation is in-plane. However the support motion δ does appear in the nonlinear terms, where it is multiplied by the modal coordinate terms $N_1\delta y_1$ and $N_2\delta y_2$ for the first and second modes respectively. This is termed *parametric* excitation. In addition there are *auto-parametric* excitation, terms in which the modal coordinate is multiplied by other modal coordinates, which can excite a response in the mode via motion in other modes. Auto-parametric excitation via the in-plane z_2 mode is possible through the $W_{12}y_1z_2^2$ and $W_{22}y_2z_2^2$ terms for the first and second modes respectively. Auto-parametric excitation is possible also between the out-of-plane modes via the $W_{12}y_1y_2^2$ and $W_{21}y_2y_1^2$ terms. Finally both out-of-plane modes have cubic stiffening terms in the form $W_{nn}y_n^3$.

When considering the response to the support excitation, if the initial conditions are zero, such that $y_1(0) = y_2(0) = z_2(0) = 0$ and $\dot{y}_1(0) = \dot{y}_2(0) = \dot{z}_2(0) = 0$, then the response will be purely in the z_2 mode and as a result the response will be governed by the equation

$$\ddot{z}_2 + 2\zeta_{z2}\omega_2\dot{z}_2 + \omega_2^2 z_2 + W_{22}z_2^3 + N_2\delta z_2 = B\ddot{\delta}.$$

No motion will be present in the out-of-plane modes since all the terms in the equations of motion for the nth out-of-plane mode contain y_n terms or time derivatives of them. Since the initial conditions are $y_n = \dot{y}_n = 0$ these terms will remain zero. However it is possible, at certain excitation levels, for the $y_n = 0$ solution to the equations of motion to become locally unstable. This means that any slight out-of-plane excitation, such as a gust of wind exciting a cable-stayed bridge, can cause a non-zero out-of-plane response.

This behaviour is demonstrated numerically, using a Runge-Kutta time-stepping simulation of Eqs. (7.66)–(7.68), in the following example. It is then addressed in more detail using the normal form analysis technique for the case where there is interaction between the second in- and out-of-plane modes in Sect. 7.5. In addition, local stability is discussed further in Sect. 8.5, for the case of a curved plate and by Gonzalez-Buelga et al. (2008) for the case of inclined cables.

Example 7.1 Three-mode model of an inclined cable

Problem Find the time series response of the three coupled cable modes given by Eq. (7.68) for a steel cable of length 1.98 m, inclined at 20° to the horizontal, of diameter 0.8 mm, mass 0.67 kg/m and static tension 205 N. The damping ratio of all three modes can be assumed to be 0.2 % and the support motion frequency, $\Omega = 0.97\omega_{z2}$. Two cases should be simulated; (i) is for an amplitude of the support motion of 2.4 mm, and (ii) for an amplitude of the support motion of 4 mm. In both cases at $t = 25$ s a disturbance, in the form of a 0.02 s, 1 mm/s amplitude pulse, should be applied to the out-of-plane modes. What is the response in the two different cases?

Solution The time series response can be computed using 4th-order Runge-Kutta numerical integration (as discussed in Example 2.1, Chap. 2). In this case *ode45* in

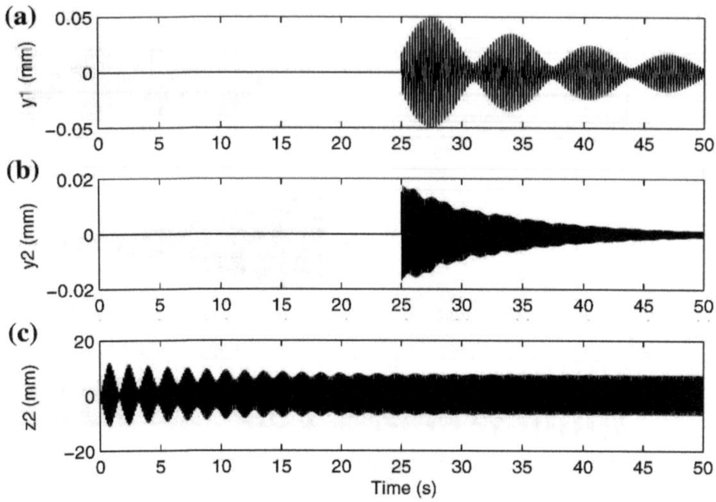

Fig. 7.7 Modal cable response to a support excitation of 2.4 mm at a frequency $\Omega = 0.97\omega_{z2}$: **a** Out-of-plane mode 1, **b** out-of-plane mode 2 and **c** in-plane mode 2

Matlab was used. For a steel cable of length 1.98 m, inclined at 20° to the horizontal, of diameter 0.8 mm, mass 0.67 kg/m and static tension 205 N the key parameters are $\omega_2 = 55.35$, $B = 0.299$ and $W_{22} = 3.783 \times 10^6$. Full details of all parameters can be found in Gonzalez-Buelga et al. (2008). Figure 7.7 shows the time response of the three modes in case (i). It can be seen that initially there is a response only in the z_2 mode. When the impulse is applied at 25 s, there is a response from both out-of-plane modes. However these responses decay away, leaving just the second in-plane response. At this amplitude of excitation, the zero solution of the out-of-plane modes is stable and so a slight disturbance does not cause a steady-state out-of-plane response.

Figure 7.8 shows the response to a larger amplitude (case (ii)) support excitation; again the second out-of-plane mode decays away after the disturbance is applied at 25 s. However, in this case the first out-of-plane mode does not decay away, instead it tends towards a steady-state response of around 10mm amplitude. This indicates that the local stability of the first out-of-plane mode about zero response has become unstable at this excitation level, leading to a non-zero response due to excitation through the parametric and auto-parametric (with the second in-plane mode) excitation. In addition, the amplitude of the second in-plane mode, z_2, decreases due to the presence of significant response in the first in-plane mode, y_1. This is due to the modal interaction term, $W_{21} z_2 y_1^2$, in the equation for z_2, Eq. (7.68). ■

7.4 Case Study of Analysis of Cable Response

In this section the dynamics of an inclined cable will be used to compare the various approximate techniques for analysing nonlinear vibrations discussed in Chap. 4.

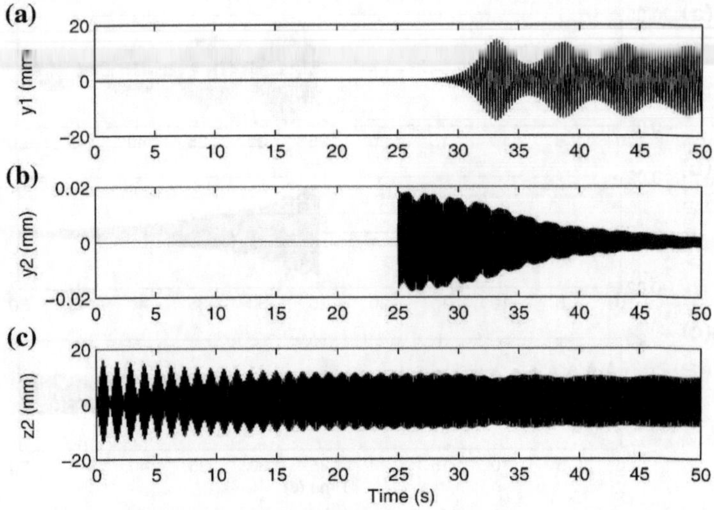

Fig. 7.8 Modal cable response to a support excitation of 4 mm at a frequency $\Omega = 0.97\omega_{z2}$, **a** out-of-plane mode 1, **b** out-of-plane mode 2 and **c** in-plane mode 2

The amplitude of response of the second in-plane mode is considered for the case where the support excitation is close to the natural frequency of the second mode. It is assumed that no modal interaction occurs between the second in-plane mode and the first two out-of-plane modes. From Eq. (7.68) the equation of motion of the second in-plane mode, assuming no other modes are present, is given by

$$\ddot{z}_2 + \omega_{z2}^2 z_2 + \varepsilon \left(2\zeta_{z2}\omega_{z2}\dot{z}_2 + W_{22}z_2^3 + N_2\delta z_2\right) = \varepsilon B\ddot{\delta}, \tag{7.69}$$

where ε is used as a book-keeping aid to indicate that the modal damping and the cubic stiffness terms are small. Note that to recover the original equations ε is simply set to unity. In addition, the forcing is assumed to be small, or order ε^1, in comparison with the response since the forcing is close to the natural frequency. The forcing is at frequency Ω with an amplitude of Δ, resulting in $\delta = \varepsilon\Delta\cos(\omega t)$.

7.4.1 Harmonic Balance

In the harmonic balance method, it is assumed that the response of the second in-plane mode is in the form $z_2 = Z_c\cos(\Omega t) + Z_s\sin(\Omega t)$. Substituting this into the equation of motion, Eq. (7.69), gives

$$(\omega_{z2}^2 - \Omega^2)(Z_c c_1 + Z_s s_1) + \varepsilon 2\zeta_{z2}\omega_{z2}\Omega(-Z_c s_1 + Z_s c_1) + \varepsilon\frac{N_2\Delta}{2}[(1 + c_2)Z_c + s_2 Z_s]$$
$$+ \varepsilon\frac{W_{22}}{4}[(3c_1 + c_3)Z_c^3 + 3(s + s_3)Z_c^2 Z_s + 3(c - c_3)Z_c Z_s^2 + (3s_1 - s_3)Z_s^3] = \varepsilon B\Delta\Omega^2 c_1,$$

where the shorthand, $c_k = \cos(k\Omega t)$ and $s_k = \sin(k\Omega t)$ is used. Applying a harmonic balance to the $\cos(\Omega t)$ and $\sin(\Omega t)$ terms respectively (see Chap. 4) gives the two equations

$$\left(\omega_{z2}^2 - \Omega^2 + \varepsilon\frac{3W_{22}}{4}Z^2\right) Z_c + (\varepsilon 2\zeta_{z2}\omega_{z2}\Omega)\, Z_s = \varepsilon B\Delta\Omega^2$$

$$-(\varepsilon 2\zeta_{z2}\omega_{z2}\Omega)\, Z_c + \left(\omega_{z2}^2 - \Omega^2 + \varepsilon\frac{3W_{22}}{4}Z^2\right) Z_s = 0,$$

where Z is the amplitude of the cable mode, $Z^2 = Z_c^2 + Z_s^2$. Squaring these two equations and adding gives an expression for the response amplitude Z in terms of the input amplitude Δ:

$$\left[\left(\omega_{z2}^2 - \Omega^2 + \varepsilon\frac{3W_{22}}{4}Z^2\right)^2 + (\varepsilon 2\zeta_{z2}\omega_{z2}\Omega)^2\right] Z^2 = \varepsilon^2 B^2\Delta^2\Omega^4. \qquad (7.70)$$

Up to now no assessment has been made regarding the relative size of the terms. Before this can be done the term $\omega_{z2}^2 - \Omega^2$ must be considered. As the excitation is close to the second natural frequency, $\Omega \approx \omega_{z2}$, it is convenient to write $\Omega = (1+\varepsilon\mu)\omega_{z2}$, where μ is a detuning parameter and ε is used to indicate that μ is small. Making this substitution into Eq. (7.70) gives

$$\left[\left(-\omega_{z2}^2(2\mu + \varepsilon\mu^2) + \frac{3W_{22}}{4}Z^2\right)^2 + \left(2(1+\varepsilon\mu)\zeta_{z2}\omega_{z2}^2\right)^2\right] Z^2 = B^2\Delta^2\omega_{z2}^4(1+\varepsilon\mu)^4,$$

where a factor ε^2 has been cancelled. Ignoring ε^1 and higher terms gives the simplified equation

$$\left[\left(-2\mu\omega_{z2}^2 + \frac{3W_{22}}{4}Z^2\right)^2 + \left(2\zeta_{z2}\omega_{z2}^2\right)^2\right] Z^2 = B^2\Delta^2\omega_{z2}^4.$$

Rearranging this equation gives

$$(9W_{22}^2)Z^6 - (48W_{22}\mu\omega_{z2}^2)Z^4 + (64\omega_{z2}^4(\zeta_{2z}^2 + \mu^2))Z^2 = 16B^2\Delta^2\omega_{z2}^4. \qquad (7.71)$$

Since the relationship between excitation amplitude and modal response amplitude is cubic, it is possible that for some excitation amplitudes there are multiple valid response amplitudes. This can be assessed by searching for turning points in the relationship between Z and Δ, i.e. points at which the gradient is zero ($\mathrm{d}\Delta/\mathrm{d}Z = 0$). Then differentiating Eq. (7.71) with respect to Z gives

$$\frac{\mathrm{d}\Delta}{\mathrm{d}Z} = \frac{Z}{16B^2\omega_{z2}^4\Delta}[(27W_{22}^2)Z^4 - (96W_{22}\mu\omega_{z2}^2)Z^2 + (64\omega_{z2}^4(\zeta_{2z}^2 + \mu^2))]. \qquad (7.72)$$

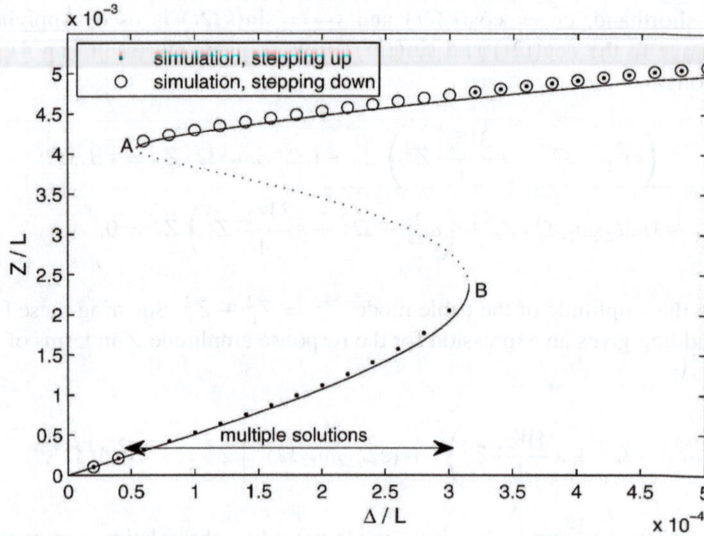

Fig. 7.9 Response amplitude for a range of forcing amplitudes and a fixed forcing frequency. L is length of the cable

From this expression the roots of Z^2 at which the gradient is zero can be found. These roots are of significance only if they are positive and real. Using the shorthand $aZ^4 + bZ^2 + c$ for the quadratic expression in Z^2 within Eq. (7.72), and noting that a and c are positive and b has the same sign as μ, a positive real root exists only if $b^2 - 4ac \geq 0$ and $\mu \geqslant 0$. This equates to the condition $\mu \geq \sqrt{3}\zeta_{2z}$ for turning points, and hence multiple possible amplitudes of response, to exist for a given excitation.

Figure 7.9 shows the predicted amplitude response curve derived from the harmonic balance technique for a steel cable with parameters defined in Example 7.1. A frequency detuning of $\mu = 0.03$ is selected; this meets the condition $\mu \geq \sqrt{3}\zeta_{2z}$ when the damping ratio is 0.2 %. As a result, multiple amplitude response solutions exist for a range of excitation values. The region of curve between points A and B is dotted to indicate that the solution is unstable (this will be shown in detail using the averaging analysis).

Figure 7.9 also shows simulation results using the Matlab variable step Runge-Kutta time-stepping routine *ode45*. In the simulation, the excitation amplitude starts at zero and is stepped up and then down gradually, ensuring that at each amplitude the simulation is run for sufficient time to ensure that a steady-state response is reached. It can be seen that the expression for the amplitude of response generated using the harmonic balance technique is a good fit to the simulation results. Now the same modal equation for the cable is considered using averaging analysis.

7.4.2 Averaging

Averaging can potentially reveal more information about the underlying dynamics of the system, compared to harmonic balance. Consider again the dynamics of an inclined cable with a support excitation frequency close to the natural frequency of the second in-plane mode, assuming no out-of-plane motion is present, Eq. (7.69)

$$\ddot{z}_2 + \omega_{z2}^2 z_2 + \varepsilon \left(2\zeta_{z2}\omega_{z2}\dot{z}_2 + W_{22}z_2^3 + N_2\delta z_2\right) = \varepsilon B\ddot{\delta},$$

with the support motion $\delta = \varepsilon\Delta\cos(\Omega t)$, where $\Omega \approx \omega_{z2}$.

This equation can be written in the standard form for averaging as

$$\ddot{z}_2 + \omega_{z2}^2 z_2 = \varepsilon X, \quad X = B\ddot{\delta} - 2\zeta_{z2}\omega_{z2}\dot{z}_2 - N_2\delta z_2 - \hat{W}_{22}z_2^3. \tag{7.73}$$

Since the forcing frequency Ω is close to the natural frequency ω_{z2} it can be written as $\Omega = \omega_{z2}(1+\varepsilon\mu)$ where, as before, μ is the small, order ε^1, frequency detuning parameter. Now time is scaled such that the forcing has frequency ω_{z2} in the new time-scale τ, hence $\tau = (1+\varepsilon\mu)t$ such that $\delta = \Delta\cos(\omega_{z2}\tau)$. Note that $\omega_{z2}\tau = \omega_{z2}(1+\varepsilon\mu)t = \Omega t$. To rewrite Eq. (7.73) in scaled time, the derivative terms (with respect to time) must be converted to derivatives with respect to τ

$$\dot{z}_2 = (1+\varepsilon\mu)z_2', \quad \ddot{z}_2 = (1+\varepsilon\mu)^2 z_2'' = z_2'' - \varepsilon 2\mu\omega_{z2}^2 z_2,$$

using $z_2'' = -\omega_{z2}^2 z_2 + \mathscr{O}(\varepsilon^1)$ and ignoring ε^2 and higher terms. Equation (7.73) can now be expressed as

$$z_2'' + \omega_{z2}^2 z_2 = \varepsilon\tilde{X},$$

where the small function $\tilde{X}$, which contains the nonlinear terms and damping, is given by

$$\tilde{X} = -B\omega_{z2}^2\Delta\cos(\omega_{z2}\tau) - 2\zeta_{z2}\omega_{z2}z_2' - N_2\Delta\cos(\omega_{z2}\tau)z_2 - W_{22}z_2^3 + 2\mu\omega_2^2 z_2.$$

Equations (4.22) and (4.23) show that the trial solutions for this example can be expressed as

$$\begin{aligned} z_2 &= z_{2c}c_2 + z_{2s}s_2, \\ z_2' &= -\omega_{z2}z_{2c}s_2 + \omega_{z2}z_{2s}c_2, \end{aligned}$$

where $c_2 = \cos(\omega_{z2}\tau)$ and $s_2 = \sin(\omega_{z2}\tau)$. With these trial solutions, expressions for z_{2c}' and z_{2s}' can be found using Eq. (4.24) to give

$$\begin{aligned} z_{2c}' = -\frac{\varepsilon}{\omega_{z2}}s_2 X \quad \rightsquigarrow \quad z_{2c}' &= -\frac{\varepsilon}{\omega_{z2}}s_2\left[-B\omega_{z2}^2\Delta c_2 - 2\zeta_{z2}\omega_{z2}^2(-z_{2c}s_2 + z_{2s}c_2)\right. \\ &\left. - N_2\Delta c_1(z_{2c}c_2 + z_{2s}s_2) - W_{22}(z_{2c}c_2 + z_{2s}s_2)^3 + 2\mu\omega_{z2}^2(z_{2c}c_2 + z_{2s}s_2)\right], \end{aligned}$$

and

$$\begin{aligned} z_{2s}' = \frac{\varepsilon}{\omega_{z2}}c_2 X \quad \rightsquigarrow \quad z_{2s}' &= \frac{\varepsilon}{\omega_{z2}}c_2\left[-B\omega_{z2}^2\Delta c_2 - 2\zeta_{z2}\omega_{z2}^2(-z_{2c}s_2 + z_{2s}c_2)\right. \\ &\left. - N_2\Delta c_2(z_{2c}c_2 + z_{2s}s_2) - W_{22}(z_{2c}c_2 + z_{2s}s_2)^3 + 2\mu\omega_{z2}^2(z_{2c}c_2 + z_{2s}s_2)\right]. \end{aligned}$$

Now the equations can be averaged over the range $\tau - \pi/\omega_{z2}$ to $\tau + \pi/\omega_{z2}$ under the assumption that $z_{2c}(\tau)$ and $z_{2s}(\tau)$ are constant over this time period. This gives

$$z'_{2ca} = -\frac{\varepsilon}{\omega_{z2}}\left[\zeta_{z2}\omega_{z2}^2 z_{2ca} - \frac{3}{8}W_{22}z_{2sa}Z_{2a}^2 + \mu\omega_{z2}^2 z_{2sa}\right], \tag{7.74}$$

$$z'_{2sa} = \frac{\varepsilon}{\omega_{z2}}\left[-\frac{1}{2}B\omega_{z2}^2\Delta - \zeta_{z2}\omega_{z2}^2 z_{2sa} - \frac{3}{8}W_{22}z_{2ca}Z_{2a}^2 + \mu\omega_{z2}^2 z_{2ca}\right], \tag{7.75}$$

where $Z_{2a}^2 = z_{2sa}^2 + z_{2ca}^2$ is the second in-plane modal amplitude of response. The steady-state response can be found by setting $z'_{2sa} = z'_{2ca} = 0$ to give

$$-\zeta_{z2}\omega_{z2}^2 z_{2ca} + \left(\frac{3}{8}W_{22}Z_{2a}^2 - \mu\omega_{z2}^2\right)z_{2sa} = 0,$$

$$-\left(\frac{3}{8}W_{22}Z_{2a}^2 - \mu\omega_{z2}^2\right)z_{2ca} - \zeta_{z2}\omega_{z2}^2 z_{2sa} = \frac{1}{2}B\omega_{z2}^2\Delta.$$

Squaring and adding the two equations gives

$$\left(\zeta_{z2}^2\omega_{z2}^4 + \left(\frac{3}{8}W_{22}Z_{2a}^2 - \mu\omega_{z2}^2\right)^2\right)Z_{2a} = \frac{1}{4}B^2\omega_{z2}^4\Delta^2, \tag{7.76}$$

which, with some algebraic rearranging, is identical to the equation found using the harmonic balance technique, Eq. (7.71).

Although the amplitude prediction is the same as that using the harmonic balance, the averaging technique can provide additional information. This is because the averaged equations, Eqs. (7.74) and (7.75), are dynamic equations for z_{2ca} and z_{2sa} and hence transient behaviour can be analysed. These equations can therefore be used to find the stability of the steady-state amplitude solutions, demonstrating that the dotted region of the solution plotted in Fig. 7.9 is unstable.

For a solution to be stable it must attract transient trajectories. This can be studied by examining a point a small distance (sometimes called a perturbation) away from the solution and seeing whether it is attracted to the solution. If so, the solution is locally stable, otherwise it is unstable (see discussion on the stability of maps in Sect. 2.5.2).

To do this, first Eqs. (7.74) and (7.75) can be expressed in matrix form:

$$\begin{Bmatrix} z'_{2ca} \\ z'_{2sa} \end{Bmatrix} = \frac{\varepsilon}{\omega_{z2}}\begin{bmatrix} -\zeta_{z2}\omega_{z2}^2 & \frac{3}{8}W_{22}Z_{2a}^2 - \mu\omega_{z2}^2 \\ -\left(\frac{3}{8}W_{22}Z_{2a}^2 - \mu\omega_{z2}^2\right) & -\zeta_{z2}\omega_{z2}^2 \end{bmatrix}\begin{Bmatrix} z_{2ca} \\ z_{2sa} \end{Bmatrix} + \begin{Bmatrix} 0 \\ -\frac{1}{2}B\omega_{z2}^2\Delta \end{Bmatrix}, \tag{7.77}$$

which is equivalent to $\mathbf{z}' = f(\mathbf{z}, \tau)$ where $\mathbf{z} = \{z_{2ca}\, z_{2sa}\}^T$.

Steady-state (or equilibrium) solutions to the equation $\mathbf{z}' = f(\mathbf{z}, \tau)$, $\bar{\mathbf{z}}$, occur when $f(\bar{\mathbf{z}}, \tau) = 0$, and are given by Eq. (7.76). Consider a point $\tilde{\mathbf{z}}$ which is perturbed a small distance away from the equilibrium position $\bar{\mathbf{z}}$. Since the perturbation is small, a Taylor series approximation can be made about the equilibrium position

$$\tilde{\mathbf{z}}' = f(\tilde{\mathbf{z}}, \tau) \approx f(\bar{\mathbf{z}}, \tau) + Df_z(\bar{\mathbf{z}}, \tau)(\tilde{\mathbf{z}} - \bar{\mathbf{z}}), \tag{7.78}$$

where $Df_z(\bar{\mathbf{z}}, \tau)$ is the Jacobian of $f(\bar{\mathbf{z}}, \tau)$. Since $\bar{\mathbf{z}}$ is an equilibrium position $\bar{\mathbf{z}}' = f(\bar{\mathbf{z}}, \tau) = 0$ by definition. Quantifying the perturbation away from equilibrium as $\mathbf{z_p}$, such that $\tilde{\mathbf{z}} = \bar{\mathbf{z}} + \mathbf{z_p}$ allows Eq. (7.78) to be written as

$$\mathbf{z_p}' = Df_z(\bar{\mathbf{z}}, \tau)\mathbf{z_p}.$$

This equation shows that, if the real parts of the eigenvalues of the Jacobian evaluated at the equilibrium position are negative, then the perturbation will decay with time and hence $\tilde{\mathbf{z}}$ will tend to $\bar{\mathbf{z}}$ such that the equilibrium solution $\bar{\mathbf{z}}$ is stable.

Using Eq. (7.77) and recalling that $Z_{2a}^2 = z_{2sa}^2 + z_{2ca}^2$, the Jacobian at $\mathbf{z}$ can be written as

$$Df_z(\mathbf{z}, t) = \frac{\varepsilon}{\omega_{z2}} \begin{bmatrix} -\zeta_{z2}\omega_{z2}^2 + \frac{3}{4}W_{22}z_{2ca}z_{2sa} & \frac{3}{8}W_{22}(3z_{2sa}^2 + z_{2ca}^2) - \mu\omega_{z2}^2 \\ -\frac{3}{8}W_{22}(z_{2sa}^2 + 3z_{2ca}^2) + \mu\omega_{z2}^2 & -\zeta_{z2}\omega_{z2}^2 - \frac{3}{4}W_{22}z_{2ca}z_{2sa} \end{bmatrix}.$$

Letting the eigenvalues of the Jacobian be λ and setting $\lambda = \gamma\varepsilon/\omega_{z2}$, results in the eigenvalue equation

$$\gamma^2 + 2\zeta_{z2}\omega_{z2}^2\gamma + (\zeta_{z2}^2 + \mu^2)\omega_{z2}^4 - \frac{3}{2}\mu\omega_{z2}^2 W_{22} Z_{2a}^2 + \frac{27}{64} W_{22}^2 Z_{2a}^4 = 0.$$

Since the coefficient of the γ^1 term is positive, the boundary between stable and unstable eigenvalues occurs at $\gamma = 0$:

$$(\zeta_{z2}^2 + \mu^2)\omega_{z2}^4 - \frac{3}{2}\mu\omega_{z2}^2 W_{22} Z_{2a}^2 + \frac{27}{64} W_{22}^2 Z_{2a}^4 = 0. \tag{7.79}$$

Of particular interest are the eigenvalues of the Jacobian evaluated at equilibrium positions $\bar{\mathbf{z}}$ defined by Eq. (7.76). The turning points on the equilibrium solution curve can be calculated using harmonic balance, by setting Eq. (7.72) to zero. The expression for the turning points matches the expression given in Eq. (7.79) and as a result, the turning points coincide with the shift between stable and unstable solutions. The real parts of the eigenvalues at $Z_{2a} = 0$ are negative, therefore along the line from $Z_{2a} = 0$ to larger values of Z_{2a} the solution is stable up until the first turning point, when it turns unstable and then restabilises after the second turning point. This justifies plotting the region of the curve between points A and B as dotted, indicating an unstable branch in Fig. 7.9.

Although the averaging technique has provided more information than the harmonic balance technique, it is difficult to extend it to include information about higher harmonics of the response. The multiple-scales method can also give information about sub and/or super harmonics, and this is considered next.

7.4.3 Multiple Scales

Recall that the equation for the second in-plane mode of an inclined cable subject to a support excitation and no out-of-plane motion, Eq. (7.69), is given by

$$\ddot{z}_2 + \omega_{z2}^2 z_2 + \varepsilon \left(2\zeta_{z2}\omega_{z2}\dot{z}_2 + W_{22}z_2^3 + N_2\delta z_2\right) = \varepsilon B\ddot{\delta},$$

with the support motion $\delta = \varepsilon\Delta\cos(\Omega t)$, where $\Omega \approx \omega_{z2}$.

To proceed, this equation will be written in the form used in the multiple-scales analysis which is given in Eq. (4.39):

$$\ddot{x} + \varepsilon 2\zeta\omega_n\dot{x} + \omega_n^2 x + \varepsilon\mathcal{N}(\dot{x}, x, t) = \varepsilon f\cos(\Omega t).$$

Written in this form, Eq. (7.69) becomes

$$\ddot{z}_2 + \varepsilon 2\zeta_{z2}\omega_{z2}\dot{z}_2 + \omega_{z2}^2 z_2 + \varepsilon\mathcal{N}(\dot{z}_2, z_2, t) = -\varepsilon B\Delta\Omega^2\cos(\Omega t), \tag{7.80}$$

where the nonlinear function is given by $\mathcal{N}(\dot{z}_2, z_2, t) = N_2\Delta\cos(\omega t)z_2 + W_{22}z_2^3$. Note that the amplitude of the forcing term in this example is a function of Ω^2, $f = -B\Delta\Omega^2$.

Now the fast and slow time-scales are defined as $\tau = \Omega t$ and $T = \varepsilon t$ and the expressions for the derivatives with respect to t can be calculated, see Eq. (4.30). The forcing is close to resonance, therefore the small (order ε^1) detuning parameter μ is used to give $\Omega = \omega_{z2}(1 + \varepsilon\mu)$. By using these relationships and substituting in the truncated power series $z_2(t) = z_{20}(\tau, T) + \varepsilon z_{21}(\tau, T)$, Eq. (7.80) can be written in a form that mirrors Eq. (4.40):

$$\begin{aligned}
&\omega_{z2}^2(1 + \varepsilon 2\mu + \varepsilon^2\mu^2)(z_{20}'' + \varepsilon z_{21}'') + 2\varepsilon\omega_{z2}(1 + \varepsilon\mu)(z_{20}'^{*} + \varepsilon z_{21}'^{*}) \\
&\quad + \varepsilon^2(z_{20}^{**} + \varepsilon z_{21}^{**}) + \varepsilon 2\zeta_{z2}\omega_{z2}(\omega_{z2}(1 + \varepsilon\mu)(z_{20}' + \varepsilon z_{21}') + \varepsilon(z_{20}^{*} + \varepsilon z_{21}^{*})) \\
&\quad + \omega_{z2}^2(z_{20} + \varepsilon z_{21}) + \varepsilon\mathcal{N}(\omega_{z2}(1 + \varepsilon\mu)(z_{20}' + \varepsilon z_{21}') + \varepsilon(z_{20}^{*} + \varepsilon z_{21}^{*}), z_{20} + \varepsilon z_{21}) \\
&\quad = -\varepsilon B\Delta\omega_{z2}^2(1 + \varepsilon\mu)^2\cos(\tau),
\end{aligned}$$

where $\bullet'$ and $\bullet^{*}$ are the partial derivatives with respect to fast and slow time respectively.

Applying a Taylor series expansion to $\mathcal{N}$, balancing the ε^0 and ε^1 terms, and ignoring higher order terms gives

$$\begin{aligned}
\varepsilon^0\!:&\quad \omega_{z2}^2 z_{20}'' + \omega_{z2}^2 z_{20} = 0, \\
\varepsilon^1\!:&\quad \omega_{z2}^2 z_{21}'' + \omega_{z2}^2 z_{21} = -\omega_{z2}^2 2\mu z_{20}'' - 2\omega_{z2}z_{20}'^{*} - 2\zeta_{z2}\omega_{z2}^2 z_{20}' \\
&\qquad - \mathcal{N}(\omega_{z2}z_{20}', z_{20}, t) - B\Delta\omega_{z2}^2\cos(\tau),
\end{aligned}$$

where the nonlinear term is given by

$$\mathcal{N}(\omega_{z2}z'_{20}, z_{20}, t) = N_2\Delta\cos(\tau)z_{20} + W_{22}z_{20}^3.$$

The solution to the ε^0 is given by

$$z_{20} = Z_{20c}(T)\cos(\tau) + Z_{20s}(T)\sin(\tau).$$

Substituting this expression into the ε^1 equation gives

$$\begin{aligned}\omega_{z2}^2 z''_{21} + \omega_{z2}^2 z_{21} = {} & (2\omega_{z2}^2\mu Z_{20c} - 2\omega_{z2}Z^*_{20s} - 2\omega_{z2}^2\zeta_{z2}Z_{20s} - \frac{3}{4}W_{22}Z_{20}^2 Z_{20c} \\ & - B\Delta\omega_{z2}^2)\cos(\tau) + (2\omega_{z2}^2\mu Z_{20s} + 2\omega_{z2}Z^*_{20c} + 2\omega_{z2}^2\zeta_{z2}Z_{20c} \\ & - \frac{3}{4}W_{22}Z_{20}^2 Z_{20s})\sin(\tau) + \frac{1}{4}W_{22}((3Z_{20s}^2 - Z_{20c}^2)Z_{20c}\cos(3\tau) \\ & + (Z_{20s}^2 - 3Z_{20c}^2)Z_{20s}\sin(3\tau)) - \frac{1}{2}N_2\Delta((1+\cos(2\tau))Z_{20c} \\ & + \sin(2\tau)Z_{20s}) \end{aligned} \tag{7.81}$$

where $Z_{20} = \sqrt{Z_{20c}^2 + Z_{20s}^2}$ is the amplitude of the z_{20} response. To eliminate secular terms,[2] and ensure that the response at the forcing frequency is captured by x_0, the amplitudes of the $\cos(\tau)$ and $\sin(\tau)$ terms on the right-hand side of the equation must be set to zero. This results in the following conditions on Z_{20c} and Z_{20s}:

$$\begin{aligned} Z^*_{20c} &= -\frac{1}{\omega_{z2}}\left(\omega_{z2}^2\mu Z_{20s} + \omega_{z2}^2\zeta_{z2}Z_{20c} - \frac{3}{8}W_{22}Z_{20}^2 Z_{20s}\right), \\ Z^*_{20s} &= \frac{1}{\omega_{z2}}\left(\omega_{z2}^2\mu Z_{20c} - \omega_{z2}^2\zeta_{z2}Z_{20s} - \frac{3}{8}W_{22}Z_{20}^2 Z_{20c} - \frac{1}{2}B\Delta\omega_{z2}^2\right). \end{aligned}$$

Since

$$Z^*_{20c} = \frac{\partial Z_{20c}}{\partial T} = \frac{\mathrm{d}Z_{20c}}{\mathrm{d}T} = \frac{\mathrm{d}Z_{20c}}{\mathrm{d}\varepsilon t} = \frac{1}{\varepsilon}\dot{Z}_{20c},$$

where the partial derivative can be converted to the full derivative since $Z_{20c} = Z_{20c}(T)$ is not a function of τ, these equations are identical to those derived using the averaging technique, Eqs. (7.74) and (7.75).

As with the averaging technique, these equations can also be used to investigate the stability of the solutions. In addition by using multiple scales, and solving the equation for z_{21}, Eq. (7.81), information regarding the response at other frequencies can be found. Since the $\cos(\tau)$ and $\sin(\tau)$ terms on the right-hand side have been set to zero, the equation simplifies to

---

[2] The definition of secular terms can be found in Sect. 4.4.2.

$$\omega_{z2}^2 z_{21}'' + \omega_{z2}^2 z_{21} = \frac{1}{4} W_{22}((3Z_{20s}^2 - Z_{20c}^2)Z_{20c}\cos(3\tau) + (Z_{20s}^2 - 3Z_{20c}^2)Z_{20s}\sin(3\tau))$$
$$- \frac{1}{2} N_2 \Delta((1+\cos(2\tau))Z_{20c} + \sin(2\tau)Z_{20s}).$$

So, for example, the component of the response at zero frequency can be written as

$$z_{21,\omega=0} = -\frac{1}{2\omega_{z2}^2} N_2 \Delta Z_{20c}$$

The multiple-scales technique produced the same results as the averaging technique. However with multiple scales it is also possible to extract information about the response at sub- and super-harmonics of the forcing frequency. It is also possible to get this type of information by using normal form analysis, and the final part of this case study is to consider how the normal form technique is applied to the inclined cable example.

7.4.4 Normal Forms

Again, the equation for the second in-plane mode of an inclined cable subject to a support excitation, with no out-of-plane motion, Eq. (7.69), is given by

$$\ddot{z}_2 + \omega_{z2}^2 z_2 + \varepsilon\left(2\zeta_{z2}\omega_{z2}\dot{z}_2 + W_{22}z_2^3 + N_2\delta z_2\right) = \varepsilon B\ddot{\delta},$$

with the support motion $\delta = \Delta\cos(\Omega t)$, where Ω is the forcing frequency.

The various steps necessary to find the normal form estimate of the near-resonant forced response of the cable are now performed, following Sect. 4.5.3. The first thing to observe is that the near-resonant forcing of the mode is of interest, the response frequency may be written as $\omega_{r1} = \Omega$.

Step 1: The first step is to rewrite the equation of motion into modal form by applying the linear modal transform $\mathbf{x} = \Phi\mathbf{q}$. Here $\mathbf{q}$ is a vector of the modal coordinates for the linear unforced, undamped system, $M\ddot{\mathbf{x}} + K\mathbf{x} = 0$. In addition Φ is a matrix of column vectors corresponding to the eigenvectors of $M^{-1}K$.

For this example the equation is already in modal form as it was generated using the Galerkin method. It can however be rewritten in the form used in Sect. 4.5.3 giving

$$\ddot{\mathbf{q}} + \Lambda\mathbf{q} + N_q(\mathbf{q}, \dot{\mathbf{q}}, \mathbf{r}) = P_q\mathbf{r},$$
$$\text{where:}\quad P_q = -B\Delta\Omega^2\left[1\ 1\right]/2, \quad \mathbf{r}^T = \left[r_p\ r_m\right] = \left[e^{i\Omega t}\ e^{-i\Omega t}\right] \qquad (7.82)$$
$$q_1 = z_2, \quad \Lambda = \omega_{z2}^2, \quad N_q = 2\zeta_{z2}\omega_{z2}\dot{q}_1 + W_{22}q_1^3 + N_2\Delta(r_p + r_m)q_1/2.$$

Note that in the examples discussed in Sect. 4.5.3, the notation for the natural frequency of a single mode system was ω_n, hence here $\omega_n = \omega_{z2}$. Also the nonlinear and damping term N_q is treated as of order ε^1 as is required for the normal form approach.

Step 1f: The next step is to apply the forcing transformation, with the aim of removing any non-resonant forcing terms. This is achieved using the transform $\mathbf{q} = \mathbf{v} + \mathbf{er}$ where for the single mode cable system $\mathbf{e}$ is a matrix of size 1×2.

The matrix $\mathbf{e}$ and the contents of the forcing vector in the transformed equation of motion, P_v, are dependent on whether each forcing term is near-resonant or not via Eqs. (4.76) and (4.77) respectively. The forcing is near-resonant in this example and so, using Eq. (4.76),

$$\mathbf{e} = \begin{bmatrix} 0 & 0 \end{bmatrix}, \quad P_v = P_q. \tag{7.83}$$

Using these, the transformed equation of motion, Eq. (4.74) is given by

$$\ddot{\mathbf{v}} + \Lambda \mathbf{v} + N_v(\mathbf{v}, \dot{\mathbf{v}}, \mathbf{r}) = P_v \mathbf{r},$$
$$\text{where: } N_v(\mathbf{v}, \dot{\mathbf{v}}, \mathbf{r}) = N_q(\mathbf{v}, \dot{\mathbf{v}}, \mathbf{r}).$$

Step 2: The nonlinear near-identity transform, which is used to remove non-resonant nonlinear terms from the equation of motion, is now sought.

As in Eq. (4.78), the nonlinear term is expanded as a series of increasingly insignificant terms. Here all the nonlinear terms are order ε hence

$$N_v(\mathbf{v}, \dot{\mathbf{v}}, \mathbf{r}) = \varepsilon \mathbf{n}_1(\mathbf{v}, \dot{\mathbf{v}}, \mathbf{r}) + \varepsilon^2 \mathbf{n}_2(\mathbf{v}, \dot{\mathbf{v}}, \mathbf{r}) + \varepsilon^3 \mathbf{n}_3(\mathbf{v}, \dot{\mathbf{v}}, \mathbf{r}) + \cdots$$
$$\text{with: } \mathbf{n}_1 = 2\zeta_{z2}\omega_{z2}\dot{v}_1 + W_{22}v_1^3 + N_2\Delta(r_p + r_m)v_1/2, \; \mathbf{n}_2 = 0, \; \mathbf{n}_3 = 0, \ldots$$

Rewriting $\mathbf{n}_1(\mathbf{v}, \dot{\mathbf{v}}, \mathbf{r})$ in terms of $\mathbf{u}$, as $\mathbf{n}_1(\mathbf{u}, \dot{\mathbf{u}}, \mathbf{r})$, and then making the substitution $\mathbf{u} = \mathbf{u}_p + \mathbf{u}_m$, such that $u_1 = u_{p1} + u_{m1}$, and expressing $\mathbf{n}_1$ in matrix form gives

$$\mathbf{n}_1(\mathbf{u}, \dot{\mathbf{u}}, \mathbf{r}) = \mathbf{n}^* \mathbf{u}^*(\mathbf{u}_p, \mathbf{u}_m, \mathbf{r}) \quad \text{where: } \mathbf{n}^* = \begin{bmatrix} W_{22} \\ 3W_{22} \\ 3W_{22} \\ W_{22} \\ N_2\Delta/2 \\ N_2\Delta/2 \\ N_2\Delta/2 \\ N_2\Delta/2 \\ \mathrm{i}2\zeta_{z2}\omega_{z2}\Omega \\ -\mathrm{i}2\zeta_{z2}\omega_{z2}\Omega \end{bmatrix}^T, \; \mathbf{u}^* = \begin{bmatrix} u_{p1}^3 \\ u_{p1}^2 u_{m1} \\ u_{p1} u_{m1}^2 \\ u_{m1}^3 \\ u_{p1} r_p \\ u_{p1} r_m \\ u_{m1} r_p \\ u_{m1} r_m \\ u_{p1} \\ u_{m1} \end{bmatrix}.$$

The general form for each term in $\mathbf{u}^*$, given by Eqs. (4.84) and (4.86) is used to find β^*

$$\beta^* = \begin{bmatrix} 8 & 0 & 0 & 8 & 3 & -1 & -1 & 3 & 0 & 0 \end{bmatrix} \Omega^2.$$

where $\omega_{r1} = \Omega$ has been used.

Step 3: Now matrix β^* can be used, term by term, to determine which of Eqs. (4.58) and (4.59) should be satisfied, hence finding the near identity transform and the resonant nonlinear terms that remain in the dynamic equation for $\mathbf{u}$

$$\beta^* = \Omega^2 \begin{bmatrix} 8 \\ 0 \\ 0 \\ 8 \\ 3 \\ -1 \\ -1 \\ 3 \\ 0 \\ 0 \end{bmatrix}^T \rightsquigarrow \mathbf{n}_u^* = \begin{bmatrix} 0 \\ 3W_{22} \\ 3W_{22} \\ 0 \\ 0 \\ 0 \\ 0 \\ 0 \\ \mathrm{i}2\zeta_{z2}\omega_{z2}\Omega \\ -\mathrm{i}2\zeta_{z2}\omega_{z2}\Omega \end{bmatrix}^T, \quad \mathbf{h}^* = \frac{1}{\Omega^2} \begin{bmatrix} W_{22}/8 \\ 0 \\ 0 \\ W_{22}/8 \\ N_2\Delta/6 \\ -N_2\Delta/2 \\ -N_2\Delta/2 \\ N_2\Delta/6 \\ 0 \\ 0 \end{bmatrix}^T.$$

Step 4: These two matrices may be used, along with $\mathbf{u}^*$, to write the transformed equation of motion and the near-identity transform, to order ε^1 accuracy, as

$$\ddot{\mathbf{u}} + \Lambda\mathbf{u} + \mathbf{n}_{u1}\mathbf{u}^* = P_u\mathbf{r} \tag{7.84}$$

$$\mathbf{v} = \mathbf{u} + \mathbf{h}^*\mathbf{u}^* \tag{7.85}$$

where Eqs. (4.80) and (4.79) have been used respectively. Using Eqs. (7.82) and (7.83), the dynamic equation, Eq. (7.84), may be written as

$$\ddot{\mathbf{u}} + 2\zeta_{z2}\omega_{z2}\dot{\mathbf{u}} + \omega_{z2}^2\mathbf{u} + 3W_{22}(u_{p1}^2 u_{m1} + u_{p1}u_{m1}^2) = -\frac{B\Delta\Omega^2}{2}(r_p + r_m), \tag{7.86}$$

where, since there is only one degree-of-freedom, $\mathbf{u} = u_1$. Substituting in the trial solution for $\mathbf{u} = \mathbf{u}_p + \mathbf{u}_m$, see Eq. (4.51), along with $\mathbf{r} = [\mathbf{r}_p \ \ \mathbf{r}_m]^T = [\mathrm{e}^{\mathrm{i}\Omega t} \ \ \mathrm{e}^{-\mathrm{i}\Omega t}]^T$ results in

$$\left(\omega_{z2}^2 - \Omega^2\right) U_1 \left(\mathrm{e}^{\mathrm{i}(\Omega t-\phi_1)} + \mathrm{e}^{-\mathrm{i}(\Omega t-\phi_1)}\right) + \mathrm{i}2\zeta_{z2}\omega_{z2}\Omega U_1 \left(\mathrm{e}^{\mathrm{i}(\Omega t-\phi_1)} - \mathrm{e}^{-\mathrm{i}(\Omega t-\phi_1)}\right)$$
$$+ \frac{3W_{22}}{4}U_1^3 \left(\mathrm{e}^{\mathrm{i}(\Omega t-\phi_1)} + \mathrm{e}^{-\mathrm{i}(\Omega t-\phi_1)}\right) = -B\Delta\Omega^2 \left(\mathrm{e}^{\mathrm{i}\Omega t} + \mathrm{e}^{-\mathrm{i}\Omega t}\right).$$

Balancing the $\mathrm{e}^{\mathrm{i}(\Omega t-\phi_1)}$ terms gives

$$\left(\omega_{z2}^2 - \Omega^2\right) U_1 + \mathrm{i}2\zeta_{z2}\omega_{z2}\Omega U_1 + \frac{3W_{22}}{4}U_1^3 = -B\Delta\Omega^2\mathrm{e}^{\mathrm{i}\phi_1}.$$

Note that balancing the $\mathrm{e}^{-\mathrm{i}(\Omega t-\phi_1)}$ would give the complex conjugate of this. Now the real and imaginary components are balanced such that

$$Re: \ \left(\omega_{z2}^2 - \Omega^2\right) U_1 + \tfrac{3W_{22}}{4}U_1^3 = -B\Delta\Omega^2\cos(\phi_1)$$
$$Im: \ 2\zeta_{z2}\omega_{z2}\Omega U_1 = -B\Delta\Omega^2\sin(\phi_1)$$

Finally squaring and adding these equations eliminates the phase, ϕ_1, to give the amplitude of response equation

$$9W_{22}^2U_1^6+24W_{22}U_1^4\left(\omega_{z2}^2-\Omega^2\right)+16\left(\omega_{z2}^2-\Omega^2\right)^2U_1^2+64\zeta_{z2}^2\omega_{z2}^2\Omega^2U_1^2=16B^2\Delta^2\Omega^4 \tag{7.87}$$

This can be shown to be identical, to order ε^1, to the equation derived using the harmonic balance, Eq. (7.71), recalling that for the harmonic balance the detuning of the forcing away from resonance was taken to be ε^1 small such that $\Omega = \omega_{z2}(1+\varepsilon\mu)$.

As with the multiple scales technique, the response at sub- and super-harmonics of the forcing frequency can be found using Eq. (7.85). In addition, as with both the averaging and the multiple scales approaches, the stability of the steady-state solutions can be found using the slowly time-varying amplitude version of the trial solution, Eq. (4.92), substituted into the dynamic equation for $\mathbf{u}$, Eq. (7.86). This leads to a first order dynamic equation in amplitudes U_{p1} and U_{m1} which can be used to find the solution stability as discussed in Sect. 4.5.4.

In summary, this case study has been used to demonstrate how the approximate analytical methods discussed in Chap. 4 can be applied to a realistic nonlinear vibration problem. The methods have been applied in order of sophistication, which also broadly corresponds to the level of complexity required for each method. However, as each method becomes more complex to apply, it also yields more useful information about the system response. Therefore, the key point is to select the method which is appropriate to the level of effort and information required for the problem at hand.

As discussed in Chap. 5, the normal form analysis can be used to give additional information on the nonlinear normal mode behaviour of the system. This will be examined in the next section.

7.5 Case Study of Modal Interaction in Cables

The potential for interaction between cable modes, first discussed in Sect. 7.3.4, is now examined in more depth using the normal form technique. This technique is chosen as it is readily formulated in a matrix form, as was presented in Chap. 4, and as such multiple degrees-of-freedom can be analysed relatively easily as a natural extension to the approach for a single degree-of-freedom system.

7.5.1 *Normal Form Analysis of Two Mode Response*

Here the interaction between the second in-plane mode, which is forced close to resonance, and the second out-of-plane, which is subject to no direct forcing, is considered. Taking just these two modes, using Eqs. (7.67) and (7.68) the equations of motion can be written as

$$\begin{aligned}\ddot{y}_2 + 2\zeta\omega_2\dot{y}_2 + \omega_2^2y_2 + W_{22}y_2(y_2^2+z_2^2) + N_2\delta y_2 &= 0\\ \ddot{z}_2 + 2\zeta\omega_2\dot{z}_2 + \omega_2^2z_2 + W_{22}z_2(y_2^2+z_2^2) + N_2\delta z_2 &= B\ddot{\delta},\end{aligned}$$

where $y_1 = 0$ (it is the interaction between the two second modes that are of interest in this analysis). It is assumed that both modes have the same damping ratio, $\zeta_{y2} = \zeta_{z2} = \zeta$, and the near-resonant vertical motion of the support, δ is taken to be $\delta = \Delta\cos(\Omega t)$, where $\Omega \approx \omega_2$. For a more detailed analysis in which both the first and second out-of-plane modes are considered see Gonzalez-Buelga et al. (2008).

Step 1: The first step of the normal form analysis is to perform a linear modal transform. This is not necessary here as the equations are already in modal form, having been derived via the Galerkin method. However they can be expressed in the standard form based on the two linear modes by writing

$$\ddot{\mathbf{q}} + \Lambda\mathbf{q} + N_q(\mathbf{q}, \dot{\mathbf{q}}, \mathbf{r}) = P_q\mathbf{r},$$

$$\text{where:}\quad \Lambda = \begin{bmatrix} \omega_2^2 & 0 \\ 0 & \omega_2^2 \end{bmatrix}, \quad P_q = \begin{bmatrix} 0 & 0 \\ -B\Delta\Omega^2 & -B\Delta\Omega^2 \end{bmatrix},$$

$$N_q(\mathbf{q}, \dot{\mathbf{q}}, \mathbf{r}) = \begin{Bmatrix} 2\zeta\omega_2\dot{q}_1 + W_{22}q_1(q_1^2 + q_2^2) + N_2\Delta(r_p + r_m)q_1 \\ 2\zeta\omega_2\dot{q}_2 + W_{22}q_2(q_1^2 + q_2^2) + N_2\Delta(r_p + r_m)q_2 \end{Bmatrix}$$

where $\mathbf{q} = [y_2 \quad z_2]$ and $\mathbf{r} = [r_p \quad r_m]$ with $r_p = \mathrm{e}^{\mathrm{i}\Omega t}$ and $r_p = \mathrm{e}^{-\mathrm{i}\Omega t}$. Note that the damping term has been included within the nonlinear vector, as discussed when Eq. (4.72) was introduced.

Step 1f: Non-resonant terms must now be removed using the force transformation $\mathbf{q} = \mathbf{v} + \mathbf{e}\mathbf{r}$. Equations (4.76) and (4.77) govern whether terms remain in the forcing matrix, P_v, in the transformed dynamic equation, or are scaled and placed into the transform matrix $\mathbf{e}$. Here, since both modes have the same linear natural frequency, ω_2 which is close to the forcing frequency Ω, all the terms are resonant. This results in $\mathbf{e} = 0$ and $\mathbf{q} = \mathbf{v}$, giving

$$\ddot{\mathbf{v}} + \Lambda\mathbf{v} + N_v(\mathbf{v}, \dot{\mathbf{v}}, \mathbf{r}) = P_v\mathbf{r}, \quad \text{with} \quad P_v = P_q,$$

$$N_v(\mathbf{v}, \dot{\mathbf{v}}, \mathbf{r}) = N_q(\mathbf{v}, \dot{\mathbf{v}}, \mathbf{r}) = \begin{Bmatrix} 2\zeta\omega_2\dot{v}_1 + W_{22}v_1(v_1^2 + v_2^2) + N_2\Delta(r_p + r_m)v_1 \\ 2\zeta\omega_2\dot{v}_2 + W_{22}v_2(v_1^2 + v_2^2) + N_2\Delta(r_p + r_m)v_2 \end{Bmatrix}.$$

Step 2: To find the near-identity transform, the nonlinear term N_v is firstly written in terms of $\mathbf{u}$ and then expressed in powers of ε, $N_v(\mathbf{v}, \dot{\mathbf{v}}, \mathbf{r}) = \varepsilon\mathbf{n}_1(\mathbf{u}, \dot{\mathbf{u}}, \mathbf{r}) + \varepsilon^2\mathbf{n}_2(\mathbf{u}, \dot{\mathbf{u}}, \mathbf{r}) + \cdots$ as in Eq. (4.78). Taking the terms in N_v to all be of order ε^1 gives

$$\mathbf{n}_1(\mathbf{u}, \dot{\mathbf{u}}, \mathbf{r}) = \begin{Bmatrix} 2\zeta\omega_2\dot{u}_1 + W_{22}u_1(u_1^2 + u_2^2) + N_2\Delta(r_p + r_m)u_1 \\ 2\zeta\omega_2\dot{u}_2 + W_{22}u_2(u_1^2 + u_2^2) + N_2\Delta(r_p + r_m)u_2 \end{Bmatrix},$$

with higher order ε^k terms, $\mathbf{n}_k$ set to zero.

Now expressing the nonlinear term in matrix form, see Eq. (4.83), and using $\mathbf{u} = \mathbf{u}_p + \mathbf{u}_m$ gives $\mathbf{n}_1(\mathbf{u}, \dot{\mathbf{u}}, \mathbf{r}) = \mathbf{n}^*\mathbf{u}^*(\mathbf{u}_p, \mathbf{u}_m, \mathbf{r})$ with

$$
\mathbf{n}^* = \begin{bmatrix}
W_{22} & 0 \\
3W_{22} & 0 \\
0 & W_{22} \\
0 & W_{22} \\
3W_{22} & 0 \\
0 & 2W_{22} \\
0 & 2W_{22} \\
W_{22} & 0 \\
2W_{22} & 0 \\
W_{22} & 0 \\
W_{22} & 0 \\
0 & W_{22} \\
0 & W_{22} \\
W_{22} & 0 \\
2W_{22} & 0 \\
W_{22} & 0 \\
0 & W_{22} \\
0 & 3W_{22} \\
0 & 3W_{22} \\
0 & W_{22} \\
N_2\Delta & 0 \\
N_2\Delta & 0 \\
N_2\Delta & 0 \\
N_2\Delta & 0 \\
0 & N_2\Delta \\
0 & N_2\Delta \\
0 & N_2\Delta \\
0 & N_2\Delta \\
\mathrm{i}2\zeta\omega_2\Omega & 0 \\
-\mathrm{i}2\zeta\omega_2\Omega & 0 \\
0 & \mathrm{i}2\zeta\omega_2\Omega \\
0 & -\mathrm{i}2\zeta\omega_2\Omega
\end{bmatrix}^T, \quad
\mathbf{u}^* = \begin{bmatrix}
u_{p1}^3 \\
u_{p1}^2 u_{m1} \\
u_{p1}^2 u_{p2} \\
u_{p1}^2 u_{m2} \\
u_{p1} u_{m1}^2 \\
u_{p1} u_{m1} u_{p2} \\
u_{p1} u_{m1} u_{m2} \\
u_{p1} u_{p2}^2 \\
u_{p1} u_{p2} u_{m2} \\
u_{p1} u_{m2}^2 \\
u_{m1}^3 \\
u_{m1}^2 u_{p2} \\
u_{m1}^2 u_{m2} \\
u_{m1} u_{p2}^2 \\
u_{m1} u_{p2} u_{m2} \\
u_{m1} u_{m2}^2 \\
u_{p2}^3 \\
u_{p2}^2 u_{m2} \\
u_{p2} u_{m2}^2 \\
u_{m2}^3 \\
u_{p1} r_p \\
u_{p1} r_m \\
u_{m1} r_p \\
u_{m1} r_m \\
u_{p2} r_p \\
u_{p2} r_m \\
u_{m2} r_p \\
u_{m2} r_m \\
u_{p1} \\
u_{m1} \\
u_{p2} \\
u_{m2}
\end{bmatrix}.
$$

The matrix $\boldsymbol{\beta}^*$, which governs whether terms are placed in the transformed nonlinear matrix or in the transform matrix, can now be calculated. This is done one element at a time using Eq. (4.86) along with Eq. (4.84), which provides a general form for each term in $\mathbf{u}^*$. Noting that $\omega_{r1} = \omega_{r2} = \Omega$, the resulting $\boldsymbol{\beta}^*$ matrix is

$$
\boldsymbol{\beta}^* = \Omega^2 \left[\begin{matrix}
8 & 0 & - & - & 0 & - & - & 8 & 0 & 0 & 8 & - & - & 0 & 0 & 8 \\
- & - & 8 & 0 & - & 0 & 0 & - & - & - & - & 0 & 8 & - & - & -
\end{matrix} \right.
$$
$$
\left. \begin{matrix}
- & - & - & - & 3 & -1 & -1 & 3 & - & - & - & - & 0 & 0 & - & - \\
8 & 0 & 0 & 8 & - & - & - & - & 3 & -1 & -1 & 3 & - & - & 0 & 0
\end{matrix} \right].
$$

Here a dash has been used in locations where the corresponding value in $\mathbf{n}^*$ is zero and so the value of these terms in $\boldsymbol{\beta}^*$ is unimportant.

Step 3: Using Eqs. (4.58) and (4.59) along with matrix $\boldsymbol{\beta}^*$ both the near-identity transform and the post transformed nonlinear terms in the dynamic equation for $\mathbf{u}$ can be found. Here the transformed equation of motion is of key interest and so the transform terms will be ignored, effectively assuming that the sub- and super-harmonics of the response are unimportant. The nonlinear terms in the equation for motion in $\mathbf{u}$ are given by

$$\beta^* = \Omega^2 \begin{bmatrix} 8 & - \\ 0 & - \\ - & 8 \\ - & 0 \\ 0 & - \\ - & 0 \\ - & 0 \\ 8 & - \\ 0 & - \\ 0 & - \\ 8 & - \\ - & 0 \\ - & 8 \\ 0 & - \\ 0 & - \\ 8 & - \\ - & 8 \\ - & 0 \\ - & 0 \\ - & 8 \\ 3 & - \\ -1 & - \\ -1 & - \\ 3 & - \\ - & 3 \\ - & -1 \\ - & -1 \\ - & 3 \\ 0 & - \\ 0 & - \\ - & 0 \\ - & 0 \end{bmatrix}^T \quad \rightsquigarrow \quad \mathbf{n}_u^* = \begin{bmatrix} 0 & 0 \\ 3W_{22} & 0 \\ 0 & 0 \\ 0 & W_{22} \\ 3W_{22} & 0 \\ 0 & 2W_{22} \\ 0 & 2W_{22} \\ 0 & 0 \\ 2W_{22} & 0 \\ W_{22} & 0 \\ 0 & 0 \\ 0 & W_{22} \\ 0 & 0 \\ W_{22} & 0 \\ 2W_{22} & 0 \\ 0 & 0 \\ 0 & 0 \\ 0 & 3W_{22} \\ 0 & 3W_{22} \\ 0 & 0 \\ 0 & 0 \\ 0 & 0 \\ 0 & 0 \\ 0 & 0 \\ 0 & 0 \\ 0 & 0 \\ 0 & 0 \\ 0 & 0 \\ \mathrm{i}2\zeta\omega_2\Omega & 0 \\ -\mathrm{i}2\zeta\omega_2\Omega & 0 \\ 0 & \mathrm{i}2\zeta\omega_2\Omega \\ 0 & -\mathrm{i}2\zeta\omega_2\Omega \end{bmatrix}^T .$$

Step 4: The equation of motion in the transformed co-ordinate system, to order ε^1 accuracy, may now be written as

$$
\begin{aligned}
&\ddot{\mathbf{u}} + \Lambda\mathbf{u} + \mathbf{n}^*\mathbf{u}^* = P_u\mathbf{r} \\
&\rightsquigarrow \ddot{u}_1 + 2\zeta\omega_2\dot{u}_1 + \omega_2^2 u_1 + 3W_{22}u_{p1}u_{m1}u_1 + 2W_{22}u_{p2}u_{m2}u_1 \\
&\quad + W_{22}(u_{p1}u_{m2}^2 + u_{m1}u_{p2}^2) = 0 \\
&\ddot{u}_2 + 2\zeta\omega_2\dot{u}_2 + \omega_2^2 u_2 + 3W_{22}u_{p2}u_{m2}u_2 + 2W_{22}u_{p1}u_{m1}u_2 \\
&\quad + W_{22}(u_{p1}^2 u_{m2} + u_{m1}^2 u_{p2}) = -B\Delta\Omega^2(r_p + r_m).
\end{aligned} \tag{7.88}
$$

Note that $P_u = P_v$, see Eq. (4.82), and that $P_v = P_q$, see **Step 1f**, and the forcing matrix has been used from **Step 1**. In addition the relationship $u_k = u_{pk} + u_{mk}$ for the response of the kth mode has been used.

These equations represent the resonant response of the in- and out-of-plane second modes. To progress, the constant-amplitude trial solution, Eq. (4.51), can be used to find the steady-state response of the two degrees of freedom. Alternatively, to analyse the stability of the response the slowly time-varying amplitude trial solution, Eq. (4.92), is used.

By inspection it can be seen that one solution for the in-plane mode, the resonant response that corresponds to u_1, is $u_1 = 0$. However due to the auto-parametric forcing terms, it is also possible for a non-zero response to occur. To understand this, firstly in the next subsection the backbones of the system will be analysed. Then in the following subsection, the response of the second in-plane mode will be considered, and in particular the conditions under which a non-zero response is observed will be found—this corresponds to points at which the $u_1 = 0$ solution looses stability.

7.5.2 Backbone Curves for the Cable System

The backbone equations for the two degree-of-freedom cable system, involving the second in- and out-of-plane modes, can now be analysed. Backbone curves, see Sects. 4.1 and 5.3.1, capture the resonant response of the unforced, undamped system. Using Eq. (7.88), the resonant response of the unforced, undamped equivalent system may be written as

$$
\begin{aligned}
&\ddot{u}_1 + \omega_2^2 u_1 + 3W_{22}u_{p1}u_{m1}u_1 + 2W_{22}u_{p2}u_{m2}u_1 + W_{22}(u_{p1}u_{m2}^2 + u_{m1}u_{p2}^2) = 0, \\
&\ddot{u}_2 + \omega_2^2 u_2 + 3W_{22}u_{p2}u_{m2}u_2 + 2W_{22}u_{p1}u_{m1}u_2 + W_{22}(u_{p1}^2 u_{m2} + u_{m1}^2 u_{p2}) = 0.
\end{aligned}
$$

As the system is unforced and has identical linear natural frequencies in the in-plane and out-of-plane directions, the system's response frequency may be written as $\omega_r = \omega_{r1} = \omega_{r2}$. Note that in the previous subsection $\Omega = \omega_{r1} = \omega_{r2}$ was used to determine β^*, this still holds true for the unforced system if Ω is treated as the response frequency and so the normal form analysis does not need to be reapplied for the unforced case.

Using the constant-amplitude trial solution, Eq. (4.51), these equations may be written as

$$\left[(\omega_n^2-\omega_r^2)\frac{U_1}{2}+\frac{3W_{22}}{8}U_1^3+\frac{2W_{22}}{8}U_2^2U_1\right]\left(e^{i(\Omega t-\phi_1)}+e^{-i(\Omega t-\phi_1)}\right)$$
$$+\frac{W_{22}}{8}U_2^2U_1\left(e^{i(\Omega t-\phi_1)}e^{i2(\phi_1-\phi_2)}+e^{-i(\Omega t-\phi_1)}e^{-i2(\phi_1-\phi_2)}\right)=0,$$
$$\left[(\omega_n^2-\omega_r^2)\frac{U_2}{2}+\frac{3W_{22}}{8}U_2^3+\frac{2W_{22}}{8}U_1^2U_2\right]\left(e^{i(\Omega t-\phi_2)}+e^{-i(\Omega t-\phi_2)}\right)$$
$$+\frac{W_{22}}{8}U_1^2U_2\left(e^{i(\Omega t-\phi_2)}e^{-i2(\phi_1-\phi_2)}+e^{-i(\Omega t-\phi_2)}e^{i2(\phi_1-\phi_2)}\right)=0.$$

Now balancing the $e^{i(\Omega t-\phi_1)}$ terms in the first equation and the $e^{i(\Omega t-\phi_2)}$ terms in the second equation gives

$$\left[\omega_n^2-\omega_r^2+\frac{3W_{22}}{4}U_1^2+\frac{2W_{22}}{4}U_2^2+\frac{W_{22}}{4}U_2^2e^{i2(\phi_1-\phi_2)}\right]U_1=0,$$
$$\left[\omega_n^2-\omega_r^2+\frac{3W_{22}}{4}U_2^2+\frac{2W_{22}}{4}U_1^2+\frac{W_{22}}{4}U_1^2e^{-i2(\phi_1-\phi_2)}\right]U_2=0. \quad (7.89)$$

Note that balancing the $e^{-i(\Omega t-\phi_1)}$ and $e^{-i(\Omega t-\phi_2)}$ terms in the first and second equations respectively would result in the complex conjugate pairs to Eq. (7.89). Considering the imaginary parts of either equation in Eq. (7.89) results in the condition

$$\sin(2(\phi_1-\phi_2))=0.$$

Hence the relative phase between the two modes must be $k\pi/2$ where k is an integer. It follows that $\cos(2(\phi_1-\phi_2))=p$ where either $p=+1$, when the relative phase is 0 or π, or $p=-1$, when the phase is $\pm\pi/2$ out-of-phase. This allows the real parts of Eq. (7.89) to be written as

$$\left\{\omega_n^2-\omega_r^2+\frac{W_{22}}{4}\left[3U_1^2+(2+p)U_2^2\right]\right\}U_1=0,$$
$$\left\{\omega_n^2-\omega_r^2+\frac{W_{22}}{4}\left[(2+p)U_1^2+3U_2^2\right]\right\}U_2=0. \quad (7.90)$$

with $p=\pm1$ to ensure that the imaginary parts of Eq. (7.89) are satisfied.

There are several solutions to these equations. Firstly there is the trivial solution $U_1=U_2=0$. Then there are two responses, labelled S_1 and S_2, relating to a response in just U_1 or just U_2 respectively

$$S1:\quad U_2=0,\quad \omega_r^2=\omega_n^2+\frac{3W_{22}}{4}U_1^2,$$
$$S2:\quad U_1=0,\quad \omega_r^2=\omega_n^2+\frac{3W_{22}}{4}U_2^2.$$

There are also solutions in which $U_1\neq0$ and $U_2\neq0$. In this case both equations in Eq. (7.89) can be written in terms of the response frequency giving

$$\omega_r^2 = \omega_n^2 + \frac{W_{22}}{4}\left[3U_1^2 + (2+p)U_2^2\right] = \omega_n^2 + \frac{W_{22}}{4}\left[(2+p)U_1^2 + 3U_2^2\right]$$

For this to be satisfied the condition

$$(1-p)U_1^2 = (1-p)U_2^2$$

is placed on U_1 and U_2. When $p = 1$, such that the two modes respond either in-phase or in anti-phase, no amplitude condition exists and the backbone curve is given by

$$S3^{\pm}:\ \omega_r^2 = \omega_n^2 + \frac{3W_{22}}{4}\left[U_1^2 + U_2^2\right],$$

where $S3^+$ and $S3^-$ correspond to the cases where the relative phase is 0 and π respectively. Finally considering $p = -1$, such that the two modes are $\pm\pi/2$ out-of-phase, results in the condition that $U_1 = U_2$ and using Eq. (7.89) gives the backbone curve

$$S4^{\pm}:\ \omega_r^2 = \omega_n^2 + W_{22}U_1^2,\quad U_1 = U_2,$$

where $S4^+$ and $S4^-$ correspond to the cases where the relative phase is $\pi/2$ and $-\pi/2$ respectively.

It is convenient to define $N = \sqrt{U_1^2 + U_2^2}$. For in-phase or anti-phase responses this corresponds to the total amplitude of the response. Using this solution $S1$, $S2$ and $S3$ can all be written as

$$S1, S2, S3^{\pm}:\ \omega_r^2 = \omega_n^2 + \frac{3W_{22}}{4}N^2,$$

The reason that these solution all reduce to the same backbone curve is that without forcing this system has axial symmetry (gravitational sag, which breaks the axial symmetry of the system, only effects the odd in-plane modes) and so the plane in which the cable vibrates is arbitrary. If the first mode was considered, where $\omega_{y1} \neq \omega_{z1}$, this would not be the case. However the anti-phase solution is different both in terms of the backbone curve and due to the condition that $U_1 = U_2$. This solution may also be written in terms of N as

$$S4^{\pm}:\ \omega_r^2 = \omega_n^2 + \frac{W_{22}}{2}N^2,\quad U_1 = U_2,$$

Figure 7.10 shows the backbone curves in the N versus ω_r plane (with the frequency normalised using the linear natural frequency ω_2. To visualise the cable motion, Fig. 7.11 shows the response in the y_2 versus z_2 plane, taken at quarter-span for each solution type. Note that the sub- and superharmonics contributions to the response have been neglected allowing the approximation that $y_2 = u_1$ and $z_2 = u_2$. These responses are plotted on an arbitrary scale that is the same in both directions. For the S_3 case the angle the response makes in the y_2 versus z_2 plane is also arbitrary (but is not 0 or 90° as for these

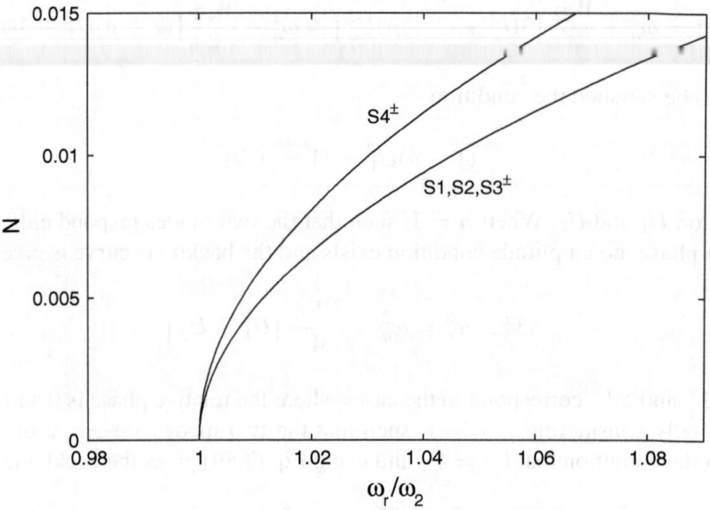

Fig. 7.10 Backbone *curves* for a model of a cable that contains the in- and out-of-plane second modes. $N = \sqrt{U_1^2 + U_2^2}$, which for the in- or anti-phase modes can be thought of at the maximum amplitude of the response, neglecting sub- and superharmonics contributions to the response. The parameter values used are the same as those in Example 7.1

angles the response consists purely of a single mode corresponding to solutions $S1$ and $S2$ respectively). It can be seen from the y_2 versus z_2 plane plots that the $S4$ solutions correspond to a whirling motion of the cable, the arrows show the direction of travel with time.

7.5.3 Autoparametric Response of the Out-of-Plane Mode

Consider the two mode model of the cable based on the second in-plane and second out-of-plane modes, as analysed in Sect. 7.5.1. Inspecting Eq. (7.88), it can be seen that one solution for the response is that $u_{p1} = u_{m1} = 0$, such that the response is confined to the second in-plane mode (corresponding to u_2). This gives

$$u_1 = 0$$
$$\ddot{u}_2 + 2\zeta\omega_2\dot{u}_2 + \omega_2^2 u_2 + 3W_{22}u_{p2}u_{m2}u_2 = -B\Delta\Omega^2(r_p + r_m).$$

which is identical to the equation derived using the normal form when considering just the second in-plane mode, see Sect. 7.4.4 and specifically Eq. (7.86) (note that in this equation u_1 referred to the second in-plane mode). The resulting amplitude of response is given by Eq. (7.87), and may be rewritten to reflect the fact that now u_1 and u_2 refer to the second out-of- and in-plane modes respectively and are expressed in terms of the quadratic in Ω^2 giving

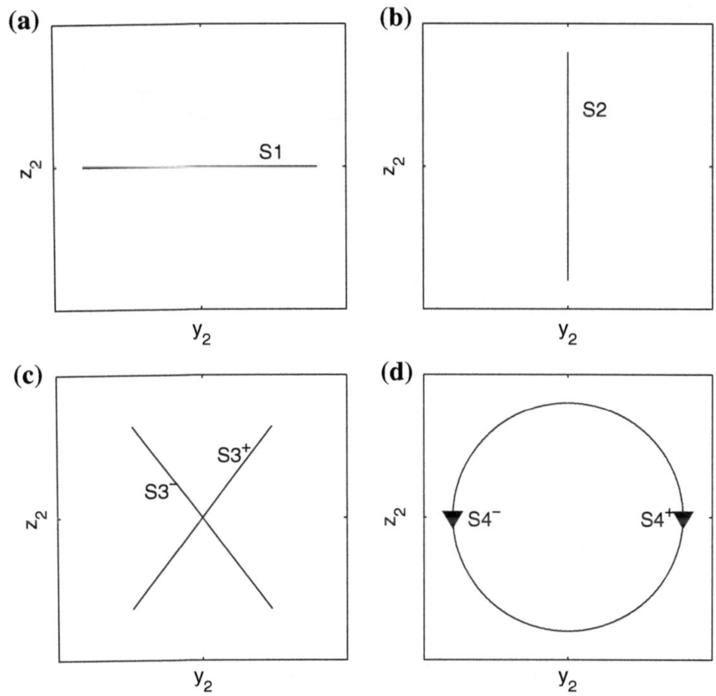

Fig. 7.11 Response of the cable at its quarter-span in the y_2 versus z_2 plane for the **a** $S1$ and **b** $S2$ single mode solutions and for the combined mode. **c** $S3$, in phase and **d** $S4$, $\pi/2$ out-of-phase solutions, plotted on a arbitrary scale that is the same in both directions

$$u_1 = 0, \tag{7.91}$$
$$9W_{22}^2U_2^6 + 24W_{22}U_2^4\left(\omega_2^2 - \Omega^2\right) + 16\left(\omega_2^2 - \Omega^2\right)^2 U_2^2 + 64\zeta^2\omega_2^2\Omega^2U_2^2 = 16B^2\Delta^2\Omega^4.$$

Figure 7.12 shows the response of the second in-plane mode when subjected to vertical support motion assuming that the other modes have zero response. This is based on the normal form prediction, Eq. (7.91), which was solved as a quadratic in Ω^2 for a range of values of U_2. It can be seen that the response follows the $S2$ backbone curve calculated in the previous analysis. The normal form prediction of the response was checked using AUTO-07p, a numerical continuation software package (see http://indy.cs.concordia.ca/auto/) and the agreement was very good. The continuation results also highlight that the steady-state solution is unstable between the two fold points. this could also be shown using the normal form, as was done for a forced Duffing oscillator in Example 4.11.

However from Sect. 7.5.1, it was seen that there is potential for the second out-of-plane mode to be excited via an autoparametric response through the second in-plane mode, see Eq. (7.88)

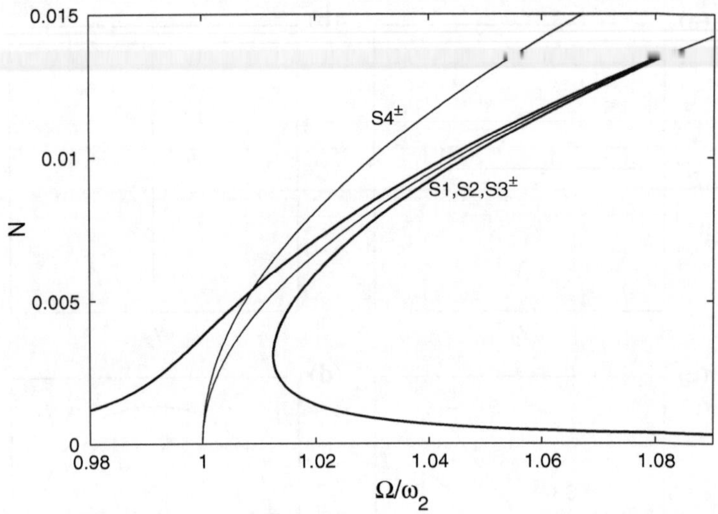

Fig. 7.12 Response of the cable, subjected to motion of the lower support with an amplitude $\Delta = 1.7 \times 10^{-4}$ and forcing frequency close to the second mode of the cable, using a model that just contains the second in-plane mode

$$\begin{aligned}
&\ddot{u}_1 + 2\zeta\omega_2\dot{u}_1 + \omega_2^2 u_1 + 3W_{22}u_{p1}u_{m1}u_1 + 2W_{22}u_{p2}u_{m2}u_1 \\
&\quad + W_{22}(u_{p1}u_{m2}^2 + u_{m1}u_{p2}^2) = 0 \\
&\ddot{u}_2 + 2\zeta\omega_2\dot{u}_2 + \omega_2^2 u_2 + 3W_{22}u_{p2}u_{m2}u_2 + 2W_{22}u_{p1}u_{m1}u_2 \\
&\quad + W_{22}(u_{p1}^2 u_{m2} + u_{m1}^2 u_{p2}) = -B\Delta\Omega^2(r_p + r_m).
\end{aligned}$$

The conditions under which this occurs will now be considered. Note that it is also possible to have a component of the response in the first out-of-plane mode, as shown in the time simulations in Sect. 7.3.4, however this possibility is not considered here. For there to be a response in u_1 (as well as in u_2), the zero response solution must become unstable. To find when this occurs, a simplified version of the u_1 equation is considered where the $3W_{22}u_{p1}u_{m1}u_1$ term is dropped giving

$$\ddot{u}_1 + 2\zeta\omega_2\dot{u}_1 + \omega_2^2 u_1 + 2W_{22}u_{p2}u_{m2}u_1 + W_{22}(u_{p1}u_{m2}^2 + u_{m1}u_{p2}^2) = 0. \tag{7.92}$$

The justification for dropping the term is that it is of magnitude U_1^3 and just at the point where the $u_1 = 0$ solution goes unstable, U_1 will be very small compared to the other terms.

As it is the solution stability that is of interest the slowly-time varying amplitude trial solution, Eq. (4.92), is used. Substituting this into the dynamic equation for u_1, Eq. (7.92), gives

$$
\begin{aligned}
&\left(-\Omega^2 U_{p1} + 2\mathrm{i}\Omega \dot{U}_{p1}\right) \mathrm{e}^{\mathrm{i}\Omega t} + \left(-\Omega^2 U_{m1} - 2\mathrm{i}\Omega \dot{U}_{m1}\right) \mathrm{e}^{-\mathrm{i}\Omega t} \\
&\quad + \mathrm{i}2\zeta\omega_2\Omega \left(U_{p1}\mathrm{e}^{\mathrm{i}\Omega t} - U_{m1}\mathrm{e}^{-\mathrm{i}\Omega t}\right) + \omega_2^2 \left(U_{p1}\mathrm{e}^{\mathrm{i}\Omega t} + U_{m1}\mathrm{e}^{-\mathrm{i}\Omega t}\right) \\
&\quad + \frac{W_{22}}{2} U_{p2}U_{m2} \left(U_{p1}\mathrm{e}^{\mathrm{i}\Omega t} + U_{m1}\mathrm{e}^{-\mathrm{i}\Omega t}\right) + \frac{W_{22}}{4} \left(U_{p1}U_{m2}^2\mathrm{e}^{-\mathrm{i}\Omega t} + U_{m1}U_{p2}^2\mathrm{e}^{\mathrm{i}\Omega t}\right) = 0,
\end{aligned}
$$

where $\omega_{r1} = \omega_{r2} = \Omega$ has been used. Balancing the $\mathrm{e}^{\mathrm{i}\Omega t}$ and $\mathrm{e}^{-\mathrm{i}\Omega t}$ terms gives

$$
\begin{aligned}
&-\Omega^2 U_{p1} + 2\mathrm{i}\Omega \dot{U}_{p1} + \mathrm{i}2\zeta\omega_2\Omega U_{p1} + \omega_2^2 U_{p1} + \frac{W_{22}}{2} U_{p2}U_{m2}U_{p1} + \frac{W_{22}}{4} U_{m1}U_{p2}^2 = 0, \\
&-\Omega^2 U_{m1} - 2\mathrm{i}\Omega \dot{U}_{m1} - \mathrm{i}2\zeta\omega_2\Omega U_{m1} + \omega_2^2 U_{m1} + \frac{W_{22}}{2} U_{p2}U_{m2}U_{m1} + \frac{W_{22}}{4} U_{p1}U_{m2}^2 = 0.
\end{aligned}
$$

This is now written in the form $\dot{\mathbf{z}} = f(\mathbf{z}, t)$ as

$$
\begin{bmatrix} \dot{U}_{p1} \\ \dot{U}_{m1} \end{bmatrix} = \frac{1}{2\Omega} \begin{bmatrix} \mathrm{i}(\omega_2^2 - \Omega^2 + W_{22}U_2^2/2)U_{p1} + \mathrm{i}W_{22}U_{p2}^2 U_{m1}/4 - 2\zeta\omega_2\Omega U_{p1} \\ -\mathrm{i}(\omega_2^2 - \Omega^2 + W_{22}U_2^2/2)U_{m1} - \mathrm{i}W_{22}U_{m2}^2 U_{p1}/4 - 2\zeta\omega_2\Omega U_{m1} \end{bmatrix}, \tag{7.93}
$$

where $U_2^2 = U_{p2}U_{m2}$ has been used.

Following the discussion on assessing solution stability in Sect. 4.5.4, the stability of a steady-state solution to $\dot{\mathbf{z}} = f(\mathbf{z}, t)$, $\bar{\mathbf{z}}$, may be assessed by considering the eigenvalues of the Jacobian of f evaluated at $\mathbf{z} = \bar{\mathbf{z}}$. In this case the steady-state solution is the zero-response solution $\bar{\mathbf{z}} = [0 \;\; 0]^T$ with the response in z_2 being governed by Eq. (7.91). Using Eq. (7.93), the Jacobian, Df_z may be written as

$$
Df_z(\mathbf{z}, t) = \frac{1}{2\Omega} \begin{bmatrix} -2\zeta\omega_2\Omega + \mathrm{i}A & \mathrm{i}W_{22}U_{p2}^2/4 \\ -\mathrm{i}W_{22}U_{m2}^2/4 & -2\zeta\omega_2\Omega - \mathrm{i}A \end{bmatrix},
$$

where the shorthand $A = \omega_2^2 - \Omega^2 + W_{22}U_2^2/2$ has been used. The eigenvalues, λ, of $Df_z(\bar{\mathbf{z}}, t)$ may be written as

$$
(2\Omega\lambda)^2 + 4\zeta\omega_2\Omega(2\Omega\lambda) + 4\zeta^2\omega_2^2\Omega^2 + (\omega_2^2 - \Omega^2)^2 + (\omega_2^2 - \Omega^2)W_{22}U_2^2 + \frac{3}{16}W_{22}^2U_2^4 = 0.
$$

Expressing this in the standard form, $a\lambda^2 + b\lambda + c = 0$, since both a and b are positive for the real part of either eigenvalue to be positive the condition $c < 0$ must be met. In other words for the zero-response solution to be unstable, and so a response in the out-of-plane mode be observed, requires the condition

$$
3W_{22}^2U_2^4 + 16(\omega_2^2 - \Omega^2)W_{22}U_2^2 + 16(4\zeta^2\omega_2^2\Omega^2 + (\omega_2^2 - \Omega^2)^2) < 0. \tag{7.94}
$$

Loss of stability of the zero-solution occurs when this inequality equals zero—this curve defines the boundary of an *Arnold tongue*. So out-of-plane motion in the response first appears when the second in-plane only response curve, governed by Eq. (7.91), enters the Arnold tongue.

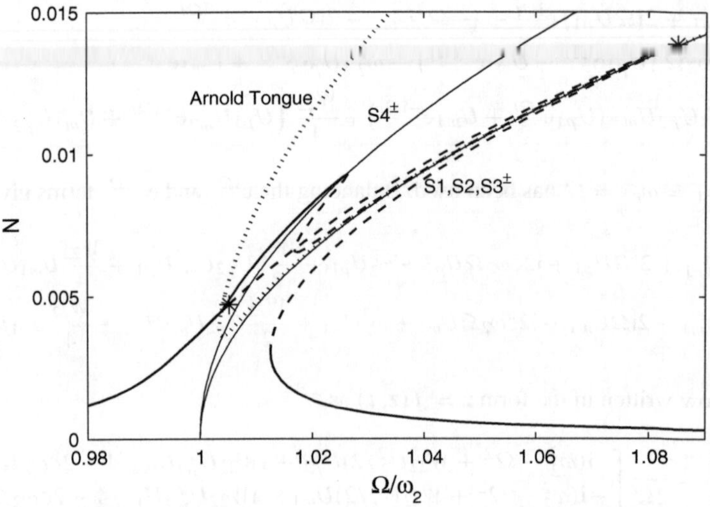

Fig. 7.13 Response of the cable, subjected to motion of the lower support with an amplitude $\Delta = 1.7 \times 10^{-4}$ and forcing frequency close to the second mode of the cable generated using numerical continuation (*thick line*), using a model that contains the second in- and second out-of-plane modes. Also shown is the Arnold tongue (*dotted line*), indicating the loss of stability of the zero-response solution in the second out-of-plane mode. Where this intersects the second in-plane response curve, shown as $*$, indicates the points at which the mixed in- and out-of-plane response will start. It can be seen that this mixed mode response follows the $S4$ backbone curve

Figure 7.13 shows this curve along with the second in-plane only response curve (previously shown in Fig. 7.12, although now generated using numerical continuation). The boundary of the Arnold tongue (shown as a dotted line) was generated by setting the inequality in Eq. (7.94) to zero and rewriting it as a quadratic in Ω^2 that can be solved for a range of N values, noting that at the point of loss of zero-response stability of the out-of-plane mode $U_1 = 0$ and hence $N = U_2$. Within the Arnold tongue the second in-plane only response is marked as unstable, denoted by the dashed line bounded by $*$ marks.

Connected to the second in-plane only response curve at the $*$ marks is the steady-state response of the combined second in- and out-of-plane response, calculated using numerical continuation. It can be seen that the start of this response coincides with the loss of stability of the zero-solution and that the response follows the $S4^{\pm}$ backbone curve indicating that whirling occurs in this region. Considering a frequency sweep, it can be seen that the response at low frequencies, up until approximately $\Omega/\omega_2 = 1.005$, the response will be in-plane. As the frequency is increased above $\Omega/\omega_2 = 1.005$ the mixed mode whirling solution will be observed up until a frequency of about $\Omega/\omega_2 = 1.026$, at which point the response will jump down to the in-plane only solution and remain there for higher frequencies. If the frequency is then reduced the jump up to the whirling solution from the fold in the in-plane only solution will occur at about $\Omega/\omega_2 = 1.013$. There is also a tiny region of stable in-plane only response between the higher $*$ marks and the upper fold of the in-plane only response solution (over frequency range $1.0850 < \Omega/\omega_2 < 1.0851$), although, due to its size, it can't be seen in the figure.

7.6 Chapter Notes

An introduction to cable vibration can be found in the texts by Irvine (1992), Krenk (2001) and Virgin (2007). Many authors have considered the nonlinear resonance phenomena which occur in cables, see for example Perkins (1992), Benedettini et al. (1995), Srinil et al. (2004), Gatulli et al. (2005), Srinil and Rega (2007), Massow et al. (2007), Srinil et al. (2007) and references therein.

The derivation for three-dimensional inclined cable vibration developed here follows the approach set out by Warnitchai et al. (1995). Using this model and Gonzalez-Buelga et al. (2008) investigated the out-of-plane response of an inclined cable when excited in the in-plane copse to the second natural frequency. This was extended by Macdonald et al. (2010), who generalised the result by considered excitation close to any natural frequency.

References

Benedettini, F., Rega, G., & Alaggio, R. (1995). Non-linear oscillations of a nonlinear model of a suspended cable. *Journal of Sound and Vibration, 182*, 775–798.

Gatulli, V., Lepidi, M., Macdonald, J., & Taylor, C. (2005). One to two global local interaction in a cable-stayed beam observed through analytical, finite element and experimental models. *International Journal of Non-linear Mechanics, 40*, 571–588.

Gonzalez-Buelga, A., Neild, S., Wagg, D., & Macdonald, J. (2008). Modal stability of inclined cables subjected to vertical support excitation. *Journal of Sound and Vibration, 318*, 565–579.

Irvine, H. M., & Caughey, T. K. (1974). The linear theory of free vibrations of a suspended cable. *Proceedings of Royal Society A, 341*(1626), 299–315.

Irvine, H. M. (1992). *Cable structures*. New York: Dover.

Krenk, S. (2001). *Mechanics and Analysis of Beams, Columns and Cables: A Modern Introduction to the Classic Theories*. Berlin: Springer.

Macdonald, J. H. G., Dietz, M. S., Neild, S. A., Gonzalez-Buelga, A., Crewe, A. J., & Wagg, D. J. (2010). Generalised modal stability of inclined cables subjected to support excitations. *Journal of Sound and Vibration, 329*, 4515–4533.

Massow, C., Gonzalez-Buelga, A., Macdonald, J., Neild, S., Wagg, D., & Champneys, A. (2007). Theoretical and experimental identification of parametric excitation of inclined cables. *7th International Symposium on Cable Dynamics, number 40 in 1* (pp. 97–104). Vienna: Austria.

Perkins, N. (1992). Modal interactions in the non-linear response of elastic cables under parametric/external excitation. *International Journal Non-linear Mechanics, 27*(2), 233–250.

Rayleigh, J. W. S. (1894a). *Theory of sound* (Vol. 1). London: Macmillan and Co.

Rayleigh, J. W. S. (1894b). *Theory of sound* (Vol. 2). London: Macmillan and Co.

Srinil, N., & Rega, G. (2007). Two-to-one resonant multi-modal dynamics of horizontal/inclined cables. part ii: Internal resonance activation, reduced-order models and nonlinear normal modes. *Nonlinear Dynamics, 48*(3), 253–274.

Srinil, N., Rega, G., & Chucheepsakul, S. (2004). Three-dimensional non-linear coupling and dynamic tension in the large-amplitude free vibrations of arbitrarily sagged cables. *Journal of Sound and Vibration, 269*(3–5), 823–852.

Srinil, N., Rega, G., & Chucheepsakul, S. (2007). Two-to-one resonant multimodal dynamics of horizontal/inclined cables. part i: Theoretical formulation and model validation. *Nonlinear Dynamics, 48*(3), 231–252.

Virgin, L. N. (2007). *Vibration of axially-loaded structures*. Cambridge: Cambridge University Press.

von Kármán, T., & Biot, M. A. (1940). *Mathematical methods in engineering*. New York: McGraw-Hill.

Warnitchai, Y., Fujino, T., & Susumpov, A. (1995). A nonlinear dynamic model for cables and its application to a cable structure-system. *Journal of Sound and Vibration*, *187*(3), 695–712.

Chapter 8
Plates and Shells

Abstract This chapter considers the nonlinear vibration of plates and shallow cylindrical shells. It starts with a description of the classical analysis of flat-plate vibration. Following on from flat plates, the vibration of a shallow curved shell is considered. Due to its curvature, this type of shell (or curved plate) naturally leads to a coupled set of nonlinear ordinary differential equations. An example in which the quadratic nonlinear terms are most significant, leading to $1/2$ subharmonic resonances, is considered. The final part of this chapter considers cylindrical shells which are *bi-stable*. This means that they have two statically stable states, both of which are in the form of a shallow cylindrical shell. To change (or *morph*) from one state to the other, the plate must be deflected past the unstable flat position via a process know as *snap-through*. The possible applications of this type of bi-stable shell to morphing structures are briefly discussed at the end of this chapter.

8.1 Vibration of Plates

Plates are planar structural elements which can carry shear, bending, torsional and axial loads in two dimensions. One way to think of a plate is as a wide beam, or as an element made up of many beams sandwiched together. In fact the analysis of plates can be developed in a similar way to beams, although the resulting equations of motion are more complex because there are additional restoring forces.

As well as being important structural elements, plates and shells are important in the study of nonlinear vibrations. This is because they often exhibit coupled modal vibrations and nonlinear resonance phenomena. This chapter will start with a review of the classical analysis of a flat plate, originally developed with small amplitude assumptions and linear vibration approximations. Developing these equations of motion depends on numerous assumptions regarding the physical behaviour of the plate, and it is important to understand what limitations these assumptions put on the final expressions for plate vibrations. With no in-plane loads and simplified mode-shape assumptions, small-deflection theory for a flat plate leads to a set of linear ordinary differential equations which govern the vibration behaviour of the plate. When axial loads, and/or large deflections, are present nonlinear equations of motion are obtained.

D. Wagg and S. Neild, *Nonlinear Vibration with Control*,
Solid Mechanics and Its Applications 218, DOI 10.1007/978-3-319-10644-1_8

8.1.1 Force Moment Relations

A plate is a three-dimensional structural element, with restoring forces which come primarily from its inherent bending stiffness. As usual in developing vibration models, a series of assumptions based on the physics is required, and these are primarily related to obtaining expressions for the restoring forces. The first assumption is that the plate is thin, which means that the plate thickness h is very small in comparison to the x and y dimensions of the plate. For linear plate equations, additional assumptions are made that the out-of-plane displacements, w, are small, often quantified by restricting w to the same order as h.

This analysis makes use of the assumptions known as the Von Kármán Plate Theory.[1] In addition to the assumptions above, it is also assumed that the slopes are small and that strain is small, so that the material is linear elastic, and that strains vary linearly with plate depth. For nonlinear analysis, Von Kármán assumed that the in-plane displacements, u and v, were negligible, and only nonlinear strain-displacement terms depending on w need be included. This will be discussed further in Sect. 8.1.2.

The forces and moments *per unit width* acting on a small element of plate, with area $\Delta x \times \Delta y$ and thickness h are shown in Fig. 8.1. Notice that in Fig. 8.1 the change in force or moment over the small element has been represented by the addition of Δ terms meaning small change (or variation) over the small distance Δx or Δy. Expanding the small change as a Taylor series and taking the limit as $\Delta x \to 0$ and $\Delta y \to 0$, the variation terms (for example with N_x) can be written as

$$N_x + \Delta N_x \approx N_x + \frac{\partial N_x}{\partial x}\mathrm{d}x, \tag{8.1}$$

using just the first-order term in the Taylor series. To develop the governing equations of motion, force equilibrium expressions are taken in the x, y and z directions, and moment equilibrium expressions are taken about the x, y axes. Moments around the z axis are assumed to be negligible.

Note that for consistency with the beams and cables chapters, N are forces acting in the x or y direction and Q are forces acting in the z direction. Due to the deflection of the plate, N and Q do not always directly correspond to the membrane and shear forces. However, in the following analysis, for convenience, N and Q are still labelled as the membrane and shear forces.

Inertia forces in the x and y directions are assumed to be insignificant compared to the inertia force in the z direction. The mass of the element is $\rho h \Delta x \Delta y$ which tends to $\rho h\mathrm{d}x\mathrm{d}y$ in the limit as $\Delta x, \Delta y \to 0$.

Notice that all the forces and moments in Fig. 8.1 are *per unit width* so they need to be multiplied by the appropriate width of the small element. Bearing this in mind, the vertical force equilibrium in the z direction including the inertia of the element gives

[1] See Amabili (2008) for a discussion of the historical aspects of plate theory.

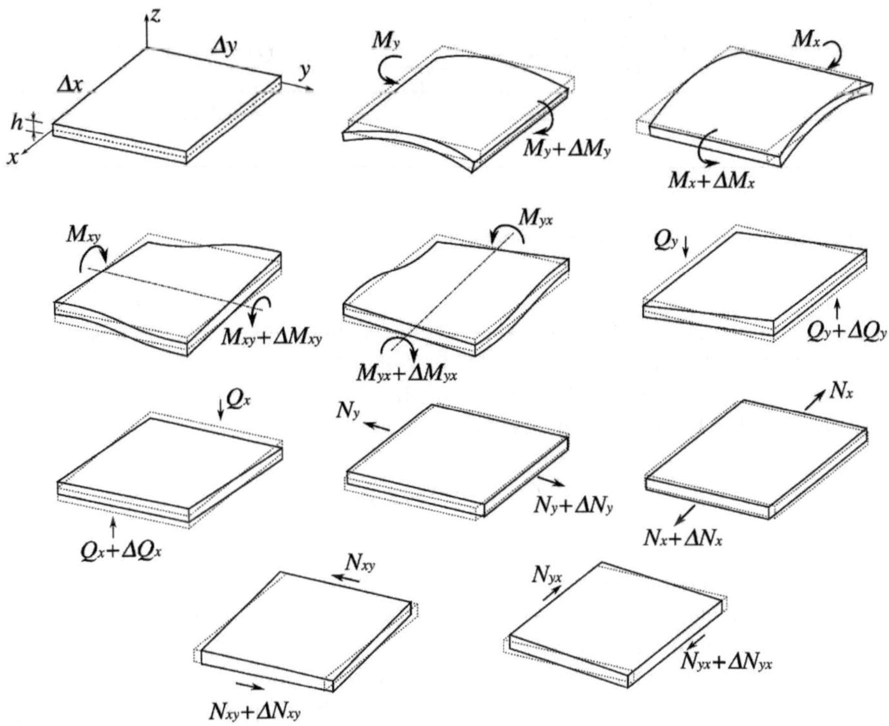

Fig. 8.1 Forces and moments acting on an element of a thin flat plate

$$\uparrow z \begin{cases} -Q_y\mathrm{d}x + (Q_y + \frac{\partial Q_y}{\partial y}\mathrm{d}y)\mathrm{d}x - Q_x\mathrm{d}y + (Q_x + \frac{\partial Q_x}{\partial x}\mathrm{d}x)\mathrm{d}y \\ -\rho h\mathrm{d}x\mathrm{d}y\frac{\partial^2 w}{\partial t^2} = 0, \end{cases}$$

which reduces to

$$\frac{\partial Q_y}{\partial y}\mathrm{d}y\mathrm{d}x + \frac{\partial Q_x}{\partial x}\mathrm{d}x\mathrm{d}y = \rho h\mathrm{d}x\mathrm{d}y\frac{\partial^2 w}{\partial t^2},$$

then dividing through by $\mathrm{d}y\mathrm{d}x$ gives

$$\frac{\partial Q_y}{\partial y} + \frac{\partial Q_x}{\partial x} = \rho h\frac{\partial^2 w}{\partial t^2}. \tag{8.2}$$

Now taking moments about an axis through the centre of the small element parallel to the y axis, while ignoring rotational inertia effects, gives

$$\circlearrowleft x \begin{cases} -M_x\mathrm{d}y + (M_x + \frac{\partial M_x}{\partial x}\mathrm{d}x)\mathrm{d}y - M_{xy}\mathrm{d}x + (M_{xy} + \frac{\partial M_{xy}}{\partial y}\mathrm{d}y)\mathrm{d}x \\ -Q_x\mathrm{d}y\frac{\mathrm{d}x}{2} - (Q_x + \frac{\partial Q_x}{\partial x}\mathrm{d}x)\mathrm{d}y\frac{\mathrm{d}x}{2} = 0, \end{cases} \tag{8.3}$$

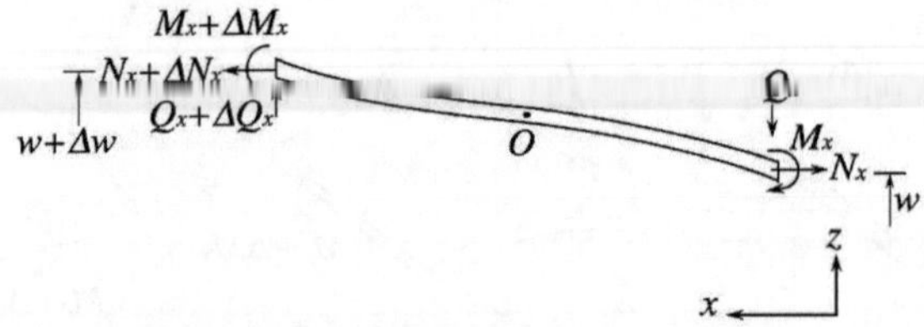

Fig. 8.2 Effect of membrane (in-plane) forces on moments in a plate

where the third-order dx^2dy term can be neglected since as Δx and Δy tend to zero the $\Delta x^2 \Delta y$ term becomes much smaller than the $\Delta x \Delta y$ terms. Note also that the effect of the in-plane forces N_x and N_{xy} has been left out for now, but is included later. Simplifying Eq. (8.3), then dividing by $dydx$ gives

$$\frac{\partial M_x}{\partial x} + \frac{\partial M_{xy}}{\partial y} = Q_x, \tag{8.4}$$

which relates the change in moment to the shear force on the plate. Equation (8.4) is the plate equivalent of Eq. (6.4) found for beams, the only difference being the additional twisting moment, M_{xy}.

But now consider what effect the in-plane forces (often called the *membrane* forces) have on the moment. This is demonstrated in the simplified two-dimensional section shown in Fig. 8.2. Here the effect of N_x on M_x can be seen. As the element is deflected, the membrane force N_x imposes a moment on the element. The deflections in the z-axis are w and $w + \Delta w$ for either side of the element and the deflection of the centre point O is approximated to be $w + \Delta w/2$, resulting in the membrane forces having moment arms of $\Delta w/2$ or in the limit $(1/2)(\partial w/\partial x)dx$.

So, taking moments about O, ignoring inertia and the y-related moments, gives

$$\circlearrowleft_x \left\{ \begin{array}{l} -M_x dy + (M_x + \frac{\partial M_x}{\partial x}dx)dy - Q_x dy\frac{dx}{2} - (Q_x + \frac{\partial Q_x}{\partial x}dx)dy\frac{dx}{2} \\ +N_x dy\frac{1}{2}\frac{\partial w}{\partial x}dx + (N_x + \frac{\partial N_x}{\partial x}dx)dy\frac{1}{2}\frac{\partial w}{\partial x}dx = 0. \end{array} \right.$$

This can be simplified by observing that $(\partial Q_x/\partial x)dx$ and $(\partial N_x/\partial x)dx$ are small in comparison to Q_x and N_x respectively (which is the same as noting that in the limit ΔQ_x and ΔN_x are small compared to Q_x and N_x, respectively) which gives us

$$\frac{\partial M_x}{\partial x} = Q_x - N_x\frac{\partial w}{\partial x}.$$

Now the change in moment is equal to the shear force minus a correction for the effect of the membrane force, N_x. An equivalent argument can be used to show the effect of N_{xy} in opposing the twist moment M_{xy}. As a result the full moment equation with the effect of membrane forces becomes

$$\frac{\partial M_x}{\partial x} + \frac{\partial M_{xy}}{\partial y} = Q_x - N_x \frac{\partial w}{\partial x} - N_{xy} \frac{\partial w}{\partial y}. \tag{8.5}$$

Using an equivalent approach, the moments around the x axis are given by

$$\circlearrowleft_y \left\{ \frac{\partial M_y}{\partial y} + \frac{\partial M_{yx}}{\partial x} = Q_y - N_y \frac{\partial w}{\partial y} - N_{yx} \frac{\partial w}{\partial x}. \right. \tag{8.6}$$

If the in-plane inertia are neglected, the force balance in both the x and y directions gives

$$\swarrow_x \left\{ \frac{\partial N_x}{\partial x} + \frac{\partial N_{xy}}{\partial y} = 0, \right. \tag{8.7}$$

and

$$\searrow_y \left\{ \frac{\partial N_y}{\partial y} + \frac{\partial N_{yx}}{\partial x} = 0. \right. \tag{8.8}$$

Equations (8.2), (8.5), (8.6), (8.7) and (8.8) are used as a basis for a model of the (small amplitude) vibration of the plate. Assuming that $M_{xy} = M_{yx}$ and $N_{xy} = N_{yx}$ leaves three moments, two shear forces and three axial forces, which is a total of eight unknowns, with five equations.

The shear forces, Q_x and Q_y can be eliminated by substituting Eqs. (8.5) and (8.6) into Eq. (8.2) to give

$$\frac{\partial^2 M_x}{\partial x^2} + 2\frac{\partial^2 M_{xy}}{\partial y \partial x} + \frac{\partial^2 M_y}{\partial y^2} + \frac{\partial}{\partial x}\left(N_x \frac{\partial w}{\partial x} + N_{xy}\frac{\partial w}{\partial y}\right)$$
$$+ \frac{\partial}{\partial y}\left(N_{xy}\frac{\partial w}{\partial x} + N_y \frac{\partial w}{\partial y}\right) = \rho h \frac{\partial^2 w}{\partial t^2}.$$

This equation can be simplified, by expanding out the differential terms in the brackets and then cancelling out terms using Eqs. (8.7) and (8.8), to give

$$\frac{\partial^2 M_x}{\partial x^2} + 2\frac{\partial^2 M_{xy}}{\partial y \partial x} + \frac{\partial^2 M_y}{\partial y^2} + N_x \frac{\partial^2 w}{\partial x^2} + 2N_{xy}\frac{\partial^2 w}{\partial y \partial x} + N_y \frac{\partial^2 w}{\partial y^2} = \rho h \frac{\partial^2 w}{\partial t^2}. \tag{8.9}$$

The result of considering the force and moment balances and then manipulating them is that three equations have been derived, Eqs. (8.7), (8.8) and (8.9)[2] which contain seven unknowns; $M_x, M_y, M_{xy}, N_x, N_y, N_{xy}$ and w (noting that $N_{xy} = N_{yx}$). To develop the model further, compatibility, strain-displacement and stress-strain, relationships must be considered for the plate element.

[2] Note that although Eqs. (8.7) and (8.8) have been used in the last step of the derivation of Eq. (8.9) they still contain useful information as none of the variables, N_x, N_y or N_{xy}, were eliminated in the simplification.

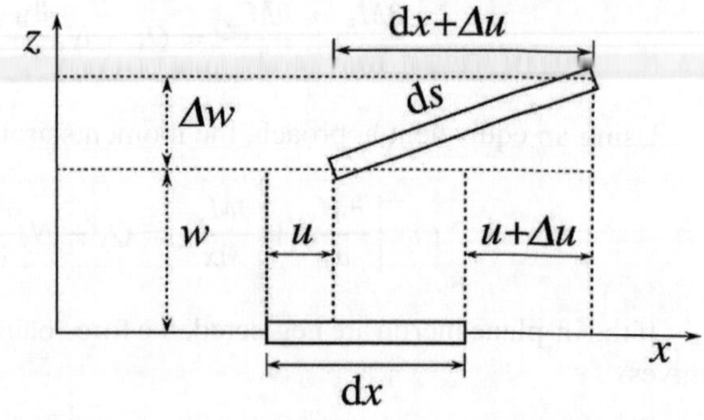

Fig. 8.3 Strain displacement relationship for a small element of plate

8.1.2 Strain-Displacement Relations

When the plate deforms, the two major strain effects are the bending strain and axial strain. Considering axial strain effects first, the strain displacement relationship for a small element of plate in the x, z plane is shown in Fig. 8.3. The axial strain in the x, z plane is equal to the change in length, $\mathrm{d}s - \mathrm{d}x$ divided by the original length $\mathrm{d}x$.

$$\frac{\mathrm{d}s - \mathrm{d}x}{\mathrm{d}x} = \frac{\mathrm{d}s}{\mathrm{d}x} - 1.$$

The variation of displacements u and w can be approximated as the first term in a Taylor series expansion, as discussed above for Eq. (8.1), so that

$$u + \Delta u = u + \frac{\partial u}{\partial x}\mathrm{d}x, \quad w + \Delta w = w + \frac{\partial w}{\partial x}\mathrm{d}x.$$

The new length of the element $\mathrm{d}s$ is then given by

$$\mathrm{d}s = \sqrt{\left(\mathrm{d}x + \frac{\partial u}{\partial x}\mathrm{d}x\right)^2 + \left(\frac{\partial w}{\partial x}\mathrm{d}x\right)^2} = \mathrm{d}x\sqrt{\left(1 + \frac{\partial u}{\partial x}\right)^2 + \left(\frac{\partial w}{\partial x}\right)^2}.$$

This means that

$$\frac{\mathrm{d}s}{\mathrm{d}x} = \sqrt{1 + 2\frac{\partial u}{\partial x} + \left(\frac{\partial u}{\partial x}\right)^2 + \left(\frac{\partial w}{\partial x}\right)^2}. \tag{8.10}$$

Expanding Eq. (8.10) as a binomial approximation, $\sqrt{1+\delta} = 1 + \frac{\delta}{2} - \frac{\delta^2}{8} + \cdots$, including only terms up to second order gives

$$\frac{ds}{dx} = 1 + \frac{1}{2}\left[2\frac{\partial u}{\partial x} + \left(\frac{\partial u}{\partial x}\right)^2 + \left(\frac{\partial w}{\partial x}\right)^2\right] - \frac{1}{8}\left[4\left(\frac{\partial u}{\partial x}\right)^2 + \mathcal{O}(\bullet)^3\right]$$
$$\approx 1 + \frac{\partial u}{\partial x} + \frac{1}{2}\left(\frac{\partial w}{\partial x}\right)^2.$$

So that, finally, the axial strain in the x direction is given by

$$v_x = \frac{ds - dx}{dx} = \frac{ds}{dx} - 1 \approx \frac{\partial u}{\partial x} + \frac{1}{2}\left(\frac{\partial w}{\partial x}\right)^2, \qquad (8.11)$$

where v is used to indicate axial strain (ϵ will be used to indicate total strain due to axial and bending deflections).

The same approach can be used to obtain the axial strains in the y direction

$$v_y = \frac{\partial v}{\partial y} + \frac{1}{2}\left(\frac{\partial w}{\partial y}\right)^2, \qquad (8.12)$$

and the strain due to plate twisting

$$v_{xy} = \frac{\partial u}{\partial y} + \frac{\partial v}{\partial x} + \frac{\partial w}{\partial x}\frac{\partial w}{\partial y}. \qquad (8.13)$$

Note that small-deflection analysis is usually developed by neglecting the last terms in each of Eqs. (8.11)–(8.13). Large deflection analysis is developed by including these terms. Equations (8.11)–(8.13) are sometimes referred to as the Von Kármán strain-displacement expressions.

These expressions may be manipulated to eliminate u and v. This is achieved by differentiating Eq. (8.13) with respect to x and y. Equation (8.11), differentiated twice with respect to y, and Eq. (8.12), differentiated twice with respect to x, are then used to eliminate u and v respectively to give

$$\frac{\partial^2 v_x}{\partial y^2} - \frac{\partial^2 v_{xy}}{\partial x \partial y} + \frac{\partial^2 v_y}{\partial x^2} = \left(\frac{\partial^2 w}{\partial x \partial y}\right)^2 - \frac{\partial^2 w}{\partial x^2}\frac{\partial^2 w}{\partial y^2}. \qquad (8.14)$$

If the deflection is assumed to be small then the right-hand side of this expression is negligible and can be set to zero. This results in the two-dimensional linear strain compatibility equation (see for example Timoshenko and Goodier (1970)). Equation (8.14), together with the bending strains (which are developed below), can be used to derive the nonlinear (i.e. large deflection) equations of motion.

The moments are assumed to depend on the bending strains. The bending strain of a plate element can be derived by considering the small element shown in Fig. 8.4. As the element bends it is assumed that the neutral axis stays at the same length, Δx, as the unstrained element. Then, the strain at point A in the beam, which is a distance

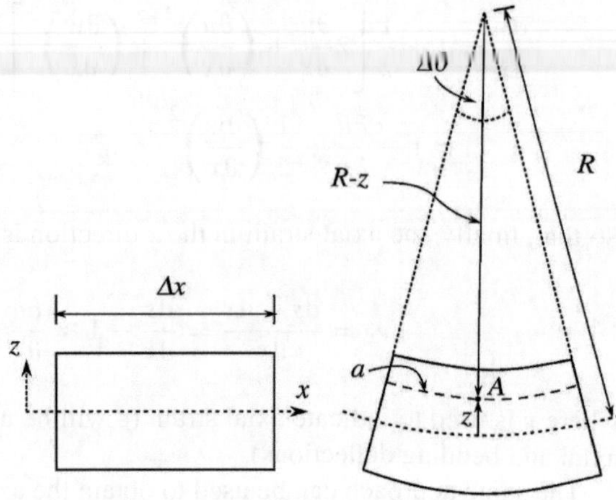

Fig. 8.4 Bending strain for a small element of plate

z from the neutral axis, can be found by defining the change in length of the circular arc a which intersects with A.

The length of the neutral axis in the deformed element is $R\Delta\theta$. The length of the arc, a, is $(R - z)\Delta\theta$, so the change in length is $(R - z)\Delta\theta - R\Delta\theta$. As a result the bending strain can be written as

$$\kappa_x = \frac{(R - z)\Delta\theta - R\Delta\theta}{R\Delta\theta} = -\frac{z}{R}. \tag{8.15}$$

Then using the approximate relationship that

$$\frac{1}{R} = \frac{\partial^2 w}{\partial x^2},$$

gives a bending strain relationship of

$$\kappa_x = -z\frac{\partial^2 w}{\partial x^2}. \tag{8.16}$$

The other two bending strains are

$$\kappa_y = -z\frac{\partial^2 w}{\partial y^2}, \tag{8.17}$$

$$\kappa_{xy} = -2z\frac{\partial^2 w}{\partial x \partial y}. \tag{8.18}$$

See Szilard (1974) for further details of the derivation of these strain relationships.

8.1.3 Stress-Strain Relations

The strain-stress relationships for a homogeneous, isotropic, elastic body in three dimensions are

$$\epsilon_x = \frac{1}{E}[\sigma_x - \nu(\sigma_y + \sigma_z)],$$
$$\epsilon_y = \frac{1}{E}[\sigma_y - \nu(\sigma_z + \sigma_x)],$$
$$\epsilon_z = \frac{1}{E}[\sigma_z - \nu(\sigma_x + \sigma_y)],$$

where ϵ is the total strain due to bending and axial loading, $\boldsymbol{\sigma}$ is the total stress, E Young's modulus and ν Poisson's ratio (see Timoshenko and Goodier (1970) for further discussion). The plate is assumed to be thin, and as a result the effect of the normal stress, $\boldsymbol{\sigma}_z$, on the x and y stresses is negligible, so that the strain expressions for x and y are

$$\epsilon_x = \frac{1}{E}(\sigma_x - \nu\sigma_y),$$
$$\epsilon_y = \frac{1}{E}(\sigma_y - \nu\sigma_x).$$

With some rearranging, these may be rewritten in terms of total stresses to give

$$\begin{aligned} \sigma_x &= \frac{E}{1-\nu^2}(\epsilon_x + \nu\epsilon_y) = \frac{E}{1-\nu^2}[\nu_x + \boldsymbol{\kappa}_x + \nu(\nu_y + \kappa_y)], \\ \sigma_y &= \frac{E}{1-\nu^2}(\epsilon_y + \nu\epsilon_x) = \frac{E}{1-\nu^2}[\nu_y + \boldsymbol{\kappa}_y + \nu(\nu_x + \kappa_x)], \end{aligned} \tag{8.19}$$

where the total strains have been written as a sum of the axial and bending strains $\epsilon_x = \nu_x + \boldsymbol{\kappa}_x$ and $\epsilon_y = \nu_y + \boldsymbol{\kappa}_y$. The total shear stress is given by

$$\boldsymbol{\tau}_{xy} = \frac{E}{2(1+\nu)}\Gamma_{xy} = \frac{E}{2(1+\nu)}(\nu_{xy} + \kappa_{xy}), \tag{8.20}$$

where the total shear strain has been written as a sum of the axial and bending strains $\Gamma_{xy} = \nu_{xy} + \boldsymbol{\kappa}_{xy}$, see Timoshenko and Goodier (1970) for a full derivation.

The objective of defining the stress-strain relationships is to find further expressions relating the seven unknowns $M_x, M_y, M_{xy}, N_x, N_y, N_{xy}$ and w. Taking the in-plane membrane expressions first, the forces are related to stresses by integrating over the depth of the plate to give

$$N_x = \int_{-\frac{h}{2}}^{\frac{h}{2}} \sigma_x \mathrm{d}z, \quad N_y = \int_{-\frac{h}{2}}^{\frac{h}{2}} \sigma_y \mathrm{d}z, \quad N_{xy} = \int_{-\frac{h}{2}}^{\frac{h}{2}} \tau_{xy} \mathrm{d}z.$$

Likewise the moments can be written in terms of the stresses as

$$M_x = \int_{-\frac{h}{2}}^{\frac{h}{2}} \sigma_x z \mathrm{d}z, \quad M_y = \int_{-\frac{h}{2}}^{\frac{h}{2}} \sigma_y z \mathrm{d}z, \quad M_{xy} = \int_{-\frac{h}{2}}^{\frac{h}{2}} \tau_{xy} z \mathrm{d}z.$$

To evaluate these integrals, firstly Eqs. (8.19) and (8.20) can be used to replace the stresses with axial and bending strains to give

$$\begin{aligned}
N_x &= \tfrac{E}{1-\nu^2} \int_{-\frac{h}{2}}^{\frac{h}{2}} [v_x + z\bar{\kappa}_x + \nu(v_y + z\bar{\kappa}_y)]\mathrm{d}z,\\
N_y &= \tfrac{E}{1-\nu^2} \int_{-\frac{h}{2}}^{\frac{h}{2}} [v_y + z\bar{\kappa}_y + \nu(v_z + z\bar{\kappa}_z)]\mathrm{d}z,\\
N_{xy} &= \tfrac{Eh}{2(1+\nu)} \int_{-\frac{h}{2}}^{\frac{h}{2}} (v_{xy} + z\bar{\kappa}_{xy})\mathrm{d}z,\\
M_x &= \tfrac{E}{1-\nu^2} \int_{-\frac{h}{2}}^{\frac{h}{2}} [v_x + z\bar{\kappa}_x + \nu(v_y + z\bar{\kappa}_y)]z\mathrm{d}z,\\
M_y &= \tfrac{E}{1-\nu^2} \int_{-\frac{h}{2}}^{\frac{h}{2}} [v_y + z\bar{\kappa}_y + \nu(v_x + z\bar{\kappa}_x)]z\mathrm{d}z,\\
M_{xy} &= \tfrac{Eh}{2(1+\nu)} \int_{-\frac{h}{2}}^{\frac{h}{2}} (v_{xy} + z\bar{\kappa}_{xy})z\mathrm{d}z,
\end{aligned} \tag{8.21}$$

where $z\bar{\kappa}_x = \kappa_x$, $z\bar{\kappa}_y = \kappa_y$ and $z\bar{\kappa}_{xy} = \kappa_{xy}$. These axial and bending strain terms are related to plate deflections through Eqs. (8.11)–(8.13) and (8.16)–(8.18) respectively. Notice that the axial strains are constant over the thickness of the plate (in other words they are not functions of z) whereas the bending strains are linearly proportional to z. This means that $\bar{\kappa}_x$, $\bar{\kappa}_y$ and $\bar{\kappa}_{xy}$ are constant over z. As a result, using the integrals in Eq. (8.21) to calculate the axial stresses, means that the bending strain terms go to zero, to give

$$N_x = \frac{Eh}{1-\nu^2}(v_x + \nu v_y), \quad N_y = \frac{Eh}{1-\nu^2}(v_y + \nu v_x), \quad N_{xy} = \frac{Eh}{2(1+\nu)} v_{xy}. \tag{8.22}$$

Similarly when using the integrals in Eq. (8.21) to calculate the moments, the stresses are multiplied by z before the integration is performed, so the axial terms go to zero while the bending terms remain giving

$$M_x = \frac{Eh^3}{12(1-\nu^2)}(\bar{\kappa}_x - \nu\bar{\kappa}_y), \quad M_y = \frac{Eh^3}{12(1-\nu^2)}(\bar{\kappa}_y - \nu\bar{\kappa}_x), \quad M_{xy} = \frac{Eh^3}{24(1+\nu)}\bar{\kappa}_{xy}.$$

Which, using Eqs. (8.16)–(8.18) may be written as

$$\begin{aligned} M_x &= -D\left(\frac{\partial^2 w}{\partial x^2} + \nu\frac{\partial^2 w}{\partial y^2}\right), \\ M_y &= -D\left(\frac{\partial^2 w}{\partial y^2} + \nu\frac{\partial^2 w}{\partial x^2}\right), \\ M_{xy} &= -D(1-\nu)\frac{\partial^2 w}{\partial x \partial y}, \end{aligned} \tag{8.23}$$

where $D = Eh^3/(12[1-\nu^2])$ is the *flexural rigidity* of the plate. This can be thought of as an equivalent to EI for beams, see Timoshenko (1940).

The compatibility expressions, Eqs. (8.22) and (8.23), along with the strain-displacement expression, Eq. (8.14), provide 7 further equations for analysing plate vibrations to add to the three force balance equations, Eqs. (8.7)–(8.9). They also introduce three additional unknowns; ν_x, ν_y and ν_{xy}, to add to the seven unknowns $M_x, M_y, M_{xy}, N_x, N_y, N_{xy}$ and w already present from the force balance equations. These combined relationships are therefore solvable as there are ten equations and ten unknowns.

8.1.4 Force Balance and Compatibility

To proceed in the analysis of plate vibrations the force balance equations must be combined with the compatibility equations. Firstly the moments M_x, M_y, M_{xy} may be eliminated from Eq. (8.9) using the compatibility expressions given by Eq. (8.23) to give

$$-D\left(\frac{\partial^4 w}{\partial x^4} + 2\frac{\partial^4 w}{\partial y^2 \partial x^2} + \frac{\partial^4 w}{\partial y^4}\right) + \left(N_x\frac{\partial^2 w}{\partial x^2} + 2N_{xy}\frac{\partial^2 w}{\partial y \partial x} + N_y\frac{\partial^2 w}{\partial y^2}\right) = \rho h\frac{\partial^2 w}{\partial t^2}. \tag{8.24}$$

This equation represents a fourth-order governing equation of motion for the plate in terms of the transverse displacement w, which has no damping included, but does include the effect of axial forces N_x, N_y and N_{xy}. The constant, D, acts as a flexural rigidity of the plate assuming homogeneous, isotropic, elastic material, h is the thickness of the plate (assumed to be thin) and ρ is the mass density of the plate.

In the case where large axial forces are present, due to pre-stressing of the plate between its boundary supports, and the vibration amplitude is small then the axial forces are often assumed to be constant and as a result Eq. (8.24) is the equation of motion for the system. This approach is very similar to the small-deflection cable dynamics discussed in Chap. 7 in which the tensile force in the x-axis is constant over the length of the cable and is discussed further in Virgin (2007).

More generally the time-dependent axial forces N_x, N_y and N_{xy} (where $N_{xy} = N_{yx}$) may be calculated using Eqs. (8.7), (8.8), (8.14) and (8.22). The first two expressions in Eq. (8.22) can be viewed as simultaneous equations in terms of ν_x and ν_y and with some rearranging can be written as

$$\nu_x = \frac{N_x - \nu N_y}{Eh}, \quad \nu_y = \frac{N_y - \nu N_x}{Eh}.$$

These two expressions along with the third expression in Eq. (8.22) can now be used to eliminate ν_x and ν_y and ν_{xy} in Eq. (8.14) to give

$$\frac{(1-\nu)}{Eh}\left(\frac{\partial^2 N_x}{\partial y^2} + \frac{\partial^2 N_y}{\partial x^2}\right) - \frac{2(1+\nu)}{Eh}\frac{\partial^2 N_{xy}}{\partial x \partial y} = \left(\frac{\partial^2 w}{\partial x \partial y}\right)^2 - \frac{\partial^2 w}{\partial x^2}\frac{\partial^2 w}{\partial y^2}.$$

Noting that all the terms containing ν cancel out using Eqs. (8.7) and (8.8) and expanding out the differential terms gives

$$\frac{\partial^2 N_x}{\partial y^2} - 2\frac{\partial^2 N_{xy}}{\partial x \partial y} + \frac{\partial^2 N_y}{\partial x^2} + Eh\left[\frac{\partial^2 w}{\partial x^2}\frac{\partial^2 w}{\partial y^2} - \left(\frac{\partial^2 w}{\partial x \partial y}\right)^2\right] = 0. \tag{8.25}$$

In the classical small amplitude linear theory, where only the first-order strain terms are taken to be significant in Eq. (8.14), both the terms in the square brackets would disappear.

Now there are four equations governing the motion of the plate, Eqs. (8.7), (8.8), (8.24) and (8.25), with the four unknowns w, N_x, N_y and N_{xy}. To reduce the number of equations further, the form of Eqs. (8.7) and (8.8) is exploited. This is done by introducing the function, $\Phi(x, y, t)$, known as the *Airy stress function*. By relating Φ to the axial forces in the following definitions[3]

$$N_x = \frac{\partial^2 \Phi}{\partial y^2}, \quad N_y = \frac{\partial^2 \Phi}{\partial x^2}, \quad N_{xy} = -\frac{\partial^2 \Phi}{\partial x \partial y}, \tag{8.26}$$

with $N_{xy} = N_{yx}$, ensures that the constraint equations Eqs. (8.7) and (8.8) are satisfied.

---

[3] The Airy stress function can be derived as a special case of the Maxwell stress function for two-dimensional stresses (Timoshenko and Goodier 1970). Here the functional form of the Airy stress function is being used to set conditions on the axial forces in the plate. For further details of this technique, see for example Ventsel and Krauthammer (2001).

Substituting the expressions from Eqs. (8.26) into (8.24) gives a governing equation

$$-D\left(\frac{\partial^4 w}{\partial x^4}+2\frac{\partial^4 w}{\partial y^2\partial x^2}+\frac{\partial^4 w}{\partial y^4}\right)+\left(\frac{\partial^2 \Phi}{\partial y^2}\frac{\partial^2 w}{\partial x^2}-2\frac{\partial^2 \Phi}{\partial x\partial y}\frac{\partial^2 w}{\partial y\partial x}+\frac{\partial^2 \Phi}{\partial x^2}\frac{\partial^2 w}{\partial y^2}\right) = \rho h\frac{\partial^2 w}{\partial t^2}. \tag{8.27}$$

This represents the governing equation of the plate, based on the force balance, in terms of the transverse displacement w, with the function Φ representing the in-plane constraint (force) effects. Here no damping is included, the constant D acts as a flexural rigidity of the plate assuming homogeneous, isotropic, elastic material, h is the thickness of the plate (assumed to be thin) and ρ is the mass density of the plate.

The compatibility equation, Eq. (8.25) may also be expressed in terms of w and Φ, using Eq. (8.26), to give

$$\frac{1}{Eh}\nabla^2\nabla^2\Phi+\frac{\partial^2 w}{\partial x^2}\frac{\partial^2 w}{\partial y^2}-\left(\frac{\partial^2 w}{\partial x\partial y}\right)^2=0, \tag{8.28}$$

where $\nabla^2=\frac{\partial^2}{\partial x^2}+\frac{\partial^2}{\partial y^2}$. For the classical small amplitude linear analysis this equation reduces to $\nabla^2\nabla^2\Phi=0$.

Equation (8.28) coupled with Eq. (8.27) are the two partial differential equations used to model the (moderately) large amplitude vibrations of plates. This is a popular approach for curved plates and shells, in which the membrane forces play a significant role in the dynamics. This will be discussed in detail in Sect. 8.4.

8.2 Small Amplitude Vibration

In the preceding section, governing equations of motion have been derived for flat plates including in-plane forces, but without damping or other external forces. Considering the case when no in-plane forces are present, using Eq. (8.24), the governing equation of motion for a plate can be written

$$D\left(\frac{\partial^4 w}{\partial x^4}+2\frac{\partial^4 w}{\partial y^2\partial x^2}+\frac{\partial^4 w}{\partial y^4}\right)+\rho h\frac{\partial^2 w}{\partial t^2}=0. \tag{8.29}$$

The first term on the left of Eq. (8.29) represents the restoring force (from bending stiffness of the plate), and the second term the inertia force. Equation (8.29) has no damping or other external forces included. Now consider an example which finds the free vibration response frequencies for a flat plate.

Example 8.1 Free vibration of a flat plate

Problem Find the free vibration response frequency of an (undamped) flat plate assuming that the transverse displacement can be assumed to be harmonic, $w(x, y, t) = \sum_1^M \sum_1^N X_m(x) Y_n(y) C e^{i\omega t}$, where $X_m(x) = \sin(\frac{m\pi x}{a})$ and $Y_n(y) = \sin(\frac{n\pi x}{b})$ are the assumed mode-shapes[4] and C is an arbitrary constant. The plate has thickness, h, and a is the x dimension, b is the y dimension. The number of x and y modes is given by M and N. The frequency of vibration response is ω and t is time.

Solution The equation of motion for an undamped flat plate is given by Eq. (8.29). Substituting for w gives

$$\sum_1^M \sum_1^N \left\{ D \left(\frac{d^4 X_m}{dx^4} Y_n + 2 \frac{d^2 X_m}{dx^2} \frac{d^2 Y_n}{dy^2} + X_m \frac{d^4 Y_n}{dy^4} \right) - \omega^2 \rho h X_m Y_n \right\} C e^{i\omega t} = 0.$$

The complex exponential term can be divided out and so for each term in the summation

$$\frac{d^4 X_m}{dx^4} Y_n + 2 \frac{d^2 X_m}{dx^2} \frac{d^2 Y_n}{\partial y^2} + X_m \frac{d^4 Y_n}{\partial y^4} - \frac{\omega^2 \rho h}{D} X_m Y_n = 0. \quad (8.30)$$

Now substituting for X_m and Y_n using the sine wave expressions given in the problem, Eq. (8.30) becomes

$$\frac{m^4 \pi^4}{a^4} + 2 \frac{m^2 \pi^2}{a^2} \frac{n^2 \pi^2}{b^2} + \frac{n^4 \pi^4}{b^4} - \frac{\omega^2 \rho h}{D} = 0. \quad (8.31)$$

Factorising, Eq. (8.31) can be written as

$$\pi^4 \left(\frac{m^2}{a^2} + \frac{n^2}{b^2} \right)^2 = \frac{\omega^2 \rho h}{D},$$

which can be rearranged to give an expression for the response frequency

$$\omega = \pi^2 \left(\frac{m^2}{a^2} + \frac{n^2}{b^2} \right) \sqrt{\frac{D}{\rho h}}. \quad (8.32)$$

Now for a given choice of M and N, the response frequencies for each of the $m = 1, 2, \ldots, M$ and $n = 1, 2, \ldots, N$ values can be estimated using Eq. (8.32). These

[4] This is based on the simplifying assumption that the boundary conditions are simply supported along both directions of the plate and that in both directions the modes take the same form as for a pinned-pinned beam.

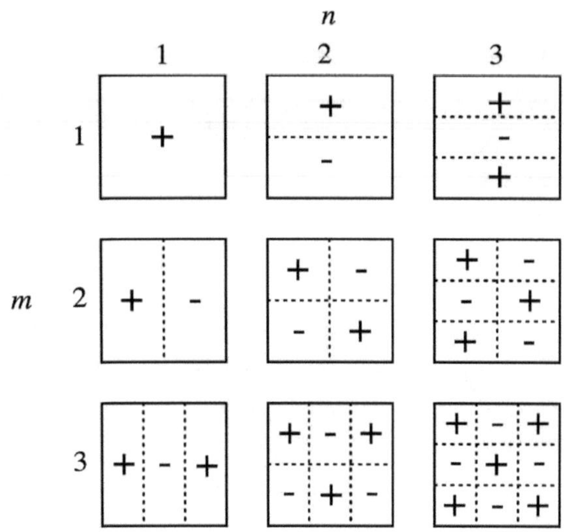

Fig. 8.5 First nine expressions for the term $\sin(\frac{m\pi x}{a})\sin(\frac{n\pi x}{b})$ which come from the assumed mode-shapes of a simply-supported square plate. See also Fig. 8.6

frequency estimates are based on the assumption that the mode-shapes are simple sine waves. The plate mode-shapes, $q_{mn}(x, y)$ are given by the combination of the sine wave terms so that $q_{mn}(x, y) = \sum_1^M \sum_1^N \sin(\frac{m\pi x}{a})\sin(\frac{n\pi x}{b})$, where m and n give the number of half sine waves in the x and y directions for each mode-shape combination. The first nine modes from the term $\sin(\frac{m\pi x}{a})\sin(\frac{n\pi x}{b})$ are shown in Figs. 8.5 and 8.6.[5] ■

Notice that in Example 8.1, the time part of the solution was approximated using a complex exponential function, which corresponds to an assumed harmonic response. A more general approach is to use the Galerkin method (see Sect. 6.1.2), to decompose the partial differential equations of motion for the plate into a set of ordinary differential equations. For beams, the Galerkin method makes use of the orthogonality properties of the mode-shapes. However, general orthogonality conditions for plates and shells do not exist. Despite this, good approximations can be made for a range of thin (von Kármán type) plate applications with restricted boundary conditions, by combining beam modes in the x and y directions.

To apply the Galerkin method, the transverse vibration is separated into three functions, such that $w(x, y, t) = \sum_1^M \sum_1^N X_n(x)Y_m(y)q_{nm}(t)$, where $q_{nm}(t)$ is a time-dependent modal coordinate. The beam mode-shapes are $X_n(x)$ and $Y_m(y)$ respectively, and can be taken from standard beam theory based on the end conditions, i.e. free, clamped or pinned etc.[6]

---

[5] Note that increasing numbers of combination modes are possible as M and N increase, see for example the discussion in Szilard (1974).

[6] Noting that this is a crude approximation to the actual shapes.

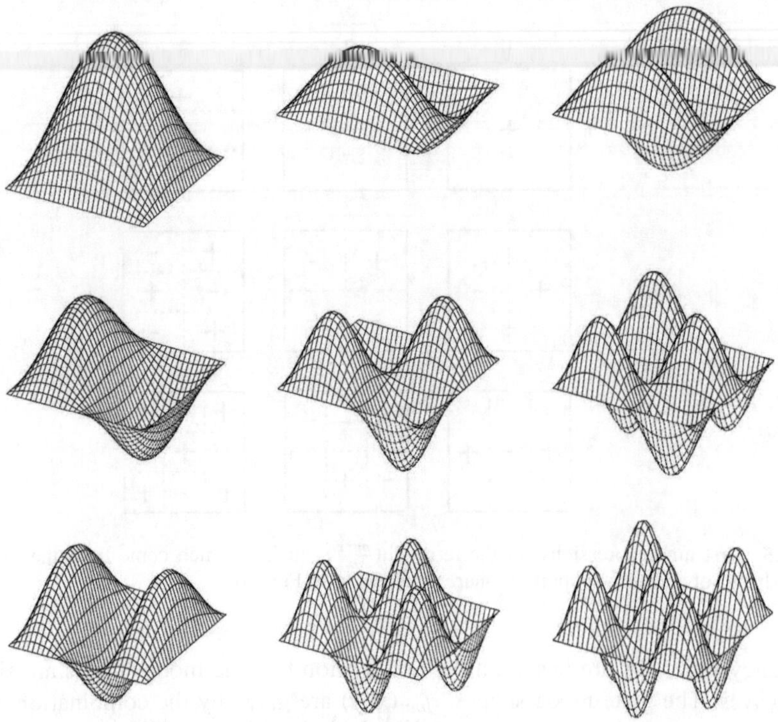

Fig. 8.6 Three-dimensional plots corresponding to the m, n values shown in Fig. 8.5

To decouple the equations of motion, the Galerkin method makes use of the orthogonality properties of the mode-shapes, X and Y. For example, using the simple sine wave mode-shapes from Example 8.1, let $X_m(x) = \sin(\alpha_m x)$ and $Y_n(y) = \sin(\beta_n y)$, where $\alpha_m = m\pi/a$ and $\beta_n = n\pi/b$. Then

$$\frac{d^4 X_m}{dx^4} = \alpha_m^4 X_m, \quad \frac{d^4 Y_n}{dy^4} = \beta_n^4 Y_n, \tag{8.33}$$

which are identical to the expressions used for orthogonality in beams.[7] The other key orthogonality condition is that

$$\int_0^a X_m(x) X_i(x) dx = \int_0^a \frac{d^2 X_m}{dx^2} \frac{d^2 X_i}{dx^2} dx = \begin{cases} 0 & \text{for } i \neq m, \\ \frac{a}{2} & i = m, \end{cases} \tag{8.34}$$

[7] In two-dimensional beams, only one equation is required as there is no y direction—see for example Weaver Jr. et al. (1990) or Inman (2006) for explanations of the beam case and Blevins (1979) for tabulated beam and plate mode-shapes.

for a pinned-pinned beam. There is an equivalent corresponding relationship for the Y_m modes. For beams, Eq. (8.33) holds for other end conditions, such as free-free and clamped-clamped. Eq. (8.34) also applies for beams, but the integral has different values for other combinations of end conditions, see for example Blevins (1979). For plates, dealing with these boundary conditions is more difficult because cross-coupling terms arise in the Galerkin decomposition. For the current analysis, the simpler pinned-pinned (i.e. simply supported on all sides of the plate) case will be used, an example of more complex boundary conditions will be discussed in Sect. 8.4.

Example 8.2 Galerkin decomposition for a flat plate

Problem Find the modal equations of motion for an (undamped) flat plate by making the Galerkin substitution, $w(x, y, t) = \sum_1^M \sum_1^N X_m(x)Y_n(y)q_{nm}(t)$, where $X_m(x) = \sin(\alpha_m x)$ and $Y_n(y) = \sin(\beta_n y)$, and $\alpha_m = m\pi/a$ and $\beta_n = n\pi/b$. The plate has thickness, h, a is the x dimension and b is the y dimension. The number of x and y modes is given by M and N respectively.

Solution The equation of motion for an undamped flat plate is given by Eq. (8.29). Substituting for w gives

$$\sum_1^M \sum_1^N \left\{ D\left(\frac{d^4 X_m}{dx^4} Y_n + 2\frac{d^2 X_m}{dx^2}\frac{d^2 Y_n}{dy^2} + X_m \frac{d^4 Y_n}{dy^4} \right) q_{mn} + \rho h X_m Y_n \ddot{q}_{mn} \right\} = 0. \tag{8.35}$$

Now multiplying by arbitrary mode-shapes X_i and Y_j and integrating over the area of the plate, Eq. (8.35) becomes

$$\sum_1^M \sum_1^N \int_0^a \int_0^b \left\{ \frac{D}{\rho h} \left(\frac{d^4 X_m}{dx^4} X_i Y_n Y_j + 2\frac{d^2 X_m}{dx^2} X_i \frac{d^2 Y_n}{\partial y^2} Y_j + X_m X_i \frac{d^4 Y_n}{\partial y^4} Y_j \right) q_{mn} \right.$$
$$\left. + X_m X_i Y_n Y_j \ddot{q}_{mn} \right\} dydx = 0. \tag{8.36}$$

By separating the integrals into x and y components, using the orthogonality conditions given by Eqs. (8.33) and (8.34) and the relationships $d^2X_m/dx^2 = -\alpha_m^2 X_m$ and $d^2Y_n/dy^2 = -\beta_n^2 Y_n$, it can be seen that the terms where $m \neq i$ or $n \neq j$ are zero, so Eq. (8.36) reduces to

$$\int_0^a \int_0^b \left\{ \frac{D}{\rho h} \left(\frac{d^4 X_i}{dx^4} X_i Y_j^2 + 2\frac{d^2 X_i}{dx^2} X_i \frac{d^2 Y_j}{\partial y^2} Y_j + X_i^2 \frac{d^4 Y_j}{\partial y^4} Y_j \right) q_{ij} + X_i^2 Y_j^2 \ddot{q}_{ij} \right\} dydx = 0.$$

Now using the $i = m$ expression in Eq. (8.34) and the equivalent expression for Y_n gives

$$\frac{D}{\rho h}\left(\alpha_i^4\frac{ab}{4}+\frac{ab\alpha_i^2\beta_j^2}{2}+\beta_i^4\frac{ab}{4}\right)q_{ij}+\frac{ab}{4}\ddot{q}_{ij}=0. \tag{8.37}$$

Equation (8.37) can be reduced to

$$\ddot{q}_{ij}+\omega_{ij}^2 q_{ij}=0, \tag{8.38}$$

for $i = 1, 2, \ldots, M$ and $j = 1, 2, \ldots, N$, where, using the definitions of α_i and β_j, the natural frequency may be written as

$$\omega_{ij}=\pi^2\left(\frac{i^2}{a^2}+\frac{j^2}{b^2}\right)\sqrt{\frac{D}{\rho h}}. \tag{8.39}$$

Now for a given choice of M and N, the governing (undamped) modal equations of motion are given by Eq. (8.38) with response frequencies for each of the $m = 1, 2, \ldots, M$ and $n = 1, 2, \ldots, N$ values estimated using Eq. (8.39) (which is the same as Eq. (8.32) in Example 8.1). As in Example 8.1, these frequency estimates are based on the assumption that the mode-shapes are simple sine waves. The governing modal equations in this case are the same as for a simple linear oscillator, also derived for Euler-Bernoulli beams in Sect. 6.1.1. ■

Note that appropriate modal damping terms should be added to Eq. (8.38) to give a physically realistic model of the plate vibration. The initial conditions and forcing can be added to the model using the same approach described for beams in Sect. 6.1.3.

Notice also that (unlike beams) it is possible to have up to $M \times N$ frequency and mode-shape combinations, and therefore the same number of modal equations. However, in this case the Galerkin method completely decouples the partial differential equation, and there are no cross-coupling terms between the $M \times N$ modal equations defined by Eq. (8.38).

Even though there are $M \times N$ modes, when the plate is harmonically forced with increasing frequency, the resonance peaks associated with each mode will appear in a particular *modal sequence*. For a square plate with simply-supported edges, this sequence is typically $(m, n) = (1, 1), (2, 1), (1, 2), (2, 2), (3, 1), (1, 3)$ for increasing forcing frequency. Modal sequences for other boundary conditions and different a/b ratios are tabulated by Blevins (1979).

8.3 Vibration with Axial Loading

The governing equations of motion for a flat plate with axial loading are given either by Eq. (8.24) if the normal forces are treated as constants, or by Eq. (8.27) with Eq. (8.28) when the forces are not treated as constant. In the first case, the Galerkin approach substitutes for the transverse vibration, w, in a similar way to the example above. However, in the latter case there are two variables, w and Φ. In this situation, the Galerkin approach can be used to obtain nonlinear ordinary differential equations of motion by making the dual substitution $w(x,y,t) = \sum_1^M \sum_1^N X_n(x)Y_m(y)q_{nm}(t)$ and $\Phi(x,y,t) = \sum_1^R \sum_1^S \boldsymbol{\Theta}_r(x)\Psi_s(y)F_{rs}(t)$, where the additional indices r, s are introduced. In this formulation the Airy function, Φ, is assumed to be separable into two space functions $\boldsymbol{\Theta}_n(x)$ and $\Psi_m(y)$ and a time function $F_{mn}(t)$. The shape functions $\boldsymbol{\Theta}_n(x)$ and $\Psi_m(y)$ are assumed to have the same orthogonality properties as the mode-shapes $X_n(x)$ and $Y_m(y)$.

The approach will be demonstrated with the example below. Two points should be noted in the example. First, the compressed summation notation will be used, such that $\sum^{M,N} = \sum_1^M \sum_1^N$. Secondly, when summations are squared, special attention must be given to the indices. For example w is approximated as a series of $M \times N$ terms so w^2 will have $(M \times N)^2$ terms, and to ensure all the *cross-coupling terms* are defined, different indices are used so that

$$w^2 = \left(\sum_1^M \sum_1^N X_n(x)Y_m(y)q_{nm}(t)\right) \times \left(\sum_1^G \sum_1^H X_g(x)Y_h(y)q_{gh}(t)\right),$$

where $G = M$ and $H = N$. This notation ensures that the correct number of terms arise from multiplying the summation.

Example 8.3 Galerkin decomposition for a flat plate with in-plane loading

Problem Find the modal equations of motion for an (undamped) flat plate with in-plane loading governed by Eqs. (8.27) and (8.28). Assume Galerkin substitution variables of the form, $w(x,y,t) = \sum_1^M \sum_1^N X_m(x)Y_n(y)q_{nm}(t)$, and $\Phi(x,y,t) = \sum_1^R \sum_1^S \Theta_r(x)\Psi_s(y)F_{rs}(t)$ where $X_m(x) = \sin(\alpha_m x)$ and $Y_n(y) = \sin(\beta_n y)$, and $\Theta_r(x) = \sin(\xi_r x)$ and $\Psi_s(y) = \sin(\eta_s y)$. The plate has thickness, h, a is the x dimension and b is the y dimension.

Solution The equation of motion for an undamped flat plate is given by Eq. (8.27). Substituting for w and Φ gives

$$\begin{aligned}
&\sum^{M,N}\left\{D\left(\frac{\mathrm{d}^4X_m}{\mathrm{d}x^4}Y_n + 2\frac{\mathrm{d}^2X_m}{\mathrm{d}x^2}\frac{\mathrm{d}^2Y_n}{\mathrm{d}y^2} + X_m\frac{\mathrm{d}^4Y_n}{\mathrm{d}y^4}\right)q_{mn} + \rho h X_m Y_n \ddot{q}_{mn}\right\} \\
&\quad - \sum^{M,N,R,S}\left\{\left(\Theta_r\frac{\mathrm{d}^2\Psi_s}{\mathrm{d}y^2}\frac{\mathrm{d}^2X_m}{\mathrm{d}x^2}Y_n - 2\frac{\mathrm{d}\Theta_r}{\mathrm{d}x}\frac{\mathrm{d}\Psi_s}{\mathrm{d}y}\frac{\mathrm{d}X_m}{\mathrm{d}x}\frac{\mathrm{d}Y_n}{\mathrm{d}y}\right.\right. \\
&\qquad \left.\left. + \frac{\mathrm{d}^2\Theta_r}{\mathrm{d}x^2}\Psi_s X_m\frac{\mathrm{d}^2Y_n}{\mathrm{d}y^2}\right)F_{rs}q_{mn}\right\} = 0. \qquad (8.40)
\end{aligned}$$

The corresponding compatibility equation is given by Eq. (8.28). Substituting for w and Φ gives

$$\sum^{R,S}\left\{\frac{1}{Eh}\left(\frac{d^4\Theta_r}{dx^4}\Psi_s+2\frac{d^2\Theta_r}{dx^2}\frac{d^2\Psi_s}{dy^2}+\Theta_r\frac{d^4\Psi_s}{dy^4}\right)F_{rs}\right\}$$
$$+\sum^{G,H,T,U}\left\{\left(\frac{d^2X_g}{dx^2}Y_hX_t\frac{d^2Y_u}{dy^2}-\frac{dX_g}{dx}\frac{dY_h}{dy}\frac{dX_t}{dx}\frac{dY_u}{dy}\right)q_{gh}q_{tu}\right\}=0, \tag{8.41}$$

where the indices $1\leq g\leq G=M$, $1\leq t\leq T=M$, $1\leq h\leq H=N$ and $1\leq U\leq H=N$ have been introduced to ensure that the correct number of terms arises when the summations are multiplied.

To apply the Galerkin method, first consider the compatibility equation. Multiply Eq. (8.41) by arbitrary mode-shapes Θ_p and Ψ_q and integrate across the area of the plate, giving

$$\sum^{R,S}\int_0^a\int_0^b\left\{\frac{1}{Eh}\left(\frac{d^4\Theta_r}{dx^4}\Psi_s+2\frac{d^2\Theta_r}{dx^2}\frac{d^2\Psi_s}{dy^2}+\Theta_r\frac{d^4\Psi_s}{dy^4}\right)\Theta_p\Psi_qF_{rs}\right\}dydx$$
$$+\sum^{G,H,T,U}\int_0^a\int_0^b\left\{\frac{d^2X_g}{dx^2}Y_hX_t\frac{d^2Y_u}{dy^2}-\left(\frac{dX_g}{dx}\frac{dY_h}{dy}\frac{dX_t}{dx}\frac{dY_u}{dy}\right)\right\}q_{gh}q_{tu}\Theta_p\Psi_qdydx=0.$$

The first summation term consists of the linear terms in the compatibility expression and so the cross-terms within this summation will go to zero when the integrals are evaluated. To achieve this the integrals are separated into x and y components and the orthogonality conditions in Eq. (8.34) along with the relationships in Eq. (8.33) are used (noting that these expressions are valid for Θ_r and Ψ_s as well as X_m and Y_n), to give

$$\frac{ab}{4Eh}\left(\xi_p^2+\eta_q^2\right)^2F_{pq}$$
$$+\sum^{G,H,T,U}\int_0^a\int_0^b\left\{\frac{d^2X_g}{dx^2}Y_hX_t\frac{d^2Y_u}{dy^2}-\left(\frac{dX_g}{dx}\frac{dY_h}{dy}\frac{dX_t}{dx}\frac{dY_u}{dy}\right)\right\}\Theta_p\Psi_qdydx\,q_{gh}q_{tu}=0.$$

This may be rewritten as

$$\Gamma_{2pq}F_{pq}+\sum^{G,H,T,U}\Gamma_{1ghtupq}q_{gh}q_{tu}=0, \tag{8.42}$$

where

$$\Gamma_{1ghtupq}=\int_0^a\int_0^b\left(\frac{d^2X_g}{dx^2}Y_hX_t\frac{d^2Y_u}{dy^2}-\left(\frac{dX_g}{dx}\frac{dY_h}{dy}\frac{dX_t}{dx}\frac{dY_u}{dy}\right)\right)\Theta_p\Psi_qdydx, \tag{8.43}$$

$$\text{and}\quad \Gamma_{2pq}=\frac{ab}{4Eh}\left(\xi_p^2+\eta_q^2\right)^2.$$

Now the Galerkin approach is applied to Eq. (8.40) but using X_i and Y_j as the arbitrary modes. This leads to a set of ordinary differential equations of the form

$$\ddot{q}_{ij} + \omega_{ij}^2 q_{ij} - \sum^{M,N,R,S} \Gamma_{3mnrsij} F_{rs} q_{mn} = 0, \tag{8.44}$$

where

$$\omega_{ij}^2 = \frac{D}{\rho h}\left(\alpha_i^2 + \beta_j^2\right)^2 \tag{8.45}$$

and

$$\Gamma_{3mnrsij} = \frac{4}{\rho hab}\int_0^a\int_0^b \left(\Theta_r \frac{\mathrm{d}^2\Psi_s}{\mathrm{d}y^2}\frac{\mathrm{d}^2 X_m}{\mathrm{d}x^2} Y_n - 2\frac{\mathrm{d}\Theta_r}{\mathrm{d}x}\frac{\mathrm{d}\Psi_s}{\mathrm{d}y}\frac{\mathrm{d}X_n}{\mathrm{d}x}\frac{\mathrm{d}Y_m}{\mathrm{d}y} + \frac{\mathrm{d}^2\Theta_r}{\mathrm{d}x^2}\Psi_s X_m \frac{\mathrm{d}^2 Y_n}{\mathrm{d}y^2}\right) X_i Y_j \mathrm{d}y \mathrm{d}x. \tag{8.46}$$

Finally, eliminating F_{rs} from Eq. (8.44) using Eq. (8.42) gives

$$\ddot{q}_{ij} + \omega_{ij}^2 q_{ij} + \sum^{M,N,R,S,G,H,T,U} \frac{\Gamma_{1ghturs}\Gamma_{3mnrsij}}{\Gamma_{2rs}} q_{gh} q_{tu} q_{mn} = 0, \tag{8.47}$$

which represents the nonlinear (cubic, i.e. Duffing-type) modal equation for the plate vibration (without damping). ■

Notice that the modal equations given by Eq. (8.47) have cross-coupling arising from the nonlinear cubic terms. Note also that the coefficients Γ_1, Γ_2 and Γ_3 contain terms which are not decoupled, and include summations over the respective indices.[8] The values of all three constants Γ_1, Γ_2 and Γ_3 depend on the mode-shapes and the values of ξ_r and η_s, which in this example $\xi_r = \alpha_r$ and $\eta_s = \beta_s$, based on the assumption that the modes Θ_r and Ψ_s satisfy the same eigenvalue equations as X_m and Y_n.

Example 8.4 Modal equations of motion for a plate with axial loading

Problem Find the modal equations of motion for a flat plate with axial loading given by Eq. (8.47) when $N = 2$. Assume the plate is pinned along all its edges and has the following properties; $a = b = 0.5\,\mathrm{m}$, $h = 0.0005\,\mathrm{m}$, $E = 70\,\mathrm{GPa}$, $\rho = 2,700\,\mathrm{kg/m^3}$.

Solution In order to find the equations of given by Eq. (8.47) the coefficients $\Gamma_{1ghturs}$, Γ_{2rs} and $\Gamma_{3mnrsij}$ need to be computed using the integral expression in Eqs. 8.43 and 8.46. These have been approximated using numerical integration, and the resulting

[8] For convenience, these, and other similar coefficients, will be used without indices in the text.

Table 8.1 Nonlinear coefficients for Example 8.4 computed using Eq. 8.47

ghtupq	*mnrsij*	$\frac{\Gamma_{1ghturs}\Gamma_{3mnrsij}}{\Gamma_{2rs}}$	$q_{gh}q_{tu}q_{mn}$
111111	111111	93.7	q_{11}^3
111112	111112	0.0	–
111121	111121	0.0	–
111122	111122	0.0	–
111211	111211	0.0	–
111212	111212	999.3	$q_{11}^2 q_{12}$
111221	111221	0.0	–
111222	111222	0.0	–
112111	112111	0.0	–
112112	112112	0.0	–
112121	112121	40.0	$q_{11}^2 q_{21}$
112122	112122	0.0	–
112211	112211	0.0	–
112212	112212	0.0	–
112221	112221	0.0	–
112222	112222	479.7	$q_{11}^2 q_{22}$
121111	121111	0.0	–
121112	121112	60.0	$q_{12}^2 q_{11}$
121121	121121	0.0	–
121122	121122	0.0	–
121211	121211	359.9	q_{12}^3
121212	121212	0.0	–
121221	121221	0.0	–
121222	121222	0.0	–
122111	122111	0.0	–
122112	122112	0.0	–
122121	122121	0.0	–
122122	122122	25.6	$q_{12}^2 q_{21}$
122211	122211	0.0	–
122212	122212	0.0	–
122221	122221	173.0	$q_{12}^2 q_{22}$
122222	122222	0.0	–
211111	211111	0.0	–
211112	211112	0.0	–
211121	211121	1498.9	$q_{21}^2 q_{11}$

(continued)

Table 8.1 (continued)

ghtupq	*mnrsij*	$\frac{\Gamma_{1ghturs}\Gamma_{3mnrsij}}{\Gamma_{2rs}}$	$q_{gh}q_{tu}q_{mn}$
211122	211122	0.0	–
211211	211211	0.0	–
211212	211212	0.0	–
211221	211221	0.0	–
211222	211222	11280.5	$q_{21}^2 q_{12}$
212111	212111	359.9	q_{21}^3
212112	212112	0.0	–
212121	212121	0.0	–
212122	212122	0.0	–
212211	212211	0.0	–
212212	212212	6927.5	$q_{21}^2 q_{22}$
212221	212221	0.0	–
212222	212222	0.0	–
221111	221111	0.0	–
221112	221112	0.0	–
221121	221121	0.0	–
221122	221122	959.5	$q_{22}^2 q_{11}$
221211	221211	0.0	–
221212	221212	0.0	–
221221	221221	9236.7	$q_{22}^2 q_{12}$
221222	221222	0.0	–
222111	222111	0.0	–
222112	222112	230.7	$q_{22}^2 q_{21}$
222121	222121	0.0	–
222122	222122	0.0	–
222211	222211	480.7	q_{22}^3
222212	222212	0.0	–
222221	222221	0.0	–
222222	222222	0.0	–

coefficients are shown in Table 8.1. Using the coefficients from Table 8.1 for a $N = 2$ model gives governing equations

$$\begin{aligned}
\ddot{q}_{11} + \omega_{11}^2 q_{11} + \alpha_1 q_{11}^3 + \alpha_6 q_{12}^3 + \alpha_{11} q_{21}^3 + \alpha_{16} q_{22}^3 &= 0, \\
\ddot{q}_{12} + \omega_{12}^2 q_{12} + \alpha_2 q_{11}^2 q_{12} + \alpha_5 q_{12}^2 q_{11} + \alpha_{12} q_{21}^2 q_{22} + \alpha_{15} q_{22}^2 q_{21} &= 0, \\
\ddot{q}_{21} + \omega_{21}^2 q_{21} + \alpha_3 q_{11}^2 q_{21} + \alpha_8 q_{12}^2 q_{22} + \alpha_9 q_{21}^2 q_{11} + \alpha_{14} q_{22}^2 q_{12} &= 0, \\
\ddot{q}_{22} + \omega_{22}^2 q_{22} + \alpha_4 q_{11}^2 q_{22} + \alpha_7 q_{12}^2 q_{21} + \alpha_{10} q_{21}^2 q_{12} + \alpha_{13} q_{22}^2 q_{11} &= 0.
\end{aligned} \tag{8.48}$$

■

As seen in Example 8.4, in most cases a significant number of the cross-coupling coefficients, Γ_1 and Γ_3, will be found to be negligibly small (depending on which mode-shapes are used). However, it is worth remembering that this model is based on the assumption of using simplified beam modes to try and capture the vibration of the plate. Therefore, when computing cross-coupling coefficients it is important to check, when possible, with experimental data. In fact, one can use the modal testing idea (Ewins 2000) and identify the Γ_1, Γ_2 and Γ_3 coefficients directly from experimental data, and this will be discussed in the case study in Sect. 8.5. Once the important cross-coupling terms have been established, the nonlinear modal oscillators can be analysed using the techniques developed in Chaps. 4 and 5.

Notice also that in the derivation, Eq. (8.44) was obtained by dividing through by $\rho h \int_0^a \int_0^b X_i^2 Y_j^2 \mathrm{d}y\mathrm{d}x = ab/4$ for a simply-supported plate. For different boundary conditions the value of this double integral can change. An example is shown in Sect. 8.5.

Plates with curvature in either one or both planar directions are referred to as shells, and the vibration of these structural elements is considered next.

8.4 Vibration of Shells

Two types of shell structure are shown in Fig. 8.7. Figure 8.7a shows a cylindrical shell, which has curvature in a single direction (x in this case), and Fig. 8.7b shows a doubly-curved shell, which has curvature in both the x and y directions. The radii of curvature for the x and y directions are denoted by R_x and R_y respectively.

The development of the equations of motion is exactly the same as for flat plates with axial loading, with the addition of terms to represent the effects of curvature. Note that shells are normally divided into *shallow* and *deep*: for shallow shells the curvature is small (radius large), and for deep shells the curvature is larger. When deriving the equations of motion, assuming the shell is shallow allows higher order terms in Taylor series expansions to be neglected. In the following derivation a shallow shell will be assumed.

Including the (first-order) effects of curvature in the x and y directions gives membrane (i.e. in-plane) strains of the form

$$\upsilon_x = \frac{\partial u}{\partial x} + \frac{w}{R_x} + \frac{1}{2}\left(\frac{\partial w}{\partial x}\right)^2,$$
$$\upsilon_y = \frac{\partial v}{\partial y} + \frac{w}{R_y} + \frac{1}{2}\left(\frac{\partial w}{\partial y}\right)^2, \qquad (8.49)$$
$$\upsilon_{xy} = \frac{\partial u}{\partial y} + \frac{\partial v}{\partial x} + \frac{\partial w}{\partial x}\frac{\partial w}{\partial y},$$

where (u, v, w) correspond to the displacements in the same rectangular coordinate system (x, y, z) as used for the flat-plate analysis.

The extra curvature term in the strain expressions can be explained with reference to the bending stiffness derivation and particularly Fig. 8.4. In the bending stiffness derivation the distance z was used as the distance from the neutral axis, leading to Eq. (8.15), $\kappa_x = -\frac{z}{R}$. If, instead, the neutral axis moves to the new position (equivalent to a in Fig. 8.4) from transverse deflection, w, then the associated strain is $\frac{w}{R_x}$ (note that the sign of the curvature is reversed to correspond to the scenario shown in Fig. 8.7). In the following derivation, to conform with the usual convention adopted elsewhere, when deriving force balance relationships for an element of shell, N and Q have been redefined as the membrane and shear forces respectively (previously N indicated a force acting in the x or y direction, and Q a force acting in the z direction). In addition it is assumed that the curvature of the plate is shallow and that the dynamic deflection in the z-axis, w, is not large.[9]

The force moment relations for an element of doubly-curved shell can be found using the same approach as for an axially-loaded flat plate and, for a shallow shell, the moment shear force relations are identical to the flat-plate expressions, being given by

$$\frac{\partial M_x}{\partial x} + \frac{\partial M_{xy}}{\partial y} = Q_x, \quad \frac{\partial M_y}{\partial y} + \frac{\partial M_{xy}}{\partial x} = Q_y, \tag{8.50}$$

where $M_{yx} = M_{xy}$ has been assumed. The equations for membrane force equilibrium (neglecting in-plane inertia) have additional terms due to curvature

$$\frac{\partial N_x}{\partial x} + \frac{\partial N_{xy}}{\partial y} + \frac{Q_x}{R_x} = 0, \quad \frac{\partial N_{xy}}{\partial x} + \frac{\partial N_y}{\partial y} + \frac{Q_y}{R_y} = 0, \tag{8.51}$$

and the equation of motion for the transverse vibration of the shell is given by

$$\begin{aligned} &-\frac{\partial Q_x}{\partial x} - \frac{\partial Q_y}{\partial y} + \frac{N_x}{R_x} + \frac{N_y}{R_y} - \frac{\partial}{\partial x}\left(N_x\frac{\partial w}{\partial x} + N_{xy}\frac{\partial w}{\partial y}\right) \\ &-\frac{\partial}{\partial y}\left(N_{xy}\frac{\partial w}{\partial x} + N_y\frac{\partial w}{\partial y}\right) + \rho h\frac{\partial^2 w}{\partial t^2} = 0. \end{aligned} \tag{8.52}$$

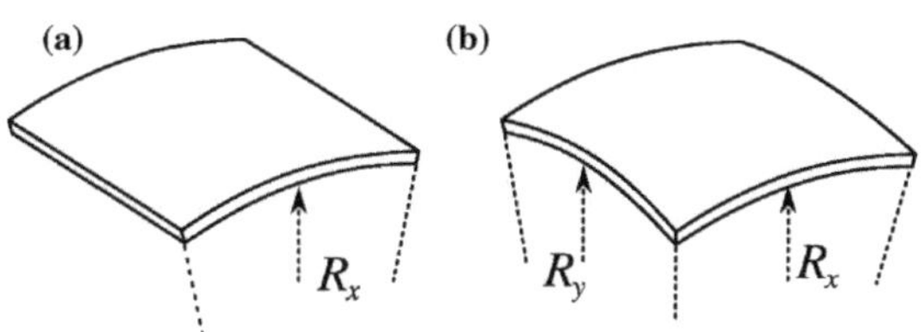

Fig. 8.7 Shell structures, showing **a** cylindrical shell (single curvature), and **b** doubly-curved shell

[9] A more detailed discussion of the assumptions used in deriving the force balance relationships is given in Amabili (2008).

Equations (8.50)–(8.52) are a simplified form of the Love equations[10] including the von Kármán nonlinearity. As with previous derivations, damping terms and external forcing will be included at a later stage.

Eliminating the shear forces, Q_x and Q_y, from Eq. (8.52) using Eq. (8.50) leads to a governing equation of the transverse vibration of the shell in the form

$$\frac{\partial^2 M_x}{\partial x^2} + 2\frac{\partial^2 M_{xy}}{\partial x \partial y} + \frac{\partial^2 M_y}{\partial y^2} - \frac{N_x}{R_x} - \frac{N_y}{R_y} + \frac{\partial}{\partial x}\left(N_x \frac{\partial w}{\partial x} + N_{xy}\frac{\partial w}{\partial y}\right) + \frac{\partial}{\partial y}\left(N_{xy}\frac{\partial w}{\partial x} + N_y \frac{\partial w}{\partial y}\right) = \rho h \frac{\partial^2 w}{\partial t^2}. \tag{8.53}$$

Equation 8.53 is the partial differential equation that governs the out-of-plane deflection of a general elastic shell with the von Kármán nonlinearity. As with the derivation of the governing equations for a flat plate, the Airy stress function, $\Phi(x, y, t)$, using Eq. (8.26), is introduced. In addition the moment relations from Eq. (8.23) are used to give

$$-D\left(\frac{\partial^4 w}{\partial x^4} + 2\frac{\partial^4 w}{\partial y^2 \partial x^2} + \frac{\partial^4 w}{\partial y^4}\right) + \left(\frac{\partial^2 \Phi}{\partial y^2}\frac{\partial^2 w}{\partial x^2} - 2\frac{\partial^2 \Phi}{\partial x \partial y}\frac{\partial^2 w}{\partial x \partial y} + \frac{\partial^2 \Phi}{\partial x^2}\frac{\partial^2 w}{\partial y^2}\right) - \left(\frac{1}{R_x}\frac{\partial^2 \Phi}{\partial y^2} + \frac{1}{R_y}\frac{\partial^2 \Phi}{\partial x^2}\right) = \rho h \frac{\partial^2 w}{\partial t^2}. \tag{8.54}$$

This equation may be compared to Eq. (8.27), the equivalent equation for a flat plate.

As with the flat-plate derivation, a second governing equation, based on the strain relationship, is required. Eliminating the u and v terms from the axial strain equations, Eq. (8.49), in exactly the same way as was done when deriving Eq. (8.14), gives

$$\frac{\partial^2 \upsilon_x}{\partial y^2} - \frac{\partial^2 \upsilon_{xy}}{\partial x \partial y} + \frac{\partial^2 \upsilon_y}{\partial x^2} = \left(\frac{\partial^2 w}{\partial x \partial y}\right)^2 - \frac{\partial^2 w}{\partial x^2}\frac{\partial^2 w}{\partial y^2} + \frac{1}{R_x}\frac{\partial^2 w}{\partial y^2} + \frac{1}{R_y}\frac{\partial^2 w}{\partial x^2}. \tag{8.55}$$

Now the Airy stress function, Eqs. (8.26), and (8.51) give

$$\frac{1}{Eh}\nabla^2\nabla^2\Phi + \frac{\partial^2 w}{\partial x^2}\frac{\partial^2 w}{\partial y^2} - \left(\frac{\partial^2 w}{\partial x \partial y}\right)^2 - \frac{1}{R_x}\frac{\partial^2 w}{\partial y^2} - \frac{1}{R_y}\frac{\partial^2 w}{\partial x^2} = 0. \tag{8.56}$$

Note that in deriving this relationship the Q_x/R_x and Q_y/R_y terms in Eq. (8.51) have been assumed to be small.

---

[10] See Love (1892) and later editions.

The modal decomposition of the shell equations follows a similar approach to that for a flat plate with axial loading. The following example demonstrates how the Galerkin method can be applied to obtain governing ordinary differential equations of motion using Eqs. (8.54) and (8.56).

Example 8.5 Galerkin decomposition for a doubly-curved shell

Problem Find the modal equations of motion for an (undamped) doubly-curved shell governed by Eqs. (8.54) and (8.56). Assume the Galerkin substitution, $w(x,y,t) = \Sigma_1^M \Sigma_1^N X_m(x) Y_n(y) q_{nm}(t)$, and $\Phi(x,y,t) = \Sigma_1^R \Sigma_1^S \Theta_r(x) \Psi_s(y) F_{rs}(t)$ where $X_m(x) = \sin(\alpha_m x)$ and $Y_n(y) = \sin(\beta_n y)$, and $\Theta_r(x) = \sin(\xi_r x)$ and $\Psi_s(y) = \sin(\eta_s y)$. The shell has thickness, h, a is the x dimension and b is the y dimension.

Solution The equation of motion for an undamped doubly-curved shell is given by Eq. (8.54). Substituting for w and Φ gives

$$\begin{aligned}\sum^{M,N}\left\{D\left(\frac{d^4X_m}{dx^4}Y_n + 2\frac{d^2X_m}{dx^2}\frac{d^2Y_n}{dy^2} + X_m\frac{d^4Y_n}{dy^4}\right)q_{mn} + \rho h X_m Y_n \ddot{q}_{mn}\right\}\\ -\sum^{M,N,R,S}\left\{\left(\Theta_r\frac{d^2\Psi_s}{dy^2}\frac{d^2X_m}{dx^2}Y_n - 2\frac{d\Theta_r}{dx}\frac{d\Psi_s}{dy}\frac{dX_m}{dx}\frac{dY_n}{dy} + \frac{d^2\Theta_r}{dx^2}\Psi_s X_m\frac{d^2Y_n}{dy^2}\right)F_{rs}q_{mn}\right\}\\ +\sum^{R,S}\left\{\left(\frac{1}{R_x}\Theta_r\frac{d^2\Psi_s}{dy^2} + \frac{1}{R_y}\frac{d^2\Theta_r}{dx^2}\Psi_s\right)F_{rs}\right\} = 0.\end{aligned} \tag{8.57}$$

The corresponding compatibility equation is given by Eq. (8.56). Substituting for w and Φ gives

$$\begin{aligned}\sum^{R,S}\left\{\frac{1}{Eh}\left(\frac{d^4\Theta_r}{dx^4}\Psi_s + 2\frac{d^2\Theta_r}{dx^2}\frac{d^2\Psi_s}{dy^2} + \Theta_r\frac{d^4\Psi_s}{dy^4}\right)F_{rs}\right\}\\ +\sum^{G,H,T,U}\left\{\frac{d^2X_g}{dx^2}Y_h X_t\frac{d^2Y_u}{dy^2}q_{gh}q_{tu} - \left(\frac{dX_g}{dx}\frac{dY_h}{dy}\frac{dX_t}{dx}\frac{dY_u}{dy}\right)q_{gh}q_{tu}\right\}\\ -\sum^{G,H}\left\{\frac{1}{R_x}X_g\frac{d^2Y_h}{dy^2}q_{gh} + \frac{1}{R_y}\frac{d^2X_g}{dx^2}Y_h q_{gh}\right\} = 0,\end{aligned} \tag{8.58}$$

where the indices $1 \le g \le G = M$, $1 \le h \le H = N$, $1 \le t \le T = M$ and $1 \le u \le U = N$ have been introduced to ensure that the correct number of terms arise when the summations are multiplied.

Now, multiply Eq. (8.58) by arbitrary mode-shapes Θ_p and Ψ_q and integrate across the area of the shell to give

$$\Gamma_{2pq}F_{pq} + \sum^{G,H,T,U}\Gamma_{1ghtupq}q_{gh}q_{tu} - \sum^{G,H}\Gamma_{4ghpq}q_{gh} = 0, \tag{8.59}$$

where $\Gamma_{1ghtupq}$ and Γ_{2pq} are given by Eq. (8.43) and

$$\Gamma_{4ghpq} = \int_0^a \int_0^b \left(\frac{1}{R_x} X_g \frac{\mathrm{d}^2 Y_h}{\mathrm{d}y^2} - \frac{1}{R_y} \frac{\mathrm{d}^2 X_g}{\mathrm{d}x^2} Y_h \right) \Theta_p \Psi_q \mathrm{d}y \mathrm{d}x. \tag{8.60}$$

Now apply the same Galerkin approach to Eq. (8.57) (using X_i and Y_j as the arbitrary modes, as in Example 8.3) to get a set of ordinary differential equations of the form

$$\ddot{q}_{ij} + \omega_{ij}^2 q_{ij} - \sum^{M,N,R,S} \Gamma_{3mnrsij} F_{rs} q_{mn} + \sum^{R,S} \Gamma_{5rsij} F_{rs} = 0, \tag{8.61}$$

where ω_{ij} is given by Eq. (8.45), Γ_3 is given by Eq. (8.46) and

$$\Gamma_{5rsij} = \frac{4}{\rho hab} \int_0^a \int_0^b \left(\frac{1}{R_x} \Theta_r \frac{\mathrm{d}^2 \Psi_s}{\mathrm{d}y^2} + \frac{1}{R_y} \frac{\mathrm{d}^2 \Theta_r}{\mathrm{d}x^2} \Psi_s \right) X_i Y_j \mathrm{d}y \mathrm{d}x. \tag{8.62}$$

Finally, F_{rs} can be eliminated from Eq. (8.61) using Eq. (8.59). First, Eq. (8.59) is rewritten as

$$F_{rs} = \frac{1}{\Gamma_{2rs}} \sum^{G,H} \left[\left(\Gamma_{4ghrs} - \sum^{T,U} \Gamma_{1ghturs} q_{tu} \right) q_{gh} \right].$$

Then substituting this equation into Eq. (8.61) gives

$$\ddot{q}_{ij} + \omega_{ij}^2 q_{ij} + \sum^{R,S,G,H} \frac{q_{gh}}{\Gamma_{2rs}} \left[\left(\Gamma_{5rsij} - \sum^{M,N} \Gamma_{3mnrsij} q_{mn} \right) \left(\Gamma_{4ghrs} - \sum^{T,U} \Gamma_{1ghturs} q_{tu} \right) \right] = 0, \tag{8.63}$$

which represents the nonlinear modal equation for the shell vibrations (without damping). ■

Equation (8.63) governs the modal vibrations of a doubly-curved shell. Note that the form of the equations includes both quadratic and cubic terms. The quadratic terms are due to the curvature effects, and the cubic terms come from the membrane force interaction (i.e. in-plane stretching). There is also an additional linear term which can have the effect of modifying the linear natural frequency for each modal equation. Modal damping and external forcing terms can be included in Eq. (8.63) to represent forced, damped vibrations of a shallow shell. Note also that the Γ coefficients have been derived for simply-supported boundary conditions in this derivation.

8.5 Case Study of Nonlinear Shell Vibration

As a case study, consider the vibration of a shallow cylindrical shell made from composite material. The use of composite material is important, as these materials are being used for an increasing number of applications, particularly those in the adaptive or smart structures area. The composite material has a more complex constitutive relationship which requires modification of the governing equations of motion. In addition to including composite material, this case study will consider free-free boundary conditions for all edges of the plate. The plate is forced by a support motion at the centre—point o in Fig. 8.8.

The first part of the case study will describe experimental tests carried out on the physical test specimen. The aim of the experimental testing is to carry out a *system identification* for the shell. For linear vibration, the main aim is to identify the main modes of vibration, the damping coefficients and the existence of important nonlinear oscillations in the response.[11]

The second part of the case study aims to derive a mathematical model which captures the key nonlinear vibration behaviour of the shell. Using the approach already outlined in this chapter, a modal decomposition using the Galerkin method is carried out, which leads to a coupled set of nonlinear modal equations. From these equations, a three-mode model can be defined which captures the key vibration behaviour of the shell.[12] The model is compared to the experimental results and key resonance features such as linear resonance peaks and a 1/2 subharmonic resonance are found in the frequency range 0–50 Hz.

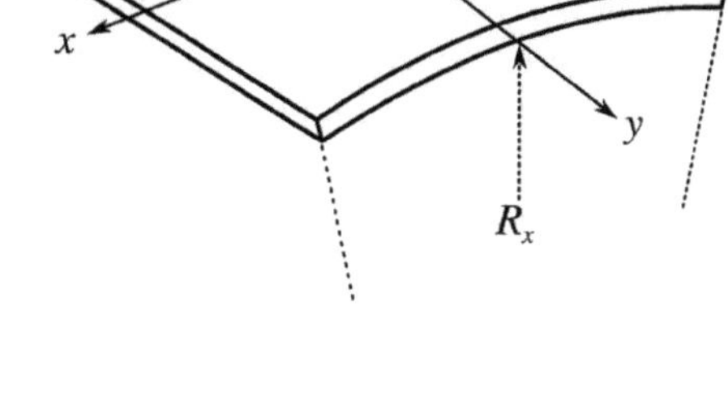

Fig. 8.8 Measured points on the shell, P_x and P_y, each point is 75 mm from o

---

[11] See Ewins (2000) for detailed descriptions of identifying linear modal properties from experimental vibration tests.

[12] The authors would like to acknowledge the work of Andres F. Arrieta, who originally developed this model and carried out the experiments shown in this section. See Arrieta et al. (2009) for further details.

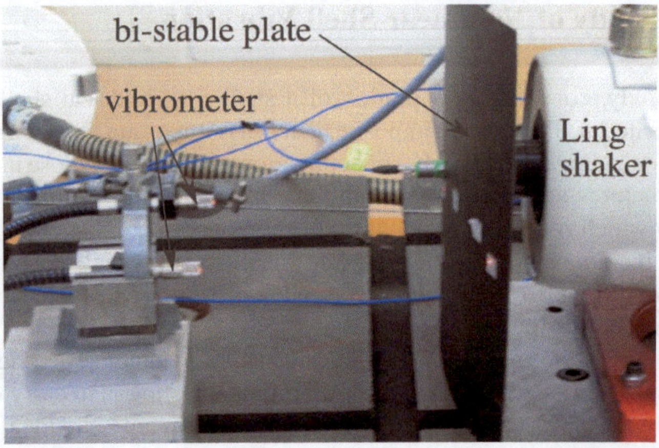

Fig. 8.9 Experimental shallow cylindrical composite shell. Reproduced with kind permission from Arrieta et al. (2009)

8.5.1 Description of Case Study

The physical system is shown in Fig. 8.9. A carbon-fibre epoxy $[0_4-90_4]_T$ 300 mm × 300 mm square shallow cylindrical shell is the experimental structure to be modelled. For this shell $1/R_x \approx 1.67\,\mathrm{m}^{-1}$ and $1/R_y \approx 0$. The shell was experimentally tested using a (Ling) vibration shaker, attached at the centre, in order to capture (or identify) the vibration characteristics of the system. The vibration response of the shell was measured for points P_y and P_x, as shown schematically in Fig. 8.8, using a differential laser vibrometer.

The general approach for obtaining frequency response diagrams has been described in Sect. 2.8.2. With this approach, stroboscopic sampling of the time series at the forcing period is used to create steady-state frequency response diagrams. In this case, the shell is excited with a sine wave and then peak-to-peak displacement measurements were recorded from the time responses for at least ten steady-state forcing periods.

An experimental frequency response diagram obtained using this approach, and a forcing amplitude $F_0 = 1.0\,\mathrm{N}$, for point P_x is shown in Fig. 8.10. Two resonance peaks appear, which correspond directly to linear modes of vibration, one at 17.45 Hz and the second at 45.4 Hz. There is additional resonance close to 19 Hz, which will be explained below.

The frequency response diagram for point P_x for a higher input force amplitude of $F_0 = 5.0\,\mathrm{N}$ is shown in Fig. 8.11. The response resembles the results obtained for Fig. 8.10 except for the region around 35 Hz. This range coincides with twice the natural frequency of the first resonance peak in the x direction, and a significant additional resonance peak occurs in the response. This additional resonance peak is due to a 1/2 subharmonic resonance.

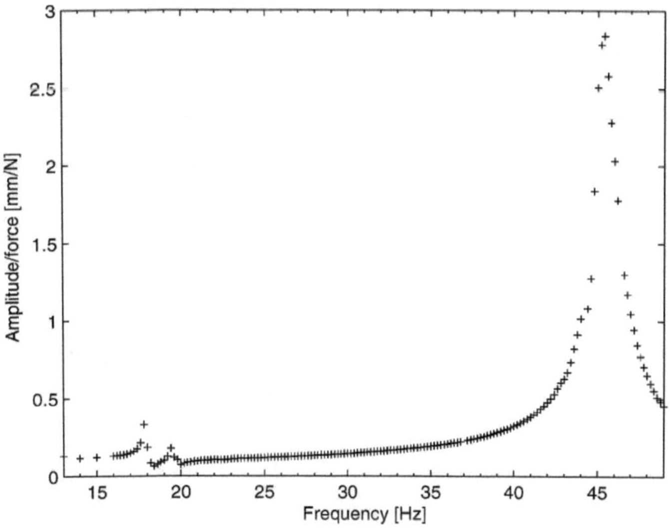

Fig. 8.10 Experimental receptance (Displacement/Force) frequency response function (FRF) for the curved (x) direction, point P_x. Forcing amplitude $F_o = 1.0$ N, frequency range $\Omega = [13, 49]$. Reproduced with kind permission from Arrieta et al. (2009)

The frequency response diagram for point P_y is shown in Fig. 8.12. In this case the dynamic response consists of a linear resonance at 19.6 Hz and a subharmonic resonance at around 38 Hz. The linear resonance peak is coupled to the x direction, and this explains the additional resonance peak which appears in the P_x plots shown in Figs. 8.10 and 8.11. At higher excitation levels, the 1/2 subharmonic resonance dominates the P_y response, as shown in Fig. 8.12.

The experimental observations indicate that at low forcing amplitudes an approximately linear response is found, although some degree of coupling occurs between resonances in the x and y directions. At higher amplitudes, a 1/2 subharmonic resonance can be observed. It is well known from the study of nonlinear differential equations that quadratic-type nonlinearities generate 1/2 subharmonic behaviours (see for example Strogatz (2001), Jordan and Smith (1999) and Cartmell (1990)). It has already been shown in Sect. 8.4 that quadratic nonlinear terms naturally arise in the modal equations for a shallow shell. In the next section, a reduced order model is derived from the general equations of a shell. The model is then used to simulate vibration behaviour, which can be compared to the experimental data.

8.5.2 Governing Equations for Composite Shells

For shells made from composite materials, the constitutive relationship becomes more complex than for homogeneous, isotropic materials. This is because compos-

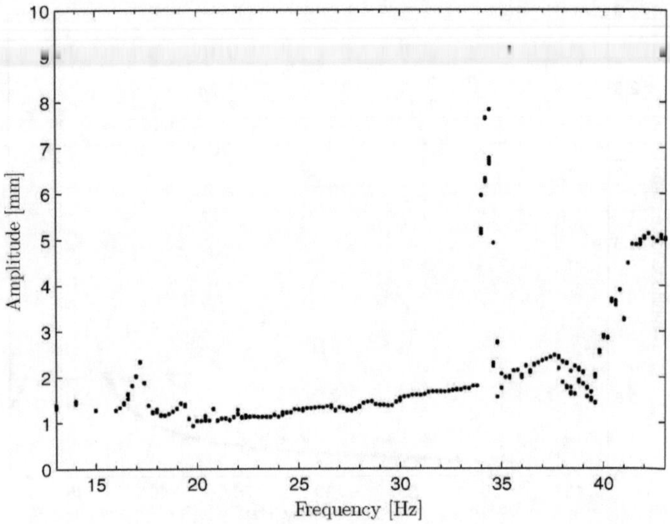

Fig. 8.11 Experimental frequency response for the curved (x) direction, point P_x. $F_o = 5.0\,\mathrm{N}$, frequency range $\Omega = [13, 43]$. Reproduced with kind permission from Arrieta et al. (2009)

ite materials can have different flexural rigidity and membrane force-displacement relationships depending on the direction. Therefore, the simplified form for the membrane forces, Eq. (8.22), and moment expressions, Eq. (8.23), cannot usually be applied. The main result of this is that the first terms on the left-hand side of both Eqs. (8.54) and (8.56) become more complex to account for this type of behaviour.

The constitutive relationships for a shallow shell made of an orthotropic[13] material, can be written as

$$
\begin{aligned}
N_x &= A_{11}\nu_x + A_{12}\nu_y,\\
N_y &= A_{21}\nu_x + A_{22}\nu_y,\\
N_{xy} &= A_{33}\nu_{xy},\\
M_x &= -D_{11}\frac{\partial^2 w}{\partial x^2} - D_{12}\frac{\partial^2 w}{\partial y^2},\\
M_y &= -D_{21}\frac{\partial^2 w}{\partial x^2} - D_{22}\frac{\partial^2 w}{\partial y^2},\\
M_{xy} &= -2D_{33}\frac{\partial^2 w}{\partial x \partial y},
\end{aligned}
\tag{8.64}
$$

[13] An orthotropic material is one that has the different materials properties or strengths in different orthogonal directions.

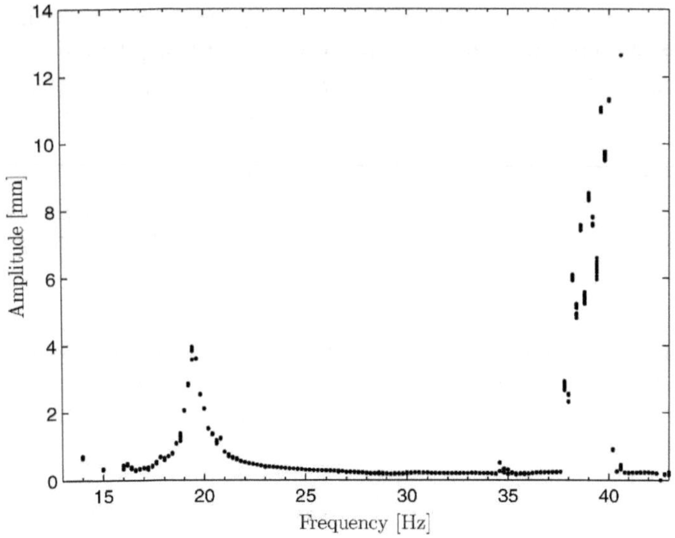

Fig. 8.12 Experimental frequency response diagram for the flat (y) direction, point P_y. Measured using stroboscopic sampling for a forcing amplitude of $F_o = 5.0\,\mathrm{N}$, frequency range $\Omega = [13, 43]$. Reproduced with kind permission from Arrieta et al. (2009)

where symmetry has been assumed, which means that $A_{21} = A_{12}$ and $D_{21} = D_{12}$. As before the objective is to rewrite the compatibility equation, Eq. (8.55), using the Airy function Φ and the constitutive relations given by Eq. (8.64). To achieve this, the membrane strains are expressed in terms of the internal forces, such that

$$\begin{aligned} v_x &= P_{22}N_x - P_{12}N_y, \\ v_y &= P_{11}N_y - P_{12}N_x, \\ v_{xy} &= P_{33}N_{xy}, \end{aligned}$$

where

$$P_{11} = \frac{A_{22}}{\Delta A}, \quad P_{12} = \frac{A_{12}}{\Delta A}, \quad P_{22} = \frac{A_{22}}{\Delta A}, \quad P_{33} = \frac{1}{A_{33}},$$

and $\Delta A = A_{11}A_{22} - A_{12}^2$. The compatibility equation for a curved plate, Eq. (8.55), can now be written as

$$P_{11}\frac{\partial^4 \Phi}{\partial x^4} + (P_{33} - 2P_{12})\frac{\partial^4 \Phi}{\partial x^2 \partial y^2} + P_{22}\frac{\partial^4 \Phi}{\partial y^4} + \frac{\partial^2 w}{\partial x^2}\frac{\partial^2 w}{\partial y^2} - \left(\frac{\partial^2 w}{\partial x \partial y}\right)^2 - \frac{1}{R_x}\frac{\partial^2 w}{\partial y^2} - \frac{1}{R_y}\frac{\partial^2 w}{\partial x^2} = 0, \tag{8.65}$$

where Eqs. (8.26) have been used to relate the membrane forces to the Airy stress function, Φ. With the moment relations given in Eq. (8.61), the governing equation for transverse vibration, Eq. (8.53), becomes

$$
\begin{aligned}
&-\left(D_{11}\frac{\partial^4 w}{\partial x^4} + 2(D_{12}+2D_{33})\frac{\partial^4 w}{\partial x^2 \partial y^2} + D_{22}\frac{\partial^4 w}{\partial y^4}\right)\\
&+\left(\frac{\partial^2 \Phi}{\partial y^2}\frac{\partial^2 w}{\partial x^2} - 2\frac{\partial^2 w}{\partial x \partial y}\frac{\partial^2 \Phi}{\partial x \partial y} + \frac{\partial^2 \Phi}{\partial x^2}\frac{\partial^2 w}{\partial y^2}\right) - \left(\frac{1}{R_x}\frac{\partial^2 \Phi}{\partial y^2} + \frac{1}{R_y}\frac{\partial^2 \Phi}{\partial x^2}\right) = \rho h \frac{\partial^2 w}{\partial t^2}.
\end{aligned}
\tag{8.66}
$$

The system of coupled partial differential equations, Eqs. (8.65) and (8.66), can now be used as the governing equations for the shallow composite shell.

The boundary conditions for a cylindrical shallow shell with free edges on all sides are defined as

$$
\begin{aligned}
&N_x(0,y) = N_x(a,y) = 0, \quad N_y(x,0) = N_y(x,b) = 0,\\
&M_x(0,y) = M_x(a,y) = 0, \quad M_y(x,0) = M_y(x,b) = 0,\\
&V_{xz}(0,s) = Q_{xz}(L_n,s) = 0, \quad V_{yz}(n,0) = V_{yz}(0,L_s) = 0,\\
&T_{xy}(0,s) = T_{xy}(L_n,s) = 0, \quad T_{xy}(n,0) = V_{yz}(0,L_s) = 0,
\end{aligned}
$$

where V_{ij} are known as the Kirchhoff effective shear stress resultants of the first kind, and T_{ij} are the Kirchhoff effective shear stress resultants of the second kind, given by

$$
V_{xz} = Q_{xz} + \frac{\partial^2 M_{xy}}{\partial x^2}, \quad V_{yz} = Q_{yz} + \frac{\partial^2 M_{yx}}{\partial y^2},
$$

and

$$
T_{xy} = N_{xy} + \frac{M_{xy}}{R_y}, \quad T_{yx} = N_{yx} + \frac{M_{yx}}{R_x}.
$$

The free-edge boundary conditions are used to define mode-shapes based on approximate beam eigenfunctions in the x and y directions (see Blevins (1979)). The mode-shapes are then used in the Galerkin method to approximate the displacement w and Airy function Φ.

The characteristic equation for a free-free beam is

$$
0 = 1 - \cosh(\lambda L)\cos(\lambda L), \tag{8.67}
$$

where L is the beam length, which will be replaced with the plate dimensions a or b as appropriate. The roots of Eqs. (8.67) (denoted λ) relate to the natural frequencies of the free-edge beam eigenfunctions, and these will be replaced with α_m, β_n, ξ_r and η_s for the shell as required. The mode-shape, in (for example) the x direction, is given by

$$X(x) = \left[(\cosh(\lambda x) + \cos(\lambda x)) - \frac{\cosh(\lambda a) - \cos(\lambda a)}{\sinh(\lambda a) - \sin(\lambda a)} (\sinh(\lambda x) + \sin(\lambda x)) \right], \tag{8.68}$$

where a is the length of the plate in the x direction. A similar approach is used to obtain the modes in the y direction. Both sets of modes are then used in a Galerkin decomposition.

8.5.3 Galerkin Decomposition

In order to reduce the governing equations, Eqs. (8.65) and (8.66), to a set of modal equations, the Galerkin method is used to separate solutions for w and Φ into space and time functions. As in Example 8.5, the Galerkin substitution, $w(x, y, t) = \Sigma_0^M \Sigma_0^N X_m(x) Y_n(y) q_{nm}(t)$, and $\Phi(x, y, t) = \Sigma_0^R \Sigma_0^S \Theta_r(x) \Psi_s(y) F_{rs}(t)$ is assumed, but in this case the mode-shapes are in the form of Eq. (8.68) and α_m, β_n, ξ_r, and η_s are the roots of Eq. (8.67) for the functions X_m, Y_n, Θ_r and Ψ_s respectively.

It is important to note that for free-free boundary conditions, $\lambda = 0$ is a valid solution of both Eqs. (8.67) and (8.68). The $\lambda = 0$ solution corresponds to the case when one (or both) of the beam modes has zero frequency, and effectively becomes a rigid body motion. Remember, however, that the complete plate mode, $W_{ij}(x, y)$, is being approximated by the combination of two (or more) beam modes, $W_{ij}(x, y) \approx X_i(x) Y_j(y)$, so that if one of the beam modes has zero frequency, then the plate mode becomes like a beam mode in the direction for which $\lambda \neq 0$. As a result the Galerkin summations are taken from 0 in this analysis to include these cases.

Substituting for w and Φ in Eqs. (8.65) and (8.66) respectively, gives

$$\begin{aligned} &\sum_0^{R,S} \left\{ \left(P_{11} \frac{d^4 \Theta_r}{dx^4} \Psi_s + (P_{33} - 2P_{12}) \frac{d^2 \Theta_r}{dx^2} \frac{d^2 \Psi_s}{dy^2} + P_{22} \Theta_r \frac{d^4 \Psi_s}{dy^4} \right) F_{rs} \right\} \\ &+ \sum_0^{G,H,T,U} \left\{ \frac{d^2 X_g}{dx^2} Y_h X_t \frac{d^2 Y_u}{dy^2} q_{gh} q_{tu} - \left(\frac{dX_g}{dx} \frac{dY_h}{dy} \frac{dX_t}{dx} \frac{dY_u}{dy} \right) q_{gh} q_{tu} \right\} \\ &- \sum_0^{G,H} \left\{ \frac{1}{R_x} X_g \frac{d^2 Y_h}{dy^2} q_{gh} + \frac{1}{R_y} \frac{d^2 X_g}{dx^2} Y_h q_{gh} \right\} = 0, \end{aligned} \tag{8.69}$$

and

$$\sum_{0}^{M,N}\left\{\left(D_{11}\frac{\mathrm{d}^4X_m}{\mathrm{d}x^4}Y_n+2(D_{12}+2D_{33})\frac{\mathrm{d}^2X_m}{\mathrm{d}x^2}\frac{\mathrm{d}^2Y_n}{\mathrm{d}y^2}+D_{22}X_m\frac{\mathrm{d}^4Y_n}{\mathrm{d}y^4}\right)q_{mn}+\rho hX_mY_n\ddot{q}_{mn}\right\}$$
$$-\sum_{0}^{M,N,R,S}\left\{\left(\Theta_r\frac{\mathrm{d}^2\Psi_s}{\mathrm{d}y^2}\frac{\mathrm{d}^2X_m}{\mathrm{d}x^2}Y_n-2\frac{\mathrm{d}\Theta_r}{\mathrm{d}x}\frac{\mathrm{d}\Psi_s}{\mathrm{d}y}\frac{\mathrm{d}X_m}{\mathrm{d}x}\frac{\mathrm{d}Y_n}{\mathrm{d}y}+\frac{\mathrm{d}^2\Theta_r}{\mathrm{d}x^2}\Psi_sX_m\frac{\mathrm{d}^2Y_n}{\mathrm{d}y^2}\right)F_{rs}q_{mn}\right\}$$
$$+\sum_{0}^{R,S}\left\{\left(\frac{1}{R_x}\Theta_r\frac{\mathrm{d}^2\Psi_s}{\mathrm{d}y^2}+\frac{1}{R_y}\frac{\mathrm{d}^2\Theta_r}{\mathrm{d}x^2}\Psi_s\right)F_{rs}\right\}=0, \tag{8.70}$$

where the indices $0\le g\le G=M,\ 0\le h\le H=N,\ 0\le t\le T=M$ and $0\le u\le U=N$ have been introduced to ensure that the correct number of terms arise when the summations are multiplied.

In order to obtain equations for each mode, Eq. (8.69) is multiplied by Θ_p and Ψ_q and Eq. (8.70) by X_i and Y_j and integrated over the surface of the shell. Then the resulting equations are simplified using the orthogonality properties of the mode-shapes (exactly as in Example 8.5). For Eq. (8.69), this procedure gives

$$\hat{\Gamma}_{2pq}F_{pq}+\sum^{G,H,T,U}\Gamma_{1ghtupq}q_{gh}q_{tu}-\sum^{G,H}\Gamma_{4ghpq}q_{gh}=0, \tag{8.71}$$

where

$$\hat{\Gamma}_{2pq}=\int_0^a\int_0^b\left(P_{11}\frac{\mathrm{d}^4\Theta_p}{\mathrm{d}x^4}\Psi_q+(P_{33}-2P_{12})\frac{\mathrm{d}^2\Theta_p}{\mathrm{d}x^2}\frac{\mathrm{d}^2\Psi_q}{\mathrm{d}y^2}+P_{22}\Theta_p\frac{\mathrm{d}^4\Psi_q}{\mathrm{d}y^4}\right)\Theta_p\Psi_q\mathrm{d}y\mathrm{d}x$$
$$=P_{11}\xi_p^4ab+\frac{\xi_p\eta_q}{ab}(P_{33}-2P_{12})(2-\xi_p)(2-\eta_q)+P_{22}\eta_q^4ab$$

and $\Gamma_{1ghtupq}$ and Γ_{4ghpq} are given by Eqs. (8.43) and (8.60) respectively. Note that in the calculation of $\hat{\Gamma}_{2pq}$, the following mode-shape relationships have been used

$$\int_0^L\phi_n^2\mathrm{d}x=L,\qquad\int_0^L\frac{\mathrm{d}^4\phi_n}{\mathrm{d}x^4}\phi_n\mathrm{d}x=\lambda_n^4L,\qquad\int_0^L\frac{\mathrm{d}^2\phi_n}{\mathrm{d}x^2}\phi_n\mathrm{d}x\approx\frac{\lambda_n}{L}(2-\lambda_n),$$

where ϕ_n is the mode-shape (X_m, Y_n, Θ_r or Ψ_s in this example), λ_n is the corresponding root of Eq. (8.67) (α_m, β_n, ξ_r, or η_s here) and L is the length (a or b here) and the approximate relationship in the second equation is taken from Blevins (1979). Using the same approach Eq. (8.70) becomes

$$\ddot{q}_{ij}+\hat{\omega}_{ij}^2q_{ij}-\sum^{M,N,R,S}\hat{\Gamma}_{3mnrsij}F_{rs}q_{mn}+\sum^{R,S}\hat{\Gamma}_{5rsij}F_{rs}=0, \tag{8.72}$$

where the natural frequency, $\hat{\omega}_{ij}$, is given by

$$\hat{\omega}_{ij}^2 = \frac{1}{\rho h}\left(D_{11}\alpha_i^4 + 2\frac{\alpha_i\beta_j}{(ab)^2}(D_{12}+2D_{33})(2-\alpha_i)(2-\beta_j) + D_{22}\beta_j^4\right),$$

and $\hat{\Gamma}_{3mnrsij} = \Gamma_{3mnrsij}/4$, where $\Gamma_{3mnrsij}$ is given by Eq. (8.46) and $\hat{\Gamma}_{5rsij} = \Gamma_{5rsij}/4$, where Γ_{5rsij} is given by Eq. (8.62). Note that the change in the constants $\Gamma_{3mnrsij}$ and Γ_{5rsij} between the original pinned-pinned derivation and the current free-free derivation is due to the fact that the equation has been divided through by $\rho h \int_0^a \int_0^b X_i^2 Y_j^2 \mathrm{d}y\mathrm{d}x$ and that this term simplifies to ρhab for the free-free mode-shapes whereas it was $\rho hab/4$ for the pinned-pinned mode-shapes.

Finally the expression for F_{rs} from Eq. (8.71) is substituted into Eq. (8.72) to obtain

$$\ddot{q}_{ij}+\hat{\omega}_{ij}^2 q_{ij}+\sum^{R,S,G,H}\frac{q_{gh}}{\hat{\Gamma}_{2rs}}\left[\left(\hat{\Gamma}_{5rsij}-\sum^{M,N}\hat{\Gamma}_{3mnrsij}q_{mn}\right)\left(\Gamma_{4ghrs}-\sum^{T,U}\Gamma_{1ghturs}q_{tu}\right)\right]=0, \quad (8.73)$$

which represents the nonlinear modal equation (without damping or forcing) for the composite plate. This is of a similar form to Eq. (8.63), but with modified coefficients $\hat{\Gamma}$ due to the change in material and boundary conditions.

It is worth pointing out that, although the Γ and $\hat{\Gamma}$ coefficients have analytical expressions, they are based on the assumption that the plate modes can be approximated as a combination of two (or more) beam modes. As a result, if Eq. (8.73) were used to model a physical system, it would *not* be expected that computed values for the Γ and $\hat{\Gamma}$ coefficients would necessarily provide the basis for a high accuracy model. In addition, because of the summation assumption in the derivation, Eq. (8.73) gives *all possible combinations* of additional linear, quadratic and cubic terms. In a physical system only a few of these terms would be significant, and so when Eq. (8.73) is used as a model, two things need to be done. Firstly, use experimental observation combined with physical reasoning to include only the terms which are required to capture the behaviour of the physical system. Secondly, use the experimental observations to identify the Γ and $\hat{\Gamma}$ coefficients, so that the model accurately simulates the experimental behaviour. This approach will be demonstrated in the next subsection, where a reduced-order, three-mode model, is derived from Eq. (8.73) and the experimental observations.

8.5.4 Three-Mode Model

Equation (8.73) can be used as the basis for a reduced order model of the physical-vibration behaviour of the shell described in Sect. 8.5.1. From the physical behaviour, no more than two modes were observed in each plate direction, so let $K = M = 1$ and $L = N = 1$ (and then of course $G = H = I = J = R = S = 1$ as well) and as the index starts at zero this will allow two modes in each direction. The experimental results show that the linear resonances and the $1/2$ subharmonic are the most significant phenomena. Therefore the simplest model will include just linear resonance terms and quadratic terms. This leaves equations of the form

$$\ddot{q}_{kl} + \omega_{kl}^2 q_{kl} + \chi_{klmn} q_{mn} q_{kl} = 0,$$

where the χ_{klmn} are constants to be identified from the experimental results. For $0 \leq k = m = l = n \leq 1$, this gives four modal equations

$$\begin{aligned}
\ddot{q}_{00} + \omega_{00}^2 q_{00} + \chi_{0000} q_{00}^2 + \chi_{0001} q_{01} q_{00} + \chi_{0010} q_{10} q_{00} + \chi_{0011} q_{11} q_{00} &= 0,\\
\ddot{q}_{01} + \omega_{01}^2 q_{01} + \chi_{0100} q_{11} q_{01} + \chi_{0101} q_{01}^2 + \chi_{0110} q_{10} q_{01} + \chi_{0111} q_{11} q_{01} &= 0,\\
\ddot{q}_{10} + \omega_{10}^2 q_{10} + \chi_{1000} q_{00} q_{10} + \chi_{1001} q_{01} q_{10} + \chi_{1010} q_{10}^2 + \chi_{1011} q_{11} q_{10} &= 0,\\
\ddot{q}_{11} + \omega_{11}^2 q_{11} + \chi_{1100} q_{00} q_{11} + \chi_{1101} q_{01} q_{11} + \chi_{1110} q_{10} q_{11} + \chi_{1111} q_{11}^2 &= 0.
\end{aligned} \tag{8.74}$$

The total out-of-plane displacement at a point x, y on the plate is then approximated by

$$w(x, y) \approx X_0(x) Y_0(y) q_{00} + X_0(x) Y_1(y) q_{01} + X_1(x) Y_0(y) q_{10} + X_1(x) Y_1(y) q_{11}.$$

However, the q_{00} displacement corresponds to a rigid body beam mode in both x and y directions, so this can be excluded from the vibration response, and the first of Eq. (8.74) can be neglected. The remaining three modes are shown schematically in Fig. 8.13, and are from top to bottom the (0,1), (1,0) and (1,1) modes. These are very close to the modes observed at the experimental resonance peaks. Specifically, at the first resonance peak (as frequency increases) in Figs. 8.10 and 8.11 the (1,0) mode is observed, at the second main resonance in Figs. 8.10 and 8.11, the (1,1) mode occurs. At the first resonance peak in Fig. 8.12 the mode (0,1) is observed.

Now consider which nonlinear terms are most important for the subharmonic resonance observed in the experimental data. First of all, the $1/2$ subharmonic resonance was observed for the first mode in the x direction, so the equation for the second resonance q_{11} will be assumed to have no nonlinear coupling terms. Second, coupling of the subharmonic was observed between the first modes in x and y (q_{10}, q_{01}) and the second mode in x (q_{11}). Finally for $1/2$ subharmonic resonance, the q_{ij}^2 terms are key to generating a subharmonic response. As a result, all the χ coefficients will be taken as zero in this example, except χ_{0101}, χ_{1010}, χ_{1011} and χ_{0111}, which means that the system of modal equations becomes

$$\begin{aligned}
\ddot{q}_{10} + \omega_{10}^2 q_{10} + \chi_{1010} q_{10}^2 + \chi_{1011} q_{11} q_{10} &= 0,\\
\ddot{q}_{11} + \omega_{11}^2 q_{11} &= 0,\\
\ddot{q}_{01} + \omega_{01}^2 q_{01} + \chi_{0101} q_{01}^2 + \chi_{0111} q_{11} q_{01} &= 0,
\end{aligned}$$

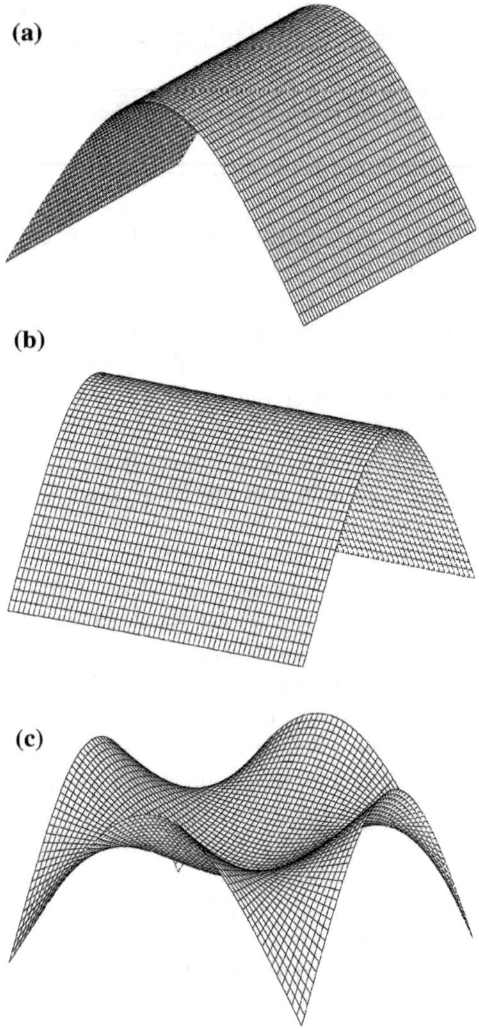

Fig. 8.13 Three-dimensional plots corresponding to the **a** (0,1), **b** (1,0) and **c** (1,1) modes for the curved composite shell

where the order of the equations is now related to the appearance of the resonance peaks in the frequency response diagram: first in x, second in x and first in y directions respectively.

Finally, to make the model physically realistic, damping and forcing terms are added to each mode, and a simplified notation is introduced such that

$$\ddot{q}_{x_1} + 2\zeta_{x_1}\omega_{x_1}\dot{q}_{x_1} + \omega_{x_1}^2 q_{x_1} + \Upsilon_1 q_{x_1}^2 + \Upsilon_2 q_{x_1} q_{x_2} = \frac{f_{x1}}{m} F_o \sin \Omega t,$$
$$\ddot{q}_{x_2} + 2\zeta_{x_2}\omega_{x_2}\dot{q}_{x_2} + \omega_{x_2}^2 q_{x_2} = \frac{f_{x2}}{m} F_o \sin \Omega t, \quad (8.75)$$
$$\ddot{q}_{y_1} + 2\zeta_{y_1}\omega_{y_1}\dot{q}_{y_1} + \omega_{y_1}^2 q_{y_1} + \Upsilon_3 q_{y_1}^2 + \Upsilon_4 q_{y_1} q_{x_2} = \frac{f_{y1}}{m} F_o \sin \Omega t,$$

where $q_{x_1} = q_{11}$ is the modal displacement for the first resonant mode in the x direction frequency response diagrams (Figs. 8.10 and 8.11) with natural frequency $\omega_{x_1} = \omega_{11} = 17.6$ Hz, $q_{x_2} = q_{12}$ is the modal displacement for the second resonant mode in the frequency response diagrams with natural frequency $\omega_{x_2} = \omega_{12} = 45.4$ Hz, $m = \rho hab$ is the mass of the plate,[14] $\Upsilon_1 = \chi_{1010}$ is the nonlinear quadratic coefficient, and $\Upsilon_2 = \chi_{1011}$ is the coupling coefficient between q_{x_1} and q_{x_2}. Similarly q_{y_1} is the modal displacement of the first resonant mode in the y direction frequency response diagrams (Fig. 8.12) with natural frequency $\omega_{y_1} = \omega_{21} = 19.6$ Hz, $\Upsilon_3 = \chi_{0101}$ is the nonlinear quadratic coefficient, and $\Upsilon_4 = \chi_{0111}$ is the coupling coefficient between q_{y_1} and q_{x_2}.

Note that forcing and damping for each mode have been added. The forcing is harmonic and F_o is the driving force amplitude, Ω is the forcing frequency, f_{x_1}, f_{x_2} and f_{y_1} are the modal participation factors for the modal equations (see Sect. 6.1.3). The damping factors for the modes are ζ_{x_1}, ζ_{x_2} and ζ_{y_1} respectively.

The coefficients in Eqs. (8.75) were selected so that the simulated model matched the experimental frequency response diagrams, Figs. 8.10, 8.11 and 8.12, using a similar modal testing approach to that described by Ewins (2000). Additional details of the method for identifying each of the coefficients is given in Arrieta et al. (2009).

Simulated frequency response diagrams can be obtained using numerical integration (as discussed in Chap. 2) combined with the stroboscopic sampling procedure. The simulated frequency response diagram obtained for point P_x for a forcing amplitude of 5 N is shown in Fig. 8.14. The simulation exhibits the same features as the experimental data (Fig. 8.11) including the two modes of vibration at the measured linear resonance frequencies, at 17.6 and 45.4 Hz and the subharmonic resonance at 35 Hz. In addition, the amplitude of the subharmonic oscillation coincides with the experimental results shown in Sect. 8.5.

The simulated frequency response diagram for point P_y is shown in Fig. 8.15. As before, the simulation captures the key dynamic features observed in the experimental response (Fig. 8.12), such as the natural frequencies and subharmonic response.

[14] Note that, in the derivation of the plate and shell equations, simply-supported boundary conditions were assumed on all sides, and the Γ constants were obtained by dividing through by $\rho h \int_0^a \int_0^b X_m Y_n X_i Y_j \mathrm{d}y\mathrm{d}x$. For a free-free plate this expression becomes ρhab.

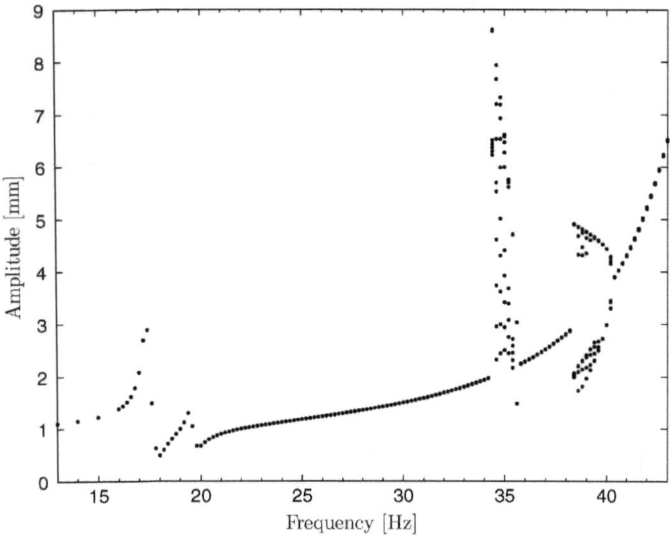

Fig. 8.14 Numerically simulated frequency response diagram for the curved (x) direction using Eq. (8.75). Reproduced with kind permission from Arrieta et al. (2009)

8.5.5 Subharmonic Resonance

The three-mode model can be used to find the behaviour of the subharmonic resonance as more than one system parameter is varied. Of particular interest is the boundary marking the onset of a subharmonic response for a range of forcing amplitudes. To do this, Eqs. (8.75) are scaled and then a first-order averaging process (as described in Chap. 4) is used to derive first-order differential equations for the response. These equations are then analysed to study at what point each mode starts to have subharmonic response when forced with harmonic excitation—further details can be found in Gonzalez-Buelga et al. (2008). The boundaries obtained are plotted in parameter space to indicate the regions where the subharmonic does not appear (corresponding to stability of the basic response) and where it does appear (thought of as instability in the basic response) for each of the vibration modes under consideration.

The equations of motion of the system, Eqs. (8.75), are first scaled so that the dynamics are dominated by the undamped linear response. This is achieved by introducing a small parameter ε and arranging the equations to take the following standard Lagrange form

$$\ddot{z}(t) + \omega^2 z(t) = \varepsilon F(z(t), \dot{z}(t), t). \tag{8.76}$$

To study the subharmonic response for mode (1,0) the forcing frequency must be close to twice the natural frequency, ω_{x_1}. This means that the forcing frequency can be expressed as $\Omega = 2\omega_{x_1}(1 + \varepsilon\mu)$, where μ is the frequency detuning and ε is a

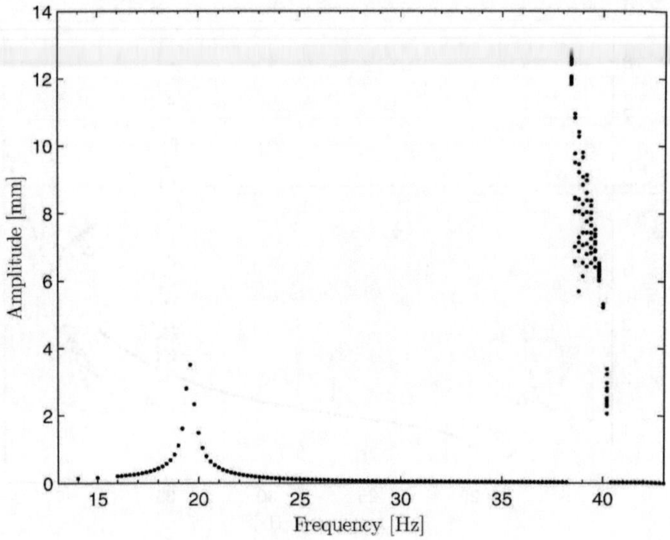

Fig. 8.15 Numerically simulated frequency response diagram for flat (y) direction. Reproduced with kind permission from Arrieta et al. (2009)

small parameter. Introducing the time transform

$$\tau = (1 + \varepsilon\mu)t$$

into Eq. (8.77) enables a scaled and transformed equation of motion for the first mode in x to be written as

$$x_1'' + \omega_{x_1}^2 x_1 + \varepsilon[2\zeta_{x_1}\omega_{x_1}x_1' + 2\mu\omega_{x_1}^2 x_1 + \Upsilon_1 x_1^2 + \Upsilon_2 x_1 x_2 - \frac{f_{x_1}}{m} Fo \sin 2\omega\,\tau] = \mathscr{O}(\varepsilon^2), \tag{8.77}$$

where, to simplify notation, $x_1 = q_{x_1}$ and $\bullet'$ represents the derivative with respect to scaled time τ.

Using the transformation

$$x(t) = x_a \cos(\omega t) + x_b \sin(\omega t),$$
$$\dot{x}(t) = -x_a\omega \sin(\omega t) + x_b\omega \cos(\omega t),$$

(see Sect. 4.3 or Verhulst (1996)) and the Lagrange standard form, Eq. (8.76), gives expressions of the form

$$\dot{x}_a = -\frac{\varepsilon}{\omega} F(\dot{x}(t), x(t), t) \sin(\omega t),$$
$$\dot{x}_b = \frac{\varepsilon}{\omega} F(\dot{x}(t), x(t), t) \cos(\omega t). \tag{8.78}$$

Approximate solutions for Eq. (8.77) can be obtained using the averaging procedure described in Chap. 4. As mode (1,0) is lightly damped, its response away from its natural frequency is very small (i.e. close to zero). Therefore, zero response is assumed for the approximate solutions away from resonance. Using Eq. (8.78) for the scaled equation Eq. (8.77) and applying averaging over the scaled time period $\frac{2\pi}{\omega_{x_1}}$ in τ (which corresponds to a time period of two cycles of frequency Ω in real-time t) the averaged equations for mode (1,0) can be written as

$$\begin{aligned} x{1'_{a_{avg}}} &= -\frac{\varepsilon}{\omega_{x1}}[\zeta_{x1}\omega_{x1}^2 x1_{a_{avg}} - \frac{\Upsilon_2}{4}x1_{a_{avg}}x2_{b_{avg}} + \frac{\Upsilon_2}{4}x1_{b_{avg}}x2_{a_{avg}} - \mu\omega_{x1}^2 x1_{b_{avg}}], \\ x{1'_{b_{avg}}} &= \frac{\varepsilon}{\omega_{x1}}[-\zeta_{x1}\omega_{x1}^2 x1_{b_{avg}} - \frac{\Upsilon_2}{4}x1_{b_{avg}}x2_{b_{avg}} - \frac{\Upsilon_2}{4}x1_{a_{avg}}x2_{a_{avg}} - \mu\omega_{x1}^2 x1_{a_{avg}}], \end{aligned} \tag{8.79}$$

where the subscript a and b refer to cosine and sine components of the Lagrange form, Eq. (8.78), respectively and again $\bullet'$ represents the derivative with respect to scaled time τ. The subscript avg indicates that the terms have been averaged over a cycle of ω_{x_1} in the scaled τ for mode (1,0).

Similar averaged equations can be obtained for the (0,1) mode (with $q_{y_1} = y_1$), where the frequency detuning $\Omega = 2w_{y_1}(1 + \varepsilon v)$ is applied, with the corresponding time transform

$$T = (1 + \varepsilon v)t,$$

giving

$$\begin{aligned} y{1^*_{a_{avg}}} &= -\frac{\varepsilon}{\omega_{y1}}[\zeta_{y1}\omega_{y1}^2 y1_{a_{avg}} - \frac{\Upsilon_4}{4}y1_{a_{avg}}y2_{b_{avg}} + \frac{\Upsilon_4}{4}y1_{b_{avg}}y2_{a_{avg}} - v\omega_{y1}^2 y1_{b_{avg}}], \\ y{1^*_{b_{avg}}} &= \frac{\varepsilon}{\omega_{y1}}[-\zeta_{y1}\omega_{y1}^2 y1_{b_{avg}} - \frac{\Upsilon_4}{4}y1_{b_{avg}}y2_{b_{avg}} - \frac{\Upsilon_4}{4}y1_{a_{avg}}y2_{a_{avg}} - v\omega_{y1}^2 y1_{a_{avg}}], \end{aligned} \tag{8.80}$$

where $\bullet^*$ represents the derivative with respect to scaled time T. Note that the averaging is now applied over the scaled time period $\frac{2\pi}{\omega_{y_1}}$ in T, equivalent to two cycles of Ω in real-time t.

The semi-trivial solution is defined as a response in mode (1,1) with no response in modes (1,0) or (0,1). The boundary of this solution can be found by considering the localized stability of each of the modes, (1,0) and (0,1), about the zero amplitude response. For mode (1,0) the averaged system, Eq. (8.79), can be expressed in matrix form as

$$\begin{pmatrix} x{1'_{a_{avg}}} \\ x{1'_{b_{avg}}} \end{pmatrix} = \varepsilon \begin{bmatrix} \frac{\Upsilon_2}{4\omega_{x_1}}x2_{b_{avg}} - \zeta_{x1}\omega_{x1} & \mu\omega_{x1} - \frac{\Upsilon_2}{4\omega_{x_1}}x2_{a_{avg}} \\ -\mu\omega_{x1} - \frac{\Upsilon_2}{4\omega_{x_1}}x2_{a_{avg}} & -\frac{\Upsilon_2}{4\omega_{x_1}}x2_{b_{avg}} - \zeta_{x1}\omega_{x1} \end{bmatrix} \begin{pmatrix} x1_{a_{avg}} \\ x1_{b_{avg}} \end{pmatrix}. \tag{8.81}$$

The localized stability about the zero response of mode (1,0) can be found by studying the eigenvalues of the matrix in Eq. (8.81), which are given by

$$\lambda^2 + 2\zeta_{x_1}\omega_{x_1}\lambda + (\zeta_{x_1}\omega_{x_1})^2 + (\mu\omega_{x_1})^2 - \frac{(\Upsilon_2 x_{2_{avg}})^2}{(4\omega_{x_1})^2} = 0, \tag{8.82}$$

where $x_{2_{avg}}^2 = x_{2_{aavg}}^2 + x_{2_{bavg}}^2$. Given that, for physically meaningful values, $\zeta_{x_1}\omega_{x_1}^2 > 0$, the boundary of local instability of the zero amplitude response occurs when $\lambda = 0$. This corresponds to the region in which a non-zero response at frequency ω_{x_1}, the subharmonic frequency, will occur. For $\lambda = 0$ the characteristic equation, Eq. (8.82), simplifies to

$$x_{2_{avg}} \geq \frac{4\omega_{x_1}^2}{\Upsilon_2}\sqrt{\zeta_{x_1}^2 + \mu^2}.$$

At the lower side of the boundary there is only a response in the (1,1) mode. The amplitude of this response can be found using Eq. (8.75), at the forcing frequency $\Omega = 2\omega_{x_1}(1+\varepsilon\mu)$. Using this (1,1) mode response amplitude, the force that triggers the appearance of the subharmonic response for mode (1,0) is found to be

$$F_o \geq \frac{4\omega_{x_1}^2}{\Upsilon_2}\frac{f_{x_2}}{m}\sqrt{\zeta_{x_1}^2 + \mu^2}\sqrt{(\omega_{x_2}^2 - \Omega^2)^2 + (2\zeta_{x_2}\omega_{x_2}\Omega)^2}. \tag{8.83}$$

The same stability analysis approach to mode (0,1) averaged equations, Eq. (8.80), shows that the $1/2$ subharmonic will appear when

$$x_{2_{avg}} \geq \frac{4\omega_{y_1}^2}{\Upsilon_4}\sqrt{\zeta_{y_1}^2 + \nu^2}.$$

Also, as with mode (1,0), the linear (1,1) modal response at which a subharmonic of mode (0,1) appears can be related to the forcing by

$$F_o \geq \frac{4\omega_{y_1}^2}{\Upsilon_4}\frac{f_{x_2}}{m}\sqrt{\zeta_{y_1}^2 + \nu^2}\sqrt{(\omega_{x_2}^2 - \Omega^2)^2 + (2\zeta_{x_2}\omega_{x_2}\Omega)^2}. \tag{8.84}$$

Equations (8.83) and (8.84) give the theoretical force amplitude required to trigger subharmonic oscillations of the (1,0) and (0,1) modes respectively. The experimental subharmonic response range for a given forcing amplitude can be directly measured from experimental frequency response diagrams for various levels of forcing. The lower and higher frequency limits for the subharmonic response are obtained and plotted for each experimental forcing amplitude for both modes (1,0) and (0,1). These results are plotted as stars and circles for modes (1,0) and (0,1) respectively in Fig. 8.16. There is good agreement between the theoretical and experimental

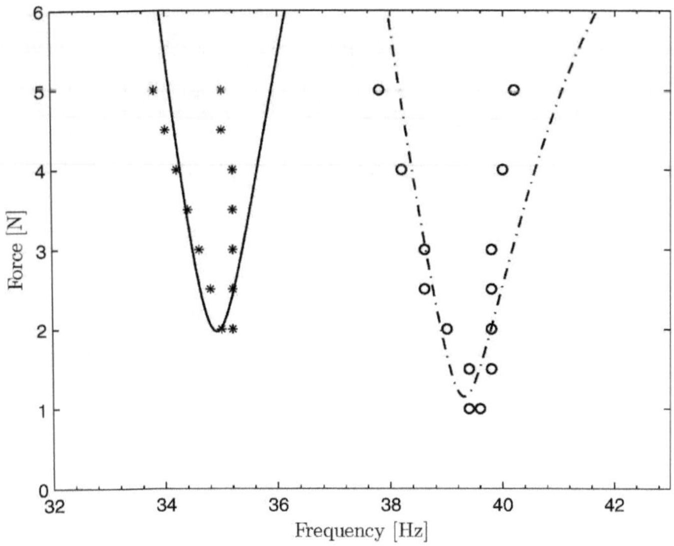

Fig. 8.16 Stability Boundaries: comparison between the theoretical predictions (*lines*) and the experimental measurements (*points*) for the subharmonic resonance of modes (1,0) and (0,1). Reproduced with kind permission from Arrieta et al. (2009)

results—see Arrieta et al. (2009) for further details. This type of *tongue* shaped curve is often referred to as an Arnold tongue, Arnold (1988).

8.6 Adaptive Structure Applications

This final section gives a brief description of how some of the techniques discussed so far are being extended to future adaptive structure applications. In particular, composite plates and shells are being developed for use in a range of adaptive structure applications. For example, vibration isolation mounts (Shaw et al. 2013), energy harvesters (Arrieta et al. 2010), and morphing aerofoils (Wildschek et al. 2010).

In this section a particular type of *bi-stable* composite shell is considered which can be used as a type of hinge, for example, potentially allowing a aerofoil structure to move a trailing edge from one position to another by deflecting the structure rather than using a rigid body hinging mechanism, see for example Diaconu et al. (2008), Kuder et al. (2013) and references therein. The actuation can be provided by a variety of means, but conventional electrical or hydraulic actuation systems offer the most practical solution (Gomis-Bellmunt and Campanile 2000). Alternatives, such as surface mounted macro-fibre composite patches with a control algorithm designed to actuate between stable states can be used for very low force applications. This type

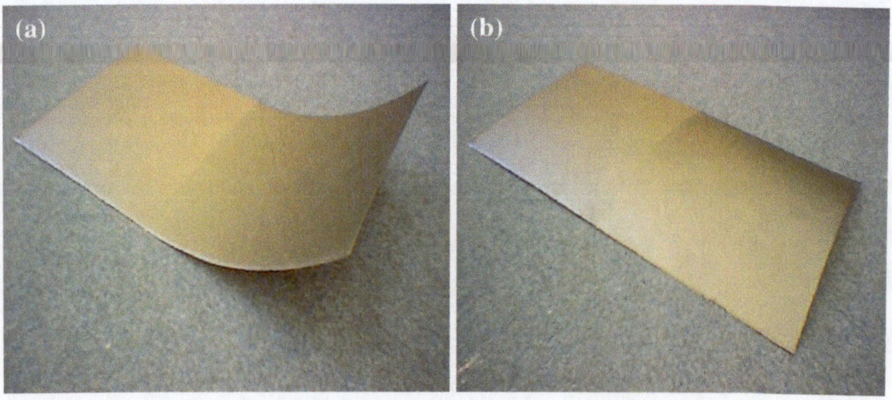

Fig. 8.17 Flat plate and bi-stable plate manufactured together, **a** bi-stable state 1, and **b** bi-stable state 2. Reproduced with kind permission from Mattioni et al. (2006)

of bi-stable composite hinge mechanism has several potential applications, of which aerospace related morphing applications have been the most widely investigated.[15]

8.6.1 Multi-form Shell Structures

The composite shell used in the case study in Sect. 8.5 is made of a polymer-matrix, fibre-reinforced composite which, during the manufacturing process is cooled from a high temperature. Thermally induced stresses occur during cooling, and as a result of these stresses the plate is a cylindrical shell shape when fully cooled. In fact, the cooled static equilibrium shape has been shown to be the shape which minimizes the potential energy of the laminate, see Hyer (1998) (see also Chap. 7 of Wagg et al. (2007)) during cooling. This process is nonlinear, and the Kirchhoff strain assumptions described in Sect. 8.1.2 are taken as an appropriate model for the shell behaviour. However, the shell can also have other potential stable configurations.[16] In the case study in Sect. 8.5, there is one other stable configuration and so the shell is said to be *bi-stable*. More generally, composite laminates can be manufactured which have *multi-stable* states.

[15] Note that morphing for aerospace structures has been investigated using other mechanisms, such as Shape Memory Alloys, see for example Calkins and Mabe (2010), Barbarino et al. (2011) and references therein.

[16] Note that bi-stability is not unique to composite materials, for example steel arches and dome-like shells can have bi-stable behaviour.

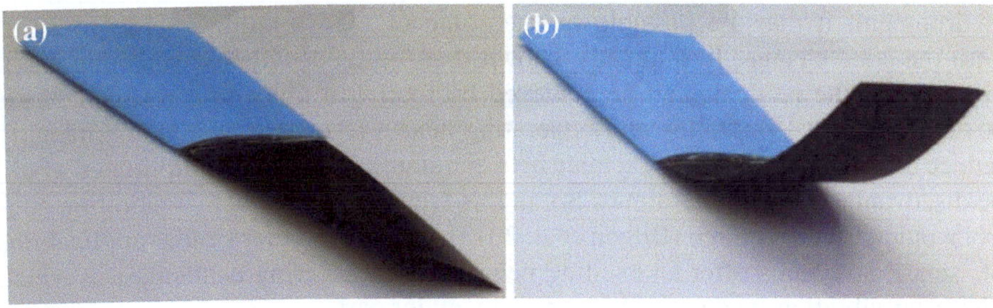

Fig. 8.18 Bi-stable plate applied to a morphing wing concept **a** winglet lowered, and **b** winglet raised. Reproduced with kind permission from Mattioni et al. (2006)

An example is shown in Fig. 8.17, where a bi-stable plate has been fabricated with a flat plate joined to the left-hand edge to form a rectangular plate-like structure.[17] In Fig. 8.17a the bi-stable plate (on the right of the sub-figure) is in the curvature up position, and in Fig. 8.17b the bi-stable plate is in the down (or flatter) position. This idea has been taken a stage further, and used to construct a small scale prototype of a morphing winglet, as shown in Fig. 8.18.

The moment required to make the bi-stable composite change state has been found to be, for example, in the range $1.186 - 1.243$ Nm for a 254 mm $\times$ 254 mm $[90_4/0_4]_T$ specimen, see Schultz and Hyer (2003) (see also Chap. 7 of Wagg et al. (2007)). When the change in state occurs, the behaviour is almost identical to the *snap-through* system discussed in Chap. 2, shown in Fig. 2.20. This is because, after a certain level of deflection, the sign of the stiffness term appears to suddenly reverse, propelling the system into the other state. This idea of negative linear stiffness is considered in Example 2.7, where the snap-through system is modelled as a Duffing oscillator. If required, the model developed for the composite shell, Eq. (8.73), can be applied to the large deflection case that occurs physically with snap-through.

The potential of this type of bi-stable material to create future adaptive structures where multiple states are required has been investigated by several authors (See Chapter references). The process of a structure changing from one shape to another is called *morphing*. In some aerospace literature this is used to describe just hinged wing aircraft, but more recently the term has become used more widely to describe any shape change in a structure. Morphing can be passive or active. A passive morphing shape change occurs as external forces alter the force distribution on the structure, see Baker and Friswell (2009). Active morphing is achieved by using active control techniques, such as those described in Chap. 3 and Gomis-Bellmunt and Campanile (2000).

Techniques for actuating bi-stable plates using both piezoelectric actuators and shape memory alloys have been described by Hyer et al. in Chap. 7 of Wagg et al. (2007). These techniques are designed to overcome the static moment required to change between the two stable states. When operating these structures in a dynamic

[17] For details see: F. Mattioni, P. M. Weaver, K. D. Potter & M. I. Friswell, Analysis of thermally induced multistable composites. *International Journal of Solids and Structures*. 45, 2008, 657–675.

environment, it may be possible to use some of the vibration energy to assist with the state change. It is already known that repeated dynamic snapping (non-periodic) of the laminate can be achieved by forcing it close to a resonance, see Arrieta et al. (2013). Deliberately operating near a resonance would significantly reduce the moment required to actuate between states, but the high amplitudes would be disadvantageous at other times. So, in this type of scenario active vibration control would also be required (Bilgen et al. 2013). These and other similar applications of 'smart' structures offer an exciting new set of engineering challenges in which nonlinear vibrations and control will play an major part.

8.7 Chapter Notes

The theory of elastic plates and shells follows from work carried out by Love (1892) and Timoshenko (1940). The classical small-deflection theories for plates and shells described here follows the approach set out by Szilard (1974) and Soedel (2004). The discussion of plates and shells with axial loading uses a similar approach to that taken by Virgin (2007). Nonlinear aspects of plate and shell theory are nicely set out in Chia (1980) and Amabili (2008), and the derivations here broadly follow this approach. For the case study, the experiments and original model derivation was carried out by Andres F. Arrieta—see Arrieta et al. (2009) for further details. The discussion on bi-stable plates comes mainly from the original work of Hyer et al. that is summarised in Chap. 7 of Wagg et al. (2007), see also Hyer (1998) and Schultz and Hyer (2003). Following this, several authors have worked on both the dynamic analysis of these types of plates and their application to morphing structures. The dynamic analysis has been studied by Arrieta et al. (2009, 2011a, 2011b, 2011c). Morphing applications are described by Schultz and Hyer (2003), Mattioni et al. (2006), Diaconu et al. (2008), Baker and Friswell (2009), Murray and Gandhi (2010), Gomis-Bellmunt and Campanile (2000), Raither et al. (2012), Arrieta et al. (2013), Kuder et al. (2013), and Bilgen et al. (2013) amongst others. Other applications for bi-stable plates have been considered, such as energy harvesting Arrieta et al. (2010) and vibration isolation mounts Shaw et al. (2013).

References

Amabili, M. (2008). *Nonlinear vibrations and stability of shells and plates*. Cambridge: Cambridge University Press.

Arnold, V. I. (1988). *Geometrical methods in the theory of ordinary differential equations*. Springer: New York.

Arrieta, A. F., Wagg, D. J., & Neild, S. A. (2009). Nonlinear dynamics of a bistable composite laminate plate with applications to adaptive structures. *Nonlinear Dynamics*, *58*, 259–272.

Arrieta, A., Hagedorn, P., Erturk, A., & Inman, D. (2010). A piezoelectric bistable plate for nonlinear broadband energy harvesting. *Applied Physics Letters*, *97*, 104102.

Arrieta, A., Spelsberg-Korspeter, G., Hagedorn, P., Neild, S. A., & Wagg, D. J. (2011a). Low order model for the dynamics of bi-stable composite plates. *Journal of Intelligent Material Systems and Structures*, *22*, 2025–2043.

Arrieta, A. F., Wagg, D. J., & Neild, S. A. (2011b). Dynamic snap-through for morphing of bi-stable composite plates. *Journal of Intelligent Material Systems & Structures*, *22*(2), 103–112.

Arrieta, A. F., Neild, S. A., & Wagg, D. J. (2011c). On the cross-well dynamics of a bi-stable composite plate. *Journal of Sound & Vibration*, *330*(14), 3424–3441.

Arrieta, A. F., Bilgen, O., Friswell, M. I., & Hagedorn, P. (2013). Dynamic control for morphing of bi-stable composites. *Journal of Intelligent Material Systems and Structures*, *24*(3), 266–273.

Baker, D., & Friswell, M. I. (2009). Determinate structures for wing camber control. *Smart Materials & Structures*, *18*(3), 035014.

Barbarino, S., Pecora, R., Lecce, L., Concilio, A., Ameduri, S., & De Rosa, L. (2011). Airfoil structural morphing based on SMA actuator series: numerical and experimental studies. *Journal of Intelligent Material Systems & Structures*, *22*(10), 987–1004.

Bilgen, O., Arrieta, A. F., Friswell, M. I., & Hagedorn, P. (2013). Dynamic control of a bistable wing under aerodynamic loading. *Smart Materials and Structures*, *22*(2), 025020.

Blevins, R. D. (1979). *Formulas for natural frequency and mode shape*. New York: Van Nostrand Reinhold.

Calkins, F. T., & Mabe, J. H. (2010). Shape memory alloy based morphing aerostructures. *Journal of Mechanical Design*, *132*(11), 111012.

Cartmell, M. (1990). *Introduction to linear, parametric and nonlinear vibrations*. London: Chapman and Hall.

Chia, C. -Y. (1980). *Nonlinear analysis of plates*. New York: McGraw-Hill.

Diaconu, C. G., Weaver, P. M., & Mattioni, F. (2008). Concepts for morphing airfoil sections using bi-stable laminated composite structures. *Thin-Walled Structures*, *46*(6), 689–701.

Ewins, D. J. (2000). *Modal testing*. Baldock: Research Studies Press.

Gomis-Bellmunt, O., & Campanile, L. F. (2010). *Design rules for actuators in active mechanical systems*. New York: Springer

Gonzalez-Buelga, A., Neild, S., Wagg, D., & Macdonald, J. (2008). Modal stability of inclined cables subjected to vertical support excitation. *Journal of Sound and Vibration*, *318*, 565–579.

Hyer, M. W. (1998). *Stress analysis of fibre-reinforced composite materials*. New York: Mc-Graw Hill.

Inman, D. J. (2006). *Vibration with control*. New York: Wiley.

Jordan, D. W., & Smith, P. (1999). *Nonlinear ordinary differential equations; An introduction to dynamical systems* (3rd ed.). Oxford: Oxford University Press.

Kuder, I. K., Arrieta, A. F., Raither, W. E., & Ermanni, P. (2013). Variable stiffness material and structural concepts for morphing applications. *Progress in Aerospace Sciences*, *63*, 33–55.

Love, A. (1892). *A treatise on the mathematical theory of elasticity*. Cambridge: Cambridge University Press.

Mattioni, F., Weaver, P. M., Potter, K., & Friswell, M. I. (2006). Multi-stable composites application concept for morphing aircraft. In M. Bernadou, J. Cagnol, & R. Ohayon (Eds.), *Proceedings of the 16th international conference on adaptive structures and technologies* (pp. 45–52).

Murray, G., & Gandhi, F. (2010). Multi-layered controllable stiffness beams for morphing: Energy, actuation force, and material strain considerations. *Smart Materials & Structures*, *19*(4), 045002.

Raither, W., Bergamini, A., Gandhi, F., & Ermanni, P. (2012). Adaptive bending-twist coupling in laminated composite plates by controllable shear stress transfer. *Composites Part A: Applied Science and Manufacturing*, *43*(10), 1709–1716.

Schultz, M. R., & Hyer, M. W. (2003). Snap-through of unsymmetric cross-ply laminates using piezoceramic actuators. *Journal of Intelligent Material Systems and Structures*, *14*(12), 795–814.

Shaw, A. D., Neild, S. A., Wagg, D. J., Weaver, P. M., & Carrella, A. (2013). A nonlinear spring mechanism incorporating a bistable composite plate for vibration isolation. *Journal of Sound and Vibration*, *332*(24), 6265–6275.

Soedel, W. (2004). *Vibrations of shells and plates*. Boca Raton: CRC Press.
Strogatz, S. H. (2001). *Nonlinear dynamics and chaos*. USA: Perseus Books Group.
Szilard, R. (1974). *Theory and analysis of plates*. Englewood Cliffs: Prentice Hall.
Timoshenko, S. P. (1940). *Theory of plates and shells*. New York: McGraw-Hill.
Timoshenko, S. P. & Goodier J. N. (1970). *Theory of elasticity*. New York: McGraw-Hill.
Ventsel, E., & Krauthammer, T. (2001). *Thin plates and shells: Theory, analysis and applications*. New York: Marcel Dekker.
Verhulst, F. (1996). *Nonlinear differential equations and dynamical systems*. New York: Springer.
Virgin, L. N. (2007). *Vibration of axially-loaded structures.* Cambridge: Cambridge University Press.
Wagg, D., Bond, I., Weaver, P., & Friswell, M., editors (2007). *Adaptive structures: Engineering applications*. New York: Wiley.
Weaver Jr. W., Timoshenko, S. P., & Young, D. (1990). *Vibration problems in engineering*. New York: Wiley.
Wildschek, A., Havar, T., & Ploetner, K. (2010). An all-composite, all-electric, morphing trailing edge device for flight control on a blended-wing-body airliner. *Proceedings of the Institution of Mechanical Engineers Part G-Journal of Aerospace Engineering*, *224*(G1), 1–9.

Chapter 9
Solutions to Problems

Chapter 2

Problem 2.1

To derive the equation of motion for the system shown in Fig. 2.20a, first define the length of the spring as $L^* = \sqrt{b^2 + x^2}$, and then the force in each spring in the x direction is equal to $f_s = k(L^* - L)\sin\theta = k(L^* - L)\frac{x}{L^*}$. Then for both springs the force in the x direction becomes

$$2f_s = 2k(L^* - L)\frac{x}{L^*} = 2kx\left(\frac{\sqrt{b^2+x^2} - L}{\sqrt{b^2+x^2}}\right) = 2kx\left(1 - \frac{L}{\sqrt{b^2+x^2}}\right).$$

This assumes that the mass is constrained to act along the horizontal centreline of the system, so that both springs are compressed or stretched the same amount. Now applying Newton's second law to the motion of the mass in the x direction the equation of motion is found to be

$$m\ddot{x} + c\dot{x} + 2kx\left(1 - \frac{L}{\sqrt{b^2+x^2}}\right) = 0.$$

The next step is to show that the stiffness term can be simplified to the same form as the Duffing oscillator equation. Note first that

$$\frac{L}{\sqrt{b^2+x^2}} = \frac{L}{\sqrt{b^2}\sqrt{1+\frac{x^2}{b^2}}} = \frac{L}{b}\left(1 + \frac{x^2}{b^2}\right)^{-\frac{1}{2}}.$$

Using the binomial expansion

$$(1+z)^n = 1 + (n)(z) + \frac{n(n-1)}{2!}(z)^2 + \cdots,$$

D. Wagg and S. Neild, *Nonlinear Vibration with Control*,
Solid Mechanics and Its Applications 218, DOI 10.1007/978-3-319-10644-1_9

it can be shown that

$$\frac{L}{b}\left(1+\frac{x^2}{b^2}\right)^{-\frac{1}{2}} \approx \frac{L}{b}\left(1-\frac{1}{2}\frac{x^2}{b^2}+\cdots\right)$$

so that the equation of motion becomes

$$m\ddot{x}+c\dot{x}+2kx\left(1-\frac{L}{b}\left(1-\frac{1}{2}\frac{x^2}{b^2}\right)\right)=0.$$

Expanding the bracket, this becomes

$$m\ddot{x}+c\dot{x}+2k\left(1-\frac{L}{b}\right)x+kL\frac{x^3}{b^3}=0.$$

because $b < L$ it is conventional to redefine the coefficient of x to be

$$\mu = 2k\left(\frac{L}{b}-1\right),$$

which gives

$$m\ddot{x}+c\dot{x}-\mu x+\alpha x^3=0,$$

where $\alpha = kL/b^3$. This equation is the same as the Duffing oscillator used in the discussion of the system shown in Fig. 2.20a. In terms of when this approximation to the system dynamics will be a valid assumption, it can be seen from the binomial expansion that the series is truncated at one term. Therefore it will be reasonable for $\frac{x^2}{b^2} \ll 1$, because in this case $(\frac{x^2}{b^2})^p$ for $p = 2, 3, 4\ldots$ will result in much smaller terms.

Problem 2.2

The normal form of the Hopf bifurcation is usually written as

$$\begin{aligned}\dot{x} &= \mu x + y - x(x^2+y^2),\\ \dot{y} &= -x + \mu y - y(x^2+y^2).\end{aligned} \tag{9.1}$$

The question asks for this to be put this into polar coordinates, so first let $r = \sqrt{x^2+y^2}$ and $\theta = \arctan(y/x)$, then $r^2 = x^2+y^2$. Dividing by two and differentiating both sides

$$\frac{\mathrm{d}}{\mathrm{d}t}\left(\frac{r^2}{2}\right) = \frac{\mathrm{d}}{\mathrm{d}t}\left(\frac{x^2+y^2}{2}\right).$$

Using the chain rule gives

$$r\dot{r} = x\dot{x} + y\dot{y} \tag{9.2}$$

and substituting for $\dot{x}$ and $\dot{y}$ gives

$$\dot{r} = \frac{1}{r}\Big(\big(\mu x^2 + xy - x^2(x^2+y^2)\big) + \big(-xy + \mu y^2 - y^2(x^2+y^2)\big)\Big),$$

$$\dot{r} = \frac{1}{r}(\mu r^2 - r^4) = r(\mu - r^2).$$

Now

$$\dot{\theta} = \frac{\mathrm{d}}{\mathrm{d}t}\left(\arctan\left(\frac{y}{x}\right)\right).$$

Let $z(t) = y(t)/x(t)$ then using the chain rule

$$\dot{\theta} = \frac{\mathrm{d}}{\mathrm{d}t}\left(\arctan(z)\right) = \frac{1}{(1+z^2)}\dot{z}$$

$$\dot{z} = \frac{\mathrm{d}}{\mathrm{d}t}\left(\frac{y}{x}\right) = \frac{x\dot{y} - y\dot{x}}{x^2}$$

$$= \frac{1}{x^2}\left(-x^2 + \mu xy - xy(x^2+y^2) - \mu xy - y^2 + xy(x^2+y^2)\right)$$

$$= -\frac{(x^2+y^2)}{x^2}. \tag{9.3}$$

So

$$\dot{\theta} = \frac{\mathrm{d}}{\mathrm{d}t}\left(\arctan(z)\right) = \frac{1}{(1+\left(\frac{y^2}{x^2}\right))}\left(-\frac{(x^2+y^2)}{x^2}\right) = -\frac{(x^2+y^2)}{(x^2+y^2)} = -1.$$

Now in polar coordinates the (considerably simpler looking) system becomes

$$\dot{r} = r(\mu - r^2)$$
$$\dot{\theta} = -1.$$

Now consider the stability of the equilibrium point at the origin $(x, y) = (0, 0) \Rightarrow$ $r = 0, \theta =$ constant. The Jacobian of Eq. 9.1 is

$$\frac{\partial(f_1, f_2)}{\partial(x_1, x_2)} = \begin{bmatrix} \mu - 3x^2 - y^2 & 1 - 2yx \\ -1 - 2xy & \mu - x^2 - 3y^2 \end{bmatrix} = D_x f.$$

Finding the eigenvalues of the Jacobian from $|D_x f - I\lambda| = 0$ gives

$$(\mu - 3x^2 - y^2 - \lambda)(\mu - x^2 - 3y^2 - \lambda) - (-1 - 2xy)(1 - 2yx) = 0,$$

which can be written as

$$\lambda^2 - 2\lambda(\mu - 3x^2 - y^2) + (\mu - 3x^2 - y^2)^2 - (4x^2y^2 - 1) = 0. \tag{9.4}$$

Focusing on the equilibrium point at $(x, y) = (0, 0)$ these values are substituted into Eq. 9.4 to give

$$\lambda^2 - 2\lambda\mu + \mu^2 + 1 = 0.$$

From this an expression for the eigenvalues can be obtained to give

$$\lambda_{1,2} = -\left(\frac{-2\mu}{2}\right) \pm \frac{1}{2}\sqrt{(-2\mu)^2 - 4(\mu^2 + 1)} = \mu \pm \mathbf{i}.$$

Now it can be seen how the eigenvalues vary as μ goes from negative to positive.

- For $\mu < 0$, $\lambda_{1,2} = -\mu \pm \mathbf{i}$, stable.
- For $\mu = 0$, $\lambda_{1,2} = \pm \mathbf{i}$, bifurcation.
- For $\mu > 0$, $\lambda_{1,2} = \mu \pm \mathbf{i}$, unstable.

So, as μ increases through zero a Hopf bifurcation takes place, and the behaviour changes from stable to unstable. Note that at $\mu = 0$ the eigenvalues are purely imaginary.

Problem 2.3

The system is already in first-order form

$$\begin{aligned} \dot{x}_1 &= x_2 = f_1, \\ \dot{x}_2 &= x_1 - x_1^2 - \mu x_2 = f_2. \end{aligned}$$

By inspection, for $\mu > 0$, the equilibrium points for this system are $\underset{\sim}{x}_a^* = (x_1 = 0, x_2 = 0)$ and $\underset{\sim}{x}_b^* = (x_1 = 1, x_2 = 0)$. The Jacobian is

$$D_{\underset{\sim}{x}} f = \frac{\partial(f_1, f_2)}{\partial(x_1, x_2)} = \begin{bmatrix} \frac{\partial f_1}{\partial x_1} & \frac{\partial f_1}{\partial x_2} \\ \frac{\partial f_2}{\partial x_1} & \frac{\partial f_2}{\partial x_2} \end{bmatrix} = \begin{bmatrix} 0 & 1 \\ 1 - 2x_1 & -\mu \end{bmatrix}.$$

For $\underset{\sim}{x}_a^* = (x_1 = 0, x_2 = 0)$, the Jacobian becomes

$$D_{\underset{\sim}{x}_a^*} f = \begin{bmatrix} 0 & 1 \\ 1 & -\mu \end{bmatrix}.$$

So for equilibrium point $\underset{\sim}{x}_a^*$, $\mathrm{tr}(A) = -\mu$ and $\det(A) = -1$ which using Fig. 2.6 is a saddle for all values of μ.

For equilibrium point $\underset{\sim}{x}_b^* = (x_1 = 1, x_2 = 0)$, the Jacobian becomes

$$D_{\underset{\sim}{x}_b^*} f = \begin{bmatrix} 0 & 1 \\ -1 & -\mu \end{bmatrix},$$

so in this case $\mathrm{tr}(A) = -\mu$ and $\det(A) = 1$, which from Fig. 2.6 means that this equilibrium point is in the upper left-hand quadrant, and so is either a stable node or spiral. To find out which, compute the eigenvalues

$$\lambda_{1,2} = -\frac{\mu}{2} \pm \frac{1}{2}\sqrt{\mu^2 - 4},$$

from which the discriminant, $\Delta = \mu^2 - 4$. In the question it states that $0 < \mu < \sqrt{4}$, which means that $\Delta < 0$ and $\lambda_{1,2}$ are complex (corresponding to the underdamped case) and the equilibrium point is a stable spiral. The local sketch for each type of fixed point is shown in Fig. 2.6, the complete state space for the system is shown in Fig. 2.7.

Problem 2.4

For the system

$$\dot{x}_1 = x_2,$$
$$\dot{x}_2 = x_1 - x_1^2 - \mu x_2,$$

the function

$$V = \frac{1}{2}x_2^2 - \frac{1}{2}x_1^2 + \frac{1}{3}x_1^3$$

is suggested as a potential Lyapunov function. However in this case V is not positive definite. How to check this? First check that $V = 0$ when $x_1 = x_2 = 0$, which is true in this case. In addition, $V > 0$ must be true for all other values of x_1 and x_2, and checking at the other equilibrium point $x_1 = 1, x_2 = 0$ then $V < 0$ it can be deduced that V is not positive definite in this case. A similar case can be seen in the upper part of Fig. 2.7 where V is plotted against x_1. At the origin, in the direction of the x_1 axis, the function V is clearly non positive definite, and to the left (negative) there is an escape to infinity.

When a suitable V function is found, then V can be differentiated with respect to t to get $\dot{V}$. Now V can be thought of as representing energy in the system, such that $\dot{V}$ represents the rate of change of energy with time:

- $\dot{V}$ positive → energy increasing → unstable
- $\dot{V}$ zero → energy stays the same →neutrally stable
- $\dot{V}$ negative → energy decreasing → asymptotically stable

The question asks for the stability of the origin to be found using V for small values of the parameters. However, if the function V is differentiated and $\dot{x}_1$ and $\dot{x}_2$ substituted from Eq. 2.36 then the relation $\dot{V} = -\mu x_2^2$ is obtained. If μ is positive, this relation will always give $\dot{V} < 0$ which seems to indicate asymptotic stability. In fact, as the relation for $\dot{V}$ only depends on x_2 this is correct, because in the x_2 direction V is a parabola, which is a positive definite relationship. However, in the x_1 direction V is not positive definite. This example demonstrates how much care is required when using Lyapunov functions to determine the stability of the origin.

Problem 2.5

Using the Jacobian computed for Problem 2.2, it can be seen that if μ changes sign from positive to negative then for equilibrium point $\underset{\sim}{x}_a^*$ the result is still a saddle point. For equilibrium point $\underset{\sim}{x}_b^* = (x_1 = 1, x_2 = 0)$, the Jacobian becomes

$$D_{\underset{\sim}{x}_b^*} f = \begin{bmatrix} 0 & 1 \\ -1 & \mu \end{bmatrix},$$

so in this case $\mathrm{tr}(A) = \mu$ and $\det(A) = 1$, which from Fig. 2.6 means that this equilibrium point is in the upper right-hand quadrant, and so is either a unstable node or unstable spiral. To find out which, compute the eigenvalues

$$\lambda_{1,2} = \frac{\mu}{2} \pm \frac{1}{2}\sqrt{\mu^2 - 4},$$

from which the discriminant, $\Delta = \mu^2 - 4$. In the question it states that $0 < \mu < \sqrt{4}$, which means that $\Delta < 0$ and $\lambda_{1,2}$ are complex (corresponding to the underdamped case) and the equilibrium point is an unstable spiral.

So at the point when $\mu = 0$ the system is crossing between the upper left and upper right hand quadrants in Fig. 2.6. In this context the system is undergoing a Hopf bifurcation and the eigenvalues become purely imaginary $\lambda_{1,2} = \pm \mathbf{i}$.

Problem 2.6

The equation of a damped unforced pendulum is given by

$$\ddot{\theta} + \delta\dot{\theta} + \omega^2 \sin\theta = 0.$$

To find the equilibrium points, first put the system into first-order form

$$\begin{aligned}\dot{x}_1 &= x_2 = f_1,\\ \dot{x}_2 &= -\delta x_2 - \omega^2 \sin x_1 = f_2,\end{aligned}$$

where $x_1 = \theta$ and $x_2 = \dot{\theta}$. By inspection, for $\delta > 0$, the equilibrium points for this system are at $x_2 = 0$ and $x_1 = 0, \pm n\pi$ with $n = 1, 2, 3 \ldots$. The Jacobian is

$$D_{\underline{x}}f = \frac{\partial(f_1, f_2)}{\partial(x_1, x_2)} = \begin{bmatrix} \dfrac{\partial f_1}{\partial x_1} & \dfrac{\partial f_1}{\partial x_2} \\ \dfrac{\partial f_2}{\partial x_1} & \dfrac{\partial f_2}{\partial x_2} \end{bmatrix} = \begin{bmatrix} 0 & 1 \\ -\omega^2 \cos x_1 & -\delta \end{bmatrix}.$$

For the equilibrium points at $x^* = (2n\pi, 0)$, where n is an integer, the Jacobian becomes

$$D_{\underline{x}^*}f = \begin{bmatrix} 0 & 1 \\ -\omega^2 & -\delta \end{bmatrix}.$$

So for these equilibrium points, $\mathrm{tr}(A) = -\delta$ and $\det(A) = \omega^2$ which from Fig. 2.6 means that this equilibrium point is in the upper left-hand quadrant, and so is either a stable node or spiral. To find out which, compute the eigenvalues

$$\lambda_{1,2} = -\frac{\delta}{2} \pm \frac{1}{2}\sqrt{\delta^2 - 4\omega^2},$$

from which the discriminant, $\Delta = \delta^2 - 4\omega^2$. In the question is states that $\delta^2 < 4\omega^2$, which means that $\Delta < 0$ and $\lambda_{1,2}$ are complex and the equilibrium points are stable spirals.

For the equilibrium points at $x^* = ((2n + 1)\pi, 0)$, where n is an integer, the Jacobian becomes

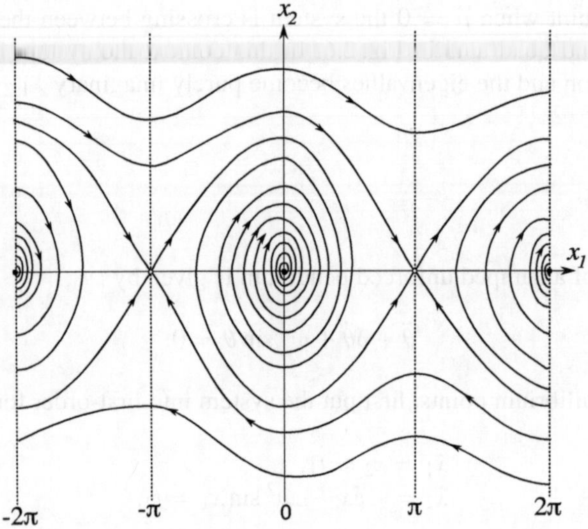

Fig. 9.1 Pendulum equation phase portrait with damping

$$D_{\underline{x}^*_d}f = \begin{bmatrix} 0 & 1 \\ \omega^2 & -\delta \end{bmatrix},$$

so in this case $\text{tr}(A) = -\delta$ and $\det(A) = -\omega^2$, which using Fig. 2.6 means that these fixed points are all saddles for all values of δ. So in the range $2\pi \leq \theta \leq 2\pi$ when $\delta^2 < 4\omega^2$ there are five fixed points (i) saddles at $x^* = (0, 0)$ and $x^* = (\pm 2\pi, 0)$, and (ii) stable spirals at $x^* = (\pm\pi, 0)$. A sketch of the phase plane is shown in Fig. 9.1.

Problem 2.7

For small angles, $\theta \ll 1$, the approximate equation is

$$\ddot{\theta} + \delta\dot{\theta} + \omega^2(\theta - \frac{\theta^3}{3!}) = 0.$$

First put the system into first-order form

$$\begin{aligned} \dot{x}_1 &= x_2, \\ \dot{x}_2 &= -\delta x_2 - \omega^2 x_1 + \frac{\omega^2}{3!}x_1^3, \end{aligned} \tag{9.5}$$

where $x_1 = \theta$ and $x_2 = \dot{\theta}$. Now consider the potential Lyapunov function

$$V = \frac{1}{2}\dot{\theta}^2 + \frac{\omega^2}{2}\theta^2 - \left(\frac{\omega^2}{3!}\right)\frac{\theta^4}{4}$$

Note that $V(0, 0) = 0$ is true in this case. But for other values this function is not positive definite. As a result, this cannot be used as a Lyapunov function to determine the stability of the origin in general. However, as the question is restricted to *small angles* $\theta = x_1 < 1$ and $\omega = 1$ it may be possible to use part of the function to determine stability *locally*. In this case, small angles implies that $\frac{x_1^2}{2} > \frac{x_1^4}{(3!)4}$, such that the stability of the fixed point at the origin $x_1 = x_2 = 0$ for these small angle conditions can be considered locally.

Using the chain rule

$$\dot{V} = x_2\dot{x}_2 + \omega^2 x_1 \dot{x}_1 - \left(\frac{\omega^2}{3!}\right) x_1^3 \dot{x}_1.$$

Substituting for $\dot{x}_1$ and $\dot{x}_2$ from Eq. 9.5 gives

$$\dot{V} = x_2(-\delta x_2 - \omega^2 x_1 + \frac{\omega^2}{3!}x_1^3) + \omega^2 x_1 x_2 - \left(\frac{\omega^2}{3!}\right) x_1^3 x_2.$$

$$\dot{V} = -\delta x_2^2 - \cancel{\omega^2 x_1 x_2} + \cancel{\frac{\omega^2}{3!} x_1^3} x_2 + \cancel{\omega^2 x_1 x_2} - \cancel{\left(\frac{\omega^2}{3!}\right) x_1^3} x_2.$$

$$\dot{V} = -\delta x_2^2,$$

which implies that the fixed point at the origin is asymptotically stable for $\delta > 0$ because $\dot{V} < 0$ for all x_2 values (except $x_2 = 0$ where it is inconclusive or at worst neutrally stable). Remember that this is only true for x_1 suitably small.

Problem 2.8

Starting with the logistic map

$$x_{n+1} = \lambda x_n(1 - x_n) = h_1(x_n),$$

the period one fixed points occur when $x_{n+1} = x_n = x_n^*$. In which case

$$x_n^* = \lambda x_n^*(1 - x_n^*), \quad \rightsquigarrow \quad x_n^* - \lambda x_n^*(1 - x_n^*) = 0, \quad \rightsquigarrow \quad x_n^* - \lambda x_n^* + \lambda x_n^{*2} = 0,$$

which has solutions

$$x_n^* = 0 \text{ and } 1 - \frac{1}{\lambda}.$$

To determine the stability compute the Jacobian which for this example is

$$\frac{\partial h_1}{\partial x_n} = \lambda - 2\lambda x_n,$$

which, as it is a scalar value, is also equal to the system eigenvalue. Now it can be seen that for $x_n^* = 0$, $\frac{\partial h_1}{\partial x_n} = \lambda$ and there is a fold bifurcation when $\lambda = 1$. For $x_n^* = 1 - \frac{1}{\lambda}$, $\frac{\partial h_1}{\partial x_n} = 2 - \lambda$ such that when $\lambda = 3$ a flip bifurcation occurs. The flip bifurcation leads to a period two behaviour as λ increases above 3. These fixed points are in the form $x_{n+2} = x_n$.

Chapter 3

Problem 3.1

The governing equation of motion for the system defined by Eq. (3.2) can be written as

$$m\ddot{x}(t) + c\dot{x}(t) + kx(t) = F_e(t) + bu(t),$$

and for velocity feedback control

$$u(t) = -\kappa \int_0^t \ddot{x} \mathrm{d}t = -\kappa \dot{x}.$$

Using the same approach as for Example 1.4 in Chap. 1 and taking Laplace transforms to give

$$\begin{aligned} ms^2X(s) + csX(s) + kX(s) = F_e(s) + bU(s) \rightsquigarrow (ms^2 + cs + k)X(s) \\ = F_e(s) + bU(s) \end{aligned}$$

and

$$U(s) = -\kappa s X(s),$$

where s the the Laplace transform variable. The control block diagram is shown in Fig. 9.2. Note that for this type of control the demand (i.e. what the the system is required to do) is zero. In other words it is required to have no vibration. As a result,

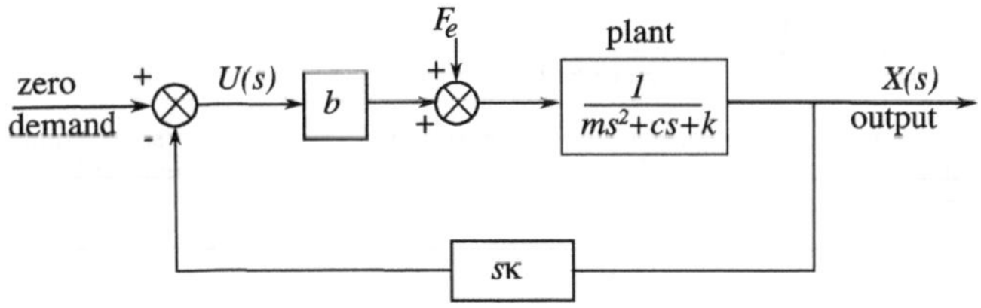

Fig. 9.2 Control block diagram for velocity feedback control

the external forcing is seen as a "disturbance", and so F_e is added into the block diagram just before the plant transfer function. Substituting $U(s)$ gives

$$(ms^2 + cs + k + \kappa sb)X(s) = F_e(s) \quad \rightsquigarrow \quad \frac{X(s)}{F_e(s)} = \frac{1}{ms^2 + [c + \kappa b]s + k}.$$

So the closed loop poles for this system are

$$s_{1,2} = -\frac{[c + \kappa b]}{2m} \pm \frac{\sqrt{[c + \kappa b]^2 - 4mk}}{2m}.$$

For stability, it is required that the real part of the poles are negative, $\mathrm{Re}\{s_{1,2}\} < 0$. In the case when $[c + \kappa b]^2 < 4mk$ the poles are complex. Then if

$$\begin{array}{ll} c + \kappa b > 0 & \text{stable} \\ c + \kappa b = 0 & \text{point of instability} \\ c + \kappa b < 0 & \text{unstable} \end{array}$$

so it is important to notice that the sign and relative size of the control parameters b and κ can affect stability.

Problem 3.2

The system shown in Fig. 3.5 has the same governing equation as that in Problem 3.1, namely Eq. (3.2) in Chap. 3. The governing equation of motion for the oscillator is given by

$$m\ddot{x}(t) + c\dot{x}(t) + kx(t) = F_e(t) + bu(t),$$

For displacement feedback $u = -\alpha x$, call this Case A.
For acceleration feedback $u = -\beta\ddot{x}$, call this Case B.

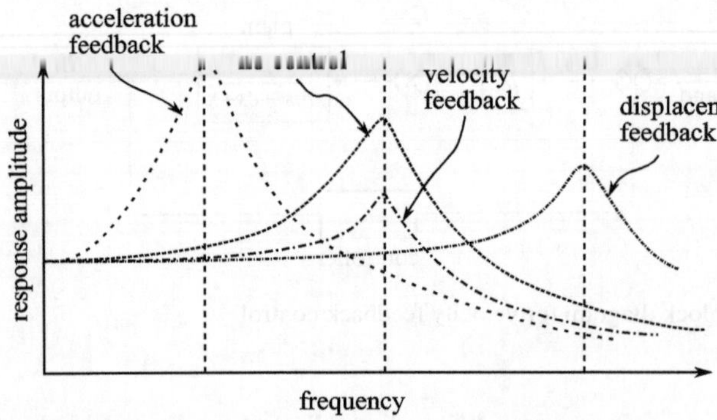

Fig. 9.3 Schematic representation of the frequency response for acceleration, velocity and displacement feedback control

Then the governing equation becomes

$$m\ddot{x}(t) + c\dot{x}(t) + (k + b\alpha)x(t) = F_e(t), \quad \text{for Case A}$$

and

$$(m + b\beta)\ddot{x}(t) + c\dot{x}(t) + kx(t) = F_e(t), \quad \text{for Case B.}$$

So instead of changing the damping, as for velocity feedback, in these cases either stiffness or mass is (effectively) being changed. This means that in both cases the effect of the feedback will be to change the apparent natural frequency of the system. So, assuming $b, \alpha, \beta > 0$, then

$$\text{for Case A} \quad \omega_n = \sqrt{\frac{k + b\alpha}{m}} > \sqrt{\frac{k}{m}},$$

$$\text{for Case B} \quad \omega_n = \sqrt{\frac{k}{m + b\beta}} < \sqrt{\frac{k}{m}}.$$

The shifts in natural frequency compared with the case where no control and velocity control are applied are shown schematically in Fig. 9.3.

Problem 3.3

The governing equation of motion for the oscillator is given by

$$m\ddot{x}(t) + (c + b\kappa)\dot{x}(t) + k_1x(t) + k_3x^3 = F_e(t).$$

For this oscillator the kinetic and potential energies are given by

$$KE = \frac{1}{2}m\dot{x}^2 \quad \text{and} \quad PE = k_1\frac{x^2}{2} + k_3\frac{x^4}{4}.$$

To do the Lyapunov function analysis, first put the governing equation of motion into first order form. Then set $F_e = 0$ and divide through by mass to give

$$\dot{x}_1 = x_2$$

$$\dot{x}_2 = -\frac{(c + b\kappa)}{m}x_2 - \frac{k_1}{m}x_1 - \frac{k_3}{m}x_1^3,$$

The question asks for the fixed point at the origin ($x_1 = x_2 = 0$) to be examined. To construct a Lyapunov function from the system energy, take the total energy of the system which in this case is

$$KE + PE = \frac{1}{2}mx_2^2 + k_1\frac{x_1^2}{2} + k_3\frac{x_1^4}{4},$$

now divide by m to give a Lyapunov function, V, for the system

$$V = \frac{1}{2}x_2^2 + \frac{k_1}{m}\frac{x_1^2}{2} + \frac{k_3}{m}\frac{x_1^4}{4}.$$

This is a positive definite function, so it can be used as a Lyapunov function for assessing the stability of $x_1 = x_2 = 0$. The next step is to differentiate V with respect to time using the chain rule to get

$$\dot{V} = x_2\dot{x}_2 + \frac{k_1}{m}x_1\dot{x}_1 + \frac{k_3}{m}x_1^3\dot{x}_1$$

Now substitute for $\dot{x}_1$ and $\dot{x}_2$ to give

$$\dot{V} = x_2\left(-\frac{(c + b\kappa)}{m}x_2 - \cancel{\frac{k_1}{m}x_1} - \cancel{\frac{k_3}{m}x_1^3}\right) + \cancel{\frac{k_1}{m}x_1}x_2 + \cancel{\frac{k_3}{m}x_1^3}x_2$$

so finally the relationship becomes

$$\dot{V} = -\frac{(c + b\kappa)}{m}x_2^2$$

which means that the origin is asymptotically stable provided that $\frac{(c+b\kappa)}{m} > 0$. This is because x_2^2 is always positive, so if the coefficient is positive, $\dot{V}$ will always be negative which implies asymptotic stability. This can easily be achieved if c, b, m and κ are made to be positive. Or that $(c + b\kappa)$ is positive, assuming that mass m will always be positive, so that $\frac{(c+b\kappa)}{m} > 0$ is true without additional restrictions on the control variables b and κ.

Problem 3.4

The governing equation of motion for the oscillator is given by

$$\ddot{x} + \delta\dot{x} - x + \alpha x^3 = 0.$$

The question asks for a Lyapunov function to be found, so the easiest starting point is to use the energy of the system. In first order form the system can be written as

$$\dot{x}_1 = x_2,$$
$$\dot{x}_2 = -\delta x_2 + x_1 - \alpha x_1^3.$$

In this case the energy is given by

$$KE + PE = \frac{1}{2}x_2^2 - \frac{x_1^2}{2} + \alpha\frac{x_1^4}{4}.$$

Note that mass m is one for this oscillator. Unfortunately, this is not a positive definite function, so it cannot be used as a valid Lyapunov function to assess the stability of $x_1 = x_2 = 0$. If the energy cannot be used the best approach is to use the parabolic function or the closest positive definite function to the energy. The first try is using

$$V = \frac{1}{2}x_2^2 + \frac{x_1^2}{2} + \alpha\frac{x_1^4}{4},$$

which is a positive definite function, and is almost the same as the system energy. Now differentiate V with respect to time using the chain rule to get

$$\dot{V} = x_2\dot{x}_2 + x_1\dot{x}_1 + \alpha x_1^3\dot{x}_1.$$

Next substitute for $\dot{x}_1$ and $\dot{x}_2$ to give

$$\dot{V} = x_2(-\delta x_2 + x_1 - \cancel{\alpha x_1^3}) + x_1 x_2 + \cancel{\alpha x_1^3 x_2},$$

so finally the relationship becomes

$$\dot{V} = -\delta x_2^2 + 2x_1 x_2$$

which means that the stability of the origin is not determined. As a result an alternative Lyapunov function needs to be found, or the origin is unstable. In fact, you should notice that this example is similar to the snap-through system given in Chap. 2, and the origin is in fact a saddle point. A saddle point is always unstable in two directions, so there will be no Lyapunov function which shows this as asymptotically stable.

Problem 3.5

The governing equation of motion for the oscillator is given by

$$\begin{bmatrix} \dot{x}_1 \\ \dot{x}_2 \end{bmatrix} = \begin{bmatrix} 0 & 1 \\ \frac{k_1}{m} & -\frac{c}{m} \end{bmatrix} \begin{bmatrix} x_1 \\ x_2 \end{bmatrix} + \begin{bmatrix} 0 \\ \frac{p}{m} u(t) - \frac{k_3}{m} x_1^3 \end{bmatrix}.$$

When

$$u(t) = \frac{1}{p} k_3 x_1^3$$

the nonlinear term is removed, but the underlying linear system is still unstable. To solve this, a term is added to the feedback control which stabilizes the underlying linear system. In this case the problem comes from the fact that the $\frac{k_1}{m}$ has a positive sign in the governing equation. If the sign of this term could be changed then the underlying linear system could be stabilised. In fact, it can be done by using the control signal

$$u(t) = \frac{1}{p}(k_3 x_1^3 - 2k_1 x_1).$$

To see how this works let's substitute for $u(t)$ in the governing equations, which gives

$$\dot{x}_1 = x_2,$$

$$\dot{x}_2 = +\frac{k_1}{m} x_1 - \frac{c}{m} x_2 + \frac{p}{m}\frac{1}{p}(k_3 x_1^3 - 2k_1 x_1) - \frac{k_3}{m} x_1^3.$$

Expanding the bracket gives

$$\dot{x}_1 = x_2,$$

$$\dot{x}_2 = \frac{k_1}{m} x_1 - \frac{c}{m} x_2 + \cancel{\frac{k_3}{m} x_1^3} - 2\frac{k_1}{m} x_1 - \cancel{\frac{k_3}{m} x_1^3}.$$

So finally

$$\dot{x}_1 = x_2,$$

$$\dot{x}_2 = -\frac{k_1}{m} x_1 - \frac{c}{m} x_2,$$

which is stable and linear. This means that a feedback linearisation control which removes the nonlinearity and leaves a stable linear system in place has now been designed.

Problem 3.6

The governing equation of motion for the oscillator is given by

$$m\ddot{x} + c\dot{x}(1 + \delta x^2) + kx = pu(t),$$

where $u(t)$ is the control input. In first order form the system is given by

$$\dot{x}_1 = x_2,$$

$$\dot{x}_2 = -\frac{c}{m}x_2(1 + \delta x_1^2) - \frac{k}{m}x_1 + \frac{p}{m}u(t).$$

The nonlinear damping term is given by

$$-\frac{c}{m}\delta x_2 x_1^2.$$

Setting

$$u(t) = \frac{1}{p}c\delta x_2 x_1^2$$

will remove the nonlinear term in the governing equation. To add more viscous damping at the same time, a velocity feedback term is added, so that

$$u(t) = \frac{1}{p}(c\delta x_2 x_1^2 - \kappa x_2).$$

Then in first order form, the system becomes

$$\dot{x}_1 = x_2,$$

$$\dot{x}_2 = -\frac{c}{m}x_2 - \frac{c}{m}\delta x_2 x_1^2 - \frac{k}{m}x_1 + \left(\frac{p}{m}\right)\frac{1}{p}(c\delta x_2 x_1^2 - \kappa x_2).$$

Expanding the bracket term

$$\dot{x}_1 = x_2,$$

$$\dot{x}_2 = -\frac{c}{m}x_2 - \xcancel{\frac{c}{m}\delta x_2 x_1^2} - \frac{k}{m}x_1 + \xcancel{\frac{c}{m}\delta x_2 x_1^2} - \frac{\kappa}{m}x_2,$$

which leaves

$$\dot{x}_1 = x_2,$$

$$\dot{x}_2 = -\frac{(c+\kappa)}{m}x_2 - \frac{k}{m}x_1,$$

which is a linear system with extra damping κ.

Problem 3.7

The governing equation of motion for the oscillator is given by

$$\begin{aligned}\dot{x}_1 &= x_2,\\ \dot{x}_2 &= -\frac{c}{m}x_2 - \frac{k_1}{m}x_1 - \frac{k_3}{m}x_1^3 + \frac{p}{m}u(t).\end{aligned}$$

An input–output controller needs to be designed for the case when the control output is $y = x_2$ (i.e. velocity). First differentiate the output so that $\dot{y} = \dot{x}_2$. Then using the governing equations gives

$$\dot{y} = \dot{x}_2 = -\frac{c}{m}x_2 - \frac{k_1}{m}x_1 - \frac{k_3}{m}x_1^3 + \frac{p}{m}u(t).$$

So, setting

$$u(t) = \frac{m}{p}\left(v(t) + \frac{c}{m}x_2 + \frac{k_1}{m}x_1 + \frac{k_3}{m}x_1^3\right),$$

gives the result that $\dot{y} = \dot{x}_2 = v(t)$ where $v(t)$ can be chosen as required to achieve the desired control task at hand.

Now, when $y = x_1$ differentiate twice to get

$$\ddot{y} = \ddot{x}_1 = -\frac{c}{m}x_2 - \frac{k_1}{m}x_1 - \frac{k_3}{m}x_1^3 + \frac{p}{m}u(t).$$

In this case the final relationship is $\ddot{y} = \ddot{x}_1 = v(t)$. In the first case the output y is x_2 (i.e. velocity) and in the second case the output y is x_1 (i.e. displacement). Both cases lead to a final relationship where the new control input $v(t)$ equals the acceleration $\dot{x}_2$. In the second case two differentiations are required instead of just one.

Problem 3.8

The governing equation of motion for the oscillator is given by

$$m\ddot{x} + c\dot{x}x + kx = pu(t).$$

In first order form, the oscillator can be expressed as

$$\begin{aligned}\dot{x}_1 &= x_2,\\ \dot{x}_2 &= -\frac{c}{m}x_2x_1 - \frac{k}{m}x_1 + \frac{p}{m}u(t).\end{aligned}$$

The question asks for an input–output linearisation control to be designed with the output $y = x_1 + x_2 = h(\underline{x})$, where $\underline{x} = [x_1, x_2]^T$ is the state vector. Also

$$f(\underline{x}) = \begin{bmatrix} x_2 \\ -\frac{c}{m}x_2x_1 - \frac{k}{m}x_1 \end{bmatrix}, \qquad g(\underline{x}) = \begin{bmatrix} 0 \\ \frac{p}{m} \end{bmatrix}.$$

The next step is to differentiate the output vector, which gives

$$\frac{\partial h}{\partial x} = [1 \quad 1].$$

Now compute the Lie derivatives

$$L_f h(\underline{x}) = [1 \quad 1]\begin{bmatrix} x_2 \\ -\frac{c}{m}x_2x_1 - \frac{k}{m}x_1 \end{bmatrix} = x_2 - \frac{c}{m}x_2x_1 - \frac{k}{m}x_1,$$

$$L_g h(\underline{x}) = [1 \quad 1]\begin{bmatrix} 0 \\ \frac{p}{m} \end{bmatrix} = \frac{p}{m}.$$

So, in this example the required control signal is

$$u = \frac{m}{p}\left(v(t) - x_2 + \frac{c}{m}x_2x_1 + \frac{k}{m}x_1\right).$$

This gives the relationship

$$\dot{y} = \dot{x}_1 + \dot{x}_2 = x_2 - \frac{c}{m}x_2x_1 - \frac{k}{m}x_1 + \frac{p}{m}u.$$

Substituting for u gives the required relationship $\dot{y} = v(t)$. In the case when $y = x_2$, $L_f h = -\frac{c}{m}x_2x_1 - \frac{k}{m}x_1$ and $\dot{y} = v(t)$. When $y = x_1$, $L_g h = 0$, so differentiate twice to get $\ddot{y} = \dot{x}_2 = v(t)$.

Problem 3.9

The governing equation of motion for the system is given by

$$\frac{\mathrm{d}}{\mathrm{d}t}\begin{bmatrix} q_1 \\ q_2 \\ \dot{q}_1 \\ \dot{q}_2 \end{bmatrix} = \begin{bmatrix} 0 & 0 & 1 & 0 \\ 0 & 0 & 0 & 1 \\ -\omega_{n1}^2 & 0 & -\zeta_1\omega_{n1} & 0 \\ 0 & -\omega_{n2}^2 & 0 & -\zeta_2\omega_{n2} \end{bmatrix}\begin{bmatrix} q_1 \\ q_2 \\ \dot{q}_1 \\ \dot{q}_2 \end{bmatrix} - \begin{bmatrix} 0 \\ 0 \\ \mu_1 q_1^2 + \delta_1 q_1 q_2 \\ \mu_2 q_2^2 + \delta_2 q_2 q_1 \end{bmatrix} + \begin{bmatrix} 0 \\ 0 \\ \alpha_1 p_1 \\ 0 \end{bmatrix} u_1 + \begin{bmatrix} 0 \\ 0 \\ 0 \\ \beta_2 p_2 \end{bmatrix} u_2.$$

First, by inspection it can be seen that setting

$$u_1 = \frac{1}{\alpha_1 p_1}(\mu_1 q_1^2 + \delta_1 q_1 q_2) \quad \text{and} \quad u_2 = \frac{1}{\beta_2 p_2}(\mu_2 q_2^2 + \delta_2 q_2 q_1)$$

will linearize each mode directly. To obtain an input–output linearisation, first take the outputs $y_1 = q_1$ and $y_2 = q_2$, differentiate twice to get the relationship with the control inputs given by

$$\ddot{q}_1 = -\omega_{n1}^2 q_1 - \zeta_1\omega_{n1}\dot{q}_1 - \mu_1 q_1^2 - \delta_1 q_1 q_2 + \alpha_1 p_1 u_1,$$
$$\ddot{q}_2 = -\omega_{n2}^2 q_2 - \zeta_2\omega_{n2}\dot{q}_2 - \mu_2 q_2^2 - \delta_2 q_2 q_1 + \beta_2 p_2 u_2.$$

Now choosing

$$u_1 = \frac{1}{\alpha_1 p_1}(v_1(t) + \omega_{n1}^2 q_1 + \zeta_1\omega_{n1}\dot{q}_1 + \mu_1 q_1^2 + \delta_1 q_1 q_2),$$

$$u_2 = \frac{1}{\beta_2 p_2}(v_2(t) + \omega_{n2}^2 q_2 + \zeta_2\omega_{n2}\dot{q}_2 + \mu_2 q_2^2 + \delta_2 q_2 q_1),$$

gives an input–output linearisation with the result that $\ddot{q}_1 = v_1(t)$ and $\ddot{q}_2 = v_2(t)$, where $v_1(t)$ and $v_2(t)$ are the new control signals, which can be chosen to give the desired linear system response.

Problem 3.10

From Eq. 3.2 the governing equation of motion for modes 2 and 3 of the beam are given by

$$\frac{d}{dt}\begin{bmatrix} q_2 \\ q_3 \\ \dot{q}_2 \\ \dot{q}_3 \end{bmatrix} = \begin{bmatrix} 0 & 0 & 1 & 0 \\ 0 & 0 & 0 & 1 \\ -\omega_{n2}^2 & 0 & -\zeta_2\omega_{n2} & 0 \\ 0 & -\omega_{n3}^2 & 0 & -\zeta_3\omega_{n3} \end{bmatrix} \begin{bmatrix} q_2 \\ q_3 \\ \dot{q}_2 \\ \dot{q}_3 \end{bmatrix} - \begin{bmatrix} 0 \\ 0 \\ \mu_2 q_2^3 \\ \mu_3 q_3^3 \end{bmatrix} + \begin{bmatrix} 0 \\ 0 \\ \alpha_2 p \\ \alpha_3 p \end{bmatrix} u,$$

where $F_c = pu$ has been assumed. First, by inspection it can be seen that because only one control signal is available, both cubic nonlinear terms cannot be removed simultaneously. However, one or the other can be cancelled by selecting

$$u = \frac{1}{\alpha_2 p}(\mu_2 q_2^3) \quad \text{or} \quad u = \frac{1}{\alpha_3 p}(\mu_3 q_3^3)$$

to linearize one of the modes mode directly. A similar situation will be found using input–output linearisation, so with this situation cancelling one modal nonlinearity is the best solution that can be obtained.

Problem 3.11

Following the same approach as in Example 3.10 gives

$$\begin{bmatrix} 2p_{21}\dfrac{k}{m} & -p_{11} + p_{21}\dfrac{c}{m} + p_{22}\dfrac{k}{m} \\ -p_{11} + p_{21}\dfrac{c}{m} + p_{22}\dfrac{k}{m} & -2p_{21} + 2p_{22}\dfrac{c}{m} \end{bmatrix} = \begin{bmatrix} q_{11} & q_{12} \\ q_{21} & q_{22} \end{bmatrix}.$$

For Q to be positive definite the conditions are that $q_{11} > 0$ and $q_{11}q_{22} - q_{21}q_{12} > 0$. Furthermore terms will be positive if $p_{11} < (p_{21}\frac{c}{m} + p_{22}\frac{k}{m})$ and $p_{21} < p_{22}\frac{c}{m}$. For P to be positive definite, the conditions are that $p_{11} > 0$ and $p_{11}p_{22} - p_{21}p_{12} > 0$.

Using the numerical values given in the question and selecting $p_{21} = 0.025$ gives $q_{11} = 0.05$, then choosing $p_{22} = 0.3$ gives $q_{22} = 0.001$. Finally, selecting $p_{11} = 0.2$ gives $q_{21} = q_{12} = 0.1025$, and the Lyapunov criteria, Eq. (3.27) is satisfied.

Now the adaptive control law can be found from Eq. (3.28). This gives

$$\begin{bmatrix} a_1 \\ a_2 \end{bmatrix} = \begin{bmatrix} a_1^* \\ a_2^* \end{bmatrix} + \int_0^t \begin{bmatrix} \gamma_{11} & 0 \\ 0 & \gamma_{22} \end{bmatrix} \begin{bmatrix} x_1 x_2 \\ x_1^3 \end{bmatrix} \begin{bmatrix} 0, & \frac{1}{2} \end{bmatrix} \begin{bmatrix} 0.2 & 0.025 \\ 0.025 & 0.3 \end{bmatrix} \begin{bmatrix} x_1 \\ x_2 \end{bmatrix} \mathbf{d}\boldsymbol{x}, \quad (9.6)$$

where the control gain matrix has been taken as diagonal, which satisfies the condition that $\Gamma^T = \Gamma$. This expression defines the adaptive gains, and the weightings γ_{11} and γ_{22} can be chosen by the control designer to give the required adaptive effort.

Chapter 4

Problem 4.1

Using the one-term response $x = X\cos(\omega_r t)$, in which ω_r is the response frequency, the equation of motion,

$$\ddot{x} + \omega_n^2 x + \alpha x^2 = 0,$$

becomes

$$-\omega_r^2 X\cos(\omega_r t) + \omega_n X\cos(\omega_r t) + \alpha X^2\cos^2(\omega_r t) = 0.$$

Using the trigonometric relationship $\cos^2(\omega_r t) = [1 + cos(2\omega_r t)]/2$ gives

$$-\omega_r^2 X\cos(\omega_r t) + \omega_n^2 X\cos(\omega_r t) + \frac{\alpha}{2}X^2[1 + cos(2\omega_r t)] = 0.$$

Now balancing the $\cos(\omega_r t)$ terms, while ignoring the other harmonics, gives

$$-\omega_r^2 X + \omega_n^2 X = 0.$$

Therefore using a one-term solution, the response frequency matches the linear natural frequency, $\omega_r = \omega_n$, for all amplitudes of response.

Problem 4.2

Based on the previous solution, a suitable three-term solution for the harmonic balance technique would include terms with frequency 0 and $2\omega_r$ as these terms were not balanced with the one-term solution. So using the solution

$$x = X_0 + X\cos(\omega_r t) + X_2\cos(2\omega_r t),$$

the equation of motion

$$\ddot{x} + \omega_n^2 x + \alpha x^2 = 0,$$

becomes

$$\begin{aligned}&-\omega_r^2 X\cos(\omega_r t) - 4\omega_r^2 X_2\cos(2\omega_r t) + \omega_n^2\,[X_0 + X\cos(\omega_r t) + X_2\cos(2\omega_r t)]\\&+\alpha\,[X_0 + X\cos(\omega_r t) + X_2\cos(2\omega_r t)]^2 = 0.\end{aligned}$$

Expanding out the squared brackets, while using the trigonometric relationship

$$2\cos(A)\cos(B) = \cos(A+B) + \cos(A-B),$$

gives

$$\begin{aligned}
&-\omega_r^2 X\cos(\omega_r t) - 4\omega_r^2 X_2\cos(2\omega_r t) + \omega_n^2\left[X_0 + X\cos(\omega_r t) + X_2\cos(2\omega_r t)\right]\\
&+\frac{\alpha}{2}\Big[2X_0^2 + 4X_0X\cos(\omega_r t) + 4X_0X_2\cos(2\omega_r t) + X^2(1+\cos(2\omega_r t))\\
&\qquad + 2XX_2(\cos(\omega_r t) + \cos(3\omega_r t)) + X_2^2(1+\cos(4\omega_r t))\Big] = 0.
\end{aligned}$$

Balancing the terms that were in the solution for X, while ignoring the other terms gives

$$\begin{aligned}
\cos(0\omega_r t):&\quad \omega_n^2 X_0 + \tfrac{\alpha}{2}(2X_0^2 + X^2 + X_2^2) = 0,\\
\cos(\omega_r t):&\quad -\omega_r^2 X + \omega_n^2 X + \alpha(2X_0X + XX_2) = 0,\\
\cos(2\omega_r t):&\quad -4\omega_r^2 X_2 + \omega_n^2 X_2 + \tfrac{\alpha}{2}(4X_0X_2 + X^2) = 0.
\end{aligned}$$

To simplify these, the relative size of the terms must be considered. Treating the nonlinear term, and so α, and the harmonic terms, X_0 and X_2, as order ε and then rewriting the equations to order ε^2 gives

$$\begin{aligned}
\cos(0\omega_r t):&\quad \omega_n^2 X_0 + \tfrac{\alpha}{2}X^2 = 0,\\
\cos(\omega_r t):&\quad -\omega_r^2 X + \omega_n^2 X + \alpha(2X_0X + XX_2) = 0,\\
\cos(2\omega_r t):&\quad -4\omega_r^2 X_2 + \omega_n^2 X_2 + \tfrac{\alpha}{2}X^2 = 0.
\end{aligned}$$

Now the $\cos(0\omega_r t)$ and $\cos(2\omega_r t)$ balance equations can be used to find X_0 and X_2 in terms of X

$$X_0 = -\frac{\alpha}{2\omega_n^2}X^2, \quad X_2 = \frac{\alpha}{2(4\omega_r^2 - \omega_n^2)}X^2.$$

Substituting these into the $\cos(\omega_r t)$ balance equation gives the response frequency-amplitude equation

$$\omega_r^2 = \omega_n^2 + \alpha^2\left(\frac{1}{2(4\omega_r^2 - \omega_n^2)} - \frac{1}{\omega_n^2}\right)X^2.$$

Since $\omega_r^2 \approx \omega_n^2$ (the right-hand term is already of order ε^2), this expression can be simplified by writing $\omega_r^2 = \omega_n^2$ in the small right-hand term to give to

$$\omega_r^2 = \omega_n^2 - \frac{5\alpha^2}{6}X^2.$$

Note that this is consistent with the ε^2 accurate normal form expression derived in Example 4.8.

Problem 4.3

To calculate the response frequency, ω_r, using the averaging method, the frequency detuning and time-scaling

$$\omega_r = \omega_n(1+\mu), \quad \tau = t(1+\mu),$$

respectively are used (note that μ is of order ε but this book-keeping aid is not used here). Using this the unforced Duffing equation,

$$\ddot{x} + \omega_n^2 x + \alpha x^3 = 0,$$

may be written in the standard form for averaging

$$x'' + \omega_n^2 x = -\alpha x^3 - (2\mu + \mu^2)x'',$$

where $\bullet'$ is the derivative with respect to τ. Since $x'' = -\omega_n^2 x$ to order ε^0 and μ^2 is of order ε^2 this can be written as

$$x'' + \omega_n^2 x = -\alpha x^3 - 2\mu\omega_n^2 x = \tilde{\mathcal{N}}.$$

Using the solution $x = x_c(\tau)\cos(\omega_n\tau) + x_s(\tau)\sin(\omega_n\tau)$ along with the relationships

$$x_c' = -\frac{1}{\omega_n}\sin(\omega_n\tau)\tilde{\mathcal{N}} \quad , \quad x_s' = \frac{1}{\omega_n}\cos(\omega_n\tau)\tilde{\mathcal{N}},$$

results in

$$x_c' = -\frac{1}{\omega_n}s_1\left[-\alpha(x_c^3c_1^3 + 3x_c^2x_sc_1^2s_1 + 3x_cx_s^2c_1s_1^2 + x_s^3s_1^3) + 2\mu\omega_n^2(x_cc_1 + x_ss_1)\right],$$
$$x_s' = \frac{1}{\omega_n}c_1\left[-\alpha(x_c^3c_1^3 + 3x_c^2x_sc_1^2s_1 + 3x_cx_s^2c_1s_1^2 + x_s^3s_1^3) + 2\mu\omega_n^2(x_cc_1 + x_ss_1)\right].$$

Here the shorthand $s_k = \sin(k\omega_n\tau)$ and $c_k = \cos(k\omega_n\tau)$ has been used. Now averaging is applied by integrating over a cycle of oscillation to give

$$x'_{ca} = -\frac{1}{\omega_n}\left[-\alpha\frac{3}{8}(x_{ca}^2 x_{sa} + x_{sa}^3) + \mu\omega_n^2 x_{sa}\right],$$

$$x'_{sa} = \frac{1}{\omega_n}\left[-\alpha\frac{3}{8}(x_{ca}^3 + x_{ca}x_{sa}^2) + \mu\omega_n^2 x_{ca}\right].$$

To find the response frequency-amplitude relationship the steady-state solution is needed. So setting $x'_{ca} = 0$ and $x'_{sa} = 0$ this gives the steady state condition

$$\mu\omega_n^2 = \frac{3}{8}\alpha(x_{ca}^2 + x_{sa}^2).$$

Since the amplitude of response, X, may be written as $X = \sqrt{x_{ca}^2 + x_{sa}^2}$ and $\omega_r = \omega_n(1+\mu)$ the backbone curve is given by

$$\omega_r^2 = \omega_n^2(1 + 2\mu) = \omega_n + \frac{3}{4}\alpha X^2.$$

Here $(1+\mu)^2 = 1 + 2\mu$ has been used as μ is of order ε.

Problem 4.4

Writing the forced Duffing oscillator in the form used when applying multiple scales gives

$$\ddot{x} + 2\zeta\omega_n\dot{x} + \omega_n^2 x + \mathcal{N}(x, \dot{x}) = f\cos(\Omega t), \quad \mathcal{N} = \alpha x^3.$$

Here the book-keeping small parameter ε has not been used, however the nonlinear term is small.

Using the fast and slow time-scales, $\tau = \Omega t$ and $T = \varepsilon t$ respectively, and introducing the small detuning parameter μ such that $\Omega = \omega_n(1+\mu)$, gives the standard results

$$\varepsilon^0 : \omega_n^2 x_0'' + \omega_n^2 x_0 = 0,$$
$$\varepsilon^1 : \omega_n^2 x_1'' + \omega_n^2 x_1 = -\omega_n^2 2\mu x_0'' - 2\omega_n x_0'^* - 2\zeta\omega_n^2 x_0' - \mathcal{N}(x_0, \omega_n x_0') + f\cos(\tau),$$

where the nonlinear term is $\gamma_x(x_0, \omega_n x_0') = \alpha x_0^3$.

The solution to the ε^0 is given by

$$x_0 = X_{0c}(T)\cos(\tau) + X_{0s}(T)\sin(\tau).$$

which when cubed, and simplified using the standard trigonometric expressions for $\cos^u(\tau)\sin^v(\tau)$ where $u + v = 3$, becomes

$$x_0^3 = \frac{3}{4}X_0^2\left(X_{0c}\cos(\tau) + X_{0s}\sin(\tau)\right) + \frac{1}{4}X_{0c}(X_{0c}^2 - 3X_{0s}^2)\cos(3\tau)$$
$$+ \frac{1}{4}X_{0s}(3X_{0c}^2 - X_{0s}^2)\sin(3\tau).$$

Here $X_0 = \sqrt{X_{0c}^2 + X_{0s}^2}$ is the amplitude of the x_0 response. Substituting this expression for x_0 into the ε^1 equation gives

$$\omega_n^2 x_1'' + \omega_n^2 x_1 = \left(2\omega_n^2 \mu X_{0c} - 2\omega_n X_{0s}^* - 2\omega_n^2 \zeta X_{0s} - \frac{3}{4}\alpha X_0^2 X_{0c} + f\right)\cos(\tau)$$
$$+ \left(2\omega_n^2 \mu X_{0s} + 2\omega_n X_{0c}^* + 2\omega_n^2 \zeta X_{0c} - \frac{3}{4}\alpha X_0^2 X_{0s}\right)\sin(\tau)$$
$$+ \frac{\alpha}{4}(3X_{0s}^2 - X_{0c}^2)X_{0c}\cos(3\tau) + \frac{\alpha}{4}(X_{0s}^2 - 3X_{0c}^2)X_{0s}\sin(3\tau).$$

Now any secular terms must be eliminated. To do this the amplitudes of the $\cos(\tau)$ and $\sin(\tau)$ terms on the right-hand side of the equation must be set to zero. This results in the following conditions on X_{0c} and X_{0s}

$$X_{0c}^* = -\frac{1}{\omega_n}\left(\omega_n^2 \mu X_{0s} + \omega_n^2 \zeta X_{0c} - \frac{3}{8}\alpha X_0^2 X_{0s}\right),$$
$$X_{0s}^* = \frac{1}{\omega_n}\left(\omega_n^2 \mu Z_{0c} - \omega_n^2 \zeta X_{0s} - \frac{3}{8}\alpha X_0^2 X_{0c} + \frac{1}{2}f\right).$$

For steady-state solutions the amplitude of oscillation is fixed and so both these equation may be set to zero giving

$$\left(\omega_n^2 \mu - \frac{3}{8}\alpha X_0^2\right) X_{0s} + \left(\omega_n^2 \zeta\right) X_{0c} = 0,$$
$$\left(\omega_n^2 \mu - \frac{3}{8}\alpha X_0^2\right) X_{0c} - \left(\omega_n^2 \zeta\right) X_{0s} = -\frac{1}{2}f.$$

Squaring and adding gives an equation relating the amplitude of response to the forcing amplitude and frequency:

$$\left(\omega_n^2 \mu - \frac{3}{8}\alpha X_0^2\right)^2 + \left(\omega_n^2 \zeta\right)^2 = \frac{1}{4}f^2,$$

where $\Omega = \omega_n(1+\mu)$ and μ can be treated as small so $\omega_n^2 - \Omega^2 = -\omega_n^2(2\mu + \mu^2) \simeq -2\omega_n^2\mu$. Using this gives

$$\left(\omega_n^2 - \Omega^2 + \frac{3}{4}\alpha X_0^2\right)^2 + \left(2\omega_n^2 \zeta\right)^2 = f^2.$$

Problem 4.5

Writing

$$\ddot{x} + \mu(x^2 - 1)\dot{x} + \omega_n^2 x = R\cos(\Omega t),$$

in the standard form for the normal form analysis results in

$$\ddot{x} + \omega_n^2 x + \tilde{\mathscr{N}}(x, \dot{x}, \mathbf{r}) = P_x \mathbf{r},$$

where $P_x = [R/2, R/2]$ and $\mathbf{r} = \{r_p, r_m\}^T = \{e^{i\Omega t}, e^{-i\Omega t}\}^T$ and $\tilde{\mathscr{N}}(x, \dot{x}, \mathbf{r}) = \mu(x^2 - 1)\dot{x}$.

Firstly the linear modal transform, which for a single degree-of-freedom system is $\mathbf{q} = \mathbf{x} = x$, is applied in **Step 1** resulting in

$$\ddot{\mathbf{q}} + \Lambda\mathbf{q} + N_q(\mathbf{q}, \dot{\mathbf{q}}, \mathbf{r}) = P_q \mathbf{r},$$
$$\text{where:} \quad \Lambda = \omega_n^2, \quad N_q(\mathbf{q}, \dot{\mathbf{q}}, \mathbf{r}) = \mu(q_1^2 - 1)\dot{q}_1, \quad P_q = P_x,$$

where q_1 is the first, and only, element in $\mathbf{q}$.

Next is **Step 1f** in which the forcing transformation, $\mathbf{q} = \mathbf{v} + \mathbf{er}$, is applied to remove non-resonant forcing terms. As the forcing is close to resonance, the response frequency is set to be the same as the forcing frequency, $\omega_{r1} = \Omega$. The forcing terms are, therefore, resonant and so the forcing transform is a unity transform, with $[\mathbf{e}] = 0$. This results in

$$\ddot{\mathbf{v}} + \Lambda\mathbf{v} + N_v(\mathbf{v}, \dot{\mathbf{v}}, \mathbf{r}) = P_v \mathbf{r},$$
$$\text{with} \quad N_v(\mathbf{v}, \dot{\mathbf{v}}, \mathbf{r}) = N_q(\mathbf{v}, \dot{\mathbf{v}}, \mathbf{r}) = \mu(v_1^2 - 1)\dot{v}_1, \quad P_v = P_q.$$

Now the near-identity transform must be found (**Step 2**). The nonlinear term is written in terms of **u** and expressed in powers of ε. As both the damping and nonlinear term are of order ε^1 gives

$$\mathbf{n}_1(\mathbf{u}, \dot{\mathbf{u}}, \mathbf{r}) = \mu(u_1^2 - 1)\dot{u}_1.$$

Using $\mathbf{u} = \mathbf{u}_p + \mathbf{u}_m$ (so $u_1 = u_{p1} + u_{m1}$) and expressing $\mathbf{n}_1$ in matrix form, $\mathbf{n}_1(\mathbf{u}, \dot{\mathbf{u}}, \mathbf{r}) = \mathbf{n}^*\mathbf{u}^*(\mathbf{u}_p, \mathbf{u}_m, \mathbf{r})$ gives

$$\mathbf{n}^* = \mu\Omega i \begin{bmatrix} 1 \\ 1 \\ -1 \\ -1 \\ -1 \\ 1 \end{bmatrix}^T, \quad \mathbf{u}^* = \begin{bmatrix} u_{p1}^3 \\ u_{p1}^2 u_{m1} \\ u_{p1} u_{m1}^2 \\ u_{m1}^3 \\ u_{p1} \\ u_{m1} \end{bmatrix}.$$

Here the general form of solution, for example $u_{p1} = (U_1/2)\mathrm{e}^{\mathrm{i}(\Omega t-\phi_1)}$ (noting $\omega_{r1} = \Omega$ for this problem), has been used. Finally, in **Step 2**, using $\mathbf{u}^*$ the matrix $\boldsymbol{\beta}^*$ is calculated as

$$\boldsymbol{\beta}^* = \begin{bmatrix} 8\Omega^2 \; 0 \; 0 \; 8\Omega^2 \; 0 \; 0 \end{bmatrix}.$$

In **Step 3** the matrix $\boldsymbol{\beta}^*$ is used to find the near-identity transform and the resonant nonlinear terms that remain in the equation of motion for $\mathbf{u}$ giving

$$\boldsymbol{\beta}^* = \begin{bmatrix} 8\Omega^2 \\ 0 \\ 0 \\ 8\Omega^2 \\ 0 \\ 0 \end{bmatrix}^T \quad \rightsquigarrow \quad \mathbf{n}_u^* = \mu\Omega\mathrm{i} \begin{bmatrix} 0 \\ 1 \\ -1 \\ 0 \\ -1 \\ 1 \end{bmatrix}^T, \quad \mathbf{h}^* = \frac{\mu\Omega\mathrm{i}}{8\Omega^2} \begin{bmatrix} 1 \\ 0 \\ 0 \\ -1 \\ 0 \\ 0 \end{bmatrix}^T.$$

Step 4 is to calculate the equation of motion in the transformed co-ordinate system and the near-identity transform. To order ε^1 these may be written as

$$\ddot{\mathbf{u}} + \mu(\frac{U_1}{4} - 1)\dot{\mathbf{u}} + \omega_n^2\mathbf{u} = R\cos(\Omega t),$$

(where $\Omega\mathrm{i}(u_{p1}u_{m1} - 1)(u_{p1} - u_{m1}) = (\frac{U_1}{4} - 1)\dot{\mathbf{u}}$ has been used) and

$$\mathbf{v} = \mathbf{u} + \frac{\mu\Omega\mathrm{i}}{8\Omega^2}\left(u_{1p}^3 - u_{1m}^3\right),$$

respectively. Note that for this problem $\mathbf{x} = \mathbf{q} = \mathbf{v}$.

Using the trial solution the equation for the dynamics may be written as

$$(\omega_n^2 - \Omega^2)\frac{U_1}{2}\left(\mathrm{e}^{\mathrm{i}(\Omega t-\phi_1)} + \mathrm{e}^{-\mathrm{i}(\Omega t-\phi_1)}\right) + \mathrm{i}\mu(\frac{U_1}{4} - 1)\Omega\frac{U_1}{2}\left(\mathrm{e}^{\mathrm{i}(\Omega t-\phi_1)} - \mathrm{e}^{-\mathrm{i}(\Omega t-\phi_1)}\right)$$
$$= \frac{R}{2}\left(\mathrm{e}^{\mathrm{i}\Omega t} + \mathrm{e}^{-\mathrm{i}\Omega t}\right).$$

Balancing the $\mathrm{e}^{\mathrm{i}(\Omega t-\phi_1)}$ terms gives

$$(\omega_n^2 - \Omega^2)U_1 + \mathrm{i}\mu(\frac{U_1}{4} - 1)\Omega U_1 = R\mathrm{e}^{\mathrm{i}\phi_1}.$$

Note that balancing $\mathrm{e}^{-\mathrm{i}(\Omega t-\phi_1)}$ results in complex conjugate of this equation and so yields no extra information. Real and the imaginary terms are now balanced to give

$$\begin{aligned} Re: &\quad (\omega_n^2 - \Omega^2)U_1 = R\cos(\phi_1), \\ Im: &\quad \mu(\tfrac{U_1}{4} - 1)\Omega U_1 = R\sin(\phi_1). \end{aligned}$$

Squaring and adding the *Re* and *Im* equations allows ϕ_1 to be eliminated to give the amplitude relationship

$$\frac{X^2}{R^2} = \frac{1}{(\omega_n^2 - \Omega^2)^2 + \mu^2\Omega^2(1 - U^2/4)^2},$$

and dividing them gives the phase relationship

$$\tan(\phi_1) = \frac{2\zeta\omega_n\Omega}{\omega_n^2 - \Omega^2 + (3/4)U_1^2}.$$

Finally the amplitude of the response at 3Ω, $X_{3\Omega}$, may be determined in terms of U_1 using the near-identity transform and the standard solution for u_1, giving

$$X_{3\Omega} = -\frac{\mu\Omega}{8\Omega^2}\frac{U_1^3}{4}\sin(3(\Omega t - \phi_1)).$$

Index

D. Wagg and S. Neild, *Nonlinear Vibration with Control*,
Solid Mechanics and Its Applications 218, DOI 10.1007/978-3-319-10644-1

MIX
Papier aus verantwortungsvollen Quellen
Paper from responsible sources
FSC® C105338

If you have any concerns about our products,
you can contact us on
ProductSafety@springernature.com

In case Publisher is established outside the EU,
the EU authorized representative is:
Springer Nature Customer Service Center GmbH
Europaplatz 3, 69115 Heidelberg, Germany

Printed by Libri Plureos GmbH
in Hamburg, Germany